Condensed MATTER THEORIES

VOLUME 2

A Continuation Order Plan is available for this series. A continuation order will bring delivery of each new volume immediately upon publication. Volumes are billed only upon actual shipment. For further information please contact the publisher.

Condensed MATTER THEORIES

VOLUME 2

Edited by

P. Vashishta and

Rajiv K. Kalia
Argonne National Laboratory
Argonne, Illinois

and

R. F. Bishop
University of Manchester Institute of Science and Technology
Manchester, United Kingdom

Plenum Press • New York and London

ISBN-13: 978-1-4612-8244-0 e-ISBN-13: 978-1-4613-0917-8
DOI: 10.1007/978-1-4613-0917-8

LC 87-656591

Proceedings of the 10th International Workshop on Condensed Matter Theories, held July 21-28, 1986, at Argonne National Laboratory, Argonne, Illinois

A Division of Plenum Publishing Corporation
233 Spring Street, New York, N.Y. 10013

Softcover reprint of the hardcover 1st edition 1987

PREFACE

The second volume of Condensed Matter Theories contains the proceedings of the 10th International Workshop held at Argonne National Laboratory, Argonne, IL, U.S.A. during the week of July 21, 1986. The workshop was attended by high-energy, nuclear and condensed-matter physicists as well as materials scientists. This diverse blend of participants was in keeping with the flavor of the previous workshops.

This annual series of international workshops was started in 1977 in Sao Paulo, Brazil. Subsequent workshops were held in Trieste (Italy), Buenos Aires (Argentina), Caracas (Venezuela), Altenberg (West Germany), Granada (Spain), and San Francisco (U.S.A.). What began as a meeting of the physicists from the Western Hemisphere has expanded in the last three years into an international conference of scientists with diverse interests and backgrounds. This diversity has promoted a healthy exchange of ideas from different branches of physics and also fruitful interactions among the participants.

The present volume is a continuation of the effort started last year when the invited papers from the 9th International Workshop were published by Plenum Press. Our only trepidation in organizing a book of this kind stemmed from the diversity of the material, which did not lend itself easily to well-defined topics. Still, the articles are loosely divided into eight categories, where the papers in each category have either a common theme or the same underlying technique.

P. Vashishta
R.K. Kalia
R.F. Bishop

ACKNOWLEDGEMENTS

For the organization of the workshop and the book, we would like to acknowledge the help of several people: Drs. Ingvar Ebbsjö, Bradley Feuston, John Clark, F. B. Malik, M. de Llano, the conference secretary--Mrs. Rose Thomas, and the conference coordinator--Mrs. Miriam Holden. We are very grateful to the Director of Argonne National Laboratory, Dr. Alan Schriesheim, for financial support and continuing encouragement for hosting the workshop. We also greatly appreciate the financial assistance and encouragement provided by Drs. Frank Fradin and Merwyn Brodsky of Materials Science Division, Argonne National Laboratory. Financial support from University of Chicago and the U.S. Department of Energy-Basic Energy Sciences is also gratefully acknowledged.

CONTENTS

I. COMPUTER SIMULATIONS

II. DENSITY FUNCTIONAL METHOD

III. CHARGED AND NEUTRAL QUANTUM FLUIDS

IV. LOCALIZATION

V. GROWTH KINETICS

VI. QUANTUM HALL EFFECT

VII. HEAVY-FERMION SYSTEMS

VIII. COUPLED CLUSTER METHOD

MOLECULAR DYNAMICS STUDIES OF GLASS TRANSITIONS:

VITRIFICATION AND AMORPHIZATION

Sidney Yip

Department of Nuclear Engineering
Massachusetts Institute of Technology
Cambridge, MA 02139

INTRODUCTION

The purpose of this talk is to briefly consider two current studies in molecular dynamics simulation of the formation of glassy states. The first problem, which I will call vitrification, is concerned with the liquid to glass transition in a one-component atomic system; my intent is to comment on the test of a recent mode-coupling theory which provides the first quantitative, dynamical description of such a transition. The second problem, which can be called amorphization, deals with the transition from crystal to glass induced by the presence of point defects. Here I would like to describe some preliminary simulation results obtained in a collaboration with A. Rahman (Argonne and Minnesota) and H. Hsieh (MIT).

One can regard, in a somewhat superficial manner, the two mentioned processes as simply different ways of producing a glassy state of matter. As shown in Fig. 1, a liquid can be cooled or compressed quickly so that the atomic configuration is structurally arrested without forming a crystal lattice. Vitrification results when nucleation and subsequent crystallization are prevented from occurring by the suddeness of the environmental change. One can also ask how a crystal can be transformed into a glass which is a metastable configuration of higher energy. One way to induce the transition is by irradiation which creates point defects in the lattice and thereby raising its energy. Amorphization would occur if the defects cannot migrate and cluster to form crystalline planes.

It is feasible to simulate by molecular dynamics the phenomena of vitrification and amorphization, and having produced these glassy states one can study in as much detail as desired their structural and dynamical properties. Because our understanding of the two problems involves quite different processes and concepts, my discussion of the simulation results will correspondingly follow different emphasis.

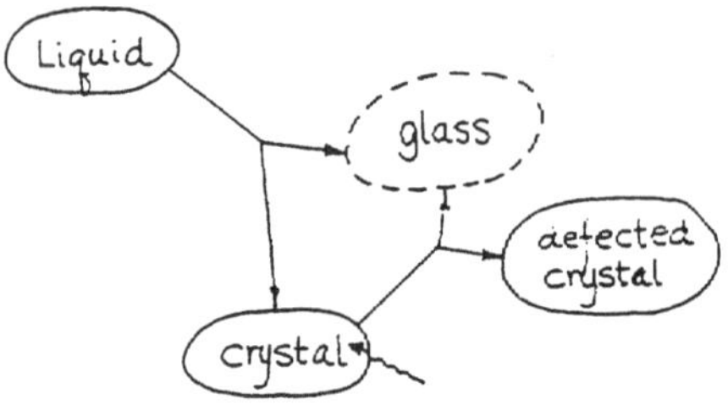

Fig. 1. Schematic showing formation of a glassy state by the sudden cooling or compression of a liquid (vitrification) or by the irradiation of a crystal (amorphization). In either case the system has an alternative path which it can follow

VITRIFICATION

It was first shown by Leutheusser [1] and by Bengtzelius, Goetze, and Sjolander [2] that a self-consistent mode-coupling approximation based on the coupling of two density fluctuation modes can lead to a dynamical equation which admits a solution of the form

$$F(k,t \to \infty) = f(k) \tag{1}$$

where F(k,t) is the density correlation function, k the wavenumber of the density fluctuation at time t, and f(k) a nonzero form factor depending on k. For such a solution to exist the density (or temperature has to reach a certain critical value n_g (or T_g). By analyzing the dynamical properties of the fluid as n approaches n_g, one finds a number of interesting behavior which indicate the onset of a transition of structural arrest at n_g. Thus, the self-diffusion coefficient D vanishes like $|\varepsilon|^{\gamma}$, with $\varepsilon = (n_g - n)/n_g$, or $(T-T_g)/T_g$, and $\gamma = 1.765$, and the shear viscosity diverges like $\eta \propto D^{-1}$. The dynamic structure factor $S(k,\omega)$, the Fourier transform of F(k,t), shows two characteristic central peaks in the frequency distribution, with both narrowing as $n \to n_g$, and the sharper component becoming an elastic line at $n = n_g$,

Subsequent analyses indicate that the exponent γ can vary somewhat [3,4], and that a glass-like transition also can be obtained from a hydrodynamic approach [4]. In addition, there exists an attempt to show the importance of coupling to the single-particle density mode [5]. The question of whether a transition can be systematically derived has been raised and discussed [6,7], and it appears that a cutoff mechanism can be found which will prevent the transition to occur in the strict sense [8].

Independent of whether a glass-like transition can be theoretically justified, it remains remarkable that the self-consistent mode-coupling approximation can lead to quantitative, dynamical calculations of dense fluids which account for all the relevant features known from molecular dynamics simulations. It is worhwhile to recall two studies as specific examples. Simulation of thermal fluctuations in dense fluids of hard spheres has revealed that the temporal behavior of F(k,t) shows a second, longer relaxation time which may be interpreted as arising from the decay of small, local clusters [9]. This feature cannot be obtained from the Enskog kinetic theory because correlated collision effects are entirely ignored, and along with the propagation of shear waves at finite k, it is regarded as a manifestation of viscoelastic behavior in a dense fluid [9,10]. It is now known that both the shear wave propagation and the second relaxation time can be quantitatively calculated by means of mode-coupling theory [11].

The second example is the hard-sphere Lorentz model of particle diffusion in a random, static medium. In this problem mode-coupling analyses have shown that a particle can be localized when the medium density reaches a critical value, and the results for the self-diffusion coefficient and the velocity autocorrelation function are found to be in satisfactory agreement with simulation data [12]. Since both theory and simulation are concerned with hard spheres, there is no adjustable parameter in any of these comparisons. It is also noteworthy that there exists no other computationally tractable method capable of analyzing problems of this kind.

Given that mode coupling is able to treat the dynamics of dense fluids and also describe particle diffusion and localization in an ergodic to nonergodic transition, it is perhaps not unreasonable to think that it can give interesting results when applied to fluids in the supercooled or mestastable regime. One should also remember that the theory has not been specifically developed to describe the glass transition, instead the transition turns out to be a natural consequence of the self-consistent treatment of the nonlinear coupling. Just as in the two prior studies, simulation can be expected to be useful in testing the validity of the theoretical predictions. In this respect it is important to realize at the outset that a number of interesting scaling properties which manifest only in the near vicinity of the transition [1,2] are out of reach because of the very long time scales involved. On the other hand, simulation provides data at finite wavenumbers and frequencies which can be calculated without being concerned about the reality of the potential function used.

In a simulation of isothermal compression of fluids interacting with a truncated Lennard-Jones potential, the onset of structural arrest could be observed and calculated along with other characteristics of the glass transition [13]. Because of the finite time interval of simulation the density at which structural arrest occurs in a strict sense could not be determined. Nevertheless, the data on F(k,t) over a range of densities can be used to directly confront the theory. Such a comparison is now possible because recently numerical results [14] have been obtained for the same system as the simulation. One finds that the theory appears to underestimate the density at which structural arrest occurs. There are two factors which are believed to contribute to this result. The first is that the numerical calculations did not take into account all the mode couplings and this is known to cause an underestimate of the self diffusion coefficient. The second factor is that activated processes are not treated in the present mode-coupling formalism; it is expected that at high densities such processes can dominate. Aside from the precise value of the transition density, the calculations give the same qualitative behavior of F(k,t) and its variation with density as the simulation. Moreover, an interesting spectral line narrowing in $S(k,\omega)$ across the transition density [13] is also seen in the calculations [14]. There are other dynamical properties obtainable from simulation [15] which can be used to test the theory, such as the transverse current correlation function, the collective mode in $S(k,\omega)$, and the shear viscosity. Comparison of these data with mode-coupling calculations should at least tell us whether the theory is qualitatively correct in the finite (k,ω) region probed by neutron and light scattering and computer simulations.

AMORPHIZATION

This phenomenon is of fundamental interest in current studies of structural transformations in crystalline materials under irradiation [16]. The molecular dynamics study of crystal to glass transition under the introduction of point defects was first carried out by Y. Limoge (Saclay) and A. Rahman. The simulations involved the introduction of vacancy-interstitial or Frankl pairs, and it was demonstrated that amorphization could be produced. A second study, involving the introduction of only interstitials and therefore no defect annihilation can occur, is now under way. In the following I will briefly describe some of the preliminary results from the latter [17].

The simulation is carried out in two stages analogous to an experiment where a sample is first irradiated and then allowed to relax. During the irradiation phase of the simulation, interstitials are inserted randomly at a constant rate into an fcc crystal maintained at constant pressure and temperature. Following the termination of defect insertion an annealing period is simulated during which various system properties are evaluated to follow the dynamical and structural evolution.

The simulation begins with 576 Lennard-Jones particles arranged on a periodic fcc lattice with the (111) planes parallel to one of the three directions of the cubic cell. An interstitial is inserted at the octahedral site which is at the center of a randomly chosen unit cell at a rate of R interstitials per unit time τ, τ being the characteristic time for a Lennard-Jones system and if one uses the parameter values for argon $\tau = 2.15 \times 10^{-12}$ s. If the neighbors of the chosen site happen to be significantly displaced from their normal lattice positions that an insertion would result in overcrowding of the particles, another site will be randomly chosen and the insertion attempt repeated. Immediately after every successful insertion the time step of simulation is reduced to prevent any violent response due to particle repulsion; this reduction is then gradually diminished so that the normal step size is recovered before the next insertion.

The system properties monitored during the relaxation stage are the density, potential energy, mean square displacement $\langle \Delta^2 r \rangle$, and the pair distribution function. Two systems which are subjected to different insertion rates will be discussed, a system B containing 320 interstitials inserted at a rate for which R = 5.92 and a system C with 190 interstitials inserted at a faster rate with R = 12.69. Throughout the simulation the system is maintained at temperature T = 0.2 (all results will be quoted in conventional reduced units appropriate to a Lennard-Jones system) and variations in system shape or volume are allowed according to the method of Parrinello and Rahman [18].

The behavior during interstitial insertion is shown in Fig. 2 where the system density is seen to initially decrease and then levels off to a value which is greater in the case of system B. The swelling is characteristic of an overcompensation in the volume expansion of the crystal; evidently with sufficient time lapse the system is able to get into configurations where additional particles can be absorbed without appreciably changing the overall density. The potential energy shows an initial increase, indicative of perturbation from the ground state, and reaches a more or less steady state value. With B raised to an energy lower than C one can infer that the latter has a less ordered structure.

The behavior during relaxation is shown in Fig. 3. One finds that there is a partial density recovery which is greater in system B. Similarly the potential energy shows an initial decrease, as the system relaxes toward a more stable configuration. From the mean square displacement one sees

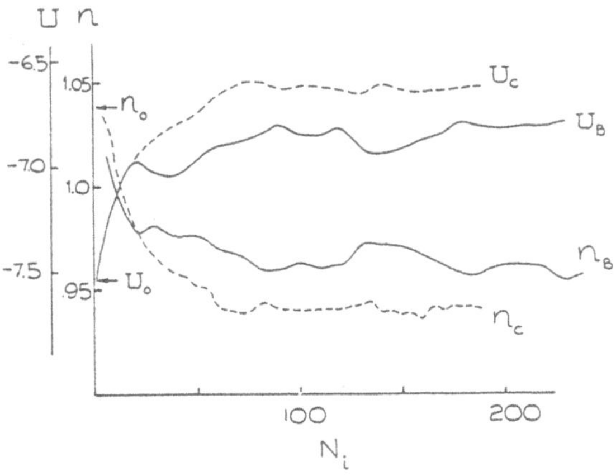

Fig. 2. Variation of density and potential energy per particle with the number of interstitials inserted. Subscript o denotes the initial fcc crystal.

that the greater density recovery of B is accompanied by considerably larger atomic displacements, whereas relatively little atomic movement takes place in C. Fig. 3 also shows that structural relaxation in the two systems apparently occurs only during an initial interval of about 3×10^3 time steps or 15τ.

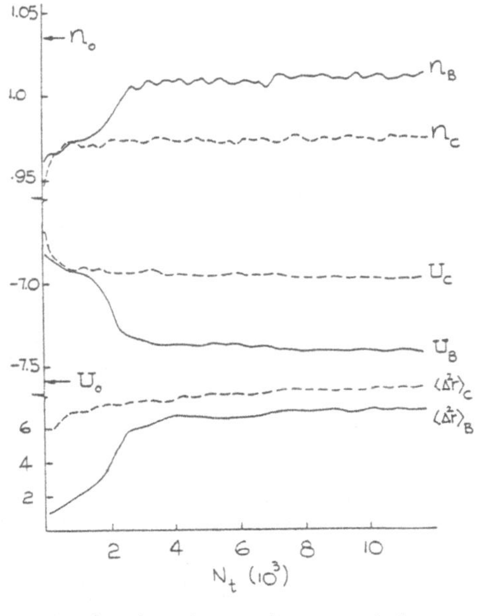

Fig. 3. Relaxation of the density and potential energy and the behavior of the mean square displacement after termination of the interstitial insertion, time in number of time steps with each step being $.005\tau$.

The results of Fig. 3 only suggest that possibly systems B and C have evolved into different structures. More direct information is provided by the pair distribution function g(r) as shown in Fig. 4. The significant feature to notice here is the intensity at the position of the second nearest neighbors for an fcc crystal at the same density. In system B an indication of a peak is already present at the start of the relaxation, and the shoulder becomes more pronounced at the end of the relaxation stage. This behavior is characteristic of crystallization. On the other hand, in system C there is no sign of any structure at the second nearest-neighbor position before and after relaxation. One may conclude that C has remained amorphous.

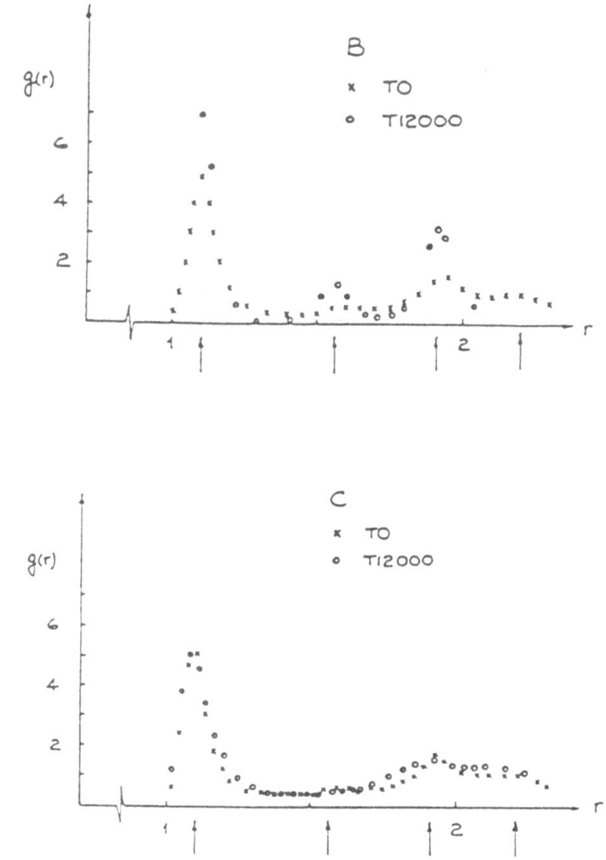

Fig. 4. Pair distribution functions of systems B and C at the start (crosses) and conclusion (circles) at time step 12000 of the relaxation phase. Arrows indicate the first four nearest-neighbor positions in an fcc crystal at the appropriate density.

These results demonstrate that amorphization does not depend on the number of point defects so much as their rate of introduction. Evidently a recrystallization mechanism exists which, if given time to activate, will allow the system to absorb a large number of defects. While the phenomenon needs to be more fully studied, we have learned through the simulation results that an athermal migration process involving elastic interactions between the interstitials, each in the form of a dumbbell in association with a lattice atom, can give rise to the formation of extra (111) planes [17].

REMARKS

My basic intent in discussing the two glass transition problems is to illustrate how atomistic simulation studies can contribute to our understanding of rather complex materials properties and behavior, either in the testing of theoretical models or in providing insights to physical phenomena. To make the present results more relevant to real systems, it is necessary to consider two atomic species so that chemical disorder is brought into play along with structural disorder. The problems then become even more rich, and both vitrification [19] and amorphization should occur under less extreme conditions of perturbation.

ACKNOWLEDGMENTS

I would like to thank A. Sjolander and U. Bengtzelius for discussions of their results on the mode-coupling calculations. I am grateful to the National Science Foundation for support and the Materials Science Division of Argonne National Laboratory for continued interest and support. I also with to express my appreciation of the hospitality of Schlumberger-Doll Research where this manuscript was written.

REFERENCES

1. E. Leutheusser, Phys. Rev. A 29, 2765 (1984).
2. U. Bengtzelius, W. Gotze, and A. Sjolander, J. Phys. C 17, 5915 (1984). See also W. Gotze, Z. Phys. B 56, 139 (1984), and in Proceedings of the 6th International Conference on Liquid and Amorphous Metals, Z. f. Phys. Chem., in press.
3. T. Kirkpatrick, Phys. Rev. A 31, 939 (1985).
4. S. P. Das, G. F. Mazenko, S. Ramaswamy, and J. J. Toner, Phys. Rev. Lett. 54, 118 (1985).
5. T. Gesti, talk given at CCP5 Meeting on the Glass Transition, Oxford, April, 1986.
6. E. Siggia, Phys. Rev. A 32, 3135 (1985).
7. S. P. Das, G. F. Mazenko, S. Ramaswamy, and J. Toner, Phys. Rev. A 32, 3139 (1985).
8. S. P. Das and G. F. Mazenko, Phys. Rev. A 34, 2265 (1986).
9. E. Alley, B. J. Alder, and S. Yip, Phys. Rev. A 27, 3174 (1984).
10. S. Yip, Ann. Rev. Phys. Chem. 30, 547 (1979).
11. E. Leutheusser, J. Phys. C 15, 2801, 2827 (1982).
12. W. Gotze, E. Leutheusser, and S. Yip, Phys. Rev. 23, 2634 (1981), 24, 1008 (1981), 25, 533 (1982).
13. J. J. Ullo and S. Yip, Phys. Rev. Lett. 54, 1509 (1985).
14. U. Bengtzelius, Ph.D. Thesis, Chalmers University of Technology, Goteborg, Sweden (1986); Phys. Rev. A, in press.
15. J. J. Ullo and S. Yip, to be published.
16. Y. Limoge and A. Barbu, Phys. Rev. B 30, 2212 (1984).
17. Y. Limoge, A. Rahman, H. Hsieh, and S. Yip, to be published.
18. M. Parrinello and A. Rahman, J. Appl. Phys. 52, 7182 (1981).
19. B. Bernu, Y. Hiwatari, and J.-P. Hansen, J. Phys. C 18, CL371 (1985), and contribution of Y. Hiwatari in this volume.

PHASE TRANSITION IN METALLIC SPIN GLASSES

Amitabha Chakrabarti and Chandan Dasgupta

School of Physics and Astronomy
University of Minensota
Minneapolis, MN 55455

ABSTRACT

Monte Carlo simulations are used to study the critical behavior of the classical Ruderman-Kittel-Kasuya-Yosida (RKKY) model of metallic spin glass both with and without the presence of weak Dzyaloshinskii-Moriya (DM) anisotropy. Finite-size scaling analyses of the data for equilibrium averages and relaxation times indicate a zero-temperature critical point for the isotropic RKKY model. When the DM interaction is included, the model exhibits a phase transition at a non-zero temperature. The values of the critical exponents at this transition are close to those obtained for the three-dimensional short-range Ising spin glass and in experiments on CuMn.

INTRODUCTION

Dilute metallic alloys of a magnetic transition metal solute in a noble metal (such as CuMn or AgMn) constitute the most widely studied class of spin-glasses. In these systems the magnetic moments interact among themselves primarily through the Ruderman-Kittel-Kasuya-Yosida (RKKY) exchange interaction. The random location of the magnetic moments and the long-range oscillatory nature of the RKKY exchange interaction prevent the system from exhibiting long-range ferromagnetic or anti-ferromagnetic order, but may induce, at low temperatures, a new phase where the spins are "frozen" in random directions. Despite extensive studies, the low temperature properties of spin-glasses are not fully understood yet. The question of whether these alloys exhibit a thermodynamic phase-transition separating a paramagnetic phase from a spin glass phase is still one of the fundamental issues in the study of spin-glasses. Most of the experiments[1] on these systems measuring thermal or transport properties exhibit smooth behavior, whereas magnetic susceptibility measurements and Mössbauer and muon spin-relaxation studies show a sharp anomaly at the so called "freezing" temperature T_g. Rapid increase of time-scales near $T = T_g$ and remanence and other history-dependent effects for $T < T_g$ are also observed in experiments. At present, no consensus exists about whether real-life spin glasses go through a sharp phase transition or a progressive freezing of the spins, although many experimental results on metallic spin-glasses have been interpreted as indications of a thermodynamic phase transition.

In their pioneering work, Edwards and Anderson (EA)[2] demonstrated theoretically the possible existence of a spin glass phase. They suggested that the frozen-in state of the spin glass phase is characterized by a non-zero value of the order parameter q (popularly known as the EA order parameter) defined by

$$q = <|<\vec{S}_i>|^2>_c, \tag{1}$$

where < > denotes a thermal average and $< >_c$ is an average over the quenched disorder. Soon afterwards Sherrington and Kirkpatrick (SK)[3] introduced a mean-field model of Ising spin glass with infinite-range interactions. This model was found to exhibit a finite-temperature transition, but the original SK solution was subsequently found to be unstable in the spin-glass phase. Later, Parisi[4] proposed an interesting new solution involving infinitely many order parameters characterized by a continuous function q(x), $0 \leq x \leq 1$. The meaning of this continuous function was subsequently clarified by the work of Sompolinsky,[5] Parisi,[6] and de Dominicis and Young.[7] Although the behavior of the infinite-range SK model is fairly well-understood now, the situation is more complicated for short-range models. There exist several theoretical arguments[8] suggesting that short-range models of spin glass should not exhibit any phase transition in three dimensions. However, recent numerical studies[9, 10] of Ising spin glass models with nearest-neighbor interactions indicate a finite transition temperature in three dimensions, whereas a zero-temperature phase transition is indicated for short-range Heisenberg models[11] in three dimensions. Although many of these models have properties that are qualitatively similar to what is observed in experiments, none of them provide a realistic description of metallic spin-glass alloys in which the RKKY exchange is the dominant interaction. The RKKY model consists of Heisenberg spins and long-range ($1/r^3$) interactions and it is worthwhile to study whether this model exhibits a thermodynamic phase transition at a finite temperature. Theoretical investigations of realistic RKKY spin glass models have mostly been confined to numerical studies. No definite conclusion about the existence of a phase transition can be drawn from the work of Ching and Huber.[12] Fernandez and Streit[13] do prescribe a transition at a finite temperature. However, their conclusion is not reliable because it is based on a method of analysis that also predicts[14] a transition for the two-dimensional short-range Ising case where actually no such transition exists.[15] The numerical work of Walstedt and Walker[16] suggests that the model with only RKKY interactions does not show any transition and a small amount of anistropy is needed to activate a clear-cut transition. These results again cannot be considered conclusive because of the following reason. The conclusion about the absence of a phase transition in the pure RKKY model was based on the observed behavior of quantities (such as the spin autocorrelation function and the components of the magnetization) which are not invariant under a uniform rotation of all the spins. Since the Hamiltonian for the RKKY spin glass is rotationally invariant, a Monte-Carlo updating process will, in general, generate uniform rotations of the spins for finite samples. Unless care is taken to account for the effects of uniform rotations, the long-time Monte-Carlo averages of the quantities mentioned above will, therefore, vanish[17] irrespective of whether a phase-transition takes place or not. This effect was not taken into account in the work of Walstedt and Walker. Also, their simulations did not provide much information about the critical behavior at the transition. Clearly, more work is needed for reliable answers to many interesting and experimentally important questions about the critical behavior of the RKKY model. In this article, we describe the results of a detailed numerical study of some of these questions.

MONTE CARLO SIMULATION OF THE RKKY MODEL

In our study[18] of the thermodynamics of the RKKY model, we used Monte Carlo simulations to calculate equilibrium averages of various rotationally invariant quantities for a wide range of sample sizes, averaged over several realizations. We also studied the temperature and sample size dependence of the different relevant time scales of this system. The Hamiltonian of the model we consider is

$$H = - J_o \sum_{j>i} [\cos (2 k_F r_{ij})/r_{ij}^3] \vec{S}_i \cdot \vec{S}_j , \tag{2}$$

where $\vec{S}_i$'s are classical Heisenberg spins of unit length randomly chosen on the sites of a fcc lattice with a concentration of 0.5 at.%, J_o is an energy constant, k_F is the Fermi wave vector of the host metal, and r_{ij} is the distance between i-th and j-th spins. The interaction parameters are chosen representing CuMn and the temperature T is measured in units of $T_o = 2\sqrt{2}\, J_o/10\, a_o^3\, k_B$ where a_o is the fcc lattice constant. The details of the Monte Carlo simulation procedure are provided in Ref. 18. Here, we discuss the main results.

We computed the single-spin autocorrelation function

$$q(t) = \langle\langle \max [\frac{1}{N} \sum_{i=i}^{N} \vec{S}_i(0) \cdot R \cdot \vec{S}_i(t)]\rangle\rangle_c , \tag{3}$$

where $\langle\ \rangle$ is the Monte Carlo (thermodynamic) average, $\langle\ \rangle_c$ is the configurational average over different realizations of the interaction, and R is a general SO(3) matrix whose inclusion corrects for any uniform rotation of the spin system.[19] The two-spin time-correlation function

$$q^{(2)}(t) = \langle \frac{2}{N(N-1)} \sum_{i>j}^{N} \langle[\vec{S}_i(0) \cdot \vec{S}_j(0)][\vec{S}_i(t) \cdot \vec{S}_j(t)]\rangle\rangle_c \tag{4}$$

which is invariant under an overall rotation of the system was also computed. The EA order-parameter q is obtained from q(t) as

$$q = \lim_{t\to\infty} q\ (t), \tag{5}$$

and

$$q^{(2)}\ (t \to \alpha) \equiv q^{(2)} \tag{6}$$

is related to the spin glass susceptibility χ_{SG} as

$$\chi_{SG} = (N-1)\ [q^{(2)} - \frac{1}{3} q^2] + 1 - q^2 . \tag{7}$$

In order to account for the sample-size dependence of $q^{(2)}(0)$ for small samples, we defined a "normalized" value of $q^{(2)}(t)$ given by

$$q^{(2)}_{norm}(t) = q^{(2)}(t)/3\ q^{(2)}(0). \tag{8}$$

Also computed was the probability distribution function for q given by

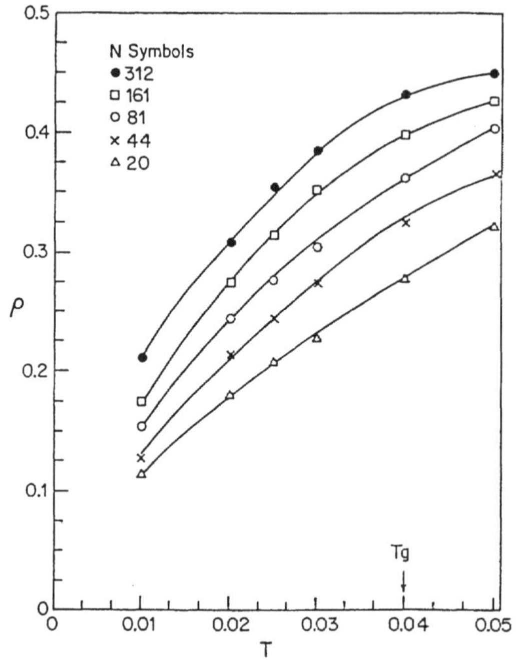

Fig. 1. Plots of ρ(defined in Eq.(10)) for the pure RKKY model vs. the temperature T (in units of T_o defined in text) for various sample sizes. Solid lines here and in the other figures are guides to the eye. The position of the experimental value of T_g is indicated by the arrow.

$$P(q) = \frac{1}{t_o} \sum_{t=\tau}^{t_o+\tau} \langle\delta(q-q(t))\rangle_c, \tag{9}$$

where τ is the equilibration time and t_o was taken between 3τ and 4τ for different sample sizes. As shown by Bhatt and Young,[9] the sample-size dependence of appropriate ratios of the moments of this distribution provides useful information about the existence of a phase transition. In particular, for a continuous phase transition at $T = T_c$, the quantity

$$\rho = [\langle q^4\rangle_{av}/(\langle q^2\rangle_{av})^2 - 1]^{1/2} \tag{10}$$

where $\langle\ \rangle_{av}$ represents an average over P(q) is expected[9 20] to have the finite-size scaling from

$$\rho(L,T) = \tilde{\rho}\ (L^{1/\nu}(T-T_c)) \tag{11}$$

where L is the sample size and ν is the correlation length exponent. Thus, curves of ρ vs. T for different L should intersect at $T = T_c$ if the system undergoes a continuous phase transition. Our results for ρ are shown in Fig. 1.

The ρ vs. T curves do not intersect each other at any point, thus indicating that no phase-transition takes place at temperatures higher than $\simeq$ 1/4 of the experimentally observed Tg. This leads us to consider the possibility that $T_c = 0$. For a zero-temperature critical point in

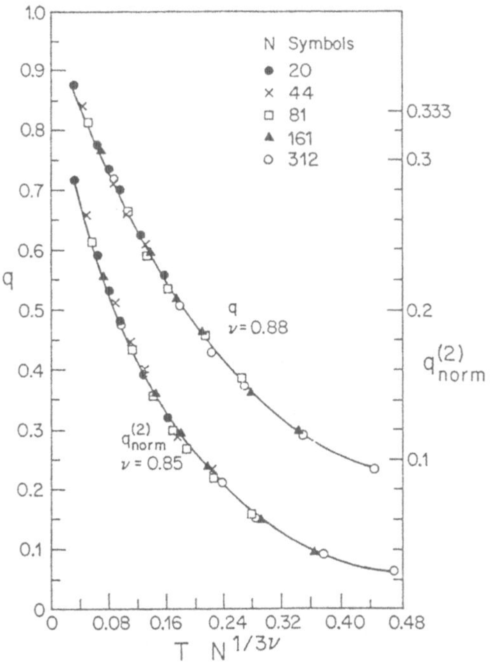

Fig. 2 Finite-size scaling plots for q and $q^{(2)}_{norm}$ for the pure RKKY model, assuming a T=0 critical point.

three dimensions, the critical exponent η is equal to -1, which implies that $\beta = 0$ and $\gamma = 3\nu$. Then finite size scaling[20] predicts that

$$q(N,T) \simeq \tilde{q}\ (TN^{1/3\nu}), \tag{12}$$

and

$$q^{(2)}_{norm}(N,T) \simeq \tilde{q}^{(2)}_{norm}(TN^{1/3\nu}), \tag{13}$$

where $N \propto L^3$ is the number of spins. As shown in Fig. 2, our results for q and $q^{(2)}_{norm}$ can be fitted very well by the above forms with a value of $\nu \simeq 0.9$.

The conclusion about the absence of any finite-temperature phase transition in the pure RKKY model is supported by the observed behavior of the various time scales. The dynamics of this system at low temperatures involve a wide distribution of relaxation times. Through the definition of a quantity

$$\delta q(t) = \frac{q(t) - q}{q(0) - q}, \tag{14}$$

we calculated the average thermal relaxation time, τ_{AV}, as

$$\tau_{AV}(T) = \int_0^\infty \delta q(t)dt. \tag{15}$$

The temperature-dependence of τ_{AV} for N = 161 is shown in Fig. 3. As can be seen, τ_{AV} does not increase as fast as $\exp(\alpha/T)$, indicating that free-energy barriers characteristic to the spin-glass phase do not

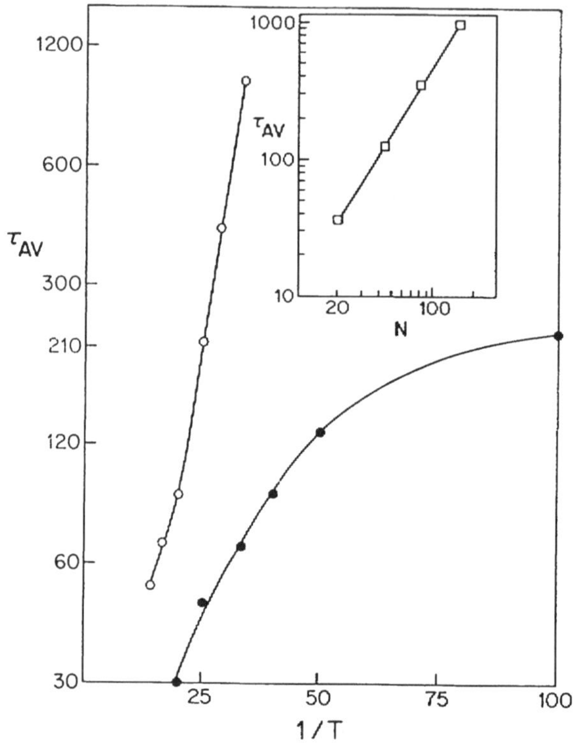

Fig. 3. Semilog plots of the average relaxation time τ_{AV} (see text) in units of Monte Carlo steps per spin for N=161 vs. the inverse of the temperature for the pure RKKY model (full circles) and for the RKKY model with DM anisotropy (open circles). The inset shows the N-dependence of τ_{AV} for the anisotropic model at T=0.03. The straight line is the best fit to the log-log plot.

develop in the RKKY system. Other relevant time scales (e.g. the longest relaxation time and the typical time between successive hops from one metastable equilibrium configuration to another) also show[18] a similar behavior.

These results, therefore, strongly indicate that the isotropic RKKY model shows a critical point at T = 0 and only RKKY exchange can not account for the experimental observations of time scales increasing faster than an Arrhenius law at low temperatures. These observations led us to carry out a numerical study of the effects of weak anisotropic interactions (which are always present in real-life spin glasses) on the critical behavior of the RKKY model. The results are described in the next section.

EFFECTS OF WEAK ANISOTROPIC INTERACTIONS

As pointed out by Fert and Levy,[21] the dominant anisotropic interaction in metallic spin-glasses is of the Dzyaloshinskii-Moriya (DM) type, arising from spin orbit scattering. There exist general arguments[22] which suggest that the presence of such anisotropy should induce a cross-over to Ising-like behavior. As simulations of short range Ising spin glass[9 10] appear to show a phase transition at a finite temperature in three dimensions, one might expect that the introduction of anisotropy in the RKKY model will bring the transition temperature to a non-zero value. Some evidence indicating the occurrence of such a phase-transition induced by the presence of dipolar anisotropy was provided by the simulation performed by Walstedt and Walker.[16] However, as discussed earlier, it is not clear whether the "freezing" observed by them in the temperature-dependence of quantities which are not invariant under overall rotations is real or is an artifact caused by spurious rotations, since the presence of small amounts of anisotropy effectively prevents uniform rotations only at very low temperatures.

The model studied by us[23] is appropriate for ternary spin glass alloys $CuMn_xT_y$, where T represents a non-magnetic impurity that mediates the DM interaction through spin-orbit scattering. In our calculation we used x = 0.5 at.% and y = 0.1 at.%. The Hamitonian for the system consists of two parts:

$$H = H_{RKKY} + H_{DM} \tag{16}$$

where H_{RKKY} is given by Eq. (2) and H_{DM} has the form derived by Fert and Levy:

$$H_{DM} = -V \sum_{i>j} \sum_{k} [\sin[k_F(r_{ij}+r_{jk}+r_{ik}) + \frac{\Pi}{10} Z_d]] \cdot \frac{\vec{r}_{ik} \cdot \vec{r}_{jk} (\vec{r}_{ik} \times \vec{r}_{jk}) \cdot (\vec{S}_i \times \vec{S}_j)}{r_{ik}^3 r_{jk}^3 r_{ij}}. \tag{17}$$

In Eq. (17), $\sum_k$ represents a sum over all sites occupied by the spin-orbit scattering impurity T, Z_d is the number of d-electrons in T and V determines the strength of the anisotropy. We took Z_d = 9.4, a value appropriate for T = Pt. For very weak anisotropy, the situation is complicated by cross-over effects which make a finite-size scaling analysis of the data for small samples very difficult. For this reason, we used the value $V/J_o = 1$, although this is several times larger than that expected[21] for Pt. Even with this relatively large value of V, there is only ≃ 7 - 8% change in the ground state energy when compared to the pure RKKY case.

In the simulations, we found that uniform rotations of substantial amounts are generated by the Monte-Carlo updating procedure at the temperatures of interest. For this reason, we concentrated our attention to $q^{(2)}(t)$ given by Eq. (4) (properly normalized to make sure that $q^{(2)}(0) = 1/3$), which is invariant under an overall rotation of the system. The probability distribution function for $q^{(2)}$,

$$P(q^{(2)}) = \frac{1}{t_o} \sum_{t=\tau}^{t_o+\tau} \langle\delta(q^{(2)} - q^{(2)}(t))\rangle_c, \tag{18}$$

was computed and, as before, we calculated an appropriate ratio of moments of $P(q^{(2)})$ as ρ' which is the same as ρ in Eq. (10), only we have used $q^{(2)}$ instead of q. Our results for ρ' vs. T for various system sizes are shown in Fig. 4. From the high temperature phase the curves for different system sizes come together at around T = 0.03, measured in units of T_o. This indicates a phase transition at a

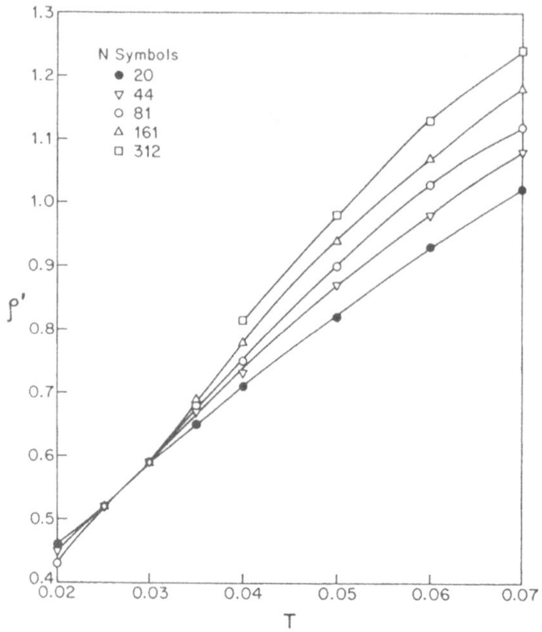

Fig. 4. Plots of ρ' (see text) for the RKKY model with DM anisotropy vs. the temperature for various sample sizes.

temperature $T_c \simeq 0.03$. From the few data points we have at lower temperatures, it appears that ρ' is nearly independent of L for T < 0.03. The data for ρ' and $q^{(2)}$ for T > 0.03 show the expected[9 10] finite-size scaling behavior with $T_c = 0.03$, $\nu \simeq 1.6$ and $\gamma \simeq 3.6$, whereas systematic deviations from scaling are found for T < 0.03. These features of the finite-size scaling behavior of the data are very similar to those observed by Bhatt and Young[9] in their simulation of the three-dimensional short-range Ising spin glass model. The values of the exponents ν and γ are slightly higher than, but consistent within the error bars with those describing the phase transition in 3-d short range Ising spin glass.[9 10 24] Our results thus provide strong support to a recent argument by Bray et al.[25] suggesting that the RKKY model with anisotropy belongs to the same universality class as the 3-d Ising spin glass with short range interactions. Our results for the exponents are also close to the values obtained experimentally[26] for CuMn. Using the value of T_o quoted in ref. 16, we estimate the experimentally observed freezing temperature of 0.5 at.% Mn in Cu to be ≃ 0.04 in units of T_o, which is not very far from the result, $T_c \simeq 0.03$, obtained from our simulation.

Our results for the temperature and sample size dependence of the average thermal relaxation time $\tau_{AV}(T)$ (defined in Eq. (15)) are displayed in Fig. 3. This time-scale clearly shows a dramatic increase near T = 0.03 when the DM interaction is present. The increase of τ_{AV} is at least as fast as exp (a/T), indicating the formation of barriers. The N dependence of τ_{AV} at T = 0.03, shown in the inset of Fig. 3, is well described by the form $\tau_{AV}(T_c) \propto N^x$ with $x \simeq 1.6$. This implies that the dynamic exponent $Z \simeq 4.8$, which is consistent with the result for 3-d short range Ising spin glass.[10]

This simulation, thus, provides strong evidence indicating that metallic spin glasses undergo an anistropy-induced phase transition in the universality class of the short-range Ising spin-glass.

REMAINING QUESTIONS

In view of the above results, the experimental fact[27] that the freezing temperature depends very weakly on the strength of the anisotropy appears puzzling. If the lower critical dimension of the isotropic RKKY model is higher than three (as suggested by our results for the pure RKKY model), then the transition temperature is expected to scale as $T_c(V) \sim [V/J_o]^{\phi}$. Agreement with experiments would then require a small value of the exponent ϕ. An alternative explanation has been proposed by Bray et al.,[25] who argue that the 3-d RKKY model is <u>at</u> the lower critical dimension. This then implies that $T_c(V) \propto 1/\surd(\ln(J_o/V)$, which would be roughly consistent with the experiments. We tried to address this question by performing the simulation with $V/J_o = 0.5$, but the finite-size scaling analysis of the data is inconclusive in this case, presumably because of crossover effects.

Another remaining question is about the nature of the low-temperature phase in the presence of anisotropy. Numerical calculations in this regime are hampered by the presence of extremely long relaxation times. It has been argued recently[28] that the nature of the ordered phase of short-range Ising model of spin glass is radically different from that of the infinite-range model. Since our simulation indicates that real metallic spin glasses belong in the same universality class as the 3-d short-range Ising spin glass, the theoretical predictions of Ref. 28 about the low-temperature behavior should apply to these systems also. It would be very interesting to determine experimentally whether this is true.

ACKNOWLEDGMENTS

This work was supported by the Alfred P. Sloan Foundation through a fellowship awarded to Chandan Dasgupta, and by the University of Minnesota Supercomputer Institute.

REFERENCES

1. For a review of the experimental results, see R. Rammal and J. Souletie, in "Magnetism of Metals and Alloys," edited by M. Cyrot (North-Holland, Amsterdam, 1982).
2. S. F. Edwards and P. W. Anderson, <u>J. Phys.</u> F5, 965 (1975).
3. D. Sherrington and S. Kirkpatrick, <u>Phys. Rev. Lett.</u> 35, 1792 (1975).
4. G. Parisi, <u>Phys. Rev. Lett.</u> 43, 1754 (1979); <u>J. Phys.</u> A 13, 1101 (1980), 13, 1887 (1980), 13, L115 (1980); <u>Philos. Mag.</u> B 41, 677 (1980).
5. H. Sompolinsky, <u>Phys. Rev. Lett.</u> 47, 935 (1981).
6. G. Parisi, <u>Phys. Rev. Lett.</u> 50, 1946 (1983).
7. C. de Dominicis and A. P. Young, <u>J. Phys.</u> A 16, 2063 (1983).

8. H. Sompolinsky and A. Zippelius in "Heidelberg Colloquium on Spin-Glasses," edited by J. L. van Hemmen and I. Morgenstern (Springer-Verlag, Berlin, 1983) and references therein.
9. R. N. Bhatt and A. P. Young, Phys. Rev. Lett. 54, 924 (1985).
10. A. T. Ogielski and I. Morgenstern, Phys. Rev. Lett. 54, 928 (1985).
11. J. R. Banavar and M. Cieplak, Phys. Rev. Lett. 48, 832 (1982); W. L. McMillan, Phys. Rev. B 31, 342 (1985).
12. W. Y. Ching and D. L. Huber, J. Phys. F 8, L 63 (1978).
13. J. F. Fernandez and T.S.J. Streit, Phys. Rev. B 25, 6910 (1982).
14. J. F. Fernandez, Phys. Rev. B 25, 417 (1982).
15. A. P. Young, Phys. Rev. Lett. 50, 917 (1983); K. Binder and I. Morgenstern, Phys. Rev. B 27, 5826 (1983).
16. R. E. Walstedt and L. R. Walker, Phys. Rev. Lett. 47, 1624 (1981); R. E. Walstedt, Physica (Amsterdam) 109 and 110 B + C, 1924 (1982).
17. D. Stauffer and K. Binder, Z. Phys. B 41, 237 (1981).
18. A. Chakrabarti and C. Dasgupta, Phys. Rev. Lett. 56, 1404 (1986).
19. H. Sompolinsky, G. Kotliar and A. Zippelius, Phys. Rev. Lett. 52, 392 (1984).
20. M. N. Barber, in "Phase Transitions and Critical Phenomena," Vol. 8, edited by C. Domb and J. Lebowitz (Academic, N.Y., 1983); K. Binder, Z. Phys. B 43, 119 (1981).
21. A. Fert and P. M. Levy, Phys. Rev. Lett. 44, 1538 (1980); P. M. Levy and A. Fert, Phys. Rev. B 23, 4667 (1981).
22. A. J. Bray and M. A. Moore, J. Phys. C 15, 3897 (1982); G. Kotliar and H. Sompolinsky, Phys. Rev. Lett. 53, 1751 (1984); B. W. Morris, S. G. Colborne, M. A. Moore, A. J. Bray and C. J. Canisius, J. Phys. C 19, 1157 (1986).
23. A. Chakrabarti and C. Dasgupta, submitted to Phys. Rev. Lett.
24. R. R. P. Singh and S. Chakravarty, Phys. Rev. Lett. 57, 245 (1986).
25. A. J. Bray, M. A. Moore and A. P. Young, Phys. Rev. Lett. 56, 2641 (1986).
26. R. Omari, J. J. Prejean and J. Souletie in "Heidelberg Colloquium on Spin Glasses," ed. J. L. van Hemmen and I. Morgenstein (Springer-Verlag, Berlin, 1983) and references therein.
27. D. C. Vier and S. Schultz, Phys. Rev. Lett. 54, 150 (1985) and references therein.
28. D. S. Fisher and D. A. Huse, Phys. Rev. Lett. 56, 1601 (1986).

STUDY OF DYNAMICAL PROPERTIES OF DENSE SOFT SPHERE FLUIDS AND GLASSES BY MOLECULAR DYNAMICS

Yasuaki Hiwatari

Department of Physics, Faculty of Science, Kanazawa University
Kanazawa, 920 Japan

and

Bernard Bernu and Jean-Pierre Hansen
Université Pierre et Marie Curie, Laboratoire de Physique Théorique des Liquides, 4 plcace Jussieu, 75230 Paris, France

INTRODUCTION

It has been found, both experimentally and through computer simulations, that liquids can be rapidly cooled far below their freezing point, and remain in a fluid state for long periods of time. If either the density of the supercooled liquid is high enough or the temperature is low enough, the liquid forms a glass. Perhaps the most striking recent theoretical development of the glass transition is due to the mode coupling theory, which predicts that the intermediate scattering function $F(k,t)$ does not decay towards zero, but has a nondecaying constant for an infinite time limit in glasses. As the glass transition is approached, e.g., the density n approaches the glass transition density $n_g (n<n_g)$, the self-diffusion constant D tends to zero following a power-law formula like $D \sim (n_g - n)^{\nu}$ with $\nu=1.8-1.9$/1/2/3/. The glass transition is a dynamical transition in the sense that the transition is signalled by a rapid drop of the diffusion constant during rapid quenching of liquids. Therefore the dynamical properties are more important for the study of the glass transition than static structures.

In this circumstance, careful molecular dynamics (MD) calculations on the dynamical properties will be useful both to test the theory and to learn about the nature of the transition without approximations. We have carried out MD simulations for the soft-sphere model, in which the temperature of the system is kept constant (isothermal MD) during annealing. Since glassy states are non equilibrium states, the system shows gradual (but very slow) relaxation during annealing. In this respect, isothermal MD is found to be superior to keep the system stable to microcanonical(constant total energy)MD.

In this article, we focus our attention on the dynamical properties, the behaviour of atomic diffusion and of the intermediate scattering function (self-part only), in liquids, supercooled liquids and glasses of the soft-sphere model.

SOFT - SPHERE MODEL

Isothermal MD simulations have been carried out for 4000 soft spheres interacting through the pair potential

$$\Phi = \varepsilon(\sigma/r)^n \; ; \; n = 12 \tag{1}$$

with periodic boundary conditions in a usual way. For the detail of our isothermal MD techniques, see Hiwatari/4/.

In Table 1 we summarize our MD results for the compressibility factors and reduced self-diffusion constants. Due to the scaling property of the soft-sphere system, only one coupling parameter is necessary to characterize thermodynamic states, for which we can take

$$\Gamma = (N\sigma^3/V)(k_B T/\varepsilon)^{-3/n} \tag{2}$$

where V is the total volume of the system, N the total number of atoms, k_B the Boltzmann constant, and T is the temperature.

The results shown in Table 1 are all obtained from MD runs over 5000 time steps in our unit of time (see Table 1). We have continued the simulations at each Γ up to fifty thousand time steps. However all these simulations have shown incipient crystallization during such annealing and we have found that final states have an fcc-like structure. Up to around five thousand time steps, all the systems in this table are found to stay in an amorphous state. Since we are concerned with amorphous systems, not with any states in crystallizing processes, we have used the MD configuration of atoms over periods such that no nucleatin occurs.

It is known that crystallization is much more easily by-passed in metallic alloys than monatomic rare gas-like substances by rapidly quenching the liquids down to low enough temperatures. For this reason binary mixtures of soft spheres with different diameters σ_α(α =1 and 2) are undertaken for a better model for typical glass-form substances, which will be reported elsewhere. Some preliminary results are found in ref. /5/.

MEAN SQUARE DISPLACEMENT (MSD) AND NON-GAUSSIAN BEHAVIOUR

Let us consider the 2n-th moment of atomic displacements;

$$\langle R^{2n}(t)\rangle = \langle|\vec{r}_i(t) - \vec{r}_i(0)|^{2n}\rangle . \tag{3}$$

For $t \to \infty$, it can be written as

$$\langle R^{2n}(t)\rangle \to (2n + 1)!/n!(Dt)^n + O(t^{n-1}), \tag{4}$$

where D is a self-diffusion constant. We define the so-called non-Gaussian parameter a(t)

$$a(t)=(3/5)\langle R^4(t)\rangle/\langle R^2(t)\rangle^2 - 1. \tag{5}$$

It follows that $a(t) \to O(t^{-1})$ for $t \to \infty$. (6)

This function is known as a measure of non-Markovian behaviour.

Table 1. Isothermal Molecular dynamics simulations for the inverse twelfth soft-sphere potential with N=4000, $\rho\sigma^3$=1.18, and the time increment $\Delta t/\tau = 5.7735\times10^{-3}(\rho\sigma^3)^{-1/3}(T^*)^{-1/2}$, where $\tau=(m\sigma^2/\varepsilon)^{1/2}$ and $T^*=k_BT/\varepsilon$. Our unit of time Δt corresponds to a constant(0.0057735) in physically meaningful time unit of $\tau_0=(\rho\sigma^3)^{-1/3}(T^*)^{-1/2}\tau$.

No.	T^*	Γ	PV/NkT	$D^*\cdot10^2$
1	4.7333	0.80	8.414	6.18
2	1.9387	1.00	13.770	3.24
3	0.9349	1.20	21.940	1.24
4	0.6788	1.30	27.357	0.59
5	0.5837	1.35	30.477	0.35
6	0.5046	1.40	34.105	0.25
7	0.4385	1.45	37.941	0.15
8	0.3829	1.50	42.499	
9	0.3358	1.55	47.448	
10	0.2958	1.60	52.753	
11	0.2615	1.65	58.739	
12	0.2321	1.70	65.220	
13	0.2067	1.75	72.488	

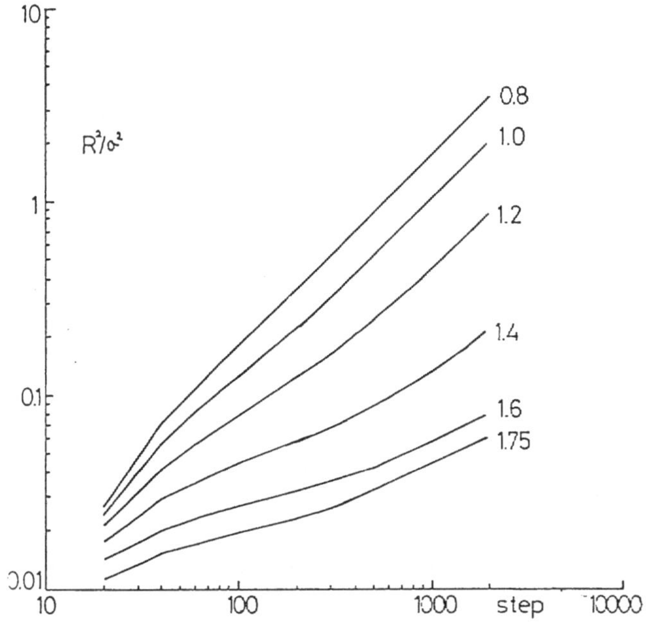

Fig. 1. The log-log plots of mean square displacement as a function of time. The unit of time step is $5.7735\times10^{-3}\tau_0$. The figures indicate coupling constants Γ.

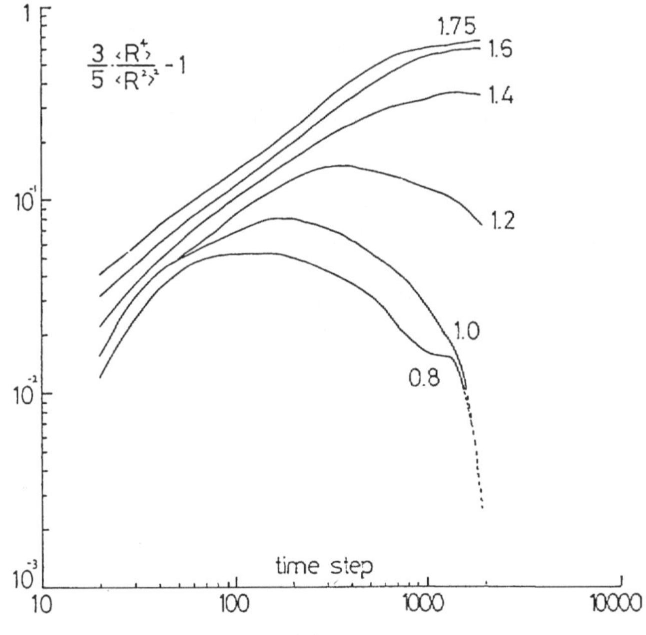

Fig. 2. The function a(t) vs. t for various copuling constants Γ.

Our MD results are summarized as follows (see Figs. 1 and 2)
(1) The MSD goes over to the Einstein limit (MSD being proportional to t) already before $t/\tau_0=1$ for liquids and supercooled liquids. On the other hand in glasses it takes a much longer time ($t/\tau_0>>10$) for the MSD to reach such an asymptotic regime; our MD result shows a roughly square root time dependence for intermediate times ($t/\tau_0>3$). This causes a serious difficulty for obtaining diffusion constants in glasses and will lead to an overestimate of the value of the diffusion constant.
(2) a(t)=0 when the Gaussian apprximation holds exactly. Figure 2 shows that in liquids a(t) is very small (maximum value is about 0.08) and it tends to nearly zero after $t/\tau_0>5$, but in glasses a(t) is always increasing as a function of time, exhibiting no sign of decrease at least up to $t/\tau_0=10$, where a(t) becomes as large as 0.6.

SELF-DIFFUSION CONSTANT

The self-diffusion constants were calculated from the mean square displacement;

$$D=\langle R^2(t)\rangle/6t \qquad \text{for } t\to\infty \tag{7}$$

The numerical values for the reduced self-diffusion constant $D^*(=\tau D(T^*)^{-5/12}\sigma^{-2})$ are listed in Table 1. D* was calculated only for $\Gamma\leq 1.45$, because for larger Γ's (glassy states) our mean square displacements do not reach the Einstein regime. As shown in Fig.3, our data for the diffusion constant can be fitted by a power-law formula, in agreement with mode coupling theory;

$$D^*=A(\Gamma^{-1}-\Gamma_g^{-1})^{\nu}; \qquad \nu=1.5 \quad \text{and} \quad \Gamma_g=1.56, \tag{8}$$

where A is constatnt, and Γ_g a copuling constant at the glass transition point for which we have taken 1.56/6/.

For the binary mixtures of soft spheres we have carefully computed the self-diffusion constant near and slightly above Γ_g by much longer calculations than here. These results show a residual diffusion in glasses due to jump processes, the details of which will be discussed in a separate paper.

INTERMEDIATE SCATTERING FUNCTION (ISF)

We have computed the self part of the intermediate scattering function Fs(k,t) from the computer generated configurations of the soft-sphere system;

$$Fs(k,t)=\langle\rho_i(\vec{k},t)\rho_i(-\vec{k},0)\rangle, \tag{9}$$

where $\rho_i(\vec{k},t)=\exp[-i\vec{k}.\vec{r}_i(t)]$. Fig.4 shows the ISF at various Γ's for $k^*=4.18$ ($k^*=k\sigma$), and Fig.5 shows ln[-ln Fs(k,t)] vs. ln t for different k* and Γ. If we assume the streched functional form for the ISF at large times and at a fixed k-value, that is,

$$Fs(k,t)=\exp[-(t/a)^{\alpha}]. \tag{10}$$

It turns out that the exponent α is equal to one for liquids, but for glasses α is clearly smaller than one (nearly one-half) on our time scale ($t/\tau_0\leq 10$). The ISF (self-part) can be expanded in powers of k as /7/,

$$Fs(k,t)=\exp[\sum_{n=1}^{\infty}(-k^2)^n\gamma_n(t)],$$

$$\gamma_1(t)=\int_0^t dt_1\int_0^{t_1} dt_2\langle v_i(t_2)v_i(t_1)\rangle=\langle R^2(t)\rangle/6,$$

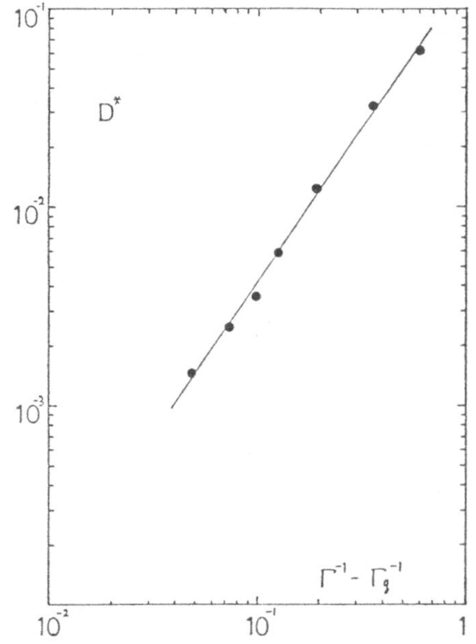

Fig. 3. The log-log plots of the reduced self-diffusion constant D^* vs. inverse coupling constant for one-component soft spheres; $\Gamma_g=1.56$ is the coupling constant at the glass transition. The straight line drawn here gives the exponent $\nu=1.5$.

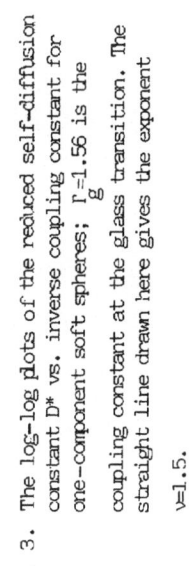

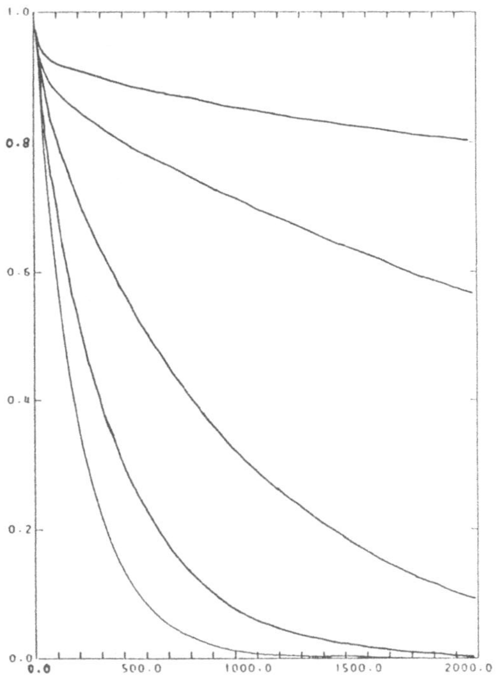

Fig. 4. Intermediate scattering function (self-part) $Fs(k,t)$ for $k^*(=k\sigma)=4.18$; from bottom to top Γ =0.8, 1.0, 1.2, 1.4, and 1.6.

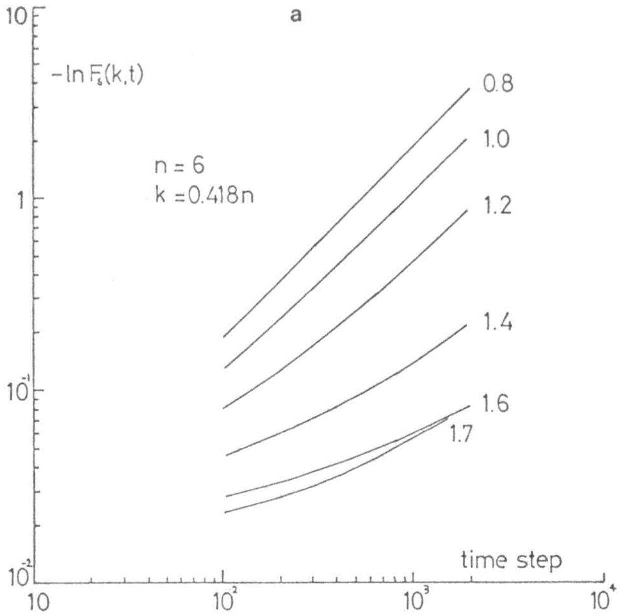

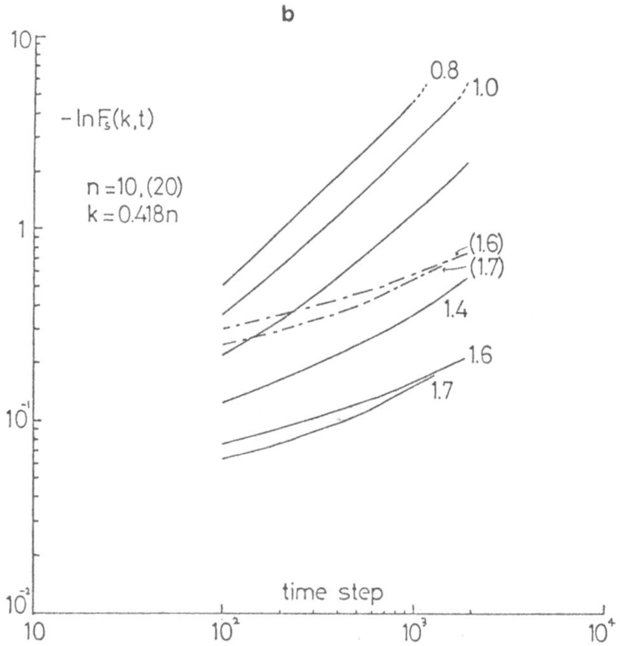

Fig. 5. ln [−ln Fs(k,t)] vs. ln t for various Γ and k*. Each function can be fitted at large times to a straight line with a slope close to one for liquids, one half for glasses. Fig. 5(a) shows results for $k^*=6k_0^*$ ($k_0^*=0.418$), while Fig. 5(b) is for $k^*=10k_0^*$ (full curves) and $k^*=20k_0^*$ (dash-dotted curves).

$$\gamma_2(t) = \int_0^t dt_1 \ldots\ldots \int_0^{t_3} dt_4 \langle v_i(t_4) \ldots\ldots v_i(t_1) \rangle - \gamma_1(t)^2/2$$

$$= (1/72)(5/3) \langle R^2(t) \rangle^2 a(t),$$

For $t \to \infty$, $\gamma_n(t) \to D_n t - c_n$, where D_n and C_n are constants. The first term of the expansion corresponds to the Gaussian approximation, and higher terms are non-Gaussian corrections. From this expansion, it turns out that the function a(t) measures non-Gaussian behaviour.

As we have seen in Fig.5. the liquid and supercooled liquid Fs(k,t) can be expressed by eq.(10) with $\alpha=1$ except in the short time region. Thus, using this analytic form at large t and original MD data at small t, we have calculated the Fourier transform of ISF, Fs(k,ω). It exhibits a single peak at ω=0. In Fig.6 is shown the half-width-at-half-maximum (HWHM) of Fs(k,ω) for various N(k*=0.418N). As Γ approaches Γ_g, Fs(k,ω) becomes very narrow for all N (the HWHM becomes very small). Such characteristic behaviour has been reported in Lennare-Jones fluids /8/ and discussed by mode coupling theory /9/.

Acknowledgement

This work was partly supported by the grant in aid for Scientific Research from the Ministry of Education, Science and Culture of Japan. One of us (Y.H.) would like to express his sincere thanks to Professor Sidney Yip for valuable discussions and for his kind hospitality during his stay at MIT.

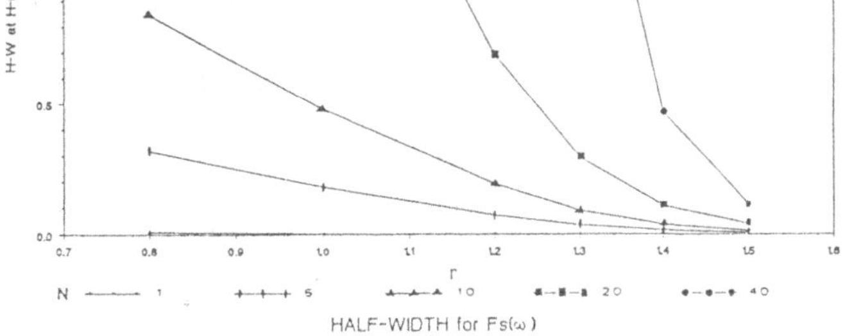

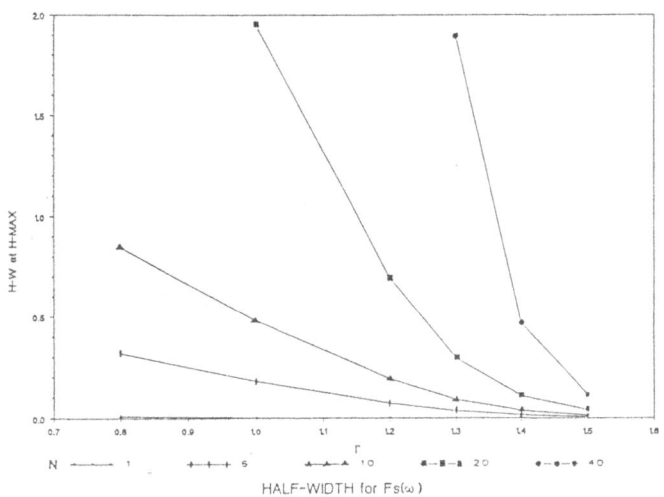

Fig. 6. The half-width-at-half-maximum (HWHM) of Fs(k,ω) vs. Γ for various wave numbers $k^*=Nk_0^*$ ($k_0^*=0.418$).

REFERENCES

1. E. Leutheusser, Phys. Rev. A29:2765 (1984).

2. T. R. Kirkpatrick, Phys. Rev. A31:939(1985).

3. U. Bengtzelins, W. Götze, and A. Sjölander, J. Phys. C 17:5915 (1984).

4. Y. Hiwatari, J. Phys. Soc. Japan. 47:733 (1979).

5. B. Bernu, Y. Hiwatari, and J.P. Hansen, J. Phys. C 18:L371 (1985); J. de Phys. C8, Supplement No. 12, 46:323 (1985).

6. Y. Hiwatari, J. Phys. C 13:5899 (1980); B. Bernu, Y. Hiwatari, and J. P. Hansen, submitted to Phys. Rev. A.

7. A. Rahman, K. S. Singwi, and A. Sjölander, Phys. Rev. 126:986 (1962).

8. J. J. Ullo and S. Yip, Phys. Rev. Letters 54:1509 (1985).

9. For example, see refs. 1 - 3.

DYNAMIC THEORY OF THE GLASS TRANSITION
IN DENSE CLASSICAL PLASMAS

Setsuo Ichimaru and Shigenori Tanaka

Department of Physics
University of Tokyo
Bunkyo, Tokyo 113, Japan

ABSTRACT

We present a new statistical theory of dynamic correlations for a classical one-component plasma (OCP) in strong Coulomb coupling, within the generalized viscoelastic formalism coupled with a fully convergent kinetic equation. The theory reproduces the existing molecular-dynamics simulation data both on the dynamic structure factor and on the coefficient of shear viscosity in the ordinary fluid state. We then extend the theory to those plasmas in the supercooled state, investigate the dynamic behaviors of the system, and predict the possibility of the glass transition. Relevance to laboratory experiment is pointed out through analyses of the metastable-state lifetimes against homogeneous nucleation of the crystalline state.

1. INTRODUCTION

The classical one-component plasma (OCP) consisting of a single species of point charges (electric charge Ze, mass m, number density n, and temperature T) embedded in a uniform neutralizing background is characterized by a single dimensionless parameter [1,2], $\Gamma = (Ze)^2/ak_BT$, where $a = (4\pi n/3)^{-1/3}$ is the ion-sphere radius. The correlation properties of such an OCP have been extensively studied [1,2] by the Monte Carlo (MC) and molecular dynamics (MD) simulation techniques as well as by numerical solutions to the integral equations. Recent experimental developments [3,4] lead us to expect that a classical OCP in strong coupling ($\Gamma >> 1$) may soon be realized in a laboratory setting.

It has been predicted [5] that the OCP may undergo a phase transition into a crystalline state (Wigner crystallization) at $\Gamma_m = 178 \pm 1$. Since the transition is of the first order, the plasma may remain in a metastable fluidlike state when it is supercooled below the corresponding transition temperature. If a sufficiently "rapid quench" is applied to such a plasma, a possibility exists that the plasma may turn into a glassy state. In light of both the theoretical significance and the practical feasibility of the experimental study, we find it essential to clarify the nature of the OCP in the supercooled state and to explore the possibility of the glass transition.

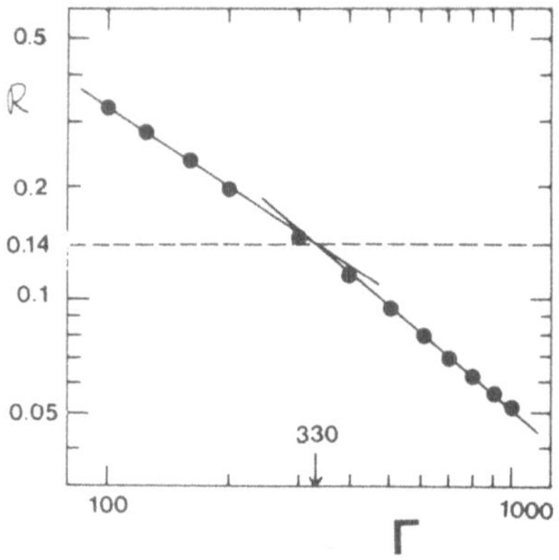

Fig. 1. The Wendt-Abraham ratio R versus Γ. The two asymptotic straight lines representing power-law fittings of R on Γ intersect at Γ = 330 and R = 0.14.

2. STATIC CORRELATIONS

The improved hypernetted chain (IHNC) scheme [6] has reproduced almost exactly the existing MC data [5] of the radial distribution function g(r) for $\Gamma \leq 160$. When we extend the IHNC calculation to those plasmas in the supercooled state ($\Gamma > \Gamma_m$)[7], a splitting of the second peak and structural developments around the third peak in g(r) are observed as Γ increases to and beyond 300 - 400. We show in Fig. 1 the Wendt-Abraham ratio [8], $R = g_{min}/g_{max}$, between the first minimum and maximum in g(r) for various Γ values, to find a slight but unmistakable kink at a Γ value around 330. Those indications may be interpreted as precursors in the static correlations to signal a dynamic glass transition at a still larger value of Γ.

3. GENERALIZED VISCOELASTIC FORMALISM AND KINETIC EQUATION

The strong Coulomb-coupling effects are described in the present theory by the dynamic local-field correction G(k,ω), which is introduced via the wave-vector $\vec{k}$ and frequency ω dependent linear-response relation between the external potential $\phi_{ext}(\vec{k},\omega)$ and the induced density fluctuations $\delta n(\vec{k},\omega)$ [2];

$$\delta n(\vec{k},\omega) = \chi_0(k,\omega)\{\phi_{ext}(\vec{k},\omega) + v(k)[1 - G(k,\omega)]\delta n(\vec{k},\omega)\} \quad . \qquad (1)$$

Here $v(k) = 4\pi(Ze)^2/k^2$ and $\chi_0(k,\omega)$ is the retarded free-particle polarizability [2]; the dynamic structure factor S(k,ω) is calculated in terms of these functions and G(k,ω) with the aid of the fluctuation-dissipation theorem [2].

Through a generalization of the viscoelastic formalism [9] into a finite k and ω regime, we find an expression [10]

$$G(k,\omega) = \frac{G(k) - i\omega\tau_m(k)I(k)}{1 - i\omega\tau_m(k)} \quad . \qquad (2)$$

Here G(k) and I(k) are the low- and high-frequency limits of G(k,ω), which are calculated [2] from the static structure factor S(k); $\tau_m(k)$ is a generalized relaxation time, which is related to the generalized shear viscosity η(k) via [10]

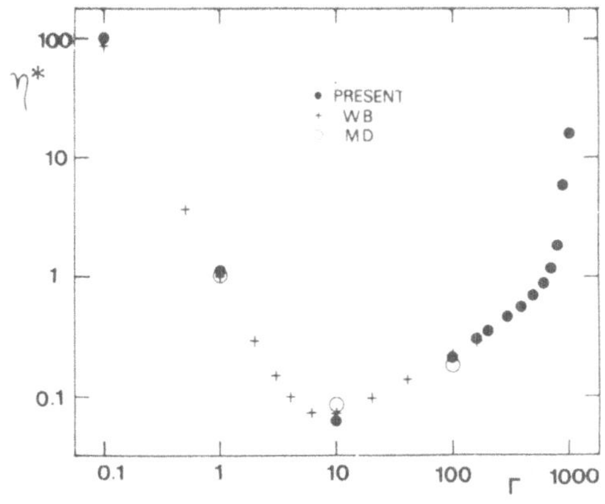

Fig. 2. The reduced shear viscosity $\eta^* \equiv \eta/mn\omega_p a^2$ calculated in the present scheme (closed circles). The crosses refer to the calculations by Wallenborn and Baus [14]; the open circles, the MD simulation result [12].

$$\eta(k)/\tau_m(k) = (3/4)n^2v(k)[G(k) - I(k)] \quad . \tag{3}$$

We assume the k-dependence of the relaxation time to be a Gaussian form, $\tau_m(k) = \tau_m \exp[-(ak/\xi)^2]$ with $\xi = 2.7$. We can thus determine $G(k,\omega)$ once the shear viscosity η at $k = 0$ is known.

In the kinetic equation for the one-particle distribution function $F(\vec{r},\vec{p};t)$, the collision term is expressed as [11]

$$\frac{\partial F(\vec{p})}{\partial t}\Big]_c = -i\int \frac{d\vec{k}}{(2\pi)^3}\int d\omega v(k)\vec{k}\cdot\frac{\partial}{\partial\vec{p}}<\delta N\delta n^*(\vec{k},\omega;\vec{p})> \tag{4}$$

with

$$\delta n(\vec{k},\omega) = \int d\vec{p}\,\delta N(\vec{k},\omega;\vec{p}) \quad . \tag{5}$$

The microscopic density fluctuations $\delta N(\vec{k},\omega;\vec{p})$ may be separated [11] into the spontaneous (free) part and the induced part. We approximate the latter as

$$\delta N_s(\vec{k},\omega;\vec{p}) = -v(k)\vec{k}\cdot\frac{\partial F}{\partial\vec{p}}\,\frac{\delta n(\vec{k},\omega)}{\omega - \vec{k}\cdot\vec{p}/m + io}[1 - G(k,\omega)] \tag{6}$$

in light of Eq. (1), leading to a fully convergent collision term [10] for the plasmas in strong Coulomb-coupling. A solution to the kinetic equation then yields an expression for η in terms of $G(k,\omega)$ [10].

We thus have a coupled set of integral equations representing a condition for self-consistency between η and $G(k,\omega)$. Solving this set of equations numerically by iteration, we find the reduced shear viscosity $\eta^* \equiv \eta/mn\omega_p a^2$ and the normalized dynamic structure factor $(\omega_p/n)S(k,\omega)$, where $\omega_p = [4\pi n(Ze)^2/m]^{1/2}$ is the plasma frequency. In the ordinary fluid regime ($\Gamma < \Gamma_m$), the computed values reproduce the existing MD results [12,13] satisfactorily both on η^* (see Fig. 2) and on $S(k,\omega)$ (see Fig. 3). For comparison we also plot in Fig. 2 the theoretical values of η^* obtained by Wallenborn and Baus [14].

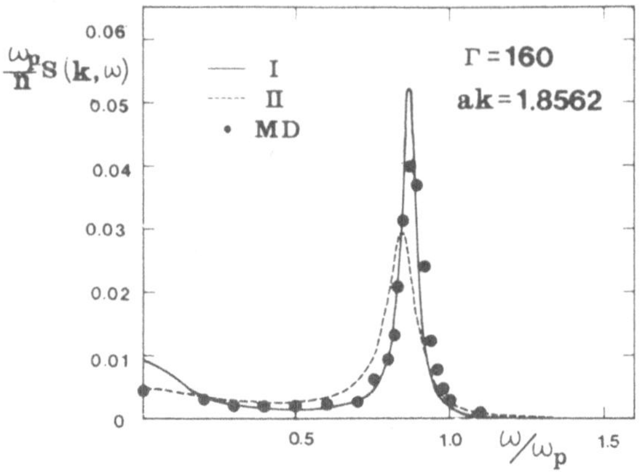

Fig. 3. The normalized dynamic structure factors at $\Gamma = 160$ for $ak = 1.8562$. The closed circles refer to the MD result [13]; the solid curve (I), the present calculation; the dashed curve (II), a modified calculation described in Ref. 10.

4. APPROACH TO THE GLASSY STATE

When the present theory is applied to those plasmas in the supercooled state, we find a steep rise in η^* at $\Gamma = 800 - 1000$, as illustrated in Fig. 2; this steep rise corresponds to a comparative increase in the relaxation time τ_m. We then evaluate the reduced self-diffusion coefficient $D^* \equiv D/\omega_p a^2$ (see Fig. 4) from η^* by evoking the Stokes-Einstein relation and the correspondence between a strongly coupled OCP and a hard-sphere system [2]; the values of D^* steeply decrease at $\Gamma = 800 - 1000$, as would be expected.

As the values of η and τ_m increase, a δ-functionlike quasi-elastic peak develops in the calculated results of $S(k,\omega)$ (see Fig. 5). This is a consequence of the use of Eq. (2) for $G(k,\omega)$; in the limit of $\tau_m \to \infty$, we find

$$S(k,\omega) \to [k_B T/v(k)][G(k) - I(k)]\delta(\omega) \quad . \tag{7}$$

This limit thus corresponds to a frozen state in which the local structure

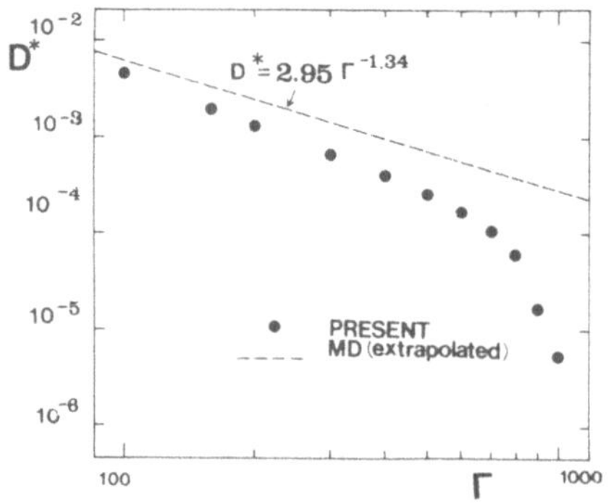

Fig. 4. The reduced self-diffusion coefficient $D^* \equiv D/\omega_p a^2$ calculated in the present scheme (closed circles). The dashed line refers to the extrapolation of the fitting formula obtained by the MD simulation for $\Gamma \leq 152.4$ [13].

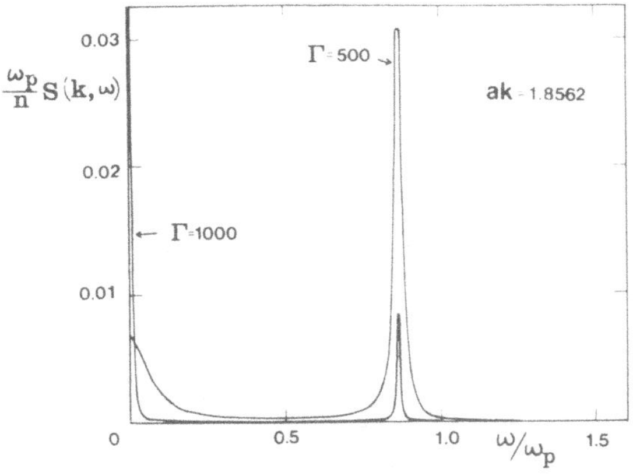

Fig. 5. The normalized dynamic structure factors at $\Gamma = 500$ and 1000 for $ak = 1.8562$ calculated in the present scheme.

does not vanish even after an infinite time or the expectation value $\langle\delta n(k,t)\delta n^*(k,0)\rangle$ stays finite for $t = \infty$. Those features in the behaviors of η^*, D^*, and $S(k,\omega)$ point to a dynamic transition to a glassy state at $\Gamma \simeq 1000$.

On the basis of a mode-coupling theory, Bengtzelius, Götze, and Sjölander [15] have derived an integral equation for

$$f(k) \equiv \lim_{t\to\infty} [\langle\delta n(k,t)\delta n^*(k,0)\rangle / nS(k)] \quad , \tag{8}$$

and discussed the glass transition in the hard-sphere and Lennard-Jones systems. If the system is in the glassy state, a nonvanishing solution of $f(k)$ appears besides a trivial solution $f(k) = 0$. We have applied their theory to the OCP and found a nonvanishing solution for $\Gamma \geq 400$ (see Fig. 6). The critical value of Γ obtained in this scheme, however, is smaller substantially than that in the present theory.

For a realization of the glassy OCP investigated above, a sufficiently rapid quench must be applied to the plasma, overcoming the nucleation of crystals in supercooled liquids. The lifetime of the metastable supercooled state can be analized on the basis of a standard statistical model [7,16] of homogeneous nucleation. The interplay between the surface energy and the free-energy difference between two phases determines the radius of critical nucleus, the minimum work needed to form the nucleus, and the probability of homogeneous nucleation. We have thus evaluated the nucleation time t_N with the aid of the computed values of D^*, and found [10] that $\omega_p t_N$ takes on a minimum value somewhere between 10^9 and 10^{13} at $\Gamma \simeq 400$. We remark that the uncertainty in $\omega_p t_N$ originates from a slight (<0.15%) ambiguity in the evaluation of the fluid-state free energy [10].

Recently, Bollinger and Wineland [4] have produced Penning-trapped, strongly coupled ($\Gamma \simeq 5 - 10$) ion plasmas by using the laser-cooling technique and maintained them stably for many hours. The theoretical limit of the possible Γ value has been estimated to be about 10^4 for their $^9Be^+$ plasmas. The minimum value of t_N takes on $2 \times (10 - 10^5)$s at $n = 10^{10}$ cm^{-3}. If the cooling can be administered sufficiently fast to overcome such a minimum value of t_N, then a highly viscous glassy OCP may be realized in the laboratory.

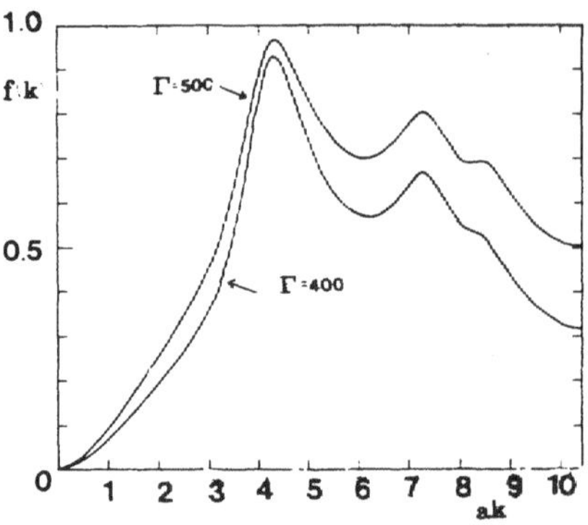

Fig. 6. The solutions to the Bengtzelius-Götze-Sjölander equation [15] for $f(k) \equiv \lim_{t\to\infty} [\langle \delta n(k,t)\delta n^*(k,0)\rangle / nS(k)]$ at $\Gamma = 400$ and 500.

REFERENCES

1. M. Baus and J.-P. Hansen, Phys, Rep. 59, 1 (1980).
2. S. Ichimaru, Rev. Mod. Phys. 54, 1017 (1982).
3. C.F. Driscoll and J.H. Malmberg, Phys. Rev. Lett. 50, 167 (1983).
4. J.J. Bollinger and D.J. Wineland, Phys. Rev. Lett. 53, 348 (1984).
5. W.L. Slattery, G.D. Doolen and H.E. DeWitt, Phys. Rev. A 26, 2255 (1982).
6. H. Iyetomi and S. Ichimaru, Phys. Rev. A 27, 3241 (1983).
7. H. Iyetomi and S. Ichimaru, Phys. Rev. A 27, 1734 (1983); S. Ichimaru, H. Iyetomi, S. Mitake and N. Itoh, Astrophys. J. (Lett.) 265, L83 (1983).
8. H.R. Wendt and F.F. Abraham, Phys. Rev. Lett. 41, 1244 (1978).
9. J. Frenkel, Kinetic Theory of Liquids (Clarendon Press, Oxford, 1946), Chap. IV.
10. S. Ichimaru and S. Tanaka, Phys. Rev. Lett. 56, 2815 (1986); S. Tanaka and S. Ichimaru, to be published.
11. S. Ichimaru, Plasma Physics: An Introduction to Statistical Physics of Charged Particles (Benjamin/Cummings, Menlo Park, Calif., 1986).
12. B. Bernu, P. Vieillefosse, and J.-P. Hansen, Phys. Lett. 63A, 301 (1977).
13. J.-P. Hansen, I.R. McDonald, and E.L. Pollock, Phys. Rev. A 11, 1025 (1975).
14. J. Wallenborn and M. Baus, Phys. Rev. A 18, 1737 (1978).
15. U. Bengtzelius, W. Götze, and A. Sjölander, J. Phys. C 17, 5915 (1984); U. Bengtzelius, Phys. Rev. A 33, 3433 (1986).
16. L.D. Landau and E.M. Lifshitz, Statistical Physics (Pergamon, Oxford, 1969), p. 471.

STRUCTURE OF A ONE-COMPONENT PLASMA IN AN EXTERNAL FIELD: A MOLECULAR DYNAMICS STUDY OF PARTICLE ARRANGEMENT IN A HEAVY-ION STORAGE RING

A. Rahman

Supercomputer Institute and School of Physics & Astronomy
University of Minnesota, Minneapolis, MN 55455

and

J. P. Schiffer

Argonne National Laboratory, Argonne, IL 60439 and
University of Chicago, Chicago, IL 60637

ABSTRACT

A one-component plasma has been studied by molecular dynamics calculations to simulate the behavior of charged particles in heavy-ion storage rings. The Hamiltonian used confines the plasma in the directions lateral to the direction of travel in the ring in the frame of reference which is moving with the beam. The results show an unexpected stratification of density in the lateral direction, and a tendency towards a first-neighbor coordination of 14(8+6) seems incipient. On each shell we observe a triangular pattern of particle arrangement.

INTRODUCTION

In heavy-ion storage rings a plasma of bare or heavily ionized nuclei, all with the same mass and positive charge, is kept circulating, with magnetic (or electric) focussing arranged so as to confine the plasma to a narrow region around the equilibrium orbit in the ring. The plasma may be cooled by several methods: among them electron cooling and the newly suggested method of laser cooling, such that the relative thermal motion of the particles with respect to each other is lowered, although the whole plasma is moving around the ring with velocities of 10^{10} cm/sec or more. The number densities in such a plasma can be between 10^5 and 10^8 ions/cm^3 and temperatures of relative motion of 1 deg.K have been reported, temperatures down to the mK range are anticipated with laser cooling. The possibility of observing condensation phenomana in such systems has been noted [1].

We have made an attempt to simulate these conditions by computer molecular dynamics [2] which, within the limitations of the Hamiltonian assumed, give fully detailed information on the structure and dynamics of the system. The basic assumption is that we have a Cartesian reference frame moving with the plasma beam and that the classical, Newtonian, equations of motion may be used to determine the dynamics of the ions. In addition, as will be seen in the next section, there are simplifying assumptions about the confining potential, which imply idealizations of the magnetic fields that contain the plasma in a real storage ring. The results obtained are rather unexpected and hence are being reported even though the more realistic aspects of the conditions prevailing in storage rings have not yet been incorporated.

THE HAMILTONIAN AND BOUNDARY CONDITIONS

We use here a periodically repeating cubic box of length L (expressed in terms of some unit of length = ξ cm) in which there are N ions with potential energy V (the summation over all periodic boxes being implicit),

$$V = \Sigma_i \Sigma_{j>i} 1/r_{ij} + \Sigma_i 1/2K(y_i^2+z_i^2),$$

where x_i, y_i, z_i (i=1,N) are the coordintates of the N particles in the box, r_{ij} is the distance between the particles i and j (all in units of ξ) and V is the energy in units of Z^2e^2/ξ. (In a typical storage ring for ions with charge Z the values of KZ^2e^2/ξ^3 range between 10^{-6} and 10^{-12} erg/cm^2.) If the value of K is large, the coordinates y_i and z_i will be confined to a narrow region around the x axis and the plasma will be spread in a cylinder along this axis, continuing to the edges of the box and joining onto the cylinder of plasma in the adjoining boxes.

In the usual manner of treating a one-component plasma with periodic boundary conditions the double summation in V is evaluated using the standard Ewald summation method first used in this context by Brush, Sahlin and Teller [3], then by Hansen and collaborators [4] in an extensive study of the one-component plasma, and then by Slattery et al. [5] in a particularly detailed study of such systems.

We note here that if the external field potential energy is modified to include the x coordinate, i.e. if it is $1/2K(x_i^2+y_i^2+z_i^2)$, one can dispense with the periodic boundary condition altogether. We shall mention this point again below.

The only parameters in our Hamiltonian are the dimensions of the box L, the number of particles N, the confining potential K and the temperature T at which the system is maintained using standard methods of molecular dynamics [2].

For such a calculation to be meaningful as an approximate simulation of conditions in a storage ring, the paramaters used must be appropriate, in other words the value of K should be sufficiently large to confine the particles in a pencil-like narrow region around the x axis with a diameter much less than the cell size L. We may then expect that the dynamics of the structure in the pencil will be only slightly perturbed by the periodic boundary conditions.

It should be noted that the present calculations only roughly approximate conditions in the storage rings that are under construction at several laboratories. The time average of the focusing forces that contain particles in a ring is proportional to the displacement from a mean equilibrium orbit. Although most storage rings have periodic strong-focusing elements, it is possible, at least in principle, to have a weak focusing ring where the restoring forces are constant in time. In weak focusing, the magnetic field that causes the particles to follow a closed orbit, is proportional to $r^{-1/2}$, where r is the bending radius, and the focusing force is equal in the horizontal (bending plane) and the vertical (perpendicular to the bending plane) directions. This idealized restoring force corresponds to the assumptions in the present calculations, with the circular motion neglected.[1]

N=2000, L=4, K=10,000, T=1/9

Under these condition the diameter of the pencil of plasma is found to be less than 0.6. Before presenting the detailed results let us consider what physical conditions such a calculation would represent.

Suppose that the unit length ξ is 0.1 cm. The beam diameter will then be <0.06 cm; the temperature would be 2°mK for Z=1 and 20°K for Z=90; while $K=2.3\times10^{-12}$ dyn/cm and 2.0×10^{-8} dym/cm for the two temperatures. The number density $\sim\xi^{-2}$ for this pencil with N=2000, 0.06 cm in diameter and 0.4 cm in length, will be 2×10^{6} cm^{-3}. These values include the parameter range of storage rings envisioned at present.

Figure 1 shows the projection of the 2000 particles onto the y-z plane. The stratification in the direction perpendicular to the beam is

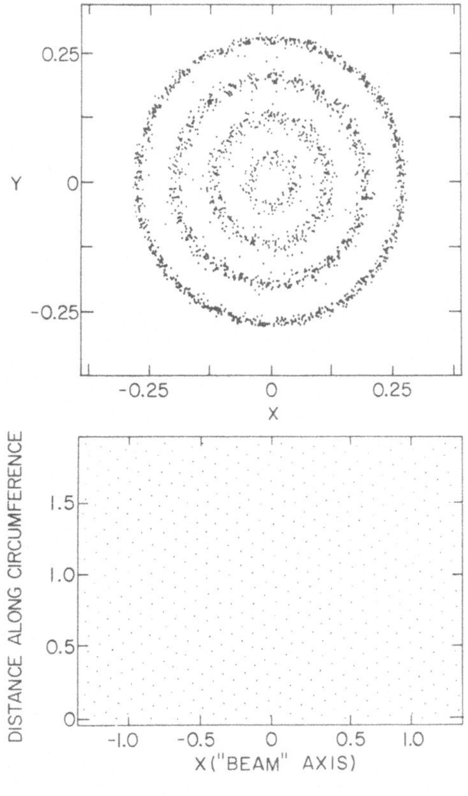

Figure 1 Upper part: Projection of 2000 particles in a molecular dynamics calculation onto the plane perpendicular to the beam (x-axis) for Γ=180. Lower part: distribution of particles in the outer shell with the shell unfolded into a plane. All but the innermost shell show a similar pattern.

quite dramatic, and immediately leads to the conclusion that there must be many more intriguing properties to be analyzed. Here we shall present a few structural properties of this system; dynamical properties will be presented elsewhere. The 3-dimensional pair correlation g(r) in the system as a whole shows a sharp peak at r=0.092 and clear but broader peaks at r=0.17 and 0.245. Since, as seen in Figure 1, a large number of the particles are in the outer shell, the overall pair correlation is distorted by the fact that particles in this shell have no neighbors on the outer side. We have therefore analyzed the 3-dimensional g(r) separately for each shell. Moreover, instead of the standard procedure of presenting g(r) as a function of r, we plot it instead in Figure 2, as a function of the coordination number n(r). It is clear that, except for the outermost shell, the first peak in g(r) has a coordination of 14.

In the lower part of Figure 2 we also show the two-dimensional pair correlation between particles in the same shell. It is clear that for all but the innermost shell, there are six neighbors under a sharp first peak at r=0.092 and 12 more neighbors under the broader second peak. In a projection of particle coordinates, which corresponds to unrolling the cylindrical shells onto a plane, the triangular pattern of particle

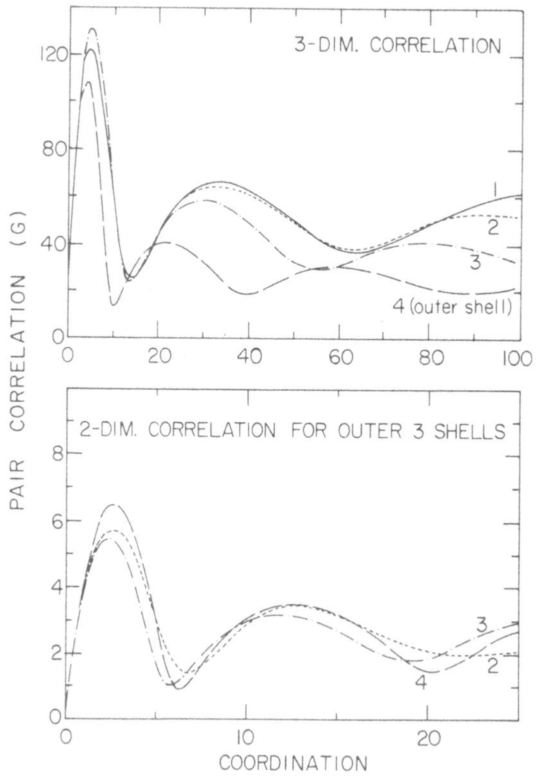

Figure 2 Upper part: Three-dimensional pair correlation function g(r) computed separately with respect to particles within the four shells shown in figure 1. It is plotted against the coordination: the number of particles included up to a given radius. Lower part: The 'two-dimensional' pair correlation function restricted to particles within one shell, for the three outer shells.

positions is clearly seen. The radii of the shells are 0.052, 0.13, 0.20, and 0.28 in the reduced units, and their population 128, 370, 616, and 886. Assuming each of these surfaces to form perfect cylinders (an assumption that becomes less valid as the radius gets smaller) and using constant surface densities for the particles with nearest neighbor distances a=0.092 and six first neighbors in a perfect triangular arrangement, one would estimate 177, 443, 682, and 955 particles.

The innermost shell has a simpler structure. There the particles form a helical pattern around the x axis, rotating about 120 degrees about that axis between successive particles. There is a tendency for the sense of rotation to maintain itself for a number of particles; it does not appear to be randomly distributed.

A calculation with only 100 particles (somewhat less than the number in the inner shell) shows a clearly defined single shell of radius 0.04 (somewhat smaller than the .058 for the inner shell seen with 128 particles) and similar ordering. When the number of particles is reduced to 40, the ions get distributed along the axis with only thermal deviations.

For the 2000 particle system the parameter Γ, which plays a central role in determining the properties of a uniform one-component plasma, can be worked out for our system. Using the volume per particle to define a radius r_s for this volume, $\Gamma=1/(r_s\ T)$. Substituting the data deduced from this case we get $\Gamma=180$.

N=2000, L=4, K=5000, T=0.411

In this calculation with higher temperature and less severe confining force, the shell structure in the lateral direction is less well defined; there are two reasonably well defined outer shells with the outer one being at a distance 0.4 from the x axis. The value of Γ for this calculation is ~40 and the presence of a two dimensional triangular arrangement on the surface of the last outer shell is still quite clearly seen. Thus "ordering": the presence of shells and a triangular two-dimensional arrangement of ions within the shells, can occur at rather low values of Γ.

ISOTROPIC CONFINEMENT WITH N=2000, L=4, K=10,000, T=1/9

This case was studied in order to ascertain whether the shell structure observed might have been induced by the presence of the periodic boundary conditions that had to be assumed for the cylindrical cases in order to allow the Ewald sums to be evaluated. With the confining potential spherically symmetric $(1/2\ K\ (x^2+y^2+z^2))$ one may remove the boundary conditions and compare the results with and without them. We find very clear spherical shell structure in both cases, with radii at 0.58, 0.48, 0.41, 0.34, 0.27, 0.21. On the outer shell the two-dimensional pair correlations show peaks at r=0.080 and 0.151 in both cases, with coordination 6 and 18. Figure 3 shows the shell structure in the case where periodic boundary conditions were not used. The figure caption explains the manner in which the results are presented. Figure 4 is the usual Mercator projection of particle positions on the outermost shell.

The close similarity of the results with and without the periodic boundary conditions leads us to the conclusion that in the isotropic case the boundary conditions are not responsible for the ordering, and thus it seems reasonable to assume that the order seen in the cylindrical case is likewise insensitive to the periodic boundary condition.

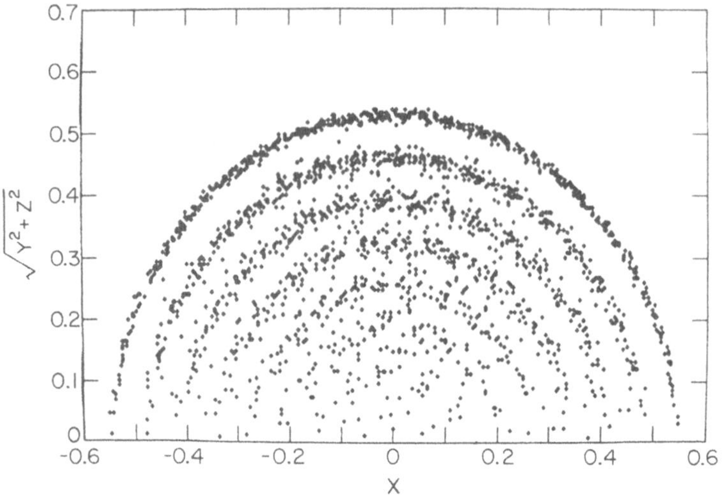

Figure 3 The abscissa is the distance along a line joining two poles of the spherical distribution obtained with 3 dimensional confinement and no periodic boundary conditions. The ordinate is the distance of the particle from this line.

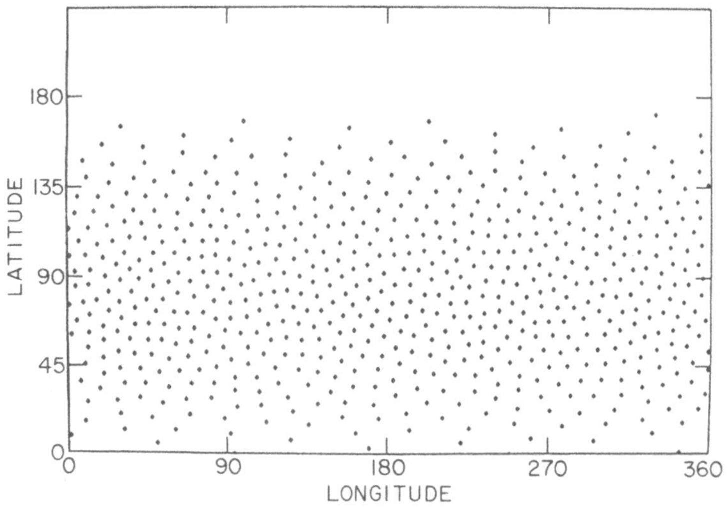

Figure 4 For the calculation mentioned in Figure 3, the Mercator projection is displayed for the positions of particles in the outermost shell.

CONCLUDING REMARKS

The calculations reported here indicate clearly that under the conditions that seem to be within reach of currently planned storage rings ordered structures can occur which are more complex and richer than the liquid-like and Wigner-solid structures that are calculated in uniform one-component plasmas. To what extent the ordering into shells and the consequent triangular ordering within the well defined surfaces will persist when the Hamiltonian is made more realistic than the one used here is now being investigated.

The assumptions made here cannot be satisfied precisely in a real storage ring. First of all, the horizontal focusing must be exerted on particles traveling in circular orbits. It is obvious, that if a beam is cooled such that all particles are travelling at accurately the same linear velocity, the order cannot be sustained if they are to each maintain their velocity and travel in circular orbits of differing radii. In other words, the horizontal focusing force has a shearing component in the direction of travel. The elastic limits againt shear of a condensed array of charged particles may be sufficient to resist slippage of rows of particles within the beam, but these limits will depend on the magnitude of the focusing forces. If slippage does occur, the beam may become heated from the frictional forces.

Another complication is the effect of a restoring force that is periodic in time, and that more nearly approximates the design of the actual focusing elements in storage rings under construction, the so called beta function. The detailed nature of the cooling process will also have to be considered. These aspects of real storage rings will need investigation before one can predict with any reasonable confidence whether this form of order might actually be achievable in currently envisioned facilities.

Finally, a theoretical basis to account for the calculational consequences of the Hamiltonian used here would be a very valuable addition to our understanding of ordering in one-component plasmas.

Work supported by the U. S. Department of Energy, Nuclear Physics Division, under Contract W-31-109-ENG-38. Some of the calculations reported were done on ERCRAY. The initial exploratory calculations were made on the Cray 2 at the Supercomputer Institute of the University of Minnesota.

References

[1] J. P. Schiffer and P. Kienle, Z. Phys. A, 321, 181 (1985); J. P. Schiffer and O. Poulson, Europhys. Let. 1, 55 (1986).

[2] A. Rahman, "Correlation Functions and Quasiparticle Interactions in Condensed Matter", NATO ASI series, Ed. J. Woods Hoslley, (1978) Plenum, NY, p. 417; A. Rahman and P. Vashista, "Physics of Superionic Conductors and Electrode Materials", Ed. J. W. Perram, (1983) Plenum, NY, p. 93.

[3] S. G. Brush, H. L. Sahlin, and E. Teller, J. Chem. Phys. 45, 2102 (1966).

[4] E. L. Pollock and J. P. Hansen, Phys. Rev. A8, 3110 (1973).

[5] W. L. Slattery, G. D. Dooley and H. E. DeWitt, Phys. Rev. A21, 2087 (1980).

CALCULATION OF ELASTIC CONSTANTS USING MOLECULAR DYNAMICS

John R. Ray[†] and Aneesur Rahman[††]

[†]Department of Physics and Astronomy
Clemson University
Clemson, South Carolina 29634-1911

[††]Supercomputer Institute and School of Physics and Astronomy
University of Minnesota
Minneapolis, Minnesota 55455

ABSTRACT

Ray and Rahman have developed a useful method to determine elastic constants in molecular dynamics computer simulations. The adiabatic elastic constants are contained in a formula involving fluctuations in the microscopic stress tensor, M_{ij}, in the microcanonical or EhN ensemble, whereas the isothermal elastic constants are contained in a fluctuation formula of the same form in the canonical or ThN ensemble. Here, E is the system energy, h is a 3x3 matrix constructed from the three vectors spanning the periodically repeating computational cell: $h=(\underline{a},\underline{b},\underline{c})$, N is the particle number, and T is the system temperature. For a potential U which depends only upon the distances between the particles (and is not necessarily pairwise additive) this formula gives the elastic constants as a sum of three terms: a fluctuation term, a kinetic term and the Born contribution which depends upon the potential U. In the static Born method of calculating elastic constants, we have only the Born term evaluated at the static lattice positions of the atoms. The fluctuation equation furnishes a practical method of calculating elastic constants which introduces temperature contributions to the static Born values, producing a significant difference.

We shall give results of our calculations for a nearest neighbor Lennard-Jones system, for which independent Monte Carlo data is available, and for silicon using the 2- and 3-body Stillinger-Weber potential.

INTRODUCTION

The main concern of this paper is the calculation of thermodynamic properties of solids using molecular dynamics; we concentrate on elastic constants. In Section II we describe the present forms of molecular dynamics which are available for such calculations. In Section III we discuss the statistical fluctuation formulas that are used to calculate the elastic constants in different ensembles. Section IV contains the results of actual calculations on a simple Lennard-Jones system which shows that certain forms of molecular dynamics may be used to calculate elastic constants accurately and efficiently. Section V concerns recent results on the calculation of the elastic constants of silicon. Section VI contains our conclusions and suggestions for future work.

FORMS OF MOLECULAR DYNAMICS

Andersen[1] has developed a procedure for carrying out molecular dynamics studies allowing the volume of a cubic molecular dynamics cell to vary in time. Andersen's theory generates the isoenthalpic-isobaric or HPN ensemble of classical statistical physics. Here H is the system enthalpy, P the pressure, and N the particle number; conventional molecular dynamics generates the microcanonical or EVN ensemble, where E is the system energy and V the volume.

Parrinello and Rahman[2,3] generalized the Andersen's theory to allow for changes in both the size and shape of the molecular dynamics cell. One of the main advantages of Parrinello-Rayman theory is that one may use molecular dynamics to study solid → solid structural phase transformations. Such studies are not possible with conventional molecular dynamics since the fixed size and shape of the molecular dynamics cell inhibits structural rearrangements in small systems.

Ray and Rahman[4] have shown how the molecular dynamics theory developed by Parrinello and Rahman can be put into accord with the theory of finite elasticity. The ensemble generated by this molecular dynamics theory is the isoenthalpic-isotension or HtN ensemble. Here H is the enthalpy of the theory of elasticity of anisotropic media and t the thermodynamic tension; see the article by Thurston[5] for a discussion of the theory of finite elasticity.

Nosé[6] has developed a canonical ensemble form of molecular dynamics. Thus, we now have available a form of molecular dynamics which generates the same ensemble as the mainstay of the Monte Carlo method. The Nosé theory generates the TtN ensemble where T is the system temperature. Ray and Rahman[7] have presented a detailed treatment of Nosé's TtN form of molecular dynamics.

If one constrains the molecular dynamics cell to remain rigid, the EhN ensemble is generated. Here h is a matrix which is constructed from the three vectors forming a parallelepiped, which spans the molecular dynamics cell. If $\underline{a}$,$\underline{b}$ and $\underline{c}$ are the three vectors forming the molecular dynamics cell, then $h=(\underline{a},\underline{b},\underline{c})$. The EhN ensemble is the microcanonical ensemble generalized to allow for an arbitrary shape and size of the molecular dynamics cell; the ThN ensemble is the canonical ensemble similarly generalized.

We have now mentioned four different forms of molecular dynamics which generate the four ensembles EhN, ThN, HtN and TtN. The EhN and ThN ensembles are related to one another by a Legendre or Laplace transform and likewise for the HtN and TtN ensemble.

The Hamiltonian for the TtN ensemble was given in Ref. (7)

$$H(\underline{s},\underline{\pi},h,\Pi,f,P) = \frac{1}{2} \Sigma\, \pi_a' \, G^{-1}\pi_a/(m_a f^2)$$

$$+ U + \mathrm{Tr}\Pi'\Pi/(2W) + V_o T_r(t\varepsilon)$$

$$+ P^2/(2M) + (F+1)k_B T_o \ln f \quad , \tag{2.1}$$

where $(\underline{s}_a,\underline{\pi}_a)$ are the scaled coordinates and conjugate momenta of particle a, U is the potential, (h_{ij},Π_{ij}) are the coordinates and conjugate momenta of the molecular dynamics cell and (f,P) are the Nosé mass scaling variable and its conjugate momentum. The constants W and M are introduced in order that h and f satisfy dynamical equations; equilibrium properties of the system of particles are independent of W and M. The prime in Eq. (2.1) represents the transpose of a matrix, m_a is the particle mass, F is the total number of particle degrees of freedom, T_o is the reservoir temperature in the canonical ensemble, ε is the strain matrix which is related to the metric tensor $G=h'h$ by

$$\varepsilon = \frac{1}{2}(h_o'^{-1}\, G\, h_o^{-1} - 1) \quad , \tag{2.2}$$

where h_o is the reference value of the h matrix; h_o is equal to the average value of h in a system under conditions of zero stress.

The equations of motion which follow from Eq. (2.1) have the form

$$m_a f^2 \ddot{s}_{ai} = - \Sigma \chi_{ab}\, s_{ab} - m_a(f^2 G^{-1}\dot{G} + 2f\dot{f})\dot{s}_a \quad , \tag{2.3}$$

$$W\ddot{h} = PA - h\Gamma \quad , \tag{2.4}$$

and

$$M\ddot{f} = 2K/f - (F+1)\, k_B T_o/f \quad , \tag{2.5}$$

where K is the particle kinetic energy $K = \frac{1}{2}\Sigma p_a^2/m_a$, $p_a = m_a f h \dot{s}_a$ being the particle momentum. The microscopic stress tensor P in Eq. (2.4) has the form

$$P_{ij} = V^{-1}[\sum_a p_{ai}p_{aj}/m_a - \sum_{a<b}\chi_{ab}x_{abi}x_{abj}] \quad , \tag{2.6}$$

where $\chi_{ab} = \frac{1}{r_{ab}}\frac{\partial U}{\partial r_{ab}}$. The area tensor A has the form $A = Vh'^{-1}$.

For simplicity, we have assumed U to be expressible in terms of distances between particles. This does not necessarily mean pairwise additivity. The physical variables for the particle, namely (x_a,p_a), are associated with the variables (s_a,π_a) by $x_a = hs_a$, $p_a = h'^{-1}\pi_a/f$. The elastic energy in Eq. (2.1) is the term $V_o T_r(t\varepsilon)$ where $V_o = \det(h_o)$ is the reference volume of the system; this elastic energy term in the Hamiltonian gives rise to the second term on the right hand side of Eq. (2.4) with $\Gamma = V_o h_o^{-1} t h_o'^{-1}$.

If we consider a variable B(x,p,h) then the time average of B along the trajectories generated by Eqs. (2.3), (2.4) and (2.5), $\bar{B}$, is equal to the ensemble average of B in the TtN ensemble

$$\bar{B} = \langle B \rangle_{TtN} \quad . \tag{2.7}$$

Equations (2.3), (2.4) and (2.5) may be used to generate the four ensembles discussed earlier by imposing various constraints. For example, if f=1 and Eq. (2.5) is dropped then we have Eq. (2.3) and (2.4) with f=1, $\dot{f}$=0; this generates the HtN ensemble. As another example, if h=const and Eq. (2.4) is dropped then Eq. (2.3) and (2.5) with h=const, $\dot{h}$=0 generate the ThN ensemble.

Note that we may also apply a uniform pressure P to the system in the presence of a tension t. The Hamiltonian Eq. (2.1) would then involve the sum $PV+V_o Tr(t\varepsilon)$. This leads to a modification of Eq. (2.4) which is

$$W\ddot{h} = (\mathcal{P} - P)A - h\Gamma \quad . \tag{2.8}$$

One has in this case the T (t and P) N ensemble. This ensemble is not a special case of the TtN ensemble because of the manner in which enthalpy is defined in the theory of finite elasticity.[4,5,7]

ELASTIC CONSTANT FORMULAS

For the HtN ensemble, the fluctuation formula for the adiabatic elastic constants was given by Parrinello and Rahman[8] and by Ray[9] and has the form

$$\begin{aligned} \delta(\varepsilon_{ij}\varepsilon_{k\ell}) &= \langle\varepsilon_{ij}\varepsilon_{k\ell}\rangle - \langle\varepsilon_{ij}\rangle\langle\varepsilon_{k\ell}\rangle \\ &= k_B T (C)^{-1}{}_{ijk\ell}/V_o \quad , \end{aligned} \tag{3.1}$$

where the compliance tensor which is the inverse of the elastic constant tensor is defined by

$$(C)^{-1}{}_{ijk\ell} = (\partial\varepsilon_{ij}/\partial t_{k\ell})_s \tag{3.2}$$

where s is the entropy. The adiabatic elastic constants are defined by

$$(C)_{ijk\ell} = (\partial t_{ij}/\partial\varepsilon_{k\ell})_s \tag{3.3}$$

The method of derivation of Eq. (3.1) presented in Ref. (9) has recently been made more rigorous by Ray and Graben.[10]

The fluctuation formula for the adiabatic elastic constants in the EhN ensemble for a general potential U is given in Ref. (17). Here, for simplicity, we state the formula for a pairwise additive central potential; later we shall give this formula for the nonadditive 2- 3-body potential. The fluctuation formula for the adiabatic elastic constants in the EhN ensemble for a pairwise additive central potential has the form

$$\begin{aligned} V_o\, h^{-1}_{oip}\, h^{-1}_{ojq}\, h^{-1}_{o\ell r}\, h^{-1}_{oms}\, C_{pqrs} &= -\frac{4}{k_B T}\,\delta(M_{ij}M_{\ell m}) \\ &+ 2Nk_B T(G^{-1}_{mi}\,G^{-1}_{j\ell} + G^{-1}_{\ell i}\,G^{-1}_{jm}) + \Big\langle \sum_{\substack{a,b\\ a<b}} g(r_{ab}) s_{abi}\, s_{abj}\, s_{ab\ell}\, s_{abm} \Big\rangle \quad , \end{aligned} \tag{3.4}$$

where M_{ij} is related to the microscopic stress tensor $\mathcal{P}_{ij}$ by

$$M = - Vh^{-1} \, Ph'^{-1}/2 \quad , \qquad (3.5)$$

and

$$g(r) = (\partial^2 U/\partial r^2 - \chi)/r^2 \quad . \qquad (3.6)$$

Equation (3.4) was given in Ref. (7).

The (TtN) ensemble for the elastic constants has the same form as Eq. (3.1) except the elastic constants are isothermal instead of adiabatic, whereas the ThN formula is the same as Eq. (3.4) except the elastic constants are isothermal.

ELASTIC CONSTANT CALCULATIONS

General Discussion

For the calculation of elastic constants we prepared an FCC crystal of 500 particles interacting with the Lennard-Jones (12-6) potential including only nearest neighbor interactions. Accurate Monte Carlo data for this potential has been published by Cowley.[11] The system was at a reduced temperature of 0.3 and at zero pressure and stress. In this case by cubic symmetry we have only nine non-zero elastic constants and only three of these are independent

$$C_{11} = C_{22} = C_{33}, \; C_{12} = C_{13} = C_{23}, \; C_{44} = C_{55} = C_{66} \quad .$$

where we have employed the Voigt notation.

HtN Results

Using HtN molecular dynamics Eqs. (2.3) and (2.4) with f=1, $\dot{f}=0$ we employed Eq. (3.1) to determine the adiabatic elastic constants. For this ensemble, we found that Eq. (3.1) did not converge to statistically significant results in molecular dynamics runs of reasonable length (140000 time steps with each time step being 0.005 $\sigma(m/\varepsilon)^{\frac{1}{2}}$ where σ and ε are the Lennard-Jones parameter. This same lack of convergence was also found by Parrinello and Rahman[12] for a different system and is implicit in the results presented by Sprik et al.[13] The unsatisfactory results obtained from Eq. (3.1) persisted at different temperatures for different potentials for different choices of the parameter W and for modified h equations. We feel this property of Eq. (3.1) deserves further study. We would expect the same type of results to occur in the TtN ensemble calculation using Eq. (3.1) for the determination of the isothermal elastic constants.

EhN AND ThN RESULTS

The use of Eq. (3.4) for the same Lennard-Jones system as discussed in the previous section leads to an accurate and efficient determination of the elastic constants as shown by Ray, Moody and Rahman[14,15]. In EhN molecular dynamics, the only dynamical equation is Eq. (2.3) with f=1, $\dot{f}=0$, h=const $\dot{h}=0$. In the ThN molecular dynamics we have Eq. (2.3) and (2.5) with h=const $\dot{h}=0$. In the EhN case, Eq. (3.4) determines the adiabatic elastic constants whereas in the ThN case, the same equation determines the isothermal elastic constants.

As an example of the data collected in this study[14,15] we show the convergence of the isothermal elastic constants in Table I. We also show the Monte Carlo values as given by Cowley[11] and the static Born values.

Table I. Comparison of isothermal elastic constants determined using molecular dynamics and Monte Carlo for a nearest neighbor Lennard-Jones (12.6) potential, FCC solid at a reduced temperature of 0.3. The quantity given is the dimensionless number $C_{ijk\ell}V_o/Nk_BT$. The symmetry averaged elastic constant $(C_{11}+C_{22}+C_{33})/3 = C_{11}$, etc., are given. The error in the molecular dynamics result is determined using the values of C_{11}, C_{22} and C_{33}, which are independently calculated, as three independent determinations of C_{11}. Also shown are the static Born values. The static Born values are calculated using only the last term in Eq. (3.4) evaluated at the static lattice positions.

ThN Molecular Dynamics

Time Steps/1000	C_{11}	C_{12}	C_{44}
5	169.6	81.0	82.4
10	166.1	77.5	81.8
15	153.9	65.2	84.2
20	155.5±1.6	65.9±0.6	83.9±1.6
	Monte Carlo (Cowley)		
25	157.1±1.0	69.3±0.9	82.2±0.2
	Static Born Values		
--	146.9	73.4	73.4

It is useful to study separately the convergence of the different terms in Eq. (3.4). We shall refer to these three terms as the fluctuation term, the kinetic term and the Born term. In Table II we show the convergence of these terms for C_{11} during the molecular dynamics run of Table I. The data for C_{12} and C_{44} shows the same behavior.

Table II. The convergence of the individual terms that appear in Eq. (3.4) in the same ThN molecular dynamics run as in Table I.

ThN Molecular Dynamics

Time Steps/1000	Fluc	Kinetic	Born	C_{11}
5	45.7	4.0	215.3	169.6
10	49.3	4.0	211.4	166.1
15	61.4	4.0	211.3	153.9
20	59.9	4.0	211.4	155.5

As Table II shows, the Born terms are fully converged at 5000 times steps. A closer study shows that the Born terms are converged after even 500 time steps! At 500 time steps for this run the Born term of Table II has the value 211.2. The calculation of the second derivatives of the potential which enter only into the Born terms is the time consuming part of the elastic constant calculation. The rapid convergence of the Born terms means that one can calculate these terms in a short calculation ∿500 time steps. Then this part of the calculation can be omitted and only the fluctuation terms need to be calculated. The fluctuation terms require only the first derivatives of the potential which are already calculated in the dynamics.

The rapid convergence of the Born terms means that the computational time involved in using Eq. (3.4) is the time for the fluctuation terms to converge to a statistically meaningful result. Since this time is apparently much shorter than the time for the strain-strain fluctuations in Eq. (3.1) to converge, Eq. (3.4) finishes a more efficient way of calculating elastic constants than Eq. (3.1).

ELASTIC CONSTANTS OF SILICON

Stillinger and Weber[6] have introduced a model potential to study the solid and liquid forms of silicon. This potential includes not only 2-atom but 3-atom contributions in the interaction potential. We have recently employed the Stillinger-Weber potential to determine the elastic constants of silicon at a number of temperatures.[17]

The microscopic stress tensor now has the form

$$P_{ij} = V^{-1}\left[\sum_a \frac{p_{ai}p_{aj}}{m_a} - \sum_{a<b} \chi_2 x_{abi} x_{abj} - \sum_{a<b} \chi_3 x_{abi} x_{abj}\right] , \qquad (5.1)$$

where $\chi_2 = (\partial U_2/\partial r_{ab})/r_{ab}$, $\chi_3 = (\partial U_3/\partial r_{ab})/r_{ab}$.

The fluctuation formula for the elastic constants in the EhN ensemble has the form

$$V_o h^{-1}_{oip} h^{-1}_{ojq} h^{-1}_{o\ell r} h^{-1}_{oms} C_{pqrs} = -\frac{4}{k_B T}\delta(M_{ij}M_{\ell m})$$
$$+ 2Nk_B T(G^{-1}_{mi} G^{-1}_{j\ell} + G^{-1}_{\ell i} G^{-1}_{jm})$$
$$+ \langle 2\text{-body Born})_{ij\ell m}\rangle + \langle(3\text{-body Born})_{ij\ell m}\rangle \ . \qquad (5.2)$$

The terms labeled 2-body Born in Eq. (5.2) depend on U_2 and have the same form as the 2-body Born terms in Eq. (3.4) whereas the terms labeled 3-body born in Eq. (5.2) depend upon U_3. In Table III we show the calculated and experimental values of the elastic constants of silicon at the temperatures 888K, 1164K and 1477K.

Table III. Elastic constants for silicon in units of 10^{11} dynes/cm^2. The experimental values are from Ref. (18). The theoretical values are calculated using the Stillinger-Weber potential and Eq. (5.2). The system was a 216 particle system at zero pressure and stress, arranged in a diamond lattice. The calculations at each temperature were for 150000 time steps with each time step being $0.01\ \sigma(m/\varepsilon)^{\frac{1}{2}} = 7.66 \times 10^{-16}$ s.

	C_{11}	C_{12}	C_{44}
		T = 888K	
Theoretical	14.14±0.01	7.52±0.00	5.24±0.84
Experimental	15.75	6.05	7.53
		T = 1164K	
Theoretical	13.73±0.03	7.43±0.01	4.57±1.14
Experimental	15.25	5.90	7.32
		T = 1477K	
Theoretical	13.32±0.01	7.39±0.04	4.20±0.83
Experimental	14.80	5.75	7.00

Although there are significant differences (25-40%) between the calculated and observed values in Table III, we note that the calculated values do show the same softening with increasing temperature as is observed. It should also be mentioned that Stillinger and Weber did not use any elastic constant data in determining their potential.

It is again useful to study the convergence of the different terms in Eq. (5.2) separately. We again find that the Born terms, both 2-body and 3-body converge within a few hundred steps. However, the fluctuation term for C_{44} converged slowly so it was necessary to carry out a long calculation (150000 time steps) in order to obtain statistically significant results.

In Table IV we show the different contributions to the elastic constants of silicon for the T=1477K simulation together with the static Born values.

In order to determine if our results were dependent on system size, we carried out similar calculations on a system of $N=2^3 \cdot 216=1728$ particles. The results of our simulation on the larger system confirms that the results for the 216 particle system is not a consequence of the small system size.

Table IV. The four contributions to the elastic constants from Eq. (5.2) in units of 10^{11} dyne/cm^2 for T=1477K. Also shown are the static Born values. Note the large value for the flucutation term in C_{44} in Table IV. The Born terms alone give a value of C_{44} of about 10×10^{11} dyne/cm^2 for C_{44} which is about 40% larger than the experimental value of 7×10^{11} dyne/cm^2. The fluctuation term modifies the Born term in the proper direction but this modification is too large so the calculated elastic constant is about 40% too small! This example shows that caution is needed when interpreting elastic constants calculated using the static Born procedure.

Term	C_{11}	C_{12}	C_{44}
2-Body Born	8.00	8.34	8.34
3-Body Born	5.65	-0.89	1.64
Kinetic	0.40	0.00	0.20
Fluc	.73	.06	5.98
TOTAL	13.32	7.39	4.20
STATIC BORN			
2-Body	9.77	9.77	9.77
3-Body	4.79	-2.39	.80
TOTAL	14.56	7.38	10.57

CONCLUSIONS

We have demonstrated how the EhN or ThN forms of molecular dynamics furnish efficient and accurate methods of calculating elastic constants using fluctuation formula. Because of the rapid convergence of the Born terms, the length of the calculation is determined by the convergence of the fluctuation of the microscopic stress tensor $\delta(M_{ij}M_{k\ell})$ in Eq. (5.2). The calculation of elastic constants in the HtN or TtN ensembles has not proven as successful due to the slow convergence of the strain-strain fluctuations in Eq. (5.1). We believe this deserves further study.

One can extend these calculational methods to other thermodynamic properties such as higher order elastic constants, specific heats . . .

Although we have mentioned that the adiabatic elastic constants are determined from Eq. (5.2) in the EhN ensemble, we can by calculating the temperature coefficients of thermodynamic tension λ_{ij} and the isostrain specific heat C_ε, using flucutuation formulas in the EhN ensemble, determine the isothermal elastic constants via the thermodynamic relation.[5]

$$C^T_{ijk\ell} = C^s_{ijk\ell} - V_o\lambda_{ij}\lambda_{k\ell}/C_\varepsilon \qquad (6.1)$$

where T indiciates isothermal and s indicates adiabatic elastic constants and $\lambda_{ij} = (\partial t_{ij}/\partial T)_\varepsilon$. This procedure was employed in Ref. (15) to determine the relative efficiency of EhN and ThN calculations of elastic constants. To the accuracy of our comparison the two ensembles are equally efficient.[15]

Our calculated result for the elastic constants of the Stillinger-Weber model of silicon, while not agreeing in absolute values, show the same softening with increasing temperature as the observed values. The three body contributions are not small especially in the case of C_{11}, see Table IV, and the fluctuation term is large for C_{44}.

Acknowledgements: Parts of this work were carried out in collaboration with Dr. M. C. Moody and M. D. Kluge. The calculations were carried out on the Clemson University Computer System and at the Supercomputer Institute at the University of Minnesota.

REFERENCES

1. H. C. Andersen, J. Chem. Phys. 72:2384 (1980).
2. M. Parrinello and A. Rahman, Phys. Rev. Lett. 45:1196 (1980).
3. M. Parrinello and A. Rahman, J. Appl. Phys. 52:7182 (1981).
4. J. Ray and A. Rahman, J. Chem. Phys. 80:4423 (1984).
5. R. N. Thurston, in Physical Acoustics, W. P. Mason, ed., Academic, New York (1964), Vol 1, Part A.
6. S. Nosé, Mol. Phys. 52:255 (1984); J. Chem. Phys. 81:511 (1984).
7. J. Ray and A. Rahman, J. Chem. Phys. 82:4243 (1985).
8. M. Parrinello and A. Rahman, J. Chem. Phys. 76:2662 (1982).
9. J. R. Ray, J. Appl. Phys. 53:6441 (1982).
10. J. R. Ray and H. W. Graben, "Fundamental Treatment of the Isoenthalpic-Isobaric Ensemble," to appear in Phys. Rev. A, and references contained therein.
11. E. R. Cowley, Phys. Rev. B 28:3160 (1983).
12. M. Parrinello and A. Rahman, unpublished results.
13. M. Sprik, R. W. Impey and M. L. Klein, Phys. Rev. B 29:4368 (1984).
14. J. R. Ray, M. C. Moody and A. Rahman, Phys. Rev. B 32:733 (1985).
15. J. R. Ray, M. C. Moody and A. Rahman, Phys. Rev. B 33:895 (1986).
16. F. H. Stillinger and T. A. Weber, Phys. Rev. B 31:5262 (1985).
17. M. D. Kluge, J. R. Ray and A. Rahman, "Molecular Dynamic Calculation of Elastic Constants of Silicon," J. Chem. Phys., to appear.
18. K. H. Hellwege, editor, Landolt-Börnstein: Crystal and Solid State Physics, Vol. 11, Springer-Verlag, Berlin (1979); p. 116.

FRAGMENTATION AND STRUCTURE OF SILICON MICROCLUSTERS

B.P. Feuston*, R.K. Kalia and P. Vashishta

Argonne National Laboratory
Argonne, Illinois 60439

*University of Cincinnati
Cincinnati, Ohio 45221

INTRODUCTION

In the past several years there has been growing interest in the properties of small clusters of atoms.[1-7] Inert gas clusters are known to form predominantly 13-, 55-, etc. atom clusters explained by stable icosahedral packing.[7] More recently there has been an effort to determine the geometrical arrangements and electronic configurations for Group IV elements, C, Si, Ge and Sn.[8-11] Silicon and germanium have both been found to form unusually stable 6- and 10-atom clusters. Si_4 and Ge_{14} are also found to be particularly stable. These unusually stable clusters are referred to as the magic numbers. Unlike Si and Ge, carbon exhibits chain-like structures due to strong π bonding with their own set of magic numbers. Bloomfield, Freeman, and Brown[8] (BFB) performed a photofragmentation experiment on ionized silicon, measuring relative cross sections and individual fragmentation channels for clusters containing upto 12 atoms. Neutral clusters were created by vaporizing bulk silicon in the presence of a stream of inert gas. These clusters were then ionized by a laser and mass selected by means of electrodes. Fragmentation occurred upon exposing the cluster to an intense beam of laser radiation. The size of the resulting charged fragment was then determined. The temperature of the cluster prior to dissociation reaches the order of the melting temperature of bulk silicon ($T \sim 2000$ K). The fragmentation channels were found to depend critically on the overlap between the laser beam and the cluster. Bloomfield et al. report relatively low total photofragmentation cross sections for $Si^+{}_4$, $Si^+{}_6$, and $Si^+{}_{10}$ in addition to the unusually common occurrence of $Si^+{}_6$ and $Si^+{}_{10}$ from the fragmentation of $Si^+{}_{7-11}$ and $Si^+{}_{12}$, respectively. The magic numbers for silicon have been determined to be 4, 6, and 10 from these experimental results.

There have been two theoretical calculations performed to explain the existence of these magic numbers. The first by Raghavachari and Logovinsky[10] (RL) employed a Hartree-Fock (HF) calculation with electron correlation included through perturbation theory. The more recent calculation was completed by Tomanék and Schlüter[11] (TS) who used both the empirical tight-binding (TS) approach and the local density approximation (LDA) in Density Functional Theory.

The two above calculations involved choosing several likely geometries as a starting configuration and minimizing the total energy to find the local potential minimum of the relaxed system. The lowest energy found from this limited search of possible configurations was identified as the ground state. The ground-state structure for Si_{2-5} as determined by all three approaches, HF, TB, and LDA, are topologically equivalent with slight variations in binding energies and bond lengths. Close-packed structures were found with significantly lower energy than the corresponding microcrystalline fragments. The most stable zero-temperature structures for Si_{3-5} are the isosceles triangle, planar rhombus and the "squashed" trigonal bipyramid, respectively. Raghavachari and Logovinsky have found the most stable configuration for Si_6 to be constructed by edge-capping a bond on the base of the trigonal bipyramid. The results of both TS calculations indicate the tetragonal bipyramid has the highest binding energy for Si_6. These two structure for Si_6, the edge-capped trigonal bipyramid and the tetragonal bipyramid, are also topologically similiar. In most cases, the charged cluster was found to have nearly the same structure as the neutral cluster. Detailed analysis of charge distributions indicated the positive charge usually resided on the least bonded atom. The ground-state geometries of clusters with seven or more atoms are found by adding face-capped atoms to the Si_6 structure. The tetracapped octahedron forms the ground-state geometry for the largest cluster considered by RL, the 10-atom cluster Si_{10}. Tománek and Schlüter determined that by capping an atom to every face of the octahedron one could form the particularly symmetric, lowest-energy structure for Si_{14}. By construction large clusters Si_N (N=7-14) can be easily dissociated into Si_6 + Si_{N-6} fragments explaining the common occurrence of Si_6 in the photofragmentation spectra. The fragmentation energy, the smallest energy necessary to break the ground-state structure of each cluster, was also calculated by RL. Their results show that Si_4 and Si_6 require relatively large energy before dissociation can occur indicating particularly stable ground-state structures. Tománek and Schlüter also found large energy gaps between bonding and anti-bonding states, similiar to typical semiconductors, for Si_6 and Si_{10}. The existence of magic numbers (N=4,6,10) is then explained by high, zero-temperature ground-state binding energies (RL and TS), high fragmentation energies relative to the ground state (RL), and the existence of relatively large energy gaps (TS).

The above approaches restrict calculations to zero-temperature ground-state structures and severely limits the sampling of possible configurations. At best the global ground-state structure may be obtained for small clusters by a well chosen initial configuration. It may be possible to determine the magic numbers and fragmentation spectra from the ground-state binding energies and structure, but the relationship between the lowest-energy zero-temperature configurations and the energetics of finite-temperature microclusters is not obvious. Recall fragmentation of Si clusters occurs at temperatures the order of the melting temperature (T~2000 K). What is needed, a first-principles finite-temperature calculation, allowing the determination of all possible structures, their corresponding binding energies, and fragmentation spectra, is not presently possible. However, a molecular dynamics calculation does allow one to study the nature of fragmentation in addition to determination of the global ground-state structure and all mechanically stable configurations underlying the finite-temperature cluster, once given an interaction potential. We present results for such a calculation for Si_{2-14} using the Stillinger-Weber[12] 3-body potential. Our results indicate that the existence of magic numbers is determined by the topology and energetics of high-energy bound structures rather than the structure and ground-state energies at zero temperature.

INTERACTION POTENTIAL, MOLECULAR DYNAMICS TECHNIQUE AND THE METHOD OF STEEPEST DESCENTS

We have chosen to use Stillinger and Weber's[12] (SW) phenomenological potential for silicon, since it is well-known that a 2-body potential in itself is insufficient to correctly describe the dynamics of a collection of covalently bonded atoms. The recently devised SW potential includes both 2- and 3-body interactions, where the 3-body contribution always increases the total potential when the angle formed by a central atom and two of its covalently bonded neighbors differs from the perfect tetrahedral angle. The 3-body term in the potential is identically zero in the perfect diamond crystal and for tetrahedrally coordinated microcrystalline fragments. The paramenters for this potential were determined from the bulk properties of silicon; cohesive energy of the diamond lattice, the melting temperature and local structure of molten silicon. The potential's range of interaction (3.77 Å) is just short of the second nearest-neighbor distance (3.83 Å) in the diamond crystal. The simulation was performed in reduced units with all energies expressed in units of ε (50 kcal = 2.17 eV), the magnitude of the minimum in the 2-body potential, and all lengths expressed in units of σ (2.0951 Å). In these reduced units a temperature (T^*) of .1 is equal to 2500 K.

Molecular dynamics[13] (MD) is intrinsically a finite-temperature method. For the given interaction potential one can determine the trajectory of the system in the 6N-dimensional phase space by numerical integration of Newton's classical equation of motion. To begin the trajectory, initial positions and velocities are chosen so that the total linear and angular momenta are zero. The accelerations can then be calculated from the SW force. Various bonding geometries, including the lowest-energy configuration of RL and TS, were used as starting configurations though the final results were independent of the initial geometry. Microcrystalline fragments were found to have the lowest-binding energies and were unsuitable starting points for large clusters Si_{8-14}.

The trajectory of individual atoms in the equilibrium state are completely determined by the MD simulation. The instantaneous positions of N atoms is represented by a 3N-dimensional vector in configuration space. As time evolves this vector traces out a trajectory with each point completely specifying the 3N coordinates of the system. By uniquely assigning each point on this trajectory in the 3N-dimensional configuration space to a potential minimum corresponding to a particular geometry, one can determine the underlying structure of the system at each time step. The assignment of each point in configuration space to a local potential minima is a mathematically well defined problem[14] that can be solved through the method of steepest descents. The steepest-descent quench (SDQ) assigns each point in configuration space to the first potential minimum encountered when descending from that point along the steepest available path. When SDQ is invoked the instantaneous positions are mapped onto a new point in configuration space where the total kinetic energy of the system is taken to be zero and the new positions correspond to the zero-force configuration of the local potential minimum. The mapping is performed in parallel with the MD simulation and does not effect the particles trajectories.

An exhaustive search was carried out to (i) enumerate all accessible mechanically stable configurations, including the global ground-state structure, and the relative probability of occupying any particular configuration at finite temperatures, (ii) find the highest energy (E_f) and temperature (T_f) that system would remain bound, and (iii) determine the fragmentation spectra, for Si_{2-14} microclusters. Operationally we have defined the fragmentation energy E_f to be the highest energy, E=T+V, that the cluster remains bound for 25,000 MD time steps before the onset of fragmentation. The fragmentation temperature T_f is determined by averaging the total kinetic energy over many time steps (typically 11,000 steps) during a MD run at the highest energy E_f.

For each initial configuration chosen the velocities were increased monotonically until fragmentation occurred. Between each velocity scaling 1000 MD steps were performed to allow the system to achieve equilibrium. A fragmented system is defined as a configuration in which some collection of atoms (usually one) no longer interacts with the parent cluster so that the original cluster is composed of two or more smaller clusters. From the heating procedure, ten to twenty bound systems of various energies were generated for each cluster. Each of these systems were allowed to continue at constant energy for 10,000 more MD steps with SDQ performed in parallel every one hundred steps to enumerate the underlying configuration as shown schematically in Fig. 1. If the system fragmented during this or any other MD run, the position prior to and immediately after fragmentation were saved for analysis of fragmentation channels and the structure from which the dissociation occurred. The ground-state energy and structure were determined from the lowest of the approximately 1000 potential minima identified for each Si cluster (see Fig. 1). Of course only a small

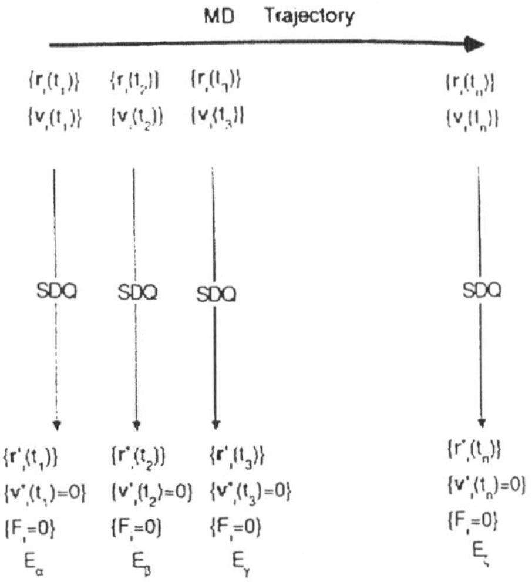

FIG. 1 Schematic representation of the steepest-descent quench (SDQ) performed in parallel with the molecular-dynamics (MD) simulation. MD generates positions $\{\mathbf{r}_i\}$ and velocities $\{\mathbf{v}_i\}$ at each time step through integration of Newton's equation of motion. The SDQ maps the instantaneous positions to the local minimum E_α where the force $\{\mathbf{F}_i\}$ on each particle vanishes and the velocities are to zero. This mapping does not in any way effect the continuation of the MD trajectory.

subset of these potential minima were numerically different corresponding to the different accessible structures. The relative probability of visiting any particular structure at finite temperatures below T_f was found through SDQ performed on systems generated from monotonic heating of the ground-state configuration. The SDQ was performed every fifth step during a 5000 MD step simulation for determining the statistics of the relative number of visits to each particular structure. In this way one finds only those states accessible from the ground state. Upon completion of this work over 500,000 MD steps and nearly 60 systems were investigated for each cluster, Si_{3-14}.

GROUND-STATE STRUCTURES

The lowest-energy structures were determined from the approximately 1000 potential minima found through the SDQ of many systems of various energies for each cluster Si_{3-14}. Steepest-descent mappings were also carried out in parallel with a MD run for Si_2 . As expected, for this model potential the distance seperating the two atoms in the dimer is exactly the nearest- neighbor distance as in crystalline silicon. The ground-state configurations of Si_3, Si_4, and Si_5 have the following symmetrical and planar geometries: the equilateral triangle (Fig. 2a) the square, and the regular pentagon (Fig. 3a), respectively. It should be noted that the isosceles triangle (Fig. 3b) consisting of only two bonds forming a perfect tetrahedral angle, is only slightly higher in energy than the ground-state structure of Si_3. The loss in the 2-body binding energy of this metastable state with respect to the lowest-energy configuration due to the decrease in the number of bonds is offset by the 3-body interaction. A more complete discussion of the high-energy structures for Si_{3-14} is given in the section on hidden structures. It is not suprising the pentagon has the lowest energy for Si_5 since the angle between adjacent bonds (108°) is only 1.5° smaller than the perfect tetrahedral angle. The "squashed" trigonal bipyramid (Fig. 3b) whose energy is higher than the ground state of the pentagon by less than 1%, was also commonly found during steepest-descent quenches from intermediate temperatures. The structure of Si_N (N=3,4,5) can be understood in terms of relaxing the edge-capped Si_{N-1} structure.

The triangle is the basic subunit in the formation of the Si_6 and Si_7 ground states. The ground-state structure of Si_6 is the first to have 3-dimensional geometry with all atoms 3-fold coordinated. The symmetrical stacking of two equilateral triangles, similar to a wedge (Fig. 4a) form the Si_6 structure while the Si_7 ground-state configuration can be thought of as capping each of the three edges on the base of trigonal pyramid. The Si_7 structure containing one equilateral and 6 nearly equilateral, isosceles triangles may be acheived from the reconstruction of an edge-capped Si_6 ground-state configuration.

One can form the symmetric ground-state geometry of Si_8, a perfect cube by either reconstructing a face-capped Si_7 structure or by attaching a Si_2 dimer to a square face of the Si_6 wedge. The addition of one atom to the edge of the Si_8 cube forms the lowest-energy structure for Si_9. Allowing this structure to relax breaks the edge-capped bond forming two identical, nearly perpendicular, non-planar pentagons. The ground-state geometry for Si_5 may be used to obtain Si_{10} in the same way the ground-state structures for Si_N (N=2,3,4) can be used to obtain those of Si_{2N} with the lowest-energy configuration for Si_{10} made of two symmetrically stacked pentagons (Fig. 5). This structure may also be formed by face-capping the Si_9 ground-state configuration or by adding a dimer parallel to one face of the Si_8 cube.

Reconstruction of an atom capped to a pentagon face on the Si_{10} ground-state structure results in the lowest-energy configuration for Si_{11}. The ground-state structure for Si_{11} is the smallest to contain a 4-fold coordinated atom. The symmetrical

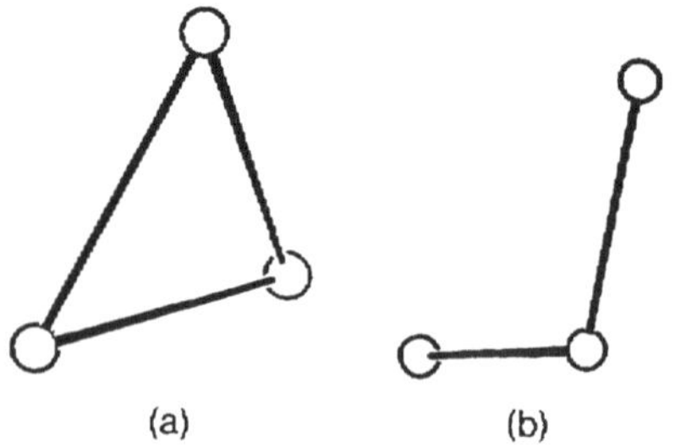

FIG. 2 Si_3 structures: (a) Ground-state structure, equilateral triangle; (b) isosceles triangle with the only two bonds forming a perfect tetrahedral angle (Si_3 chain).

ground-state geometry for Si_{12} has four identical pentagons bound in stacked pairs orientated perpendicular to one another and pointing in opposite directions. Similar to Si_{11}, the ground-state structure of Si_{13} contains one 4-fold coordinated atom. The configuration is best described by the edge-sharing of four identical pentagons to create a 4-sided cone with the small end forming a square and the open end capped by an atom connected to each of the four protruding points. The Si_{14} ground-state geometry similar to all even-numbered clusters, is very symmetrical consisting of six identical pentagons and three perfect squares with all atoms lying on the surface having only three bonds.

Several features appear common to all Si_N (N=3-14) microclusters interacting through the SW 3-body potential. First one observes the existence of three basic "building blocks" (triangle, square and pentagon) in the ground-state energy structures for each cluster Si_{3-14}. These not only form the lowest-energy configurations for Si_3, Si_4, and Si_5 but also by trivially stacking these three planar figures one obtains the ground-state configurations of Si_6, Si_8, and Si_{10}, respectively. All metastable configurations for large clusters are formed by 3-, 4-, and 5-membered rings which are the distorted ground-state configurations for Si_3, Si_4, and Si_5. The concept of basic building blocks underlying complex structures has been previously used by Chadi[15] to construct new crystalline forms for Si and Ge.

The present results compare favorably with those of RL and TS. Our calculated binding energies are consistently lower than the quantum chemical calculations due to the SW potential's inability to correctly evaluate the interaction energy between underbonded atoms. Yet the lowest-energy configurations determined by the two previous calculations do routinely appear in these MD results with the additional possibility of lower-energy structures for Si_5 and Si_6. The ground-state configurations for Si_{7-14} determined from the present MD simulation can be constructed by the addition of atoms to the lowest-energy Si_6 structure similar to the previous work. Consequently this model potential could also predict the natural occurrence of Si_6 fragments from the 6+N (N=1-6) structure if dissociation occurred from the ground state. The plot of ground-state energy per atom (E_0) versus number (N) in Fig. 6a does not show any indication for the existence of the magic numbers

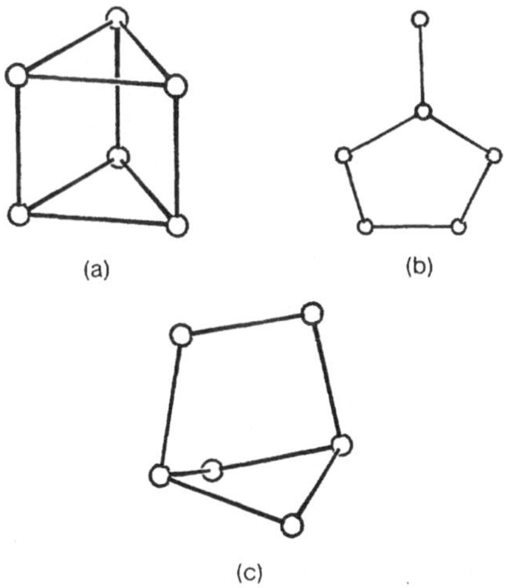

FIG. 4 Si_6 structures: (a) Ground-state geometry can be formed by symmetrically stacking two equilateral triangles; (b) Corner-capped pentagon; (c) Reconstructed face-capped pentagon.

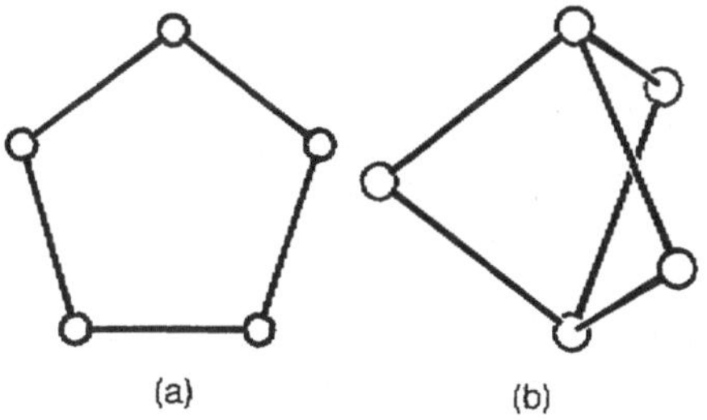

FIG. 3 Si_5 structures: (a) Ground-state structure, pentagon; (b) Trigonal bipyramid.

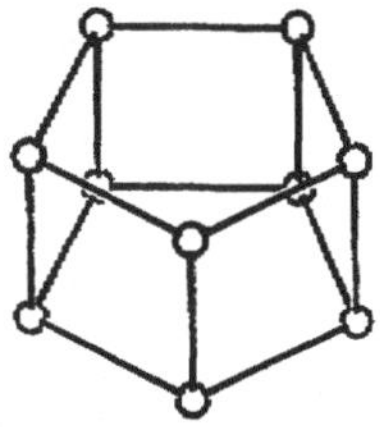

FIG. 5 Si_{10} structure ground-state geometry can be formed form by symmetrically stacking two regular pentagons.

(N=4,6,10) that appear in the plots by RL (Fig. 6b) and TS (Figs. 6c-6d). In the previous calculations, a limited search of possible ground-state geometries resulted in relatively stable structures for $Si_{4,6,10}$. The unusually common occurrence of these clusters in the photo- fragmentation spectra of silicon was then explained by their particularly stable ground-state configurations. We have found structures topologically equivalent to the ground-state structures of RL and TS though for Si_5 and Si_6 new structures of higher binding energies were also obtained. From this simulation the existence of magic numbers (N=4,6,10) can not be explained by the ground-state energies (Fig. 6a) or structures. In the present work, even-numbered clusters have relatively stable ground-state structures due to their symmetrical forms which tends to maximize the number of bonds while minimizing the 3-body potential energy. Si_6 is the only exception due to the large contribution of the two equilateral triangles to the 3-body energy.

In contrast to RL and TS, we are unable from the zero-temperature ground-state energies and structures to make any claims to the presence of magic numbers even though our calculated structures for silicon microclusters are in accord with the previous work. The present zero-temperature results indicate all even-numbered clusters are relatively stable particularly Si_8. The ground-state configurations for Si_N (N=7-14) can be constructed from Si_{N-1} with Si_6 as the basic subunit but that is a weak argument for the mechanism of fragmentation and the explanation for the common occurrence of Si_6 fragments especially when fragmentation occurs at such high temperatures. The structures underlying the high-temperature cluster must play an important role in the determination of fragmentation spectra and they can only be studied by investigating finite-temperature systems.

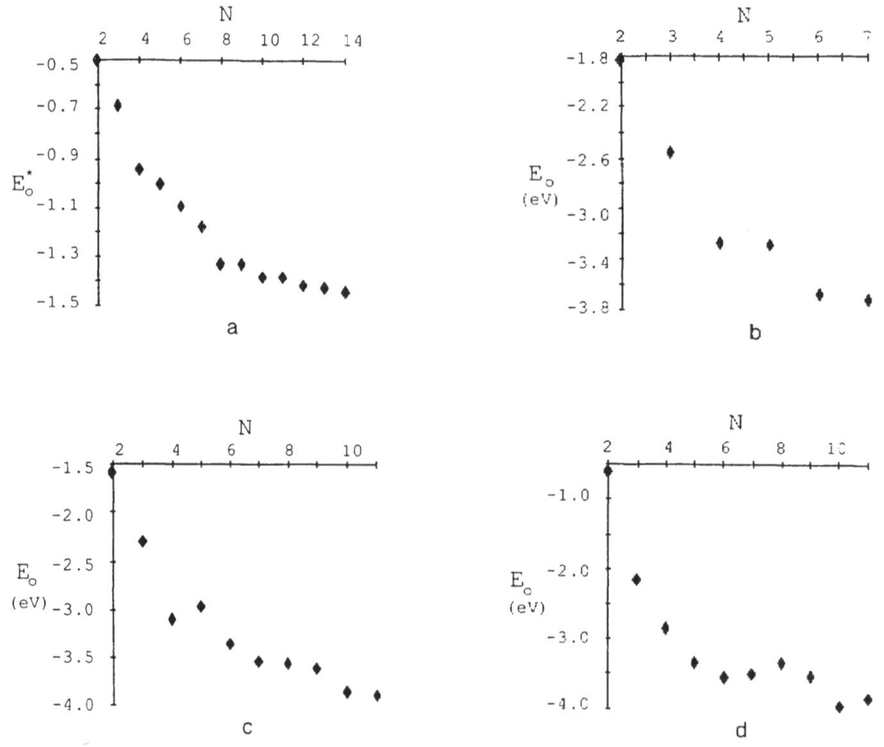

FIG. 6 Ground-state energy per atom (E_o) versus Number (N) for: (a) Present results, in the reduced units of energy* (ε); (b) Raghavachari and Logovinsky (RL), Hartree-Fock calculation; (c) Tománek and Schlüter (TS), Tight-Binding calculation; (d) Tománek and Schlüter (TS), Local Density Approximation . The unusually common occurrence of $Si_{4,6,10}$ in the photofragmentation spectra obtained by BFB is explained by RL and TS as due to the relative stability of their ground-state configurations. In the present calculation the presence of magic numbers can not be explained by ground-state energies or structures but only by the topology and energetics of finite-temperature clusters.

Temperature Dependence of the Total Internal Energy

Starting from the ground-state configuration each cluster was monotonically heated by uniformly increasing all of the atomic velocities. After scaling the velocities a constant energy MD run of 11,000 time steps was performed for determination of the average temperature. The curves of total internal energy per atom (E) versus temperature (T) were plotted for each cluster $Si_{3\text{-}14}$. There exist small jumps in these curves similar to structural transformations, particularly for small finite-temperature clusters $Si_{3\text{-}6}$, see Fig. 7. These steps which appear in the curve of E vs T are indicative of the change in the majority of underlying mechanically stable states from lower- to higher-energy structures rather than a single structural transition. An investigation of the clusters lacking this feature in the E vs T curve has shown them to possess a large number of accessible structures with small differences in energies that span the range between low- and high-energy configurations. Since the transition from the lowest to the highest-energy structures involves many intermediate configurations, the shift in the majority of underlying structures occurs smoothly and a sharp transition can not be seen in E vs T. In some cases many states are accessible but are grouped into subsets of structures of nearly equal energies with each group seperated by relatively large energy differences such that the transition between different subsets also appears as a small step in E vs T. As an example, Si_6 has 16 different configurations which can be grouped into 3 subsets of nearly equal binding energies giving rise to 2 small jumps in E vs T (Fig. 7b). For the energy distribution of the underlying structures giving rise to these jumps in E vs T for Si_6 see Fig. 10.

Hidden Structures

Of the systems generated in the monotonic heating from the ground state of each cluster, four were chosen for a 5000 time step MD run with SDQ performed in parallel every fifth step (see Fig. 1) to enumerate the underlying mechanically stable configuration and to determine the relative probability of visiting each structure. Due to the large number of accessible configurations only the ground-state structure and the most interesting higher-energy structures are given for $Si_{6\text{-}10}$. The choice of which high energy configurations were most pertinent was based on the relative number of visitations found by SDQ. The number of possible configurations given below for $Si_{6\text{-}14}$ are for only those structures that appeared more than once in a given MD run where 1000 configurations were determined. Allowed structures which were not visited or seldom found are considered to be statistically unimportant. To simplify the following discussion, we will define "a chain of atoms" to be a bound cluster with all atoms except the two end atoms exactly two-fold coordinated and all angles perfect tetrahedral with the three-body contribution to the total potential energy identically zero. In reduced units the energy of the Si_N chain is (N-1)/N setting an lower limit on possible binding energies. There are several bound configurations for Si_N with (N-1) 2-body bonds and only perfect tetrahedral angles, all degenerate in energy. The chain is the most stable of these structures since it has only two end atoms that are singly coordinated. These highest-energy structures are visited below E_f in $Si_{3\text{-}6}$ and can be identified in

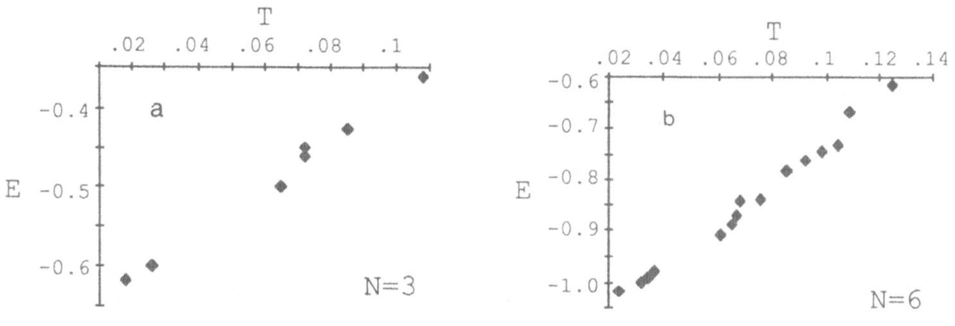

FIG. 7 The total internal energy per atom versus temperature curves are shown for Si_3 and Si_6 in (a) and (b), respectively. Each cluster was started in its ground-state configuration and monotonically heated by multipling the velocities by 1.1. After each scaling the temperature was determined by averaging the kinetic energy over 11,000 MD steps. A reduced temperature (T^*) of .1 corresponds to 2500 K.

superheated clusters upto Si_{11}. It should be noted that all of these high-energy structures can be considered fragments of the perfect diamond crystal. All possible structures are enumerated below for Si_{3-5} while only the statistically important structures are given for the larger clusters Si_{6-14}. There are probably more configurations for these larger cluster than were determined from the SDQ. Since these structures did not appear in the longest MD runs, they are considered statistically unimportant and will not affect the results.

N=3 The two previously discussed triangles were the only mechanically stable structures available to Si_3 (Fig. 2). Without ambiguity the equilateral triangle and isosceles triangle, consisting of only two bonds forming a perfect tetrahedral angle, may be referred to below as the triangle and chain

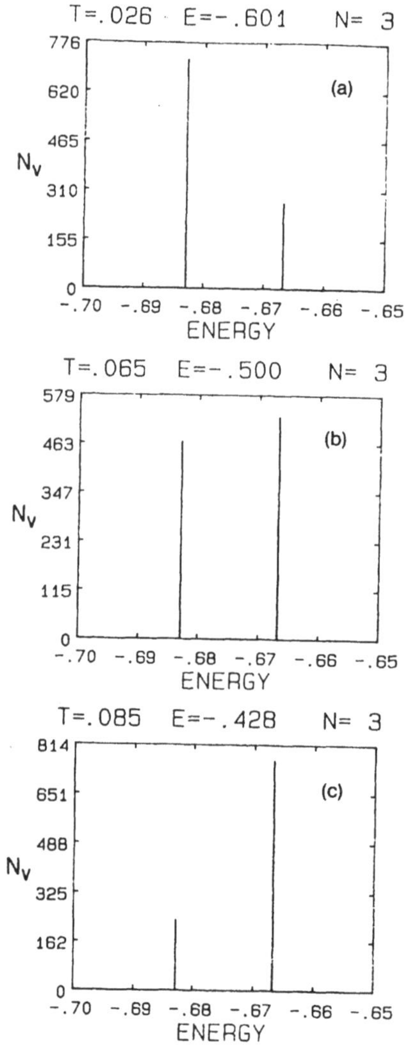

FIG. 8 Histograms of the visited potential minima for Si_3 determined by 1000 SDQ mappings performed in parallel with a 5000 MD step simulation (see Fig. 1). Number of visitations N_v vs energy per atom (E) after SDQ are given for three different temperatures; (a) T^*=.026, (b) T^*=.065 and (c) T^*=.085.

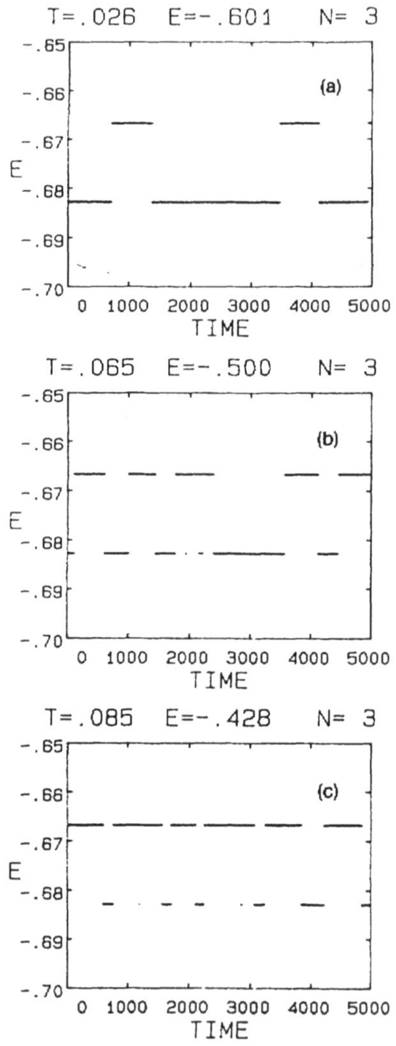

FIG. 9 The underlying potential minimum is plotted against the number of MD steps for the same three simulations depicted in Fig. 8. Not only does the majority of underlying state shift towards higher-energy configurations but the frequency of transitions also increase with increasing temperature.

respectively. At low temperatures both structures are visited while the relative distribution of underlying structure shifts more toward the chain with increasing temperatures as can be seen in Fig. 8 where the number of visitations (N_v) versus underlying potential energy minima (E) is plotted for three finite temperature systems. In Fig. 9, the local potential-energy minimum resulting from the SDQ is plotted against the MD time steps to show the frequency of transitions between the two structures. From this figure, one can see that not only does the majority of underlying states shift towards higher-energy configurations but the frequency of transitions also increases with increasing temperature. At T^*~.065 both structures are equally likely to be visited. A step, similar to a structural transition, appears near this energy in the graph of E vs T (Fig. 7a). Recall T^*=.1 corresponds to a temperature of 2500 K.

N=4 Four structures and three energies, the highest one being doubly degenerate, are accessible to Si_4. Excluding the ground state the structures listed in the order of increasing energy are the corner-shared triangle, the chain, and the trigonal pyramid. The latter two geometries, fragments of the diamond structure, have three bonds with only tetrahedral angles and are therefore degenerate in energy. Corner-capped atoms added to the center or the end of the Si_3 chain form the pyramid or chain respectively. The two small jumps in the curve of E vs T appear at T^*~.07 and T^*~.11 which indicate the change in the most common underlying structure, first from the ground state to the corner-capped triangle and then to the two high-energy structures.

N=5 Of the seven energies found for configurations of Si_5 the energy of the "squashed" trigonal bipyramid (Fig. 3b) is the closest to that of the ground-state structure (Fig. 3a) a regular pentagon. Another intermediate-energy configuration can be obtained by corner-capping the square. Three high-energy structures can be viewed as corner-capping a triangle with a two atom chain, corner-sharing two triangles and capping two atoms on the same corner of a triangle, both minor reconstructions of the crystalline tetrahedron. These last two structures are the first to appear with a 4-fold coordinated atom. The chain and the tetrahedral fragment of the diamond lattice have the same highest energy with four bonds but only the chain has been observed. The center atom of the tetrahedron is 4-fold coordinated but the remaining four atoms have only one bond and therefore the structure is unstable with respect to thermal fluctuations. Unlike Si_3 and Si_4 the ground-state structure is dominant in the underlying structures even upto energies just below E_f. In E vs T a slight jump appears at T^*~.095 corresponding to the clusters ability to access the high-energy structures for the first time.

N=6 Sixteen different potential-energy minima were found for Si_6, but only subsets of these structures were actually visited at any particular temperature. The lowest-energy structure, two equilateral triangles stacked in the form of a wedge (Fig. 4a) was not observed in systems near the fragmentation energy where more open structures were commonly found. A distorted octahedron, similiar to that of RL and TS, is found though a corner-capped pentagon (Fig. 4b) is closer in energy to the ground-state and more frequently visited. Even at relatively low temperatures T^*~.037 the cluster undergoes many transitions between configurations of nearly equal energies. These low-energy structures can be generated from the wedge by breaking one or two bonds and allowing the configuration to relax. The increase in the 2-body potential energy is again balanced by the reduction in the 3-body contribution. Two notable structures in this set are the reconstructed face-capped pentagon (Fig. 4c) and two edge-sharing squares forming a tetrahedral angle. Construction of the intermediate structures begins by breaking bonds of these latter two configurations. The most commonly found geometries at .95E_f as determined by SDQ were three such structures, the corner-capped pentagon, a 2-atom chain attached to the corner of the square and the capping of diagonally opposite corners of the square to form a distorted tetragonal bipyramid. It has been noted previously[8] that the tetragonal bipyramid is topologically similiar to the 6-membered "chair" ring though the energy corresponding to this microcrystalline fragment was seldom found from the steepest-descent mappings. The Si_6 chain was found only at the highest temperatures. The appearance of two small steps in the figure for E vs T (Fig . 7b) at T^*~.07 and T^*~.11 concur with the histograms in Fig. 10 indicating the move of the majority of underlying states to higher-energy structures. These transitions are between three groups of structures, shown in Fig. 10c and not just between single structures.

N=7 The most commonly found low-energy configuration for Si_7, only .1% higher in energy than the ground state may be obtained by reconstructing a bicapped trigonal bipyramid. The other 22 structures found can be described by various edge-sharings and cappings of the triangle, square and the pentagon. At energies near E_f a dimer attached across a distorted pentagon's face and a square sharing an edge with a pentagon accounted for over 50% of the underlying mechanically stable structures. The E vs T curve has none of the interesting features exhibited by the smaller clusters due to the large number of accessible configurations with an even distribution of binding energies.

<u>N=8</u> The Si_8 ground-state geometry, a perfect cube, is a particularly stable structure since all atoms are 3-fold coordinated and all angles are at 90° differing by 19° from the perfect tetrahedral angle. The high degree of connectivity coupled with relatively little frustration due to 3-body interactions enables the ground-state configuration to be particularly stable. However in our calculation (Figs. 11a-11b) and in the experimental results it is obvious N=8 is not a magic number. This is an example of our assertion that E_0 vs N (Fig. 6) cannot predict the existence of magic numbers. For temperatures on the order of $.5T_f$, the atomic vibrations around the equilibrium positions of the ground-state becomes too large and the first structural transformations occur. Due to the overall symmetry at least two bonds must be broken before the other 27 observed structures become accessible.

<u>N=9</u> With eight 3-fold and one 2-fold coordinated atoms the ground state structure is the only one visited for energies upwards to $.5E_f$ and remains the dominant underlying structure for all energies under E_f. Another structure commonly found near E_f is formed by four

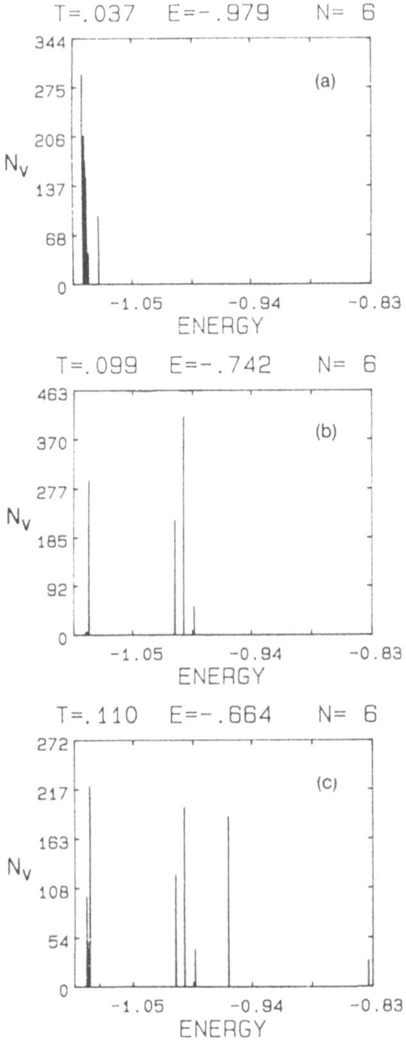

FIG. 10 Histograms of the visited potential minima for Si_6 at: (a) T^*=.037, (b) T^*=.099 and (c) T^*=.110. The structures accessible to Si_6 can be grouped into sets of configurations of nearly equal energies. Transitions between subgroups occur with increasing temperature, indicated by the small jumps in E vs T (Fig 7b).

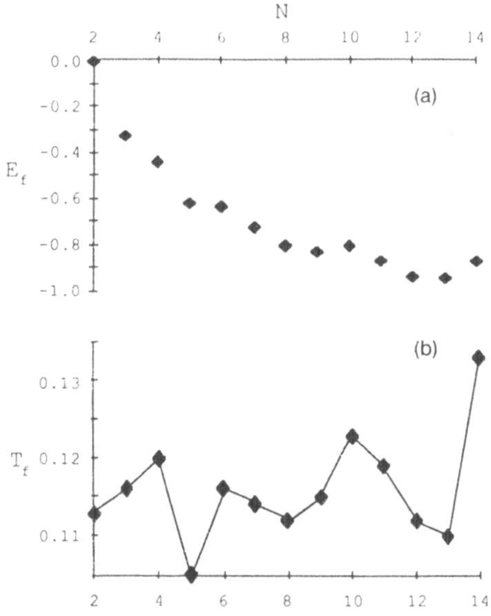

FIG. 11 (a) Fragmentation energy per atom (E_f) vs number (N); (b) Fragmentation temperature (T_f) vs number (N). Peaks in these curves at N=4,6,10 (magic numbers) clearly indicate the particular stability of the $Si_{4,6,10}$ microclusters at high energies and temperatures. Fragments of these sizes were reported to be unusually common in the photofragmentation experiment of Bloomfield, Freeman, and Brown.

5-membered rings and one 4-membered ring.

N=10 The ground state configuration of two stacked pentagons (Fig. 5) is visited in all systems of energies below E_f, though its presence decreases from 70% at $.5E_f$ to 8% at $.95E_f$ as determined from the SDQ. A particularly open structure containing one singly bonded atom was also present in 7% of the visited structures. Near E_f all of the 59 visited structures are observed. A 5-membered ring, an irregular pentagon, is the basic subunit underlying all these structures.

N=11 Seventy-four potential minima have been identified for Si_{11} though none are particularly preferred at energies near E_f. At these high energies the two lowest-energy structures account for only 2% of the underlying structures while the most common geometry is found less than 6% of the time. The high-temperature cluster frequently alters its form between the numerous available structures. Most if not all of these structures may be constructed by various surface packings of 4- and 5-sided nonplaner faces.

N=12 Only 54 potential-energy minima were identified for Si_{12} with the ground state forming a common underlying structure for all energies below E_f. Even though all atoms are 3-fold coordinated in this structure, fully bonded 4-fold coordinated atoms appear in some of the higher-energy configurations. All stable configurations are accessible at $.95E_f$ with the ground state remaining the most common accounting for 20% of the steepest-descent mapped minima.

N=13 Two low-energy structures other than the ground state are commonly found at finite temperatures. The more stable of these two high-energy structures has two fully bonded 4-fold coordinated atoms with two 5-membered rings and one 3-membered ring in addition to the six 4-membered rings while the other structure has only three bonds per atom with a surface formed by six 5-membered rings and two 4-membered rings. These three low-energy configurations account for 50% of the underlying structures at $.75E_f$ and 5% at $.95E_f$. Close to E_f nearly 70 of the 95 identified potential minima are visited but none seem to be preferred with the most common appearing less than 6% of the time. These structures can be obtained from closed surfaces made up of various combinations of 4- and 5-membered rings.

N=14 For energies on the order of $.25E_f$ all atoms vibrate about the Si_{14} ground-state equilibrium positions but for higher energies structural transformations occur and the lowest-energy configuration accounts for less than 1% of the visited structures. In total 156 potential-energy minima have been found that span the energy range of $.24\varepsilon$ between lowest and highest energies. The majority of these structures are best seen as the many possible combinations of taking 4- and 5-sided figures to form a closed structure.

MAGIC NUMBERS

The results of the present MD simulation indicate that 4-, 6- and 10-atom clusters are relatively stable structures at temperatures on the order of the melting temperature. Curves of fragmentation energy per atom (E_f) versus number (N) and fragmentation temperature (T_f) versus number (N) are shown in Fig. 11. The fragmentation energy, the highest attainable energy of a bound system, was determined from approximately 60 systems generated for each cluster Si_{3-14}. In practice, the highest energy observed before the occurrence of the first of twenty or more fragmentations was defined to be E_f. The fragmentation temperature was found from the average kinetic energy of the system with energy E_f. Peaks in the curve for E_f vs N clearly indicate 6- and 10-atom microclusters have particularly stable high-energy configurations for silicon, while the peak at N=4 is not so apparent. However sharp peaks do appear in the T_f vs N curve for the three experimentally determined magic numbers N=4,6 and 10. It is also obvious in both curves that Si_{14} is very stable at high temperatures with respect to Si_{13}. There has not been an experimental investigation of the relative stability of Si_{14} but Martin and Schaber[13] have found N=14 to be a magic number for Ge. In Table 1 the ground-state energy per atom (E_o) found through the SDQ mappings are given with the fragmentation energies per atom (E_f) and temperatures (T_f) for Si_{2-14}.

TABLE 1 The zero-temperature ground-state energy E_o per atom, fragmentation energy per atom E_f and the corresponding fragmentation temperature T_f for Si_{2-14}.

NUMBER	MINIMUM ENERGY	FRAGMENTATION ENERGY	TEMP
N	E_o	E_f	T_f
2	-0.5000	0.000	.113
3	-0.6828	-0.326	.116
4	-0.9386	-0.449	.120
5	-0.9996	-0.620	.105
6	-1.0906	-0.632	.116
7	-1.1788	-0.726	.114
8	-1.3223	-0.807	.104
9	-1.3271	-0.835	.115
10	-1.3797	-0.803	.123
11	-1.3829	-0.870	.119
12	-1.4178	-0.934	.112
13	-1.4234	-0.945	.110
14	-1.4455	-0.866	.133

TABLE 2 Primary channels of fragmentation for Si_{2-14} as determined from 20-30 events for each size cluster. Fragmentation from Si_N to $Si_{(N-2)}+Si_2$ indicates $Si_{4,6,10}$ are particularly common fragments while the fragmentation from Si_N to $Si_{(N-3)}+Si_3$ supports the common occurrence of Si_4 and Si_6 fragments.

N	(N-1)+1	(N-2)+2	(N-3)+3
2	1.00		
3	1.00		
4	0.90	0.10	
5	0.82	0.18	
6	0.64	0.25	0.11
7	0.62	0.19	0.19
8	0.55	0.35	0.05
9	0.64	0.12	0.16
10	0.74	0.22	0.00
11	0.65	0.15	0.10
12	0.70	0.25	0.05
13	0.76	0.12	0.04
14	0.76	0.14	0:00

In general, a N+1 atom system has a higher binding energy than a N atom system and one should expect a continuously increasing curve in $|E_f|$ vs N until the evaporation point of the bulk is reached. Since fragmentation energy and temperature increase with increasing N as a qualitative argument it will be sufficient to explain the presence of the peaks at N=4, 6 and 10 in Fig. 11 by accounting for the relative drop in the curves at N=5, 7 and 11. An argument for the peak at N=6 in the fragmentation curves follows this line of reasoning. At energies near E_f, Si_4 and Si_6 have a clear majority of the underlying structures containing singly bonded atoms. One of the three most commonly found structures at $.95E_f$ for Si_{10} also contains a singly bonded atom. With the exception of the two largest clusters, Si_{13} and Si_{14}, all other clusters are usually mapped to potential minima corresponding to closed structures. Both Si_{13} and Si_{14} have a large number of accessible minima near E_f but none are particularly preferred. Compact structures are known to have high binding energies and assumed to be stable at low temperatures but it is not clear that they can remain stable at high temperatures. Our results indicate that these closed, tightly-bound configurations are not stable with respect to large thermal fluctuations. As can be seen from the high fragmentation energy for Si_2 singly bonded atoms are quite stable with respect to large thermal vibrations. The most common structures of Si_6 and Si_{10} at energies near E_f both contain a single atom capped to a pentagon. Since the 3-body interaction is minimized in the pentagon, each atom sees a local environment similar to an atom in a chain. Each atom in the pentagon has considerable degree of movement like members of the high-energy chain but with a small additional contribution to the potential energy due to the 3-body interaction. It is the ability of the Si_6 and Si_{10} cluster to access these low-energy, loosely bound structures containing one single atom bonded to 5-membered rings that enables them to remain stable at relatively high temperatures. In contrast to Si_6, Si_7 could only form a similar stable structure with two noninteracting atoms corner-capping a pentagon, but for this small cluster the two capped atoms invariably do interact forming a closed structure. Recall the most common underlying structures for the high temperature Si_7 clusters is a pentagon with a dimer connected to one edge or across its face. This argument may explain why Si_6 is stable with respect to Si_5 and Si_7 but a similar explanation for Si_4 or Si_{10} is not so easily obtained. It is unlikely that a general topological argument can be found to explain the existence of the magic numbers since fragmentation occurs at high temperatures where a large number of structures of varying topology are visited.

Fragmentation occurs when a subset of atoms no longer interacts with the remaining atoms in the parent cluster. This definition restricts our MD simulation of fragmentation to events where only single bonds are broken. The most common channel for all Si_N (N=2-14) is the fragmentation into Si + Si_{N-1}.

When the parent cluster separates into two or more compact subunits connected by single bonds before fragmentation then nontrivial fragments can result. As an example, Si_4 can only fragment into $Si_2 + Si_2$ by severing the center bond of the Si_4 chain. The relative probability of the cluster Si_N dissociating into the fragments $Si_{N-M} + Si_M$ (M=1,2,3) is given in Table 2.

Evidence for the stability of the magic numbers is also found in the nontrivial fragmentation channels N-->2+(N-2) and N-->3+(N-3). The $Si_{6,8,12}$ clusters were observed to have the highest probability of nontrivial fragmentation with at least 25% of the observed events resulting in $Si_2 + Si_{4,6,10}$ respectively, see Table 2, column titled "(N+2)-2." $Si_{7,9}$ were also found to have a large probability of dissociating into Si_3 and $Si_{4,6}$, Table 2 column titled "(N-3)-3." These relatively high probabilities for nontrivial fragmentation into 4-, 6- and 10-atom clusters are in good agreement with the experimental fragmentation spectra and support the claim for unusually stable $Si_{4,6,10}$ microclusters.

Our results indicate that the knowledge of the zero-temperature ground-state structure and energies is not sufficient to explain the existence of unusually stable clusters for covalently-bonded systems. Rather it is the topology and energetics of these high-temperature clusters which determine its relative stability with respect to the number of atoms. The fragmentation energy (E_f) and temperature (T_f) indicate $Si_{4,6,10}$ are particularly stable clusters at high temperatures, the order of the melting temperature. Though we have only investigated the primary channel of fragmentation, the fragmentation spectra N-->2+(N-2) and N-->3+(N-3) indicates the unusually common occurrence of $Si_{4,6,10}$ fragments in the dissociation of silicon microclusters. The results of this MD simulation agree quite well with the BFB photofragmentation experiment and gives us good reason to believe that N=4,6,10 are the magic numbers for silicon. We believe that the SW 3-body potential describes the essential features of the energetics and fragmentation of silicon microclusters. It can also be argued that many of the results described above will be common to Group IV semiconductors, in particular germanium clusters. There will certainly be details which will differ when a more accurate interaction potential becomes available or a fully self-consistent finite-temperature first-principles electronic structure calculation becomes possible. The details and full results of the present MD study will be published elsewhere.[16] †Work supported by the U.S. Department of Energy, BES-Materials Sciences

References

[1] R. Van Hardeveld and F. Hartog, Adv. Catal. 22, 75 (1975)
[2] B.K. Rao and P. Jena, Phys. Rev. B 32, 2058 (1985)
[3] P.A. Montano, G.K. Shenoy, E.E. Alp, W. Schulze, and J. Urban, Phys. Rev. Lett. 56, 2076 (1986)
[4] M.R. Hoare and P. Pal, Adv. Phys. 20, 161 (1971)
[5] F.F. Abraham, Rep. Prog. Phys. 45, 1113 (1982)
[6] I.A. Harris, R.S. Kidwell, and J.A. Northby, Phys. Rev. Lett. 53, 2390 (1984)
[7] J. Farges, M.F. de Feraudy, B. Raoult, and G. Torchet, J. Chem. Phys. 78, 5067 (1983)
[8] L.A. Bloomfield, R.R. Freeman, and W.L. Brown, Phys. Rev. Lett. 54, 2246 (1985)
[9] T. P. Martin and H. Schaber, J. Chem. Phys. 83, 855 (1985)
[10] K. Raghavachari and V. Logovinsky, Phy. Rev. Lett. 55, 2853 (1985)
[11] D. Tomanek and M.A. Schluter, Phys. Rev Lett. 56, 1055 (1986)
[12] F.H. Stillinger and T.A. Weber, Phys.Rev. B 31, 5262 (1985)
[13] A. Rahman in "Correlation Functions and Quasiparticle Interaction in Condensed Matter" edited by J. Woods Halley, NATO Advanced Studies Series Vol. 35 (New York: Plenum Press, 1977)
[14] F H. Stillinger and T.A. Weber, Science 225, 983 (1984)
[15] D.J. Chadi, Phys. Rev. B 32, 6485 (1985)
[16] B.P.Feuston, R.K.Kalia, and P.Vashishta, Phys. Rev. B (to be published - 1987)

PHYSICS OF STRONGLY COUPLED ROTATION-TRANSLATION SYSTEMS

S. D. Mahanti

Department of Physics and Astronomy
Michigan State University
East Lansing, MI 48824

ABSTRACT

We review some of the interesting physical properties of simple molecular solids which arise from a strong coupling between the rotational and translational degrees of freedom of the constituent atoms or molecules.

I. INTRODUCTION

Molecular solids have been the subject of intensive experimental and theoretical research in condensed matter physics during the last fifteen years. For the purpose of this paper we will be interested in one important characteristics of these solids; they are classic examples of systems exhibiting a strong coupling between large amplitude rotational and translational motions. This coupling leads to, among other things, ferro-elastic (FE) structural phase transitions accompanied by anomalous thermo-elastic softening[1-5] and interesting dynamics[6-10] in the orientationally disordered plastic phase.

We will discuss two classes of systems, alkali cyanides in 3-dimension and diatomic Lennard-Jones molecules lying flat on a uniform 2d substrate, the latter closely resembling the δ-phase of a submonolayer of O_2 molecules adsorbed on a graphite substrate.[11] The reason for choosing these systems is to understand qualitatively and in some cases quantitatively the effects of the competition between (1) direct and indirect intermolecular interactions, the latter arising from the rotation-translation (RT) coupling and (2) intersite and intrasite (cation cage in 3d and substrate corrugation in 2d) interactions on the thermodynamic and dynamic properties of these systems.

The strength of the RT coupling in these systems can be visualized from the following typical example. In the cubic fcc phase of KCN, if one fixes the centers of mass of all the particles at the average cubic sites then the barrier height for CN^- rotation is ~300-400K depending on the values of the molecular parameters. This rigid barrier height can be reduced drastically by as much as a factor of 2-3 if one allows the centers of mass to relax i.e. allows for the RT coupling. One therefore expects this strong RT coupling to significantly affect the physical properties of these solids.

The outline of the paper is as follows. In Sec. II we discuss the physics of anomalous thermoelastic softening in alkali cyanides. Sec. III deals with a Landau theory analysis of the FE phase transition in 3d Cyanides and also gives the results of a constant pressure molecular dynamics (MD) simulation of the FE transition for the 2d Lennard-Jones system. In Sec. IV, we discuss the rotational dynamics of a single molecular impurity coupled strongly to a polarizable medium using both Mori theory (with a simple approximation to the memory function) and MD simulation. Finally in Sec. V, we summarize our current understanding of these systems.

II. ELASTIC SOFTENING IN ALKALI CYANIDES

Careful ultrasonic[1,3] and neutron[2,3] scattering measurements have shown that the shear elastic constants C_{44} and $C_{11} - C_{12}$ soften dramatically in the orientationally disordered cubic phase of the alkali cyanides with decreasing temperature (T) and the systems finally undergo first order FE phase transition to an orientationally ordered noncubic phase. In contrast, the longitudinal elastic constant C_{11} does not show appreciable softening. In Fig. 1 we show experimental results for KCN and RbCN. In all the alkali cyanides, C_{44}, the elastic constant of t_{2g} symmetry, softens most and before it becomes zero, the systems undergo a first order transition. However the molecular orientations (as referred to the symmetry directions of the cubic phase) in the low-T orientationally ordered phase are different for the different cyanides. In KCN and NaCN, molecules orient parallel to the (110) direction whereas in RbCN and CsCN they orient parallel to the (111) direction. This difference in the orientational order also leads to different lattice structures.

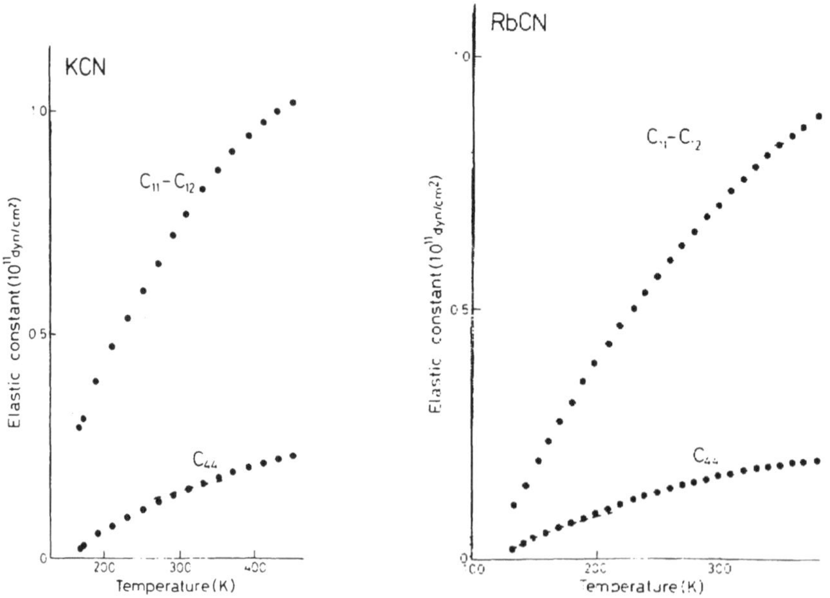

Fig. 1. Temperature dependence of elastic constants.

The observed elastic softening can be understood in terms of the coupling between the phonons of the cubic solid (bare phonons) in which the molecular ions, due to their rotation, can be treated as spherical and the nonspherical orientational fluctuations.[4,5] To account for this coupling one takes a simple Hamiltonian consisting of a bare phonon part, a pure rotational part, and a linear coupling between bare phonons and the rotational variables associated with the molecular orientation. The effect of this coupling can be calculated within random phase approximation and choosing appropriate symmetry directions and phonon polarizations, one obtains the renormalized elastic constatnts in the orientationally disordered cubic phase.[5,12] These are

$$C_{11} = C_{11}^{0} - a_{11} A_{eff}^{2} \chi_{e_g}(T) \quad (1.a)$$

$$C_{44} = C_{44}^{0} - a_{44} B_{eff}^{2} \chi_{t_{2g}}(T) \quad (1.b)$$

$$C_{12} = C_{12}^{0} - a_{12} A_{eff}^{2} \chi_{e_g}(T) \quad (1.c)$$

where C_{11}^{0} etc. are the elastic constants in the absence of RT coupling, the coefficients $a_{11}=(8/a,2/a)$, $a_{44}=(2/a,0.5/a)$, and $a_{12}=(4/a,1/a)$ are for the NaCl and CsCl structures respectively. Here 2a is the lattice constant. The isothermal susceptibilities and the coupling constants χ_{e_g}, $\chi_{t_{2g}}$ and A_{eff}, B_{eff} are associated with e_g and t_{2g} symmetries respectively. The main physical features of these systems can be summarized as follows: (a) The RT coupling constants have two major contributions which tend to oppose each other, one from the short range repulsion (A_R and B_R) and the other from the interaction between the electric quadrupole moment (Q) of the CN^- ion and the fluctuating electric field gradient (A_Q and B_Q).[5] In cyanides, B_Q is larger than B_R and B_{eff} which controls the C_{44} softening has the same sign as B_Q. The sign of B_{eff} can be directly probed by looking at the anisotropy of elastic neutron diffraction and experiments substantiate the above finding.[13] (b) Elastic softening depends sensitively on the strength and the T-dependence of $\chi(T)$ which is in turn determined by the renormalized single-site potential $V(\theta,\phi)$ and the direct interaction

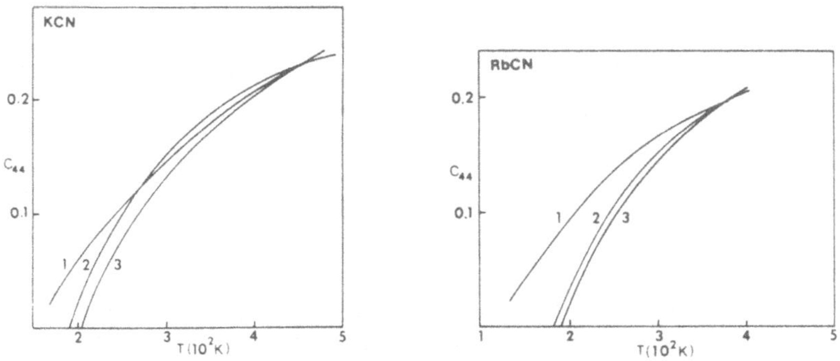

Fig. 2. Temperature dependence of elastic constant C_{44} (in units of 10^{11} dyn/cm^2). 1: experiment, theory with (2) and without (3) anharmonicity.

between the CN^- molecules. If one ignores all the multipole moments excepting the quadrupole Q of the CN^- molecule in the calculation of the elastic constants then one has to reduce Q from its bare ion value by ~40% to fit the experimental data reasonably well in cyanides with NaCl structure (see Fig. 2) and as much as 85% in CsCN. This reduction in Q is not borne out by quantum mechanical calculations[14] and furthermore higher multipole moment contributions to the single site potential are quite important as they tend to oppose the short range repulsive potential.[15,16] One can understand the elastic softening data semiquantitatively if one includes all the multipole moments and the self energy contribution to the single site potential and allows for a small but systematic reduction in the multipole moments in the solid state environment.[15,17]

III. FERROELASTIC ORDER AND STRUCTURAL PHASE TRANSITION

We will discuss the 3d (cyanides) and the 2d (diatomic LJ molecules on a smooth substrate) systems separately because for the 3d systems a Landau theoretical (mean field theory) description is adequate whereas in the 2d systems fluctuation effects are very important and one has to go beyond simple mean field theory.

Starting from a linearly coupled RT Hamiltonian, one can derive a variational form for the Landau free energy for the cyanides which is given by[18]

$$F=F_0+ (1/2) \sum_{\alpha\beta} [D_{\alpha\beta}(\vec{Q} = 0) + I_{\alpha\beta}(\vec{Q} \to 0) - \sum_{Q} I_{\alpha\beta}(\vec{Q}) - \chi_{\alpha\beta}^{o\,-1}] \eta_{\alpha\beta}\eta_{\beta}$$

$$-(1/6)[A_{3,1}\, \eta_1(\eta_1^2 - 3\eta_2^2) + A_{3,2}\eta_3\eta_4\eta_5 + A_{3,3}\{\eta_1(2\eta_3^2 - \eta_4^2 - \eta_5^2)$$

$$+ \sqrt{3}\, \eta_2(\eta_5^2 - \eta_4^2)\}] + O(\eta^4) \qquad (2)$$

where η_α's are the orientational order parameters of e_g (η_1,η_2) and $t_{2g}(\eta_3,\eta_4,\eta_5)$ symmetries respectively. Several points should be made here. (a) For NaCl structure, the indirect contribution $I_{\alpha\beta}$ dominates the FE phase transition whereas for the CsCl structure the direct contribution $D_{\alpha\beta}$ can account for the FE order. (b) The observed (110) ordering in KCN and NaCN can be understood in terms of the third order Landau terms which couple order parameters of e_g and t_{2g} symmetries. (c) It is necessary to include, as in the case of elastic softening, all the multipole and self energy contributions to the single site potential in calculating the Landau coefficients even for a qualitative understanding of the ferroelastic order in these systems. (d) For a quantitative understanding of both the FE transition temperature and the orientational order parameter in the FE phase the linear coupling approximation to the RT Hamiltonian may not be adequate.[19]

In contrast to the case of 3d alkali cyanides where the competition between short range repulsive and electrostatic forces make it difficult to understand their physical properties quantitatively (even qualitatively in some cases), 2d LJ systems are relatively simple. The major problem for these systems is the inadequacy of mean field theory. We have therefore studied the nature of FE transition and elastic softening in this system using a constant pressure molecular dynamics simulation.[20] The system consists of diatomic oxygen molecules lying flat on a smooth 2d substrate. At T=0, the system is orientationally ordered and the molecular centers of mass form a distorted triangular lattice.[21] At high T, the molecules are orientationally disordered and the lattice symmetry is triangular. One can describe this structural phase transition in terms of a two component order parameter associated with the two components of the strain tensor i.e. $\psi_1 = \varepsilon_{xy}$ and $\psi_2 = (\varepsilon_{xx} - \varepsilon_{yy})/2$. The Ginzburg-Landau-Wilson (GLW) Hamiltonian has the form:

$$H_{GLW} = (1/2) \sum_i |\nabla\psi|_i^2 + (1/2)\, r_2(T) \sum_i \psi_i^2 + w\,(\psi_2^3 - 3\psi_1^2\psi_2) + O(\psi^4) \qquad (3)$$

Although the mean field theory for the above Hamiltonian gives a first order transition, the Renormalization Group (RG) calculation predicts a continuous transition belonging to the same universality class as the three state ferromagnetic Potts model.[22] In contrast to this continuous transition, constant pressure MD simulation gave a first order transition. The T-dependence of the orientational and structural order parameters are shown in Fig. 3. We believe that the underlying reason for this first order transition is the presence of low-lying excitations of local herringbone (HB) ordering of the diatomic molecules on a triangular lattice and these excitations are not included in the above GLW description of the system Hamiltonian. The underlying physical mechanism that drives the structural transition first order can be explored through an effective XY spin Hamiltonian.[20] At the transition, one finds a sudden unbinding of a large number of vortex-antivortex pairs which are localized defects with HB-like structure.

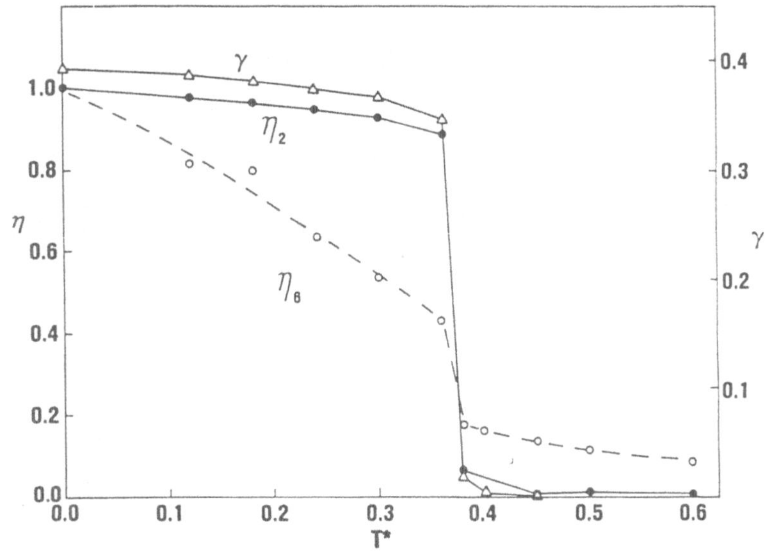

Fig. 3. Lattice anisotropy parameter γ and orientational order parameters η_2 and η_6 vs. reduced temperature T*; $\eta_p = \langle \mathrm{Cos}\,p\theta \rangle$, when θ gives the molecular orientation.

The elastic constants have been calculated in the orientationally disordered triangular phase which we believe is a 2d 'solid' phase due to the absence of translational diffusion. We find that $C_{11} - C_{12} = C_{44}$, $C_{11} = C_{22}$, and $C_{44}/C_{11} = \lesssim 0.1$. Thus the high temperature phase is extremely soft towards transverse distortions and almost liquid-like. Experimentally[11,23], for O_2/ graphite submonolayer coverage one finds a first order transition from the distorted triangular δ-phase to an orientationally disordered liquid phase. At the present time it is not clear whether the soft 2d solid phase (plastic crystal phase) that we see in our MD simulation appears as a liquid-like phase in the experiment or it is absent completely due to the effects not incorporated within our model. More careful simulation studies including the effects of substrate potential and out-of-plane motion of the molecules are needed to clarify this situation.

IV. ORIENTATIONAL DYNAMICS (SINGLE IMPURITY)

For the purpose of this talk I will confine my discussion to the rotational dynamics of a single molecular impurity coupled strongly to a polarizable medium. Our study[16] of real systems such as CN^- impurity in KBr shows that the adiabatic potential in which the molecule rotates has a significant contribution from the RT coupling. One therefore expects to see the effect of this coupling on the rotational dynamics. To understand the nature of rotational dynamics in a strongly coupled RT system as function of different physical parameters such as the moment of inertia of the rotor, characteristic frequency scale of the polarizable medium, temperature, and RT coupling constant etc. , we have chosen a simple RT Hamiltonian H_{RT} which incorporates the essential physics. H_{RT} describes an one dimensional rotor coupled to N harmonic oscillators and is given by

$$H_{RT} = (1/2)\mu\dot{\theta}^2 - \cos\theta + (1/2)\sum_i(\dot{X}_i^2 + \Omega_i^2 X_i^2)$$

$$-2\sqrt{(\alpha/N)} \quad \sum_i X_i \sin(\theta/2) \quad , \qquad (4)$$

where all the quantities are given in a dimensionless form.[25] The energy is measured in units of D where the bare rotor potential is -D Cosθ. The frequencies are measured in units of ω_D, the maximum frequency of the bath oscillators. $\mu = I\omega_D^2/D$ is a measure of the moment of inertia of the rotor. The above choice of the RT coupling leads, through a canonical transformation, to an effective adiabatic potential for the rotor of the form $-D(1-\alpha)\cos\theta$ which makes the rotor adiabatically 'free' in the 'strong coupling limit' $\alpha=1$.

We have studied the rotational dynamics using MD and Mori theory with different approximations to the Mori memory function for a large range of values of α, μ, T. For the present, I will discuss the results for the strong coupling limit and for the simplest approximation to the memory function.

In Mori theory one calculates the relaxation function Φ_{UU} associated with a particular dynamical variable U, where

$$\Phi_U(z) = \beta\langle U,U\rangle \left[z - \langle\omega^2\rangle/(z + \textstyle\sum_U(z))\right]^{-1}, \qquad (5)$$

where $\beta = 1/k_B T$, $\langle U,U\rangle$ is the equilibrium correlation function, $\langle\omega^n\rangle$ is the nth moment and $\sum_v(z)$ is the Mori memory function. In principle Mori theory is exact. However one makes approximations in obtaining the memory function. One particularly simple approximation that has been used in the past for this problem (in the absence of coupling to a polarizable medium) is[24]

$$\sum_U(z) = -\frac{1}{\langle\omega^2\rangle}\;\frac{\left(\langle\omega^4\rangle - \langle\omega^2\rangle^2\right)}{z + i\left[\langle\omega^4\rangle/\langle\omega^2\rangle\right]^{1/2}} \quad . \tag{6}$$

The above approximation gives the moments of the spectral weight function $P_U(\omega) = -T/\pi \;\; \mathrm{Im}\;\Phi_U(\omega + i\,0)$ exactly up to order 4. We have chosen $U = \mathrm{Sin}(\theta/2)$, $\mathrm{Cos}(\theta/2)$ and the rotational velocity $\dot{\theta}$ and have calculated P_U for all three using MD simulation and have compared MD results with those of Mori theory. Our basic findings can be summarized as follows:

(a) In Mori theory with the above approximation to the memory function, one finds a three peak structure in general, one central peak and two side bands.[16,24] The frequency scale is determined by $\sqrt{\langle\omega^2\rangle}$. The spectral weights, the widths and the frequency shifts of the side bands are determined by a single dimensionless parameter R which is given by

$$R = [\langle\omega^4\rangle \;/\; \langle\omega^2\rangle^2]^{1/2} \quad . \tag{7}$$

In particular, the spectral weights of the side bands X and the width of the central peak Γ decrease with increasing R.[16]

(b) From MD simulations[25], we find that the weight of the central peak A and its width Γ are very sensitive to the density of oscillator freqencies which is missing in the Mori theory for P_S and P_C. Similarly, the μ-dependence of P_C and P_S , in particular the appearance of a narrow central peak over a broad relaxational peak for P_C, is absent in the Mori scheme where the parameter R associated with the above two dynamical variables is independent of μ (see Table I).

Spectral parameters $\sqrt{\langle\omega^2\rangle/\omega_D^2}$ and $R = [\langle\omega^4\rangle/\langle\omega^2\rangle^2]^{1/2}$ for the three dynamic variables appearing in Mori theory, $\langle\Omega^2\rangle$ is the average of Ω^2, where Ω is the dimensionless bath (oscillator) frequency; $0 \leq \Omega \leq 1$.

TABLE I

U	$\mathrm{Sin}\theta/2$	$\mathrm{Cos}\theta/2$	$\dot{\theta}$
$[\langle\omega^2\rangle/\omega_D^2]^{\frac{1}{2}}$	$\frac{1}{2}\sqrt{T/\mu}$	$\frac{1}{2}\sqrt{T/\mu}$	$\sqrt{1/2\mu}$
R	$\sqrt{3(1+\frac{1}{T})}$	$\sqrt{3(1+\frac{1}{3T})}$	$\sqrt{\frac{3}{2}(1+\frac{T}{3}+\frac{4}{3}\mu/\langle\Omega^2\rangle)}$

(c) Another weak point that we have found in the above approximation scheme is in the T-dependence of Γ for P_V, the velocity (θ) power spectrum. Γ increases with decreasing T because R decreases with T. In contrast, MD simulation gives just the opposite behavior. We have recently developed an alternate approximation to the memory function which can correct for this particular shortcoming.[26]

It is not clear at the present time whether it is possible within Mori approximation to explain all the qualitative features of the spectral response in the strong coupling limit of a RT system. It is however clear that one has to go up to atleast the 8th moment which will incorporate the translational dynamics in the calculation of the rotational dynamics. Another possible way of approaching the dynamics of these strongly coupled rotation translation system is to develop a theory similar to the strong coupling superconductivity theory for electron phonon systems.

V. SUMMARY

In summary, the strongly coupled rotation-translation systems show an interesting range of thermodynamic and dynamic behavior. In particular, a theoretical analysis of the temperature dependence of the elastic constants brings out the sensitive nature of the cancellation between the short range steric and relatively longer ranged electrostatic forces in 3d ionic molecular solids. Molecular dynamics simulations in realistic systems are able to bring out not only the true nature of the orientational dynamics but also are able to shed some light on the adequacy of the values of different physical parameters describing the system. Our study shows that simple truncation schemes within Mori formalism are not able to reproduce the orientational dynamics over the broad range of parameter space, particularly in the strong coupling limit.

ACKNOWLEDGEMENTS

I should like thank Dr. G. Kemeny for many stimulating discussions and my graduate students, D. Sahu, P. Murray, and S. Tang who did much of the work described here. Full details can be found in the references cited. I should also like to thank N.S.F. (Grant # DMR 81-17297) for financial support.

REFERENCES

1. S. Haussühl, Sol. St. Comm., 32, 181 (1979) and references therein; W. Rehwald, Phys. Lett. 87A, 245 (1982).
2. J. M. Rowe, J. J. Rush, M. Vegelatos, D. L. Pryce, D. J. Hinks, and S. Susman, J. Chem. Phys. 62, 455 (1975).
3. A. Loidl, S. Haussühl, and J. K. Kjems, Z. Phys. B, Cond. Matter 50, 187 (1983); A. Loidl, K. Knorr, J. K. Kjems, and S. Haussühl, J. Phys. C, Sol. St. Phys. 13, L349 (1980).
4. K. H. Michel and J. Naudts, J. Chem. Phys. 67, 547 (1977).
5. D. Sahu, and S. D. Mahanti, Phys. Rev. B26, 2981 (1982); Sol. St. Comm., 47, 207 (1983).
6. J. M. Rowe, J. J. Rush, N. J. Chesser, K. H. Michel, and J. Naudts, Phys. Rev. Lett. 40, 455 (1978); B. DeRaedt and K. H. Michel, Discuss. Faraday Soc. 69, 88 (1980).
7. A. Loidl, K. Knorr, J. Daubert, W. Dultz, and W. J. Fitzerald, Z. für Physik B38, 153 (1980).

8. M. L. Klein and I. R. McDonald, J. Chem. Phys. 79, 2333 (1983); M. Ferrario, I. R. McDonald and M. L. Klein, J. Chem. Phys. 84, 3975 (1986).
9. R. M. Lynden-Bell, I. R. McDonald, and M. L. Klein, Mol. Phys. 48, 1093 (1983).
10. M. L. Klein, Y. Ozaki, and I. R. McDonald, J. Phys. C, Sol. St. Phys. 15, 4993 (1982).
11. P. A. Heiney, P. W. Stephens, S.G.J. Mochrie, J. Akimatsu, and R. J. Birgeneau, Surf. Sci. 125, 539 (1983).
12. D. Sahu, Ph. D thesis, Mich. St. University (1983).
13. J. M. Rowe and S. Susman, Phys. Rev. 29, 4727 (1984).
14. R. LeSar and R. G. Gordon, J. Chem. Phys. 77, 3682 (1982).
15. P. Murray and S. D. Mahanti (to be published).
16. S. D. Mahanti, P. Murray, and G. Kemeny, Phys. Rev. 32, 3263 (1985); S. D. Mahanti and G. Kemeny, Phys. Rev. B30, 7362 (1985).
17. For a discussion on the reduction of molecular multipole moments in the solid state environment, see M. Ferrario et al. (ref.8).
18. D. Sahu and S. D. Mahanti, Phys. Rev. B29, 340 (1984).
19. K. H. Michel and J. M. Rowe, Jour. de Chimi Physique, 82, 199 (1985).
20. S. Tang, S. D. Mahanti, and R. K. Kalia, Phys. Rev. Lett. 56, 484 (1986); S. Tang, Ph. D thesis, Mich. State Univ. (1985).
21. R. D. Etters, R. Pan, and V. Chandrasekharan, Phys. Rev. Lett. 45, 645 (1980); R. Pan, R. D. Etters, K. Kobashi, and V. Chandrasekharan, J. Chem. Phys. 77, 1035 (1982).
22. E. Domany and E. K. Riedel, Phys. Rev. B19, 5817 (1979).
23. M. F. Toney, R. D. Diehl, and S. C. Fain, Phys. Rev. B27, 6413 (1983);
24. B. DeRaedt and K. H. Michel, Phys. Rev. B 19, 767 (1979).
25. G. Kemeny, S. D. Mahanti, and J. Gales, Phys. Rev. B 33, 3512 (1986).
26. G. Kemeny and S. D. Mahanti, Bull. Am. Phys. Soc. 30, 320 (1985).

COMPUTER SIMULATION OF "SPECIAL" GRAIN BOUNDARIES IN METALS AND IONIC MATERIALS*

D. Wolf

Materials Science Division
Argonne National Laboratory, Argonne, IL 60439

ABSTRACT

In recent years a set of lattice statics computer codes has been developed which permits one to determine the energies, structures, and point-defect properties of grain boundaries in metals and predominantly ionic ceramic materials. In contrast to metals, long-range Coulomb forces between the ions have to be considered in ionic bicrystals. The selection criteria for the identification of "special" (i.e., low-energy) grain boundaries (GBs) is discussed with particular emphasis on (a) the role of the GB plane, (b) the influence of the planar unit-cell area, and (c) the effect of the interatomic potential on the predicted energies and structures of GBs in metals and ionic materials.

INTRODUCTION

Within the framework of the geometrical GB models based on Bollmann's coincident-site-lattice (CSL) theory[1] it is usually assumed that the lowest-energy grain boundaries are associated with the highest densities, Σ^{-1}, of CSL sites. As a consequence, for a given misorientation of two crystal lattices characterized, for example, by the parameter Σ, the lowest-energy GB plane is thought to be the symmetrical configuration, i.e., the densest plane in the CSL.

Much of the experimental evidence is in conflict with this model. Thus, in their thermal-grooving experiments on <110> tilt GBs in aluminum, Hasson et al.[2] observed energy cusps for the $\Sigma3$ (111) orientation but not for the $\Sigma3(211)$ GB. Also, whereas the $\Sigma11$ (113) boundary gave rise to an energy cusp, the $\Sigma9(221)$ and $\Sigma9(114)$ GBs did not. Hasson et al.[2] also report that in many cases their bicrystals preferred asymmetrical plane configurations, in contrast to what is expected within the framework of the CSL model. Very recent high-resolution TEM results by Merkle et al.[3-5] and Krakow et al.[6] on NiO

*Work supported by the U.S. Department of Energy, Basic Energy Sciences-Materials Science, under contract W-31-109-Eng-38.

and Au, respectively, show asymmetrical GB-plane orientations in many instances, even for low-Σ boundaries.

It is the purpose of this brief overview to attempt to reconcile the experimental and theoretical situations by reviewing the physical criteria established recently[7-9] for the identification of "special" (i.e., low-energy interfaces). Two more extensive overviews over computer-simulation methods applied to GBs in metals and predominantly ionic ceramic materials were presented recently.[10-11] Hence, in what follows, only the most recent developments will be highlighted.

2. GEOMETRICAL CHARACTERIZATION OF GRAIN BOUNDARIES

As is well known, eight geometrical parameters are necessary--in addition to the crystal structure and lattice parameter--to fully characterize a planar grain boundary. These eight degrees of freedom (dof) may be subdivided into macroscopic and microscopic ones.[12] The latter represent the three translational dof of the two crystals with respect to one another parallel and perpendicular to the GB plane. They are usually characterized by the translation vector $\vec{T}$. By the very nature of $\vec{T}$, only experiments which can measure the translation on an atomic scale are capable of determining the three components of $\vec{T}$.

The five macroscopic dof characterize the misorientation of the two crystal lattices relative to each other and the GB plane. In the conventional GB characterization scheme,[12] the misorientation is characterized by the rotation axis, $\vec{n}_R$ (a unit vector, hence constituting only two dof), and the rotation angle, θ_R, whereas the GB plane is defined by the plane normal, $\vec{n}$, with respect to either of the two crystal coordinate systems. The eight dof of a GB may hence be summarized as follows:

$$\vec{n}_R,\ \theta_R,\ \vec{n},\ \vec{T}. \qquad (1)$$

This choice of parameters emphasizes the fact that a bicrystal may be thought of as having been generated by a single rotation, $R(\vec{n}_R,\theta_R)$, of two identical interpenetrating crystal lattices relative to each other with the subsequent removal of corresponding material to accommodate the particular GB plane chosen.[1,12] It suffers a certain disadvantage in that symmetry properties of GBs are not readily apparent from this choice of the five macroscopic parameters. For example, to the unexperienced reader it may not be obvious that the GB characterized by $\vec{n}_R,\theta_R,\vec{n}$ = <100>, 36.87°, (013) is a <u>symmetrical</u> tilt boundary which, in fact, is characterized by only <u>three</u> independent macroscopic parameters.

A simple alternate choice of the eight dof which avoids this difficulty is the following[7] (see Fig. 1):

$$\vec{n}_1,\vec{n}_2,\Theta,\vec{T}, \qquad (2)$$

where $\vec{T}$ is the same translation vector as in scheme (1). The unit vectors $\vec{n}_1$ and $\vec{n}_2$ are now chosen to represent the GB-plane normal relative to the principle coordinate systems of crystals 1 and 2, respectively, whereas Θ is the twist angle about this plane normal.

In this scheme, a symmetrical GB is readily recognized by the condition $\vec{n}_1 = \pm\,\vec{n}_2$. For example, the above (013) twin GB in the fcc lattice may be characterized as the 180° twist GB,[7] (013),(013),180°, from which it is apparent that only three variables are needed for its

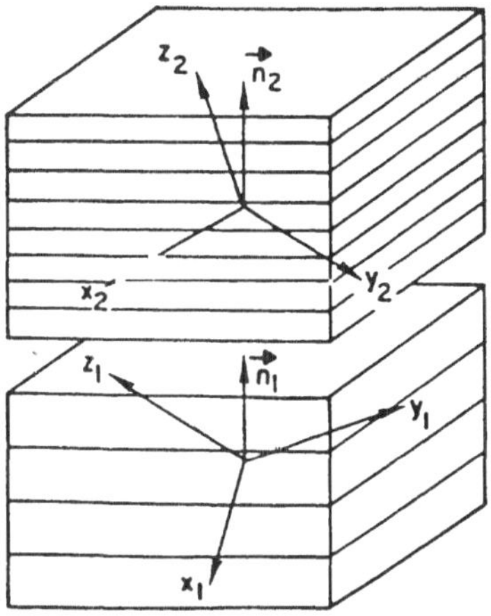

Fig. 1. GB-plane normals $\vec{n}_1$ and $\vec{n}_2$ in the principle coordinate systems of crystals 1 and 2, respectively.

macroscopic characterization, namely the GB plane (two dof) and the twist angle. Also, the separation of tilt and twist components for a general GB is readily accomplished: Whereas Θ represents the angle of a twist rotation about the GB-plane normal $\vec{n}$ (= $\vec{n}_1$ or $\vec{n}_2$, depending on which coordinate system is chosen), the tilt component is readily extracted from $\vec{n}_1$ and $\vec{n}_2$, considering that the tilt axis is perpendicular to both normals. Finally, the determination of the total misorientation, $R(\vec{n}_R, \theta_R)$, requires merely a multiplication of the rotation matrices for the twist and the tilt rotation. For further details see Ref. 13.

3. TWO CRITERIA FOR THE SELECTION OF LOW-ENERGY SYMMETRICAL GBs

According to the conventional classification scheme (1), symmetrical GBs may be either of a pure twist or a pure tilt type. On the other hand, however, according to the second scheme, such boundaries are characterized by $\vec{n}_1 = \pm \vec{n}_2$, with the only remaining parameter being the twist angle Θ in (2). Hence, as illustrated in Ref. 7, in crystal lattices with only one atom in the basis, all symmetrical tilt GBs may be considered as 180° twist GBs. Consequently, the distinction between symmetrical tilt and twist GBs is unnecessary and somewhat arbitrary. For the purpose of investigating the conditions for which symmetrical GBs with particularly low energy are obtained only twist GBs, hence, need to be considered. According to (2), this requires the investigation of the role of the GB plane normal, $\vec{n}$, and the twist angle, Θ, on the GB energy.[7]

3.1 Role of the GB Plane

When creating a planar defect on a given crystallographic plane (with normal $\vec{n}$), the perfect stacking of lattice planes in the $\vec{n}$ direction is destroyed. As a consequence, the perfect coordination of atoms near the defect plane is destroyed; i.e., atoms which were, for example, nearest neighbors in the perfect crystal may have been forced much more closely together as a result of the defect creation. One might ask about the role played by $\vec{n}$ on the closest distance of approach between any two such perturbed atoms.

The $\vec{n}$ direction in the perfect crystal is characterized by a certain number of staggered lattice planes in the repeat unit, referred to as the

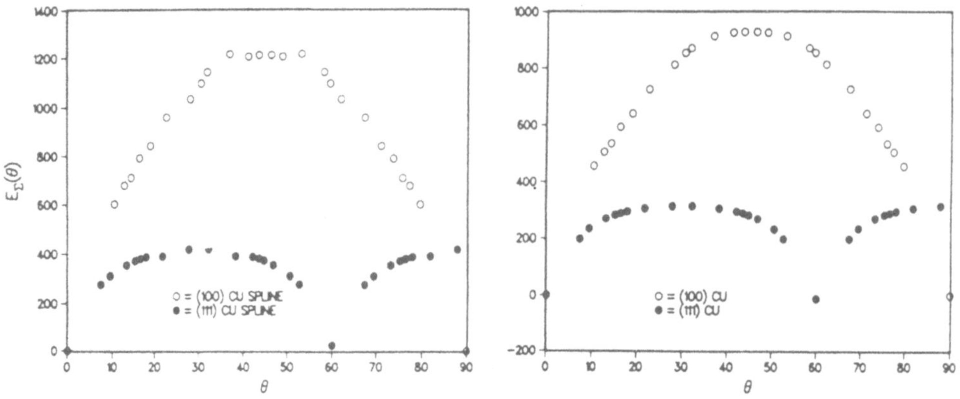

Fig. 2. Relaxed energies (in erg/cm^2) of (100) and (111) twist GBs in Cu computed by means of two different interatomic potentials. (For details see Ref. 10.)

period, $P(\vec{n})$. For example, in the fcc lattice three planes form the repeat unit in the (111) direction; hence P(111) = 3, whereas P(100) = P(110) = 2. In the unrelaxed interface, the closest distance between any two atoms facing each other across the interface plane is the interplanar spacing, $d(\vec{n})$. In fact, in the energetically worst case, two identical lattice planes which were staggered relative to one another prior to the creation of the interface may end up being right on top of each other at the distance $d(\vec{n})$. Since, in general, $d(\vec{n})$ may be substantially less than the nearest-neighbor distance, d_{NN}, in the perfect crystal, all atoms facing each other across such an interface repell each other substantially.

As a consequence of Pauli's principle, all interatomic interaction potentials are strongly repulsive (with an approximately exponential dependence on the interatomic separation) for short distances of approach between two atoms. The GB energy is therefore expected to increase substantially with decreasing separation, $d(\vec{n})$, of the lattice planes facing each other across the GB plane. This effect was confirmed in recent calculations for (100)[14,15] and (111) twist GBs[7,10] (see Fig. 2) and for <110> symmetrical tilt GBs[7] (see Fig. 3) in the fcc lattice. These calculations illustrate that--practically independent of the interatomic interaction potential chosen--the (100) twist GB energies are larger, typically by a factor of two to three, than the energies of (111) twist GBs, although the interplanar spacings differ by less than 20% [d(111) = 0.578a, d(100) = 0.5a, where a is the lattice parameter]. A similar connection between $d(\vec{n})$ and the GB energies exists for the <110> symmetrical tilt GBs [see Figs. 3(a) and (b)].

The fact that the planar atom density is proportional to the interplanar spacing leads to the conclusion[7] that in a given crystal structure the densest planes are the most favored lattice planes for the accommodation of low-energy symmetrical GBs.

3.2 Role of the Unit-Cell Area

One of the most pronounced structural features of GBs is the translation, $\vec{T}$, of the two grains for which a state of minimum energy is obtained. As illustrated in Ref. 7, the GB-energy variation for translations parallel to the GB plane is particularly sensitive to the size of

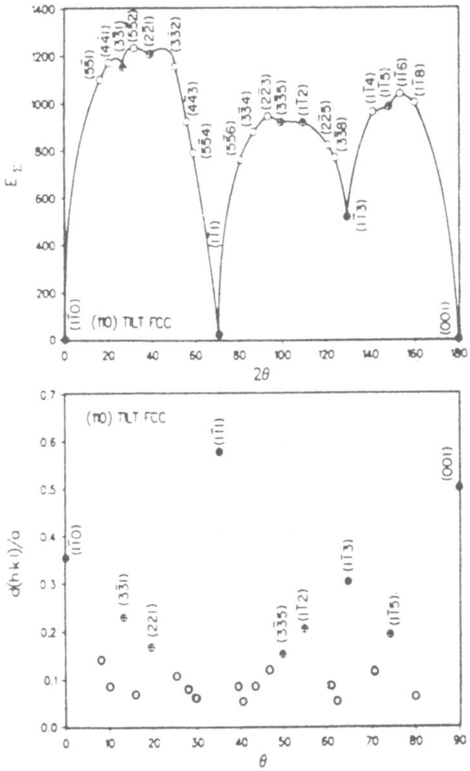

Fig. 3. Relaxed energies (in erg/cm^2) of <110> symmetrical tilt GBs in Cu vs tilt angle 2θ determined with the same potential as the energies in the left half of Fig. 2. In the bottom half the corresponding interplanar lattice spacings parallel to the GB plane are shown (in units of the lattice parameter a).

the planar unit cell (for coherent interfaces). More precisely, it is the number of atoms per lattice plane in the unit cell which controls how much energy can be gained by translating from the least to the most favorable configuration (see, e.g., Fig. 4 in Ref. 7). Since the symmetrical tilt GB on a given lattice plane contains only one atom per lattice plane in the unit cell, the translational gain of energy is greater for this "special" GB than for any other (twist) boundary on this plane,[7] thus giving rise to its particularly low energy. The basic conclusion drawn in Ref. 7 may then be formulated as follows: The translational gain of energy favors the GBs with the smallest numbers of atoms in the planar unit cell.

4. ASYMMETRICAL GBs

As mentioned in the Introduction, for a given misorientation of two crystal lattices the CSL model postulates the lowest-energy GB plane to be the symmetrical configuration; i.e., the plane with the highest density of CSL sites. The conditions under which this assumption holds were elucidated in a recent article[9] in which the energies of some asymmetrical tilt GBs were determined.

As discussed in more detail in Ref. 9, for most lattice misorientations an asymmetrical GB-plane orientation may be found such that the effective interplanar spacing at the GB,

$$d_{eff} = [d(\vec{n}_1) + d(\vec{n}_2)]/2, \tag{3}$$

is greater than for the corresponding symmetrical configurations. Here $d(\vec{n}_1)$ and $d(\vec{n}_2)$ denote the spacing of lattice planes parallel to the GB plane in the ideal semi-infinite crystals 1 and 2, respectively, which form the bicrystal. According to Sec. 3.1, this increase in the interplanar spacing is expected to give rise to a decrease in the GB energy. However, the simultaneous increase in the planar unit-cell area as one goes from a symmetrical to an asymmetrical GB-plane orientation—and the corresponding increase in GB energy (see Sec. 3.2)—often compensates for this energy gain. The net result is a delicate trade-off between the increased d-spacing and the increased unit-cell area associated with asymmetrical plane orientations. Hence, the relative stability of symmetrical and asymmetrical configurations has to be investigated in each case, and it may, to some degree, depend on the detailed nature of the interatomic interaction potentials.[9]

5. ROLE OF THE INTERATOMIC POTENTIAL IN METALS

The effect of the shape and cut-off radius of the interatomic potential on the predicted GB energies and structures was investigated recently in some detail.[10,14,15] The main conclusions are based on the observation that whereas the GB energy is governed by the very short-range part of the interatomic potential, V(r), (i.e., for distances less than the nearest-neighbor distance) the detailed GB structure is influenced by the detailed shape of the potential out to distances of the order of 1.5a to 2a. These conclusions may be summarized as follows:

(a) The absolute value of the GB energy for a given GB depends strongly on the potential since the magnitude of the short-range (practically exponential) increase in V(r) varies from one potential to another.

(b) The relative magnitudes of the energies of different GBs (i.e., energy vs misfit-angle curves such as the ones shown in Fig. 2) are rather insensitive to a change in potential.

(c) The detailed atomic structure depends strongly on the detailed shape of V(r); in particular, whether or not charge-density oscillations are included in the potential may have a major effect on the computed atomic structure of a given GB.

(d) The lowest-energy translational state of a given GB depends strongly on the shape of V(r).

6. COMPARISON BETWEEN IONIC CRYSTALS AND METALS

As discussed, for example, by Catlow et al.,[16] the strongly ionic bonding in most oxides with NaCl structure greatly simplifies the calculation of defect properties by allowing one to adopt a fully ionic model with central-force (two-body) interactions. In addition to the short-range interactions between the ions, long-range Coulomb interactions then have to be considered in the interionic potentials with the general form[16]

$$V_{jk}(r_{jk}) = \frac{q_j q_k}{r_{jk}} + A_{jk} e^{-r_{jk}/\rho_{jk}} - \frac{C_{jk}}{r_{jk}^6}, \qquad (4)$$

where q_j is the charge of ion j. A_{jk} and ρ_{jk} are parameters associated with the Born-Mayer repulsion between the ion cores, whereas the Van der Waals term governed by C_{jk} allows for an effectively covalent attraction primarily between the O^{2-} ions in metal oxides. As pointed out earlier,[10] the short range part in (4) (including the Born-Mayer and Van der Waals terms) is qualitatively very similar in nature to the interatomic potentials in metals with the exception that due to their lack of charge-density oscillations the ionic-crystal potentials are much smoother functions of r. It is therefore not very surprising that the GB energy vs misfit-angle curves calculated for ionic bicrystals bear strong resemblence to the corresponding results for metals. In particular, this resemblence was observed for (i) (100) twist GBs in MgO,[17,18] the transition-metal oxides NiO, CoO, FeO, and MnO[19,20] and in many alkali halides,[21] (ii) for <100> tilt[22] and (iii) for <110> tilt GBs in NiO.[23] As in metals, no energy cusps were observed for the (100) twist boundaries,[17-21] and similar to the results in Fig. 3, a deep energy cusp near the (111) twin orientation was observed for the <110> symmetrical tilt boundaries.[23]

Owing to the presence of the Coulomb term in the potentials (4), however, some differences between GBs in ionic and metallic bicrystals are expected. These have been investigated in detail for the (100) twist GBs.[21] The Coulomb interaction was shown to result in a net repulsion between the neutral (100) planes which leads a substantial volume expansion at these GBs which is much larger than what one obtains for a metal. However, as shown more recently[24,25] this Coulomb repulsion may be reduced significantly by removal of pairs of oppositely charged ions in energetically unfavorable lattice sites, i.e., by the introduction of Schottky disorder into the GB region.

REFERENCES

1. W. Bollmann, "Crystal Defects and Crystalline Interfaces," Springer-Verlag, New York (1980).
2. G. Hasson, J. Y. Boos, I. Herbeuval, M. Biscondi, and C. Goux, Surf. Sci. 31:115 (1972).
3. K. L. Merkle, J. F. Reddy, and C. L. Wiley, Ultramicroscopy 18:281 (1985).
4. K. L. Merkle and D. J. Smith, Ultramicroscopy (to be published).
5. K. L. Merkle, J. F. Reddy, C. L. Wiley, and D. J. Smith (in these proceedings).
6. W. Krakow, J. T. Wetzel, and D. A. Smith, Phil. Mag. A 53:739 (1986).
7. D. Wolf, J. de Physique Colloque C4, 46:C4-192 (1985).
8. D. Wolf, Materials Research Society Symposium Proceedings 40:341 (1985).
9. D. Wolf, Proc. Intl. Conf. on "Ceramic Microstructures '86: Role of Interfaces," Berkeley, CA, 1986 (to be published).
10. D. Wolf, Physica 131B:53 (1985).
11. D. M. Duffy and P. W. Tasker, Physica 131B:46 (1985).
12. C. Goux, Canadian Metall. Quarterly 13:9 (1974).
13. D. Wolf, Phil. Mag. A (to be submitted).
14. D. Wolf, Acta Metall. 32:245 (1984).
15. D. Wolf, Acta Metall. 32:735 (1984).
16. C. R. A. Catlow, I. D. Faux, and M. J. Norgett, J. Phys. C 9:419 (1976).
17. D. Wolf, J. Am. Ceram. Soc. 67:1 (1984).
18. D. Wolf, J. de Physique Colloque C6:45 (1982).
19. D. Wolf and R. Benedek, Adv. Ceram. 1:107 (1981).
20. D. Wolf, Adv. Ceram. 10:290 (1984).
21. D. Wolf, Phil. Mag. A 49:823 (1984).
22. D. M. Duffy and P. W. Tasker, Phil. Mag. A 47:817 (1983).
23. D. M. Duffy and P. W. Tasker, Phil. Mag. A 48:155 (1983).
24. P. W. Tasker and D. M. Duffy, Phil. Mag. A 47:L45 (1983).
25. D. Wolf, Mater. Res. Soc. Symp. Proc. 24:47 (1984).

ORDER AND CHAOS IN NEURAL SYSTEMS

K. E. Kürten
Institut für Theoretische Physik
Universität zu Köln
D-5000 Köln, West Germany

J. W. Clark
McDonnell Center for the Space Sciences
and Department of Physics
Washington University, St. Louis, Missouri 63130, U.S.A.

ABSTRACT

The occurrence of chaos in continuous-time nerve-net models is demonstrated in randomly connected networks of 26 and 80 neurons. For nets of sizeable dimensions one can conclude that chaos is a quite common occurrence; this may have important biological implications.

INTRODUCTION

A challenging problem in describing the activity of neural networks is to select a model of single nerve cells which is simple enough to permit extensive analysis, yet complex enough to include the important characteristics of real neurons. In 1973 R. B. Stein *et al.*[1] developed a continuous-time model of networks of neuron-like elements from a mathematical, computational, and physiological point of view. In Refs. 1 the dynamical behavior of the model has been investigated in detail for networks consisting of up to 3 neurons. In particular, conditions for the existence of stable and unstable steady-state solutions and for the occurrence of sustained (stable) periodic oscillations have been derived analytically. However there has been no explicit consideration of the possibility of the existence of chaotic solutions, although in recent years chaotic motions have been found in a variety of nonlinear systems relevant to fluid dynamics, chemical reactions, and nonlinear optics (lasers).[2,3] Moreover, within the theory of nonlinear dynamical systems the existence of unusually complex solutions is itself of considerable interest.

THE MODEL

The microscopic dynamics of a nerve net of N interconnected neurons is described with some realism by a set of $2N$ coupled first-order nonlinear differential equations due to Stein *et al.*[1] Within this mathematical model, expermentally observed properties and response characteristics of individual neurons are reproduced. The corresponding autonomous system of differential equations is of the form

$$a_i^{-1}\frac{dx_{1i}}{dt} = -x_{1i} + S\left[f_i + \sum_{j=1}^{N} c_{ij}x_{1j} + b_i x_{2i}\right] \quad , \tag{1a}$$

$$\frac{dx_{2i}}{dt} = x_{1i} - p_i x_{2i} \quad , \tag{1b}$$

where S, a function of sigmoid shape, is chosen as $S(u) = [1 + \exp(-u)]^{-1}$, and i runs from 1 to N.

Equation (1a) relates the basic or output variable x_{1i}, the normalized firing rate of neuron i at time t, to the total input of the cell, which is given by the argument of the sigmoid function. The auxiliary variable x_{2i} describes the phenomenon of adaption of the neuron. The first term in the sigmoid function, f_i, is the external input; the second term represents the total synaptic input from all cells of the net; and the last term incorporates adaption, allowing for a kind of self-excitation ($b_i > 0$) or self-inhibition ($b_i < 0$). The parameters $a_i > 0$ and $p_i > 0$ are rate constants, characteristic of the individual neurons or of specific types of neurons.

The strengths of the interactions among the neurons are specified by the efficiency matrix (c_{ij}), wherein c_{ij} measures the efficacy of the activity of neuron j in producing input to neuron i. A positive (negative) coupling coefficient c_{ij} means that the postsynaptic effect of j on i is excitatory (inhibitory). If neuron j has no synapses on i, c_{ij} is set equal to zero.

In the work to be reported, neurons are assumed to have identical intrinsic parameters, which correspond to the standard case studied by Stein *et al.*, namely $a_i = a = 100\ \text{sec}^{-1}$, $b_i = b = -200$, and $p_i = p = 100\ \text{sec}^{-1}$. These values produce a reasonable match of observed neuronal responses. The individual neuronal elements are inherently stable.

The efficiency matrix (c_{ij}) is chosen with the aid of a random-number generator, sampling uniform distributions. (i) Each neuron is assigned exactly m nonzero c_{ij}'s, where $0 < m \le N$, so that each neuron has the *same* number of inputs from network neurons. The m neurons which feed neuron i are selected randomly from all the N neurons of the net. (ii) The magnitudes of the nonzero c_{ij}'s are chosen by sampling a uniform distribution on $(0,L]$. (iii) A fraction h ($0 < h < 1$) of the Nm nonzero c_{ij}'s are supposed to be negative; the rest, positive. The couplings which are to be negative are also selected randomly, *without* requiring that all neurons have the same number of inhibitory inputs.

The network is stimulated by constant external inputs f to a randomly-chosen set of N_s neurons, where $0 < N_s \le N$. Thus $f_i = f$ for all neurons in the privileged set, and $f_i = 0$ for the others. In the cases cited below we choose $N_s/N = 0.5$.

COMPUTER SIMULATIONS

Although for small N we were able to find quite complex solutions, no example with chaotic behavior was found. In fact, the first behavior resembling chaos was encountered for a system with $N = 26$ neurons during a systematic investigation of the stability of the steady states of randomly connected nets. No further examples of chaotic solutions were uncovered for other systems of comparable size. In the example to be described below, each of the $N = 26$ neurons has $m = 7$ incoming connections.

A thorough numerical search, using a damped Newton-Raphson method, has revealed only one fixed point (steady state), which is unstable, while in some of the other 26-neuron nets considered we are able to find as many as 30-40 fixed points. Upon varying the external input f, which is regarded as a control parameter, we observe instances of putative chaotic output of a subset of neurons (some 9 in number) in the range $36 \lesssim f \lesssim 43$. These neurons, nontrivially active, are dubbed eligible, the remainder of the neurons being locked into a condition of either total inactivity (with firing rate zero) or saturated activity (with firing rate unity). Increasing the stimulus f past the lower critical boundary, the indicated subset of neurons undergoes a transition from simple periodic motion to a condition resembling intermittency. Increasing f through the upper boundary, there is a transition from apparent chaos to periodic motion. By $f = 42.8$ the solutions are already periodic, with exceptionally long transients in the existing runs.

Figure 1 gives sample numerical solutions, in both ordered and chaotic domains, for the firing rate of a representative eligible neuron.

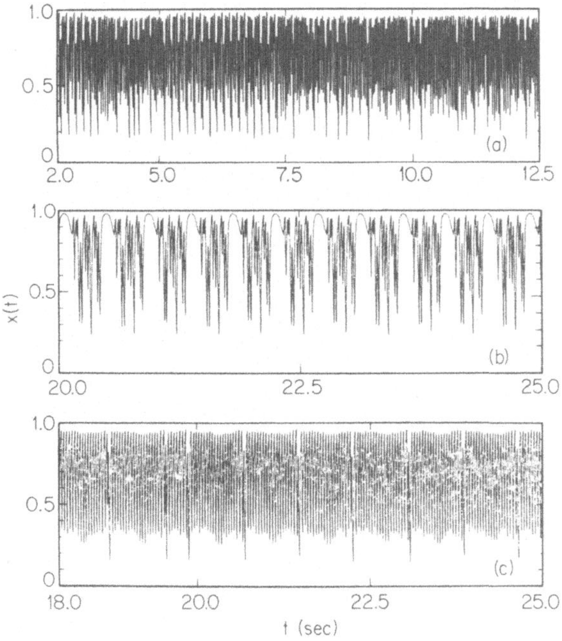

Figure 1: Time course of the firing rate of an eligible neuron at three values of the external input: (a) $f = 41.8$ (chaotic case), (b) $f = 46.8$ (periodic case) and (c) $f = 36.3$ (chaotic case).

For some systems it might happen that although the numerical solution looks chaotic, one may in reality just be seeing a very long transient, or a periodic mode of extremely long period.[4] Ideally, one would prefer to have available the spectrum of Lyapunov exponents, which provides the most incisive dynamical diagnostic for chaotic systems. Lyapunov exponents are the average exponential rates of divergence or convergence of neighboring trajectories in phase space. Any motion having at least one positive Lyapunov exponent is by definition chaotic, and the magnitude of the positive exponent reflects the time scale on which the evolution of the system ceases to be predictable. In practice, the Lyapunov exponents may be directly determined by integration of the linearized equations of motion.[5] However, since each equation has to be integrated for N different initial conditions, defining an arbitrarily-oriented frame of N orthonormal vectors, we have to solve a set of $2N(2N+1)$ differential equations. Work in this direction is in progress. Here we must be content with the examination of power spectra and the estimation of the fractal dimension of the attractor.[6] While the Lyapunov spectrum is closely related to the fractal dimension of the associated strange attractor through the information dimension, calculation of the latter requires knowledge of all but the most negative Lyapunov exponents.[5]

Two sample Poincare maps, in which the output of one eligible neuron is plotted against the simultaneous output of another eligible neuron, are presented in Figure 2. Representative power spectra, based on 8000 data points, are compared in Figure 3. At $f = 36.3$ it appears that we are just within the boundary of the chaotic regime, while the solution at $f = 46.8$ is definitely periodic. For $f = 40.8$, in a run very similar to that at input 41.8, chaos prevails. This has been verified convincingly by implementing the algorithm proposed in Ref. 6 for the characterization of an attractor using a single-variable time series. The correlation exponent ν is found to be 3.1, whereas the expected integer value (unity) is obtained in the periodic case at $f = 46.8$. Further evidence of the chaotic nature of the $f = 40.8$ and $f = 41.8$ solutions is furnished by the associated phase portraits and especially by the power spectra we have derived. Referring to Fig. 3a, we see the characteristic broad continuum at low frequency, in obvious contrast with the periodic example ($f = 46.8$) of Fig. 3b, which is distinguished by the occurrence of sharp peaks.

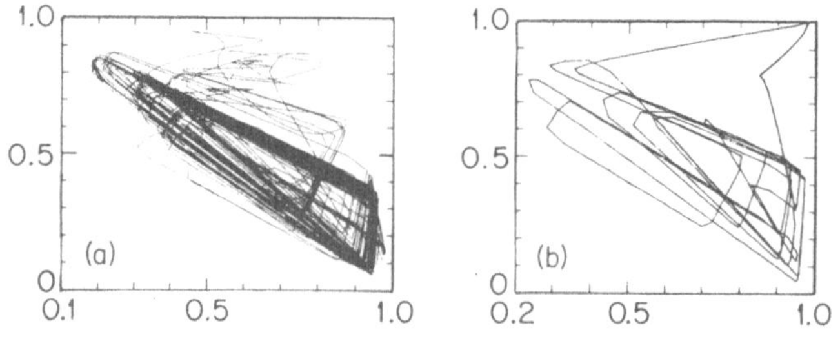

Figure 2: Phase portraits for two values of external input: (a) $f = 41.8$ (chaotic case) and (b) $f = 46.8$ (periodic case). The firing rate $x_1(t)$ of cell 1 is plotted against the firing rate $x_2(t)$ of cell 2. The initial transient is omitted from (b).

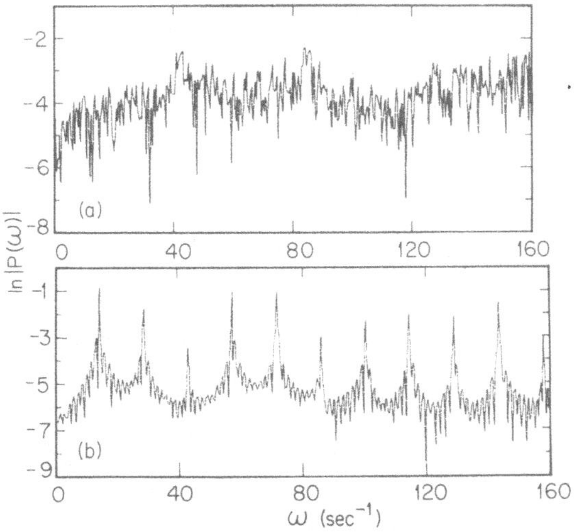

Figure 3: Power spectra for two values of external input: (a) $f = 41.8$ (chaotic case) and (b) $f = 46.8$ (periodic case). Here, $P(\omega) = \int_0^T \hat{x}(t)\exp(i\omega t)dt$ where $\hat{x}(t) = x(t) - T^{-1}\int_0^T x(u)du$ and T is the upper limit of the calculated time series.

Numerous examples of chaos have been discovered for a system of 80 neurons with $m = 40$ incoming connections per neuron. Figure 4 offers a selection of solutions which rather obviously exhibit chaotic behavior. However, the final certification has still to be made on the basis of the estimated fractal dimension, or, more definitively, the Lyapunov spectrum.

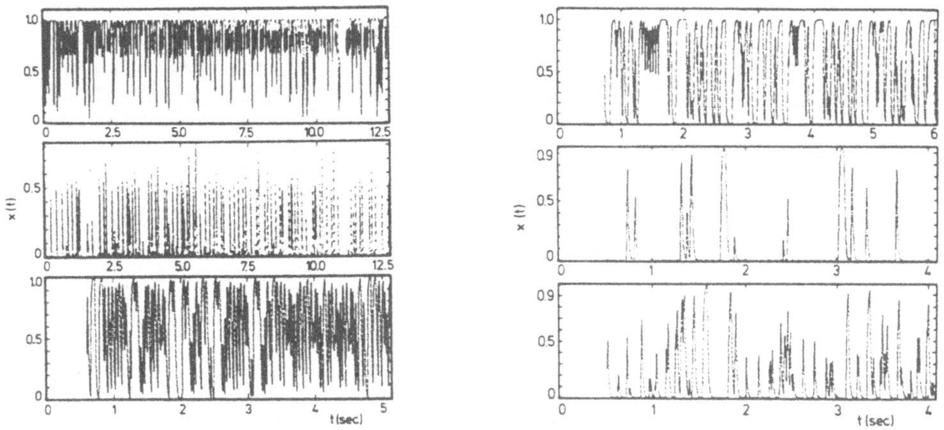

Figure 4: Representative chaotic solutions for the firing rate $x(t)$ versus time t of an eligible neuron, for several quasirandom networks containing 80 neurons.

There remains the important question of the biological implications of chaos in neural networks. In most situations a tendency to chaos would surely be harmful, an impediment to reliable response and to organized behavior conducive to survival. Regarding performance of the mature animal, there would accordingly be a need to suppress chaotic modes if the synaptic interactions and other conditions of structure and stimulus permit such behavior, by modifying the network parameters in some way. On the other hand, during development of the animal there would appear to be a need for some mechanism for avoiding, insofar as is possible, the patterns of synaptic and neuroanatomical organization which are favorable to chaos. The latter end is presumably achieved, at least in part, by genetic programming. While chaos would usually be harmful, we should not exclude the possibility that in some circumstances it could be beneficial, particularly in the context of higher mental processing where it is advantageous to have the potential of breaking out of rigid, stereotyped behavior. Thus chaos, like neuronal spontaneity, opens access to a wider variety of responses and behavior, and may, if mechanisms exist for keeping it in check, be a valuable attribute of complex biological systems. For extensive and perceptive commentary on the occurrence of chaos in neural systems and its role in neurobiology, the reader should consult Refs. 7.

ACKNOWLEDGMENTS

This work was supported in part by the Condensed Matter Theory Program of the Division of Materials Research of the U.S. National Science Foundation under Grant No. DMR-8519077 and in part by BRSG S07 RR07054-20 awarded by the Biomedical Research Support Grant Program, Division of Research Resources, National Institutes of Health. We thank J. W. Chen, U. an der Heiden, H. Moraal, K. Niklaus, H. G. Schuster and H. A. Wischmann for informative discussions.

REFERENCES

1. R. B. Stein, K. V. Leung, D. Mangeron, and M. N. Oguztöreli, Kybernetik **15**, 1 (1974); R. B. Stein, K. V. Leung, M. N. Oguztöreli, and D. W. Williams, Kybernetik **14**, 223 (1974); and references cited therein.
2. H. G. Shuster, *Deterministic Chaos: An Introduction* (Physik-Verlag, Weinheim, 1984).
3. K. E. Kürten and J. W. Clark, Phys. Lett. **114A**, 413 (1986).
4. S. Newhouse, Publ. Math. IHES **50**, 101 (1980).
5. A. Wolf, J. B. Swift, H. L. Swinney, and J. A. Vastano, Physica **16D**, 285 (1985).
6. P. Grassberger and I. Procaccia, Physica **9D**, 189 (1983).
7. E. Harth, IEEE Trans. Systems Man Cybernetics **SMC-13**, 782 (1983); M. R. Guevara, L. Glass, M. C. Mackey, and A. Shrier, IEEE Trans. Systems Man Cybernetics **SMC-13**, 790 (1983).

WHAT'S RIGHT AND WHAT'S WRONG WITH THE DENSITY-GRADIENT EXPANSIONS FOR THE EXCHANGE AND CORRELATION ENERGIES?

John P. Perdew

Department of Physics and Quantum Theory Group
Tulane University
New Orleans, Louisiana 70118, U.S.A

ABSTRACT

In a system of slowly-varying electron density, the exchange and correlation hole surrounding an electron may be expanded in gradients of the density. The zero-order term of this expansion, the local density approximation, satisfies important properties of the exact hole: The density of the exchange hole is everywhere negative, and integrates to a deficit of one electron, while the density of the correlation hole integrates to zero. Here the second-order term of the expansion is shown to improve the description of the hole near the electron, at the cost of worsening the description far away and violating these exact properties. The gradient expansion may be used to improve upon the local density approximation for the energy of a real system with long-range interactions, but only after this expansion has been appropriately cut off. The cut-off or generalized gradient approximation, which involves only the local density and its first derivative, provides qualitative and quantitative information about many-body effects in inhomogeneous systems.

INTRODUCTION

The density functional theory of Hohenberg, Kohn and Sham[1,2] provides a description of the ground-state density and energy of a many-electron system which is universal, simple, practical and (in principle) exact. The density $n(\underset{\sim}{r})$ is constructed from Kohn-Sham orbitals $\psi_i(\underset{\sim}{r})$ obtained from the solution of a self-consistent Schrödinger equation, in which $\delta E_{xc}/\delta n(\underset{\sim}{r})$ plays the role of an exchange-correlation potential. The exchange-correlation energy $E_{xc}[n]$ is an unknown functional of the electron density which is usually replaced by its local density approximation[2] (LDA):

$$E_{xc}^{LDA}[n] = \int d^3r \; n \; \epsilon_{xc}(n) \; , \tag{1}$$

where $\epsilon_{xc}(n)$ is the exchange-correlation energy per particle for an electron gas of uniform density n.

All of the true many-body effects are subsumed in the exchange-correlation energy $E_{xc}[n]$, which will be discussed here. Some of the numerical applications of LDA have been reviewed recently by Gunnarsson and Jones[3]. (Actually the local spin-density approximation is used for systems with unpaired electronic spins; for the spin-density generalizations of the expressions in this paper, see Refs. 4 and 5.) Although the successes of LDA have been impressive, this approximation still falls tantalizingly short of the "chemical accuracy" one would like to have for the description of binding and bonding in atoms, molecules and solids.

The simplest generalization of LDA is the gradient expansion approximation[1,2] (GEA)

$$E_{xc}^{GEA}[n] = \int d^3r \; n \, \epsilon_{xc}(n) + \int d^3r \; C_{xc}(n) \, |\nabla n|^2/n^{4/3} , \quad (2)$$

which is truncated at second order in the gradient. Table 1 compares the LDA, GEA and exact exchange-correlation energies for several atoms. Although the LDA errors are only about 5%, the GEA errors are worse. What is wrong with the gradient expansion? And what can be salvaged from the wreckage? The author's views will be presented here. These views are similar in essence to those proposed in the pioneering work by Ma and Brueckner[6], and, especially, by Langreth and co-authors[7-10].

The basis for this discussion will be an exact expression[11,12] for the exchange-correlation energy

$$E_{xc}[n] = \tfrac{1}{2}\int d^3r \int d^3R \; n(\mathbf{r}) \, n_{xc}(\mathbf{r},\mathbf{r}+\mathbf{R})/R , \quad (3)$$

where $n_{xc}(\mathbf{r},\mathbf{r}+\mathbf{R})$ is the density at position $\mathbf{r}+\mathbf{R}$ of the exchange-correlation hole surrounding an electron at position $\mathbf{r}$:

$$n_{xc}(\mathbf{r},\mathbf{r}+\mathbf{R}) = \int_0^1 d\lambda \Big\{ \langle [\hat{n}(\mathbf{r})-n(\mathbf{r})][\hat{n}(\mathbf{r}+\mathbf{R})-n(\mathbf{r}+\mathbf{R})] \rangle_\lambda - n(\mathbf{r}) \, \delta(\mathbf{R}) \Big\} / n(\mathbf{r}) . \quad (4)$$

Table 1. Exchange-correlation energies of atoms, in hartrees. Exchange energies from Ref. 4, and correlation energies from Ref. 5. LDA: local density approximation. GEA: gradient expansion approximation. GGA: generalized gradient approximation.

Atom	LDA	GEA	GGA	Exact
He	- 0.996	- 0.882	- 1.077	- 1.068
Be	- 2.53	- 2.27	- 2.77	- 2.76
Ne	-11.77	-10.99	-12.61	-12.50
Ar	-29.28	-27.75	-31.09	-30.97

The hole is an average over the coupling-constant λ for a series of systems with fixed density $n(\mathbf{r}) = \langle \hat{n}(\mathbf{r}) \rangle_\lambda$ and increasing electron-electron interaction λ/R. Since the total number of electrons is fixed,

$$\int d^3R \; n_{xc}(\mathbf{r}, \mathbf{r}+\mathbf{R}) = -1, \tag{5}$$

i.e., the hole represents a deficit of one electron. The concentration of the hole around its electron describes the tendency of the other electrons to avoid the one at $\mathbf{r}$, as a consequence of Coulomb repulsion and/or the Pauli exclusion principle.

Gunnarsson and Lundqvist[12] have argued that the LDA works decently, even beyond its formal domain of validity (very slow density variation), because it obeys Eq. (5). Unlike the exact hole, the LDA hole is spherically-symmetric about the electron it surrounds. However, as Gunnarsson and Lundqvist[12] also pointed out, the exchange-correlation energy of Eq. (3) depends only upon the spherical average

$$\langle n_{xc}(\mathbf{r}, R) \rangle_{sph} \equiv \int \frac{d\hat{R}}{4\pi} n_{xc}(\mathbf{r}, \mathbf{r}+\mathbf{R}). \tag{6}$$

It will be shown here that the GEA violates Eq. (5), and another property of the exact hole as well. To demonstrate the second property, it will be necessary to separate Eqs. (3) and (4) into exchange and correlation contributions. The exchange energy $E_x[n]$ dominates in the high-density limit, and is intrinsically simpler than the correlation energy $E_c[n]$.

EXCHANGE ENERGY

The exchange hole is just Eq. (4) with the expectation value evaluated for a non-interacting (λ=0) system. Thus

$$n_x(\mathbf{r}, \mathbf{r}+\mathbf{R}) = -\frac{1}{2} |\rho(\mathbf{r}, \mathbf{r}+\mathbf{R})|^2/n(\mathbf{r}), \tag{7}$$

$$\rho(\mathbf{r}, \mathbf{r}+\mathbf{R}) \equiv \sum_i \psi_i^*(\mathbf{r}+\mathbf{R}) \, \psi_i(\mathbf{r}) \, \theta(\mu-\epsilon_i). \tag{8}$$

The exchange energy is Eq.(3) with $n_{xc} \to n_x$, i.e., it is the standard Fock integral evaluated with Kohn-Sham (not Hartree-Fock) orbitals. From Eqs. (7) and (8), it follows immediately that

$$n_x(\mathbf{r}, \mathbf{r}) = -n(\mathbf{r})/2 \tag{9}$$

$$n_x(\mathbf{r}, \mathbf{r}+\mathbf{R}) \leq 0. \tag{10}$$

Since every attainable exchange hole density must be negative and must integrate to -1, the exchange energy will be more negative for a hole which is concentrated closer to the electron it surrounds. It is clear[13] that the LDA, which obeys Eqs. (9), (10) and (5), can never give <u>too</u> bad a value for the exchange energy.

Now consider the GEA for exchange. Each position $\mathbf{r}$ in an electronic system has a density n, an inhomogeneity wavevector $|\nabla n|/n$, and a Fermi wavevector $k_F = (3\pi^2 n)^{1/3}$. The gradient expansion is formally valid[1,2] when the dimensionless ratios

$$s = |\nabla n|/2k_F n, \qquad |\nabla_i \nabla_j n|/2k_F|\nabla n| \tag{11}$$

are much smaller than unity. Fig 1 displays these ratios for the neon atom. The parameter s of Eq. (11) is typically less than or about equal to one over the interior of the atom, but diverges exponentially as the density decays into the vacuum.

The Dirac density matrix of Eq. (8) and thus the exchange hole density of Eq. (7) may be expanded to second order in the gradients of $n(\mathbf{r})$, following the operator-algebra formalism of Kirzhnits[14] and of Gross and Dreizler[15]. The result for $n_x^{GEA}(\mathbf{r},\mathbf{r}+\mathbf{R})$, which involves both first and second derivatives of the density, was presented and discussed by Perdew.[13] The "double" nature of the gradient expansion is evident: each additional power of the gradient brings with it another power of R, suggesting that the gradient expansion is trustworthy only for small R.

In order to eliminate the second derivatives of the density, Perdew and Wang[4] integrated by parts on $\mathbf{r}$ in Eq.(3), thereby replacing $n_x(\mathbf{r},\mathbf{r}+\mathbf{R})$ by $\tilde{n}_x(\mathbf{r},\mathbf{r}+\mathbf{R})$, a representation of the hole involving only $n(\mathbf{r})$ and its first derivative ∇n. Fig. 2 shows the spherically-averaged exchange hole $\langle\tilde{n}_x(\mathbf{r},R)\rangle_{sph}$ in the LDA and GEA for inhomogeneity s=1. Both approximations satisfy the exact property of Eq. (9), but the GEA violates Eq. (10). In fact the GEA hole density displays the undamped oscillation $-\,ns^2\cos(2k_F R)/18$ in the limit $R\rightarrow\infty$. The $\mathbf{R}$ integrals of Eqs. (3) and (5) are not even defined in the usual sense. However, with the help of a convergence factor $e^{-\alpha R}$ $(\alpha\rightarrow 0^+)$, the GEA exchange hole density integrates to -1 and

$$E_x^{GEA}[n] = A_x\int d^3r\, n^{4/3} + C_x\int d^3r\, |\nabla n|^2/n^{4/3}, \tag{12}$$

where A_x=-0.73856 is the usual LDA (α=2/3) coefficient[2], and C_x = -0.001667 is the gradient coefficient first derived by Sham[16] and re-derived by Gross and Dreizler[15]. (Kleinman[17], however, has obtained a coefficient which is 8/7 of Sham's.)

Does the gradient expansion for the exchange hole make any sense at all? To see that it does, consider Fig. 3, which shows two sections through the GEA exchange hole <u>before</u> spherical averaging. The first section plots the hole density along a line through the electron perpendicular to the density gradient. Along this line, the hole is symmetrical about the electron,. The second section is along a line through the electron parallel to the density gradient, and shows that the deepest part of the hole has been shifted away from the electron toward the higher-density side. To linear order in R, $\tilde{n}_x^{GEA}(\mathbf{r},\mathbf{r}+\mathbf{R}) = -\frac{1}{2}n(\mathbf{r}+\mathbf{R})$. This makes good physical sense, since it is the density $n(\mathbf{r}+\mathbf{R})$ that the hole is "dug out of".

Although the gradient terms worsen the large-R behavior of the hole, they improve its small-R behavior. Thus Perdew and Wang[4], following the real-space cutoff approach of Perdew[13], have proposed a generalized gradient approximation (GGA): Start with the GEA exchange hole density $\tilde{n}_x^{GEA}(\mathbf{r},\mathbf{r}+\mathbf{R})$, and cut away everything that is most pathological. First set the hole density to zero for all $\mathbf{R}$ values at which $\tilde{n}_x^{GEA}(\mathbf{r},\mathbf{r}+\mathbf{R})$ is positive. Next set the hole density to zero for $R>R_c$, where

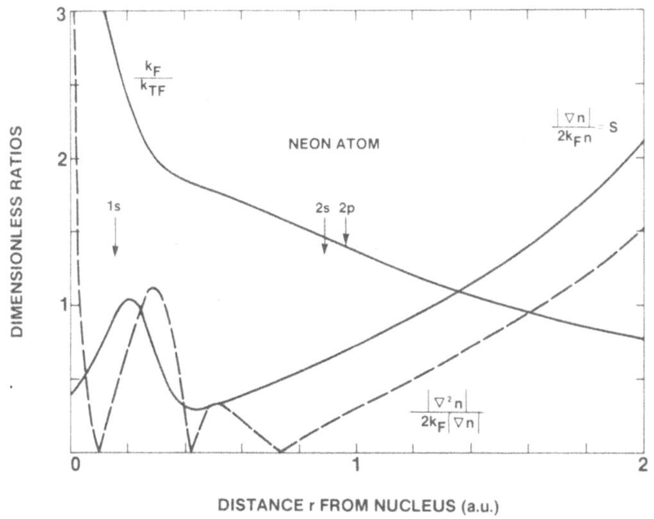

Fig. 1. The dimensionless ratios of Eq. (11) for the neon atom, together with the ratio of the Fermi to the Fermi-Thomas wavevector.

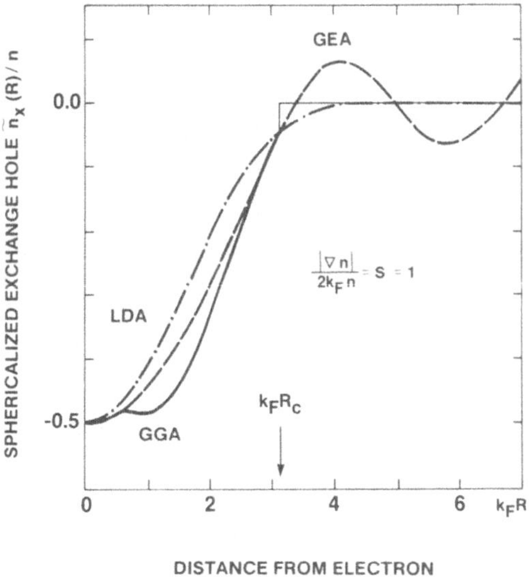

Fig. 2. The spherically-averaged exchange-hole density for inhomogeneity s=1. The electron is at R=0.

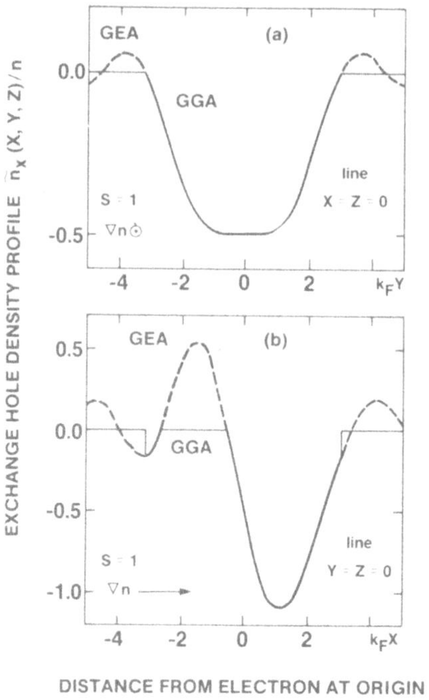

Fig. 3. The unaveraged exchange-hole density for inhomogeneity s=1, along lines passing through the electron in directions perpendicular (a) and parallel (b) to the density gradient.

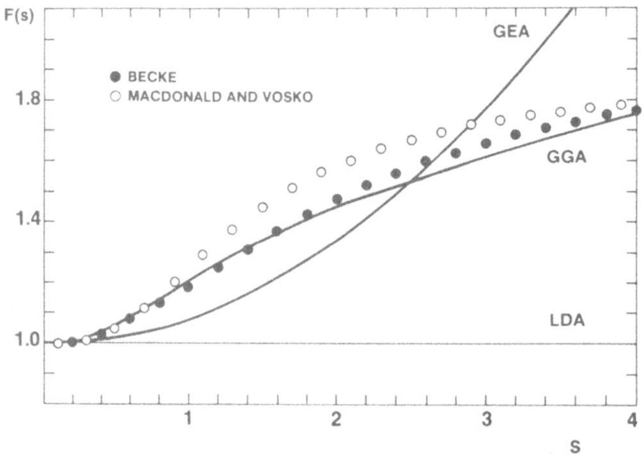

Fig. 4. The exchange function F(s) for Eq. (13).

$R_s(k_F, s)$ is a cutoff radius chosen to satisfy the sum rule of Eq. (5). The resulting exchange hole density is the solid curve (GGA) in Fig. 3, which after spherical averaging becomes the solid curve (GGA) in Fig. 2.

Finally the exchange energy is constructed from Eq. (3):

$$E_x^{GGA}[n] = A_x \int d^3r \; n^{4/3} \; F^{GGA}(s) \quad , \tag{13}$$

which scales homogeneously like the exact exchange energy. The function F(s) was evaluated numerically and then fitted accurately to the analytic form

$$F^{GGA} = (1+1.296s^2+14s^4+0.2s^6)^{1/15} \quad , \tag{14}$$

which reduces to the GEA for s<<1: $F^{GEA} = 1+0.0864s^2$. Fig. 4 compares the GEA and GGA functions F(s), which are very different for intermediate and large s. The form of Eq. (13) has also been used recently by Becke[18], who fitted a Pade approximant to the exact exchange energies of atoms, and by Macdonald and Vosko[19], who similarly fitted the exchange energy - density. As Fig. 4 shows, the a priori form of Eq. (13) is very similar to these "semi-empirical" forms, especially Becke's.

Calculated exchange energies for several atoms are presented in Table 2. For exchange alone, the GEA provides a partial improvement upon the LDA, but the GGA gives a much more substantial improvement. The GGA exchange energies are in error by less than 1%.

CORRELATION ENERGY

The Fermi-Thomas wavevector or inverse screening length is $k_{FT} = 2(k_F/\pi)^{1/2}$. As is evident from Fig. 1, the dimensionless ratio k_F/k_{FT} is close to unity in the valence shell of an atom, but is substantially bigger in the core. The gradient expansion for the correlation energy is formally valid when the dimensionless ratios of Eq. (11), and their analogs with $k_F \rightarrow k_{FT}$, are all much smaller than unity. These conditions are violated severely in an atom, and the gradient expansion therefore gives a poor account of the correlation energy (Table 3).

The correlation hole density $n_c(\underset{\sim}{r}, \underset{\sim}{r}-\underset{\sim}{R})$ integrates over $\underset{\sim}{R}$ to zero, so it must have both positive and negative parts. The analog of Eq. (9) is the cusp condition[20] on the correlation hole density at R=0. Manifestly the LDA obeys these exact conditions.

Fig. 5 shows the LDA correlation hole density for r_s=2, where $k_F r_s = (9\pi/4)^{1/3}$. The curve has been plotted from the pair distribution function of Contini et al[21], but using the crude approximation $\int_0^1 d\lambda g(\lambda) = [g(1)+g(0)]/2$ for the coupling-constant integration.

Table 2. Exchange energies of atoms, in hartrees (from Ref. 4).

Atom	LDA	GEA	GGA	Exact
He	- 0.884	- 0.970	- 1.033	- 1.026
Be	- 2.31	- 2.50	- 2.68	- 2.67
Ne	- 11.03	- 11.55	- 12.22	- 12.11
Ar	- 27.86	- 28.86	- 30.29	- 30.18
Zn	- 65.63	- 67.36	- 69.93	- 69.7
Kr	- 88.6	- 90.7	- 93.8	- 93.9
Xe	-170.6	-173.9	-178.6	-179.1

The gradient coefficient for the correlation energy has been computed and parametrized by Rasolt and Geldart[22]

$$C(n) = 0.001667 + \frac{(0.002568+0.023266r_s+7.389\times10^{-6}r_s^2)}{(1+8.723r_s+0.472r_s^2+0.07389r_s^3)} . \quad (15)$$

Very similar results have been obtained by Hu and Langreth.[8]

The GEA for the correlation energy ought to be improved by a real-space analysis similar to that of the preceding section: Construct the spherically-averaged correlation hole density $\langle \tilde{n}_c^{GEA}(\underset{\sim}{r},R)\rangle_{sph}$, and cut it off outside a radius chosen so that the cut-off hole density integrates to zero. However, even if an accurate representation of $\langle \tilde{n}_c^{GEA}(\underset{\sim}{r},R)\rangle_{sph}$ were available, the resulting correlation energy functional would still be defined only numerically; the failure of correlation to scale homogeneously would block the construction of a simple analytic functional. The wavevector analysis of Langreth and coworkers[7-10], which historically preceded the real-space analysis[4,13], overcomes this barrier.

The wavevector analysis[11] of the exchange-correlation energy is obtained from Eq.(3) when the Coulomb interaction 1/R is replaced by its Fourier decomposition. The result is

$$E_{xc} = \int d^3r\, n(\underset{\sim}{r}) \int_0^\infty dk\, \frac{1}{\pi} \langle n_{xc}(\underset{\sim}{r},k)\rangle_{sph} , \quad (16)$$

where

$$\langle n_{xc}(\underset{\sim}{r},k)\rangle_{sph} = \int d^3R\, e^{i\underset{\sim}{k}\cdot\underset{\sim}{R}} \langle n_{xc}(\underset{\sim}{r},R)\rangle_{sph} . \quad (17)$$

The wavevector analysis of the GEA correlation hole is known[7] within the random phase approximation (RPA). It is approximately[5,8]

$$\langle \tilde{n}_c^{GEA}(\underset{\sim}{r},k)\rangle_{sph} = \langle \tilde{n}_c^{LDA}(\underset{\sim}{r},k)\rangle_{sph} + \frac{2\pi\sqrt{3}}{k_{FT}}\, C(\infty)\, \frac{|\nabla n|^2}{n^{7/3}} \exp\left[\frac{-2\sqrt{3}\, C(\infty)k}{k_{FT}\, C(n)}\right] . \quad (18)$$

This representation of the GEA is better for small wavevectors

Table 3. Correlation energies of atoms and ions, in hartrees (from Ref. 5).

Atom	LDA	GEA	GGA	Exact
H	-0.022	0.033	-0.003	0
He^{+1}	-0.030	0.092	0.002	0
Li^{+2}	-0.034	0.154	0.004	0
He	-0.112	0.08	-0.044	-0.042
Li^{+1}	-0.134	0.201	-0.045	-0.044
Be^{+2}	-0.150	0.320	-0.049	-0.044
Be	-0.224	0.233	-0.094	-0.094
Ne^{+6}	-0.333	0.965	-0.136	-0.18
Ne	-0.74	0.56	-0.39	-0.39
Ar	-1.42	1.11	-0.80	-0.79
Kr	-3.27	2.12	-2.01	—
Xe	-5.18	3.26	-3.31	—

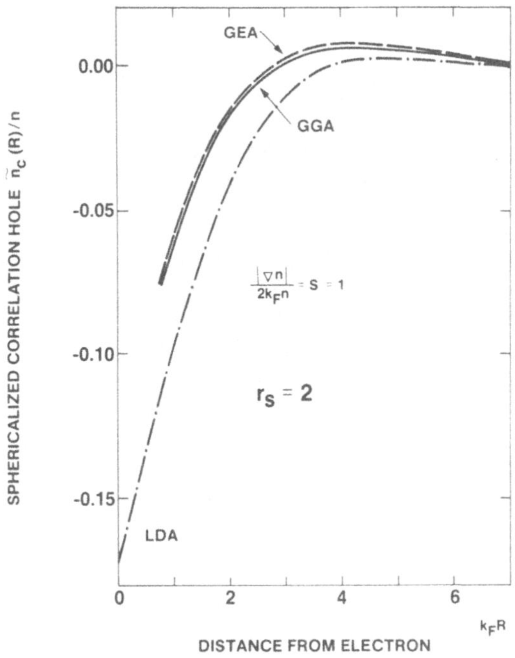

Fig. 5. The spherically-averaged correlation-hole density for $r_s=2$ and inhomogeneity s=1.

k than for $k \gtrsim 2k_F$, and better for high densities n (say, $r_s \lesssim 2$) than for low.

While the LDA correlation hole integrates to zero (i.e., $\langle \tilde{n}^{LDA}(\underset{\sim}{r}, k=0) \rangle_{sph} = 0$), it appears from Eq. (18) that the GEA hole integrates to $0.7s^2/\sqrt{r_s}$. The exponential peak around k=0 in Eq. (18) is a characteristic feature of the GEA, but one which is not realistic in the limit $k \to 0$. In this limit, the density fluctuations responsible for the exchange-correlation energy are of very long wavelength, and thus not sufficiently localized to sense either the local density or its gradient.

These considerations led Langreth and Perdew[7], Langreth and Mehl[8], and Perdew[5] to make a wavevector-space cutoff. The gradient contribution to Eq. (18) is replaced by zero for $k < k_c$, where $k_c = \tilde{f}|\nabla n|/n$ is a cutoff proportional to the inhomogeneity wavevector. Thus the generalized gradient approximation to the correlation energy is[5]

$$E_c^{GGA}[n] = \int d^3r \; n \; \epsilon_c(n) + \int d^3r \; e^{-\Phi} \; C(n)|\nabla n|^2/n^{4/3} \; , \quad (19)$$

$$\Phi = 1.745 \; \tilde{f} \; [C(\infty)/C(n)] \; |\nabla n|/n^{7/6} \; . \quad (20)$$

The electron-gas inputs $\epsilon_c(n)$[23] and C(n) (Eq.(15)) have been taken from beyond-RPA theories[8], and the cutoff parameter $\tilde{f}$ = 0.11 chosen to fit the exact correlation energy of the neon atom. The correlation energies of the other atoms and ions are then accurately reproduced, as Table 3 shows.

The GEA and GGA correlation holes are shown in Fig. 5. Note that the GEA adds a positive "bump" to the LDA correlation hole, which (in the R-region plotted) is only slightly modified by GGA. At large R, the GEA hole (in the approximate representation of Eq. (18)) falls off with a nonoscillatory R^{-4} behavior, while the GGA hole is a sum of purely oscillatory terms, the dominant one being $\cos(k_c R)/R^2$.

It is worthwhile to point out why a small-wavevector cutoff would <u>not</u> improve the GEA for exchange: The bad large-R behavior of the GEA exchange hole arises, not from the $k \to 0$ region of its Fourier transform, but from strong distributional singularities[7] at $k = 2k_F$. Such singularities are also present in the wavevector analysis of the GEA correlation hole[7], but in this case they are very weak, especially at high densities.

SUMMARY

Inhomogeneity of the density makes the exchange energy more negative, and the correlation energy less so, than the local density approximation predicts.

The gradient expansions for the exchange and correlation holes surrounding an electron seem to require not only that the dimensionless ratios of Eq.(11) be small, but also that $R|\nabla n|/n$ be small, where R is the distance from the electron. When the spurious large-R contributions are cut off to satisfy known properties of the exact holes, reasonably accurate density functionals for real electronic systems are generated.

ACKNOWLEDGEMENT

This work was supported in part by the National Science Foundation under Grant No. DMR84-20964.

REFERENCES

1. P. Hohenberg and W. Kohn, Phys. Rev. 136, B864 (1964).
2. W. Kohn and L. J. Sham, Phys, Rev. 140, A1133 (1965).
3. O. Gunnarsson and R. O. Jones, Phys. Rev. B31, 7588 (1986).
4. J. P. Perdew and Wang Yue, Phys. Rev. B33, 8800 (1986).
5. J. P. Perdew, Phys. Rev. B33, 8822 (1986); erratum to appear.
6. S.-K. Ma and K. A. Brueckner, Phys. Rev. B165, 18 (1968).
7. D. C. Langreth and J. P. Perdew, Phys. Rev. B21, 5469 (1980).
8. D. C. Langreth and M. J. Mehl, Phys. Rev. B28, 1809 (1983).
9. C. D. Hu and D. C. Langreth, Phys. Scr. 32, 391 (1985).
10. C. D. Hu and D. C. Langreth, Phys. Rev. B33. 943 (1986).
11. D. C. Langreth and J. P. Perdew, Phys. Rev. B15, 2884 (1977); Solid State Commun. 17, 1425 (1975); Phys. Rev. B26, 2810 (1982); Phys. Lett. 92A, 451 (1982).
12. O. Gunnarsson and B. I. Lundqvist, Phys. Rev. B13, 4274 (1976).
13. J. P. Perdew, Phys. Rev. Lett. 55, 1665 (1986).
14. D. A. Kirzhnits, Field Theoretical Methods in Many-Body Systems, Pergamon, Oxford (1967).
15. E. K. U. Gross and R. M. Dreizler, Z. Phys. A302, 103 (1981).
16. L. J. Sham, in Computational Methods in Band Theory, edited by P. M. Marcus, J. F. Janak and A. R. Williams, Plenum, NY. (1971).
17. L. Kleinman, Phys. Rev. B30, 2223 (1984).
18. A. D. Becke, J. Chem. Phys. 84, 4524 (1986).
19. L. D. Macdonald and S. H. Vosko, private communication. See also this volume.
20. See, for example, A. K. Rajagopal, J. C. Kimball and M. Banarjee, Phys. Rev. B18, 2339 (1978).
21. V. Contini, G. Mazzone and F. Sacchetti, Phys. Rev. B33, 712 (1986).
22. M. Rasolt and D. J. W. Geldart, Phys. Rev. B34, 1325 (1986). See also Phys. Rev. B13, 1477 (1976).
23. J. P. Perdew and A. Zunger, Phys. Rev. B23, 5048 (1981).

EXCHANGE-ONLY ENERGY FUNCTIONALS FROM ATOMIC EXCHANGE ENERGY DENSITIES

S. H. Vosko[†‡] and L. D. Macdonald[†]

Department of Physics,[†]
University of Toronto,
Toronto, Canada M5S 1A7

Serin Physics Laboratory,[‡]
Rutgers University,
Piscataway, N. J. 08854, U. S. A.

ABSTRACT

An exchange energy functional for use in metals has been obtained that gives better densities than previous gradient corrections to the local density approximation. This was done by fitting atomic exchange energy densities to a function which is a product of the exchange energy density for a uniform system and a function of density and its gradient with the constraint that it have the proper limit for slowly varying densities. These improvements are at the expense of substantially larger energy errors (typically 1% of the exchange energy) but generally improved energy differences.

INTRODUCTION

The seminal works of Hohenberg, Kohn, and Sham [HK64,KS65] not only established the fundamentals of density functional theory (DFT), but also put forward a systematic procedure for improving the local density approximation (LDA) for the exchange-correlation energy functional $E_{xc}[n]$ by use of gradients of the density. By the mid 1970's the utility of these improvements fell out of favour for a number of reasons: (*i*) The Ma and Brueckner [MB68] calculation of the gradient corrections (to first order in e^2) to the LDA for the correlation energy of atoms produced values of the wrong sign. (*ii*) The analysis of Herman *et al.* [HDO69/70] for the gradient correction to the LDA for the exchange energy functional $E_x[n]$ (the so called $X_{\alpha\beta}$ method with β the coefficient of the gradient term) found that to obtain good agreement with the Fock exchange energy for atoms β had to vary by approximately 50%, thus destroying the universality of the functional. Moreover, the a priori calculation (β_S) by Sham [S71] (to first order in e^2) gave a value 2 to 3 times smaller than required. (*iii*) A number of proofs [K74/75,RR75,GRA75] were given that the gradient expansion for the Hartree-Fock (HF) version of DFT does not exist past the first order in e^2. (*iv*) Although Geldart and Rasolt [GR76] showed that the gradient expansion exists for exchange plus correlation, the resulting coefficient has the wrong sign to improve the LDA.

Recently, there has been renewed interest in the gradient expansion for the exchange energy [K84,P85,B85/86,PW86]. This has largely been due to the recognition

[SGP82] that the Optimized Potential Model (OPM) approximation to HF [TS76] is a DFT and that a gradient expansion exists for its $E_x[n]$. This OPM is commonly refered to as the "exchange-only" DFT. Also, Langreth, Perdew, and Mehl [LP80,LM83] have shown how to modify the gradient expansion for correlation so that it produces sensible and reasonably accurate results. The remaining major error in their approximate $E_{xc}[n]$ is in the exchange part.

The present investigation was stimulated by the work of Becke [B85/86] who accurately fit the Fock exchange energies of many atoms, by means of two parameters (β and γ), (to about 0.2% compared to 4% from the lowest order term of the gradient expansion and LDA's 5 to 17%) by a simple universal functional of $n(\vec{r})$ and $\vec{\nabla}n(\vec{r})$, which he has coined the $X_{\alpha\beta\gamma}$ method because of its similarity to the $X_{\alpha\beta}$ procedure. It should be emphasized that, given an accurate $E_x[n]$, DFT should provide accurate densities as well as energies. To this end the densities for the eleven zero spin spherical atoms He, Be, Ne, Mg, Ar, Ca, Zn, Kr, Sr, Cd, and Xe (later refered to as the data base) were calculated self-consistently and compared to both the Clementi and Roetti [CR74] HF and the OPM densities from Aashamar *et al.* [ALT78]. A representative example (Xe) of the improvement over the LDA is shown in Figure 1. It should be observed that the OPM density, (which really is the objective for the gradient correction to the LDA), is very close to the HF density while the improvement from Becke's functional is not nearly as dramatic as occurs in the energy. (The Becke densities typically have two thirds of the LDA error in the atomic core compared to the Becke energies which have a couple of percent of the LDA error.) A natural question that arises from the density comparisons is, "Are total energies in atoms a good criterion for the validity of an approximate exchange energy functional?".

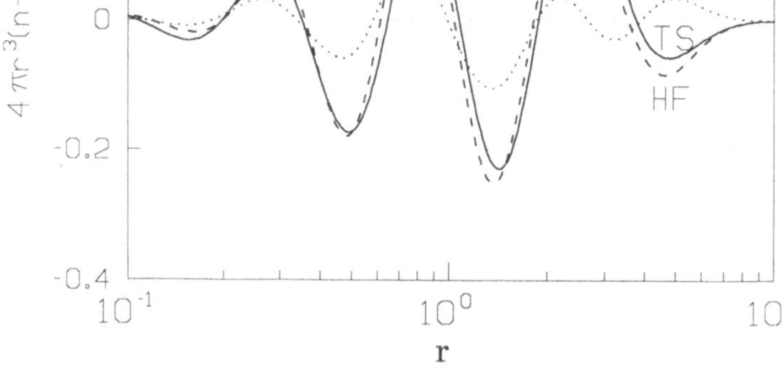

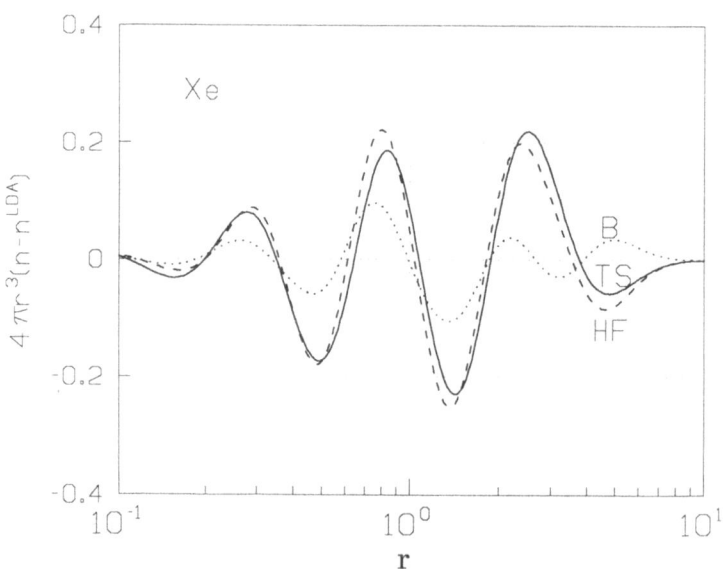

Fig. 1. Comparison of the Becke (B), OPM (TS), and HF densities with respect to the LDA for Xe.

Our primary objective is to find an $E_x[n]$ that gives both significantly better density and energy than the LDA for metals in which the electron density is composed of atomic-like cores surrounded by an interstitial region where the density is slowly varying. Hence, to obtain the correct limit in the interstitial region, the lowest gradient correction should be constrained to the exact theoretical form. (Kleinman [K84] has recently obtained a value for the gradient coefficient (β_K) that is 8/7 of Sham's β_S. This difference is small in comparison to Becke's β which is $\approx 2.7\beta_S$. We will use β_S for convenience since our analysis using atomic densities cannot distinguish between β_K and β_S.) To insure that improved density and energy is obtained in the atomic-like core region we fit the atomic exchange energy densities from our database. A second objective is to gain some insight into how a simple functional of $n(\vec{r})$ and $\vec{\nabla}n(\vec{r})$, (without $\nabla^2 n$ and higher gradients) can give such accurate values for $E_x[n]$.

GRADIENT EXPANSIONS AND EXCHANGE-ONLY ENERGY DENSITY

Herman *et al.* [HDO69/70] made a general analysis of the gradient series for $E_x[n]$ and showed that the first few terms of the exchange energy density could be written as

$$e_x(n, \vec{\nabla}n, \nabla^2 n) = e_x^L(n) g(\xi, \eta)$$

that is $E_x[n] = \int d\vec{r}\, e_x(n, \vec{\nabla}n, \nabla^2 n)$, where the LDA exchange energy density $e_x^L(n)$ is $-A_x n^{4/3}(\vec{r})$ with $A_x = \frac{3}{4}(3/\pi)^{1/3}$,

$$g(\xi, \eta) = \sum_{i,j=0}^{2} c_{ij}\xi^i \eta^j$$

$c_{00} = 1$, $\xi \equiv |\vec{\nabla}n|^2/n^{8/3}$, and $\eta \equiv \nabla^2 n/n^{5/3}$. They also point out that the linear terms are equivalent in that they could be transformed into each other by an integration by parts. It is worth noting that, for atoms, the quadratic (and higher order) terms lead to divergent contributions to $E_x[n]$. These observations emphasize the difficulty of obtaining a universal expression for the exchange energy density from atoms. On the one hand the linear (convergent) terms only give $\sim 40\%$ of the required correction to the LDA; on the other hand the coefficients of the quadratic terms are unknown and even if they were known, atoms could not be used to test the rate of convergence of the series. Clearly, a resummation of the gradient terms is a necessity, for example a Chisholm approximant [C73]. Thus we are faced with the difficulty of resumming a series for which only the linear coefficient is known! This is more difficult than in the case of the kinetic energy gradient expansion which is known to 6th order in the gradient (which diverges for atoms), and for which the linear term gives a much larger contribution to the error in its LDA.

Becke's form for e_x is

$$e_x^B(n, \vec{\nabla}n) = e_x^L(n) - n^{4/3}(\vec{r})\beta \frac{\xi}{1+\gamma\xi} \tag{1}$$

with $\beta = 2^{1/3}(0.0036) = 0.0045$, and $\gamma = 2^{2/3}(0.004) = .0063$. (Note that the values of β and γ shown in parentheses are those of Becke for Spin-DFT which is related to the above by replacing $n(\vec{r})$ in the above by $2n_\sigma(\vec{r})$, and taking half the sum over $\sigma = \uparrow / \downarrow$.) Recall that $\beta_S = 7/432\pi(3\pi^2)^{1/3} = 0.00167$. Equation (1) can be rewritten as a quotient of polynomials in ξ,

$$e_x^B(n, \vec{\nabla}n) = e_x^L(n) \left\{ \frac{1 + a_1\xi}{1 + b_1\xi} \right\} \tag{2}$$

where $a_1 = \gamma + \beta/A_x$ and $b_1 = \gamma$ or more generally, with the assumption that the η dependence can be neglected, (*i.e.* $e_x = e_x^L f(\xi)$ which following Perdew is refered to as the generalized gradient approximation (GGA)) the factor in {} of Eq. (2) can be interpreted as a [1/1] Padé approximant to the function $f(\xi)$. The function $f(\xi)$ can be constrained to the proper $\xi \to 0$ limit by increasing the order of the Padé approximant. The constrained [2/2] Padé for $f(\xi)$ is

$$f_{[2/2]}(\xi) = \frac{1 + (b_1 + \frac{\beta_S}{A_x})\xi + a_2\xi^2}{1 + b_1\xi + b_2\xi^2} \tag{3}$$

The coefficients a_2, b_1, and b_2 can be obtained in a variety of ways. In analogy with Becke, a fit to the exchange energies (*i.e.* $\sum_a \{(E_x^{[2/2]} - E_x)/E_x\}_a^2$) has been made for the atoms $\{a\}$ in the data base. The sign of the $\xi \to 0$ constraint forced the parameters to become very large and ill determined. Typical values are $a_2 = 15.5$, $b_1 = 1230.$, and $b_2 = 8.0$. (Since there are no small ξ components in the atomic data base this result doesn't mean that there is a contradiction between the sign of the $\xi \to 0$ constraint and the data base.) The resulting $f(\xi)$ and the corresponding densities were indistinguishable from that of Becke.

To appreciate the source of the difficulty in obtaining improved densities, recall that in the OPM-DFT that the potential in the Kohn-Sham single particle equations contain the functional derivative of $E_x[n]$. For the GGA this has the form

$$v_x^{GGA}[n;\vec{r}] = \frac{\delta E_x^{GGA}[n]}{\delta n(\vec{r})} = v_x^L(n)\left\{f(\xi) - \frac{3}{2}\left(\frac{\partial f}{\partial \xi}\eta + \frac{\partial^2 f}{\partial \xi^2}\tau\right)\right\} \tag{4}$$

where $\tau \equiv \vec{\nabla} n \cdot \vec{\nabla}\xi/n^{5/3}$, and the LDA exchange potential is $v_x^L(n) = -A_x \frac{4}{3} n^{1/3}(\vec{r})$. It is clear from (4) that accurate values of $f(\xi)$ and its first and second derivatives are required for v_x^{GGA} in contrast to E_x^{GGA} which depends on an integral over $e_x^L f(\xi)$. If more terms in the gradient series were known, more constraints could be put into a Padé approximant to $f(\xi)$ and the coefficients could then be determined by fitting a set of total E_x's. The resultant $f(\xi)$ and its derivatives would be better defined. The main point is that more detailed information of the ξ dependence of $f(\xi)$ is needed. Our approach is to make a reasonable ansatz for an exchange energy density functional $e_x[n;\vec{r}]$ and calculate it for the atoms in the data base to obtain this information and more insight into its internal structure.

The OPM-DFT exchange energy is

$$E_x^{OPM}[n] = -\frac{1}{2}\int d\vec{r} \int d\vec{r}\,' \frac{|n_1(\vec{r},\vec{r}\,')|^2}{|\vec{r}-\vec{r}\,'|} \tag{5}$$

where the Dirac one particle density matrix $n_1(\vec{r},\vec{r}\,') = \sum_\nu^{occ} \phi_\nu(\vec{r})\phi_\nu^*(\vec{r}\,')$ with the single particle wavefunctions $\phi_\nu(\vec{r})$ generated by the optimized *local* potential $v_{OPM}(\vec{r})$ in the Schrödinger equation that minimizes $E_v[n] = T_s[n] + V[n] + E_H[n] + E_x[n]$ which is the sum of the kinetic, external, Hartree, and exchange energy functionals respectively. (The OPM atomic energies are less than 0.005% above the HF when E_x is calculated by (5).)

There is no unique transformation of this exchange energy (5) into the exchange energy density because of the arbitrary divergence term that may be added to e_x. Hence, it is necessary to choose a definition of the exchange energy density that is fit well by the LDA and doesn't have large oscillations about the LDA which would arise from the divergence term that could be present. Two plausible choices for $e_x[n;\vec{r}]$ are:

$$e_x[n;\vec{r}] = -\frac{1}{2}\int d\vec{r}\,'\frac{|n_1(\vec{r},\vec{r}\,')|^2}{|\vec{r}-\vec{r}\,'|}; \tag{6}$$

$$\tilde{e}_x[n;\vec{r}] = -\frac{1}{2}\int d\vec{R}\frac{|n_1(\vec{r}+\frac{\vec{R}}{2},\vec{r}-\frac{\vec{R}}{2})|^2}{R}.$$

Folland [F71] has shown that there is substantial structure in both $e_x^L(n)$ and $e_x[n;\vec{r}]$ of Eq. (6) and that the structures coincide very closely. It is seen in Figures 2-4 that this definition does not oscillate significantly around the LDA and that an arbitrary divergence term is not dominant. (Since it was found (Fig. 1 for example) that the OPM gives densities that are very close to HF compared to the LDA and since Clementi and Roetti [CR74] give HF wave functions in analytical form, the HF exchange energy density is used instead of the OPM for convenience.)

Also from these graphs it can be seen that ξ ($e_x^{S_1} = e_x^L - \beta_S n^{4/3}\xi$) is more closely in phase with the difference between $e_x[n;\vec{r}]$ and the LDA than η ($e_x^{S_2} = e_x^L - 3\beta_S n^{4/3}\eta$) is. These graphs explain why E_x can be fit accurately by using the form $e_x^L(n)f(\xi)$, with a simple form for $f(\xi)$. Hence, in common with Becke we use a function of ξ to provide a gradient correction. Admittedly, there is a component of η, and perhaps significant higher order gradients, in this definition of $e_x[n;\vec{r}]$ and at least the η dependence would be needed for a more accurate fit. In fact, the correction to the LDA is not a single-valued function of ξ in the region of the atom where the correction is large. The spread of this multi-valuedness in our database is about 100%

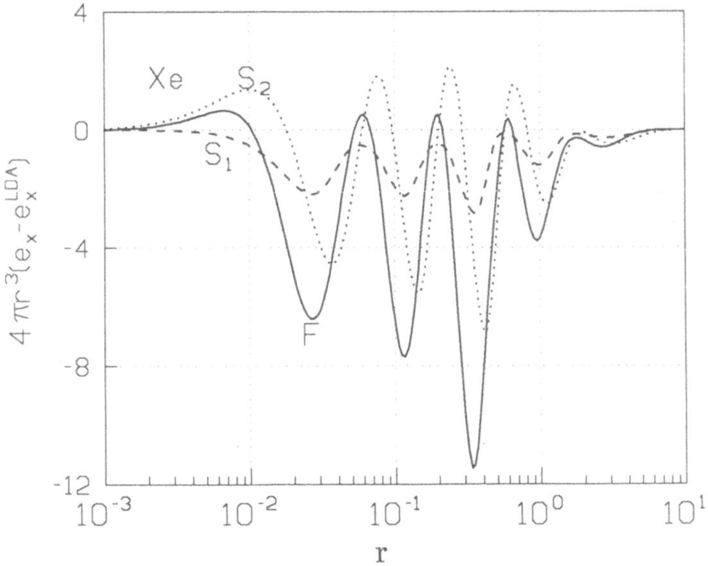

Fig. 2. Comparison of the Sham correction using ξ (S_1), using η (S_2), and HF (F) exchange energy densities (in Ry) with respect to the LDA for Xe.

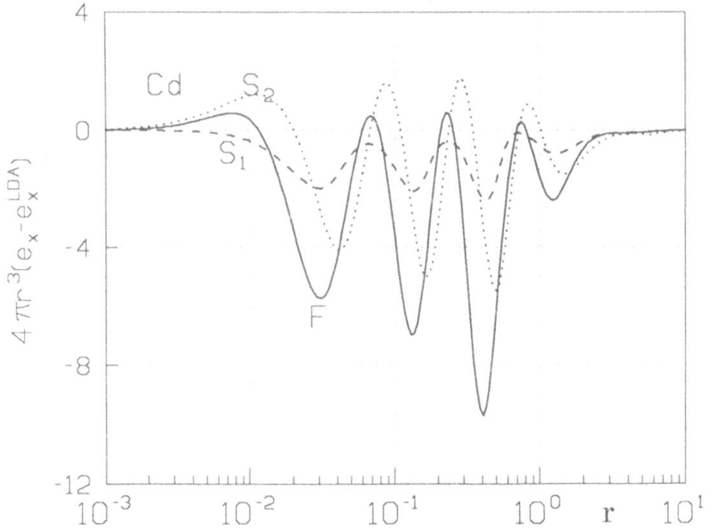

Fig. 3. Comparison of the Sham correction using ξ (S_1), using η (S_2), and HF (F) exchange energy densities (in Ry) with respect to the LDA for Cd.

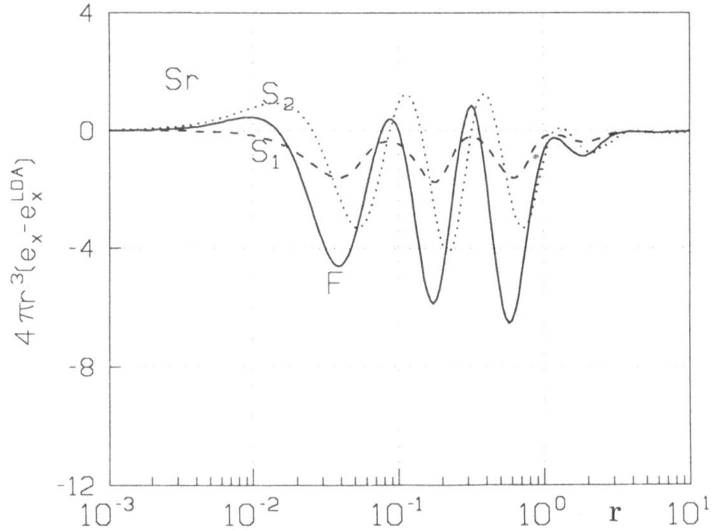

Fig. 4. Comparison of the Sham correction using ξ (S_1), using η (S_2), and HF (F) exchange energy densities (in Ry) with respect to the LDA for Sr.

of the correction to the LDA! Hence, there is a limit on how well the correction can be fit. The effective contribution of the η and higher order gradients will be absorbed in some sense into the parameters that we use in the fit. Since these contributions are likely to differ from atom to atom and because they are so large, according to the large spread in the multi-valuedness of the correction as a function of ξ alone, fits to the exchange energy densities of individual atoms are not likely to be universal functions of ξ. Finally, it can be noted that ξ tends to grow exponentially in any exponential density region and so tends to increase just outside each shell of electrons in the atoms. In the interior of the atom these are precisely the positions at which the corrections shown in Figures 2-4 to the LDA are needed and in turn contribute very closely to the structure observed in the exchange part of the OPM potential [TS76] through Equation (4) however, the region of the outer shell of electrons cannot be corrected using an expansion in ξ since it has begun to diverge there.

ENERGY DENSITY FIT

To fit the energy density with a function of ξ when the correction to the LDA is not single-valued and yet get reasonable exchange energies it is necessary to weight the fit according to the contribution to the exchange energy. A least squares fit is computationally simpler than other powers of the residuals. Hence the following expression was minimized with the positions r_i on a uniform logarithmic mesh

$$\sum_a \sum_i \left\{ \frac{r_i^3 \left(e_x^F(r_i) - e_x^L(r_i) f_{[2/2]}(\xi_i) \right)}{E_x^F} \right\}_a^2 . \qquad (7)$$

Fits were done using both a [2/2] Padé approximant of ξ (Eq. 3) multiplying the LDA and a function of the form of Perdew and Wang's [PW86]. There was an indistinguishable difference between the two fits so only one is shown on Figure 5. (The parameters found for $f_{[2/2]}(\xi)$ are $a_2 = .001064$, $b_1 = .0641$, and $b_2 = .000459$, and for the PW form of $f(\xi) \rightarrow (1 + a_1\xi + a_2\xi^2 + a_3\xi^3)^\epsilon$ are $a_1 = .0273$, $a_2 = .00352$, $a_3 = .000109$, $\epsilon = .0826$.)

The atomic E_x from Eq. (7) typically have errors of $\approx 1\%$ (except in the iron series where it is $\approx 0.1\%$) which is almost an order of magnitude more than Becke's GGA, significantly more than PW's GGA, and about a quarter of the error from the lowest order gradient correction. The self-consistent energies (see Table 1) for the LDA and the GGA's are dominated by this error from their respective exchange energy functionals except up to about Kr in Becke's. However, energy differences such as ionization potentials are generally slightly improved over the other GGA's but not as good as the LDA, and the iron series interconfiguration energies are better than the other GGA's which are better than the LDA. Figures 6-8 show the improvement in densities towards the OPM compared to the LDA and the [PW86] GGA. These representative cases show that the improvement is significant out to $r \approx 2$. This is the point at which ξ grows exponentially and cannot be expected to be good expansion parameters in this region. This also causes the gradient corrected exchange potential to have an exponential tail similar to the LDA potential and not the exact $-e^2/r$ tail contained in the OPM.

CONCLUSION

It has been found that a gradient correction can be inferred from exchange energy densities that gives improved densities for $r < 2$ in atoms over those obtained from such GGA's as result from either fits to total exchange energy or to fits to

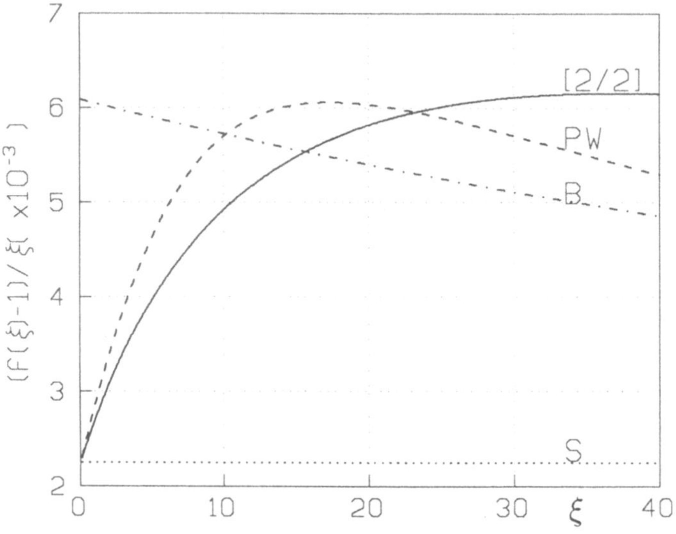

Fig. 5. Comparison of the Sham (S), Becke (B), Perdew Wang (PW), and the [2/2] fit from equation (7) effective β as a function of ξ.

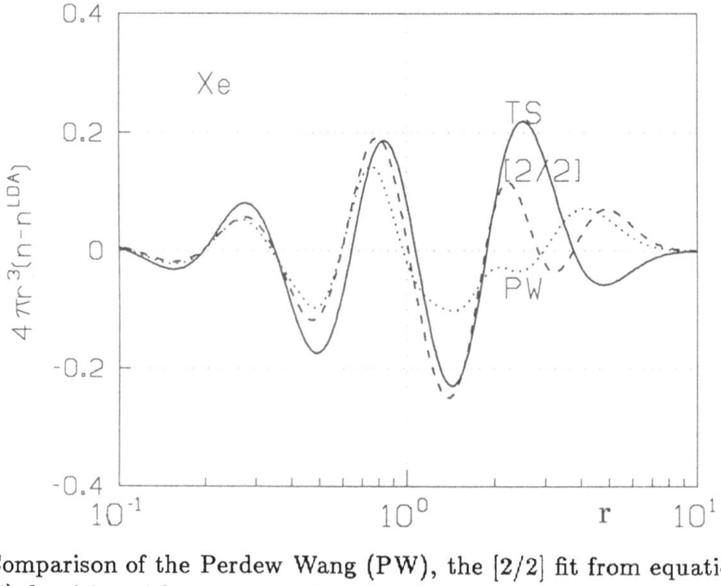

Fig. 6. Comparison of the Perdew Wang (PW), the [2/2] fit from equation (7), and OPM (TS) densities with respect to the LDA for Xe.

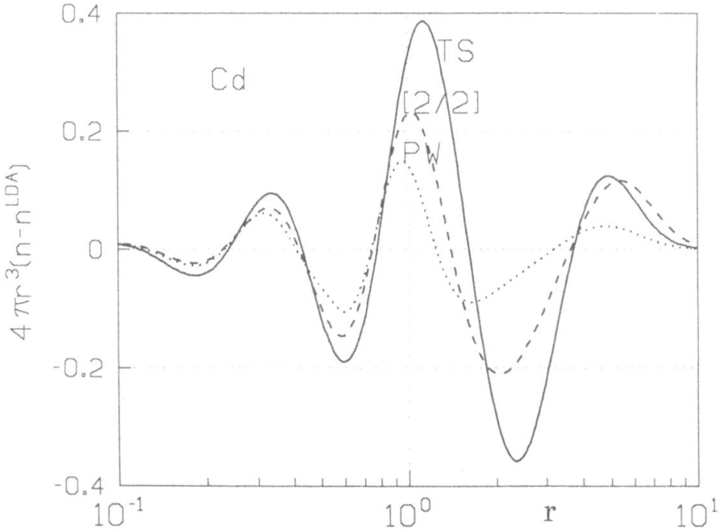

Fig. 7. Comparison of the Perdew Wang (PW), the [2/2] fit from equation (7), and OPM (TS) densities with respect to the LDA for Cd.

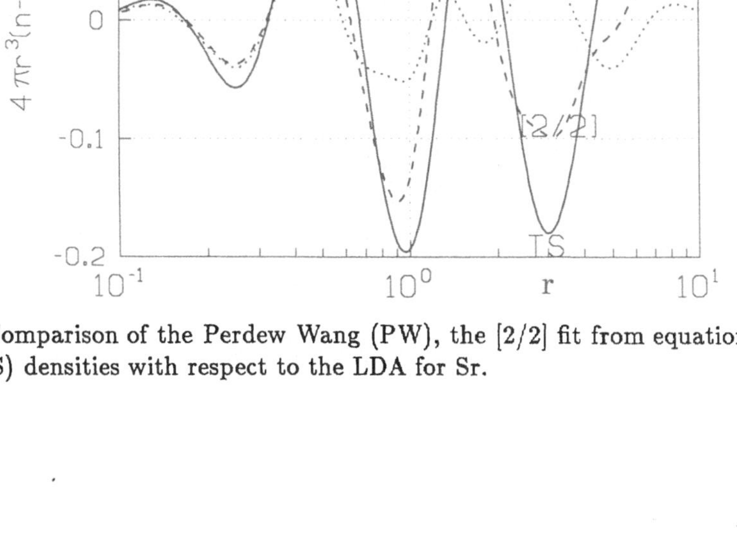

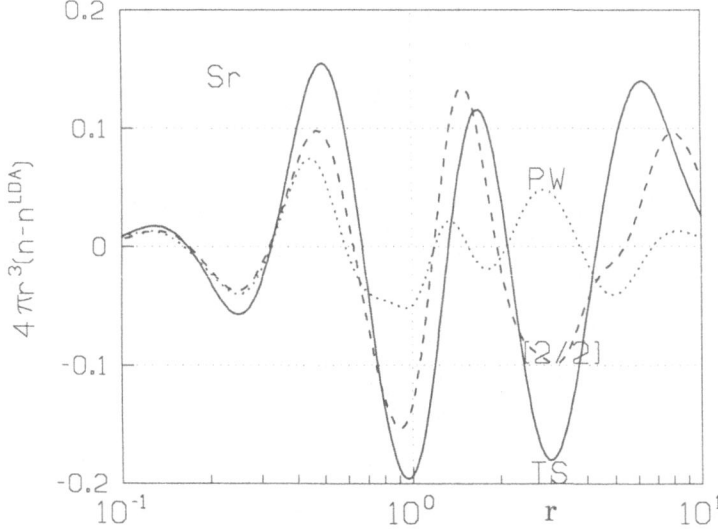

Fig. 8. Comparison of the Perdew Wang (PW), the [2/2] fit from equation (7), and OPM (TS) densities with respect to the LDA for Sr.

Table 1. Self-consistent energies and Ionization Potentials (IP) are compared to HF (RHF) [CR74]. (OPM energies are less than 0.005% above HF.) The values shown for the LDA, Becke (B), Perdew-Wang (PW), and the [2/2] Padé fit from Eq. (7) have the RHF energy subtracted off.

Energies and Ionization Potentials (mRy)					
Atom	E_{RHF}	LDA	B	PW	[2/2]
H	−1000.	86.	4.	0.	−16.
H +	0.	0.	0.	0.	0.
IP	1000.	−86.	−4.	0.	16.
He	−5723.	276.	−6.	−20.	−73.
He^+	−4000.	175.	10.	3.	−29.
IP	1723.	−101.	16.	23.	44.
Li	−14865.	479.	1.	−21.	−112.
Li^+	−14473.	456.	8.	−14.	−97.
IP	393.	−23.	7.	8.	14.
Be	−29146.	699.	−5.	−36.	−168.
Be^+	−28555.	670.	6.	−25.	−148.
IP	591.	−30.	11.	11.	20.
Ne	−257094.	2113.	−117.	−257.	−379.
Na	−323718.	2429.	−82.	−237.	−357.
Na^+	−323354.	2423.	−64.	−213.	−335.
IP	364.	−6.	18.	23.	21.
Mg	−399229.	2732.	−76.	−249.	−364.
Mg^+	−398743.	2722.	−54.	−223.	−337.
IP	486.	−10.	21.	26.	27.
Ar	−1053635.	4600.	16.	−238.	−232.
K	−1198329.	4906.	14.	−254.	−226.
K +	−1198035.	4909.	33.	−228.	−207.
IP	294.	3.	20.	25.	20.
Ca	−1353516.	5196.	−11.	−295.	−243.
Ca^+	−1353140.	5197.	11.	−267.	−218.
IP	376.	1.	22.	28.	24.
Cr	−2086710.	6164.	−241.	−529.	−303.
Cr^+	−2086278.	6247.	−154.	−432.	−233.
IP	433.	83.	87.	97.	70.
Mn	−2299731.	6565.	−175.	−463.	−219.
Mn^+	−2299297.	6571.	−138.	−419.	−176.
IP	434.	6.	37.	44.	44.
Cu	−3277926.	7447.	−547.	−802.	−346.
Cu^+	−3277455.	7534.	−451.	−691.	−264.
IP	470.	86.	97.	112.	82.
Zn	−3555695.	7876.	−446.	−693.	−192.
Zn^+	−3555133.	7934.	−371.	−602.	−128.
IP	562.	59.	75.	92.	64.
Kr	−5504109.	10377.	100.	−41.	708.
Rb	−5876694.	10727.	115.	−7.	776.
Rb^+	−5876414.	10729.	132.	15.	791.
IP	280.	2.	17.	23.	15.
Sr	−6263076.	11080.	128.	20.	842.
Sr^+	−6262734.	11087.	154.	51.	868.
IP	342.	7.	25.	32.	26.
Mo	−7951068.	12403.	129.	136.	1109.
Mo^+	−7950636.	12474.	198.	211.	1159.
IP	432.	71.	69.	75.	49.
Tc	−8409551.	12806.	170.	197.	1196.
Tc^+	−8409166.	12826.	217.	252.	1246.
IP	384.	21.	47.	55.	50.
Ag	−10395370.	14214.	206.	366.	1521.
Ag^+	−10394939.	14284.	282.	455.	1581.
IP	432.	71.	76.	89.	60.
Cd	−10930144.	14500.	142.	327.	1526.
Cd^+	−10929731.	14650.	303.	501.	1674.
IP	414.	150.	161.	174.	148.
Xe	−14464260.	16945.	530.	908.	2399.

a gradient expansion of the exchange energy using a cutoff exchange hole. These improvements in density are at the expense of substantially larger errors in energies than in these other GGA's but it's energy differences are generally slightly better as indicated by the IP's in Table 1. Since this gradient correction also includes the $\xi \to 0$ limit it should be a useful improvement over the LDA in metals. With more accurate form factors becoming available (Hansen *et al.* [HSL84] for example) improvements in the densities are necessary (for example Pindor *et al.* [PVU86]). It is not clear at this point which of the GGA's will be most useful for describing the properties of molecules and solids and more tests will have to be made to learn what compromise in accuracy between energy and density will be most useful.

ACKNOWLEDGEMENTS

We wish to thank David C. Langreth, John P. Perdew, and Dale D. Koelling for a number of helpful discussions. This work was supported in part by the Natural Sciences and Engineering Research Council of Canada (A6294) and the National Science Foundation (DMR 8304210).

REFERENCES

ALT78 K. Aashamar, T. M. Luke, and J. D. Talman, At. Data Nucl. Data Tables **22**, 443 (1978)
B85/86 Proceedings of Conference on *Density Matrices and Density Functionals*, Kingston, Canada, edited by R. M. Erdahl and V. H. Smith, Jr. Aug. 1985; A. D. Becke, J. Chem. Phys. **84**, 4524 (1986)
C73 J. S. R. Chisholm, Math. Comp. **27**, 841 (1973)
CR74 E. Clementi and C. Roetti, At. Data Nucl. Data Tables **14**, 177 (1974)
F71 N. O. Folland, Phys. Rev. **3**, A1535 (1971)
GR76 D. W. Geldart and M. Rasolt, Phys. Rev. **13**, B1477 (1976)
GRA75 D. W. Geldart, M. Rasolt, and C. O. Almbladh, Solid State Commun. **16**, 243 (1975)
HSL84 N. K. Hansen, J. R. Schneider, and F. K. Larsen, Phys. Rev. **29**, B917 (1984)
HDO69/70 F. Herman, J. P. Van Dyke, and I. B. Ortenburger, Phys. Rev. Lett. **22**, 807 (1969); Int. J. Quant. Chem. **3S**, 827 (1970)
HK64 P. Hohenberg and W. Kohn, Phys. Rev. **136**, B864 (1964)
KS65 W. Kohn and L. J. Sham, Phys. Rev. **140**, A1133 (1965)
K74/75 L. Kleinman, Phys. Rev. **10**, B2221 (1974); **12**, B3512 (1975)
K84 L. Kleinman, Phys. Rev. **30**, B2223 (1984)
LM83 D. C. Langreth and M. J. Mehl, Phys. Rev. **28**, B1809 (1983)
LP80 D. C. Langreth and J. P. Perdew, Phys. Rev. **21** B5469 (1980)
MB68 S.-K. Ma and K. A. Brueckner, Phys. Rev. **165** 18 (1968)
P85 J. P. Perdew, Phys. Rev. Lett. **55**, 1665 (1985); see also this volume
PW86 J. P. Perdew and Wang Yue, Phys. Rev. **33**, B8800 (1986)
PVU86 A. J. Pindor, S. H. Vosko, and C. J. Umrigar, J. Phys. **16**, F1207 (1986)
RR75 A. K. Rajagopal and S. Ray, Phys. Rev. **12**, B3129 (1975)
SGP82 V. Sahni, J. Gruenebaum, and J. P. Perdew, Phys. Rev. **26**, B4371 (1982)
S71 L. J. Sham, in *Computational Methods in Band Theory*, edited by P. M. Marcus, J. F. Janak, and A. R. Williams, (Plenum, New York 1971), p. 458
TS76 J. D. Talman and W. F. Shadwick, Phys. Rev. **14**, 36 (1976)

DENSITY FUNCTIONAL THEORY AND f ELECTRON SYSTEMS

Michael R. Norman

Argonne National Laboratory
Materials Science Division
Argonne, IL 60439

Density functional theory within the local density approximation has proven to be an extremely successful way to elucidate the electronic structure and ground state properties of a wide variety of electronic systems [1]. Because of the discovery of exotic behavior in f electron systems such as the mixed valent effect and heavy fermion phenomenon, the applicability of the above theory has been questioned [2,3]. Recently, dramatic progress has been made in determining more exact density functionals which have been extensively tested in atomic systems. The purpose of this paper is to speculate what impact these new methods will have on the f electron question. In the first section, a brief review of density functional theory and the local density approximation (LDA) will be given, as well as an equally brief review of results on f electron solids. In the second section, the new density functional techniques will be discussed and results for atoms reviewed, and potential extension of these methods to the case of the lattice, with emphasis on f electron solids, will be presented. Finally, some speculations will be mused over in the final section.

INTRODUCTION

The two theorems by Hohenberg and Kohn [4] allow one to transform the ground state problem for a many-body system to an effective one-particle form. The transformation, though, involves the unknown energy functional for the effective one-particle system. The paper by Kohn and Sham [5] reduces this problem to that of finding the exchange-correlation energy functional (that part of the energy due to the non-direct part of the Coulomb interaction). Also in that paper, they presented an approximation to that functional known as the local density approximation (LDA) which was based on previous work by Slater [6]. This functional assumes that the quasiparticles are completely screened and behave locally like an electron gas. Such an approximation should only be valid for slowly varying charge densities, but the LDA does quite well for a variety of systems including atoms, molecules, bulk solids, and surfaces. There are several reasons why this is so. It can be shown on dimensional grounds that the exchange-correlation energy density is equal to a shape function times the density to the one-third power [7]. The LDA merely involves replacing this shape function by a constant. The reason that works so well is that only spherical averages enter the total energy expression [8]. Another way to see this is to note that the exchange-correlation energy represents the interaction of the electron with its exchange-correlation hole (the hole is due to Pauli exclusion and Coulomb correlation effects keeping other electrons away). Although the LDA does a poor job on the shape of the hole, it does fairly well on the spherical average, which is what enters the total energy expression. Moreover, it satisfies the correct sum rule (that is, the electron plus its hole is a neutral object).

A spin-generalization to LDA exists and does quite well for predicting magnetic behavior and moment size for itinerant magnets [9]. But a generalization involving orbital degrees of freedom has not been devised yet (at least to the author's knowledge). This is the first problem to consider when dealing with f electron systems, since no orbital degeneracies (besides basic spin-orbit splitting) can be dealt with. The problem, though, is deeper than that. One implicitly assumes within Kohn-Sham theory that the density is constructed from a single Slater determinant of (fictitious) one-particle orbitals. Thus, no multiplet structure can be resolved. Use of the spin generalization of the theory leads to an incorrect multiplet structure [10]. The only way to get around this is to express the multiplet energies as linear combinations of energies corresponding to single determinant states [10]. Clearly, one has difficulties extending such a procedure to the lattice.

Although the LDA does well for energy-related problems, the eigenvalue spectrum which comes from the effective one-particle equations physically do not mean anything [11]. Even in the exact density functional theory, only the highest occupied level has physical significance (in a solid, this is the chemical potential), since an excitation spectrum is not a ground state property [12]. The problem is exasperated in LDA since the potential used in the effective one-particle equations is a derivative of the energy, and thus samples the shape of the exchange-correlation hole, which is very bad in LDA for localized or inhomogeneous systems. In practice, one oftens gets good results when comparing experimental photoemission results to the LDA eigenvalue spectra in solids [1]. This is because for itinerant systems, the effect of an excitation is small, and thus a ground state eigenvalue comparison does well. Such is not the case for localized systems. In particular, one sees a breakdown in this comparision beginning in cases such as nickel, and a complete failure in the case of cerium. Such a failure, though, does not imply anything about the ground state properties being given incorrectly.

In fact, this is a good time to review some of the local density work on f electron compounds. For a more complete discussion, the reader is referred to Ref. 13. First, the lattice constants determined from total energy minimization are within 5% of experiment, but are always an underestimate (this is for mixed valent systems; for localized systems, the LDA does very well for lattice constants if the f electrons are treated as core states [14]). The reason for the underestimate is twofold. First, the f states are placed too low relative to the other states by the LDA (the estimate is 40 mRy in TmSe [15]). This is a well-known problem, though, in that the LDA places higher angular momentum states too low relative to lower ones. This has been most analyzed for d states in transition metals (the d bands are too low in V and Nb, for instance [16]). This problem has been worked on in atoms for many years without much success, and will be discussed further in the next section. Second, the amount of f hybridization is being over predicted by the LDA. This is an extremely important result which will be dwelt on in the next section.

Magnetic calculations within the LDA have proven to be rather successful for itinerant f systems, as long as one includes orbital-moment effects [17]. This is surprising since within the LDA only the spin-part drives the system (the orbital part of the moment is a passive participant). Despite this, we are in need of a complete theory for moment-polarized systems.

Because of the highly localized response of f electron systems to photoemission, it is difficult to see any type of f electron band dispersion. A better technique that gets one closer to the ground state is use of the deHaas-vanAlphen effect to map out the Fermi surface. Comparisons of itinerant uranium f electron dHvA data to band calculations is amazingly good [18]. Even for mixed valent $CeSn_3$, the LDA band calculations are in rather surprising agreement with experiment, indicating true coherent f bands at low temperatures [19]. For CeSb, though, one only gets agreement with experiment when one treats the f electrons as unhybridized states [20]. Thus the LDA description of the f electrons as band states breaks down somewhere between these two cases. Unfortunately, dHvA data is hard to obtain for mixed valent/heavy fermion systems. Recently, though, data was obtained on heavy fermion $CeCu_6$[21]. This system may prove to be the link between the two cases mentioned above.

Temperature dependent quantities can be extracted from zero temperature band calculations under the assumption that they are due to thermal broadening of the bands. This is clearly not the case in most of the interesting systems. For instance, the band calculations cannot explain the susceptibility curve for $CeSn_3$ [22]. Physically, the curve has a "Kondo" maximum at 150 K representing a crossover between itinerant behavior at low T and free moment behavior at high T [2]. Finite temperature spin entropy calculations have been performed on Fe within the LDA with some success [23]. Techniques similar to those used there might explain the behavior of $CeSn_3$.

In conclusion, the LDA works reasonably for several quantities, but not so well for others. More importantly, the band-like description for the f electrons, coming from the use of Fermi-Dirac statistics on effective single-particle states, clearly breaks down for most rare-earth systems. There is, therefore, plenty of ground for improvements.

NEW DENSITY FUNCTIONAL METHODS

In this section, we will discuss several of the new density functional methods and their possible impact on the f electron question. Within the past several years, a series of new gradient corrections to LDA have been proposed [24]. They are a significant improvement over previous such corrections for reasons too complicated to go into here. As yet, they have not been extensively tested in solids (they work very well in atoms, leading to significant improvements in the exchange-correlation energy and the charge densities). We found in vanadium that the problem of the d bands being too low was only corrected partly by these corrections [25]. In $CeSn_3$, these corrections had virtually no effect on the Fermi surface [26]. The reason for this is that the direct part of the Coulomb interaction dominates over any type of electron gas exchange-correlation term for such a localized f system. For this reason, we do not expect the gradient corrections to lead to any qualitative changes, although it is possible that the ground state properties will be improved by the use of such functionals.

Besides the gradient corrections, there is one other major correction to the LDA that has been extensively worked on. This is the self-interaction correction (SIC). The reason there is a self-interaction error in LDA is that when one replaces the Fock exchange operator by its local density counterpart, the exact cancellation of self-Coulomb and self-exchange terms no longer occurs. Unfortunately, one cannot get rid of this error in a unitarily invariant fashion, thus one is left with a representation-dependent functional. In fact, such corrections are zero for Bloch states in an infinite solid, but non-zero for Wannier states, so the extension of the correction to a solid is ambiguous. There are many reasons, though, why inclusion of such a correction is desirable [27]. First, the correct -1/r tail to the potential occurs for neutral atoms, and the stability of negative ions is enhanced over LDA. Second, the eigenvalue spectrum is a good approximation to physical removal energies (this does not occur for LDA or for Hartree-Fock). Third, several properties of the exact density functional theory that are not reproduced by LDA are reproduced by SIC. This includes derivative discontinuities [28,29] (the potential changes discontinuously across an excitation gap) and the correct dependence of several key quantities for fractional occupation of orbitals. In the exact theory [28], the eigenvalues should be constant with respect to the occupation number of the orbital, jumping discontinuously at integer occupation values. Moreover, the exchange-correlation hole should no longer integrate to one in the fractional case [28]. SIC reproduces these effects [28,29], and in fact stabilizes correctly at integer occupation values for atoms (in the case of Fe, SIC gives d^6s^2 [29] whereas LDA gives $d^{6.4}s^{1.6}$). This latter deficit for LDA is probably why LDA does not localize the f electrons into integer occupation configurations for most rare earth metals. Finally, Harrison has calculated the full non-spherical SIC and applied it to transition-metal atoms [30]. The problem of the d electrons being too low in energy with respect to the s electrons is apparently resolved by this correction.

Given these promising effects, what work has been done with SIC in solids? SIC has been applied successfully to insulators, the primary success being the reproduction of the experimental band gaps [31,32]. The method of Ref. [32] has the advantage of being simple and being applied within a Bloch formalism. Because of this, though, the author doubts whether it will give the desired effect for f electron systems (tendency of the f electrons to decouple from the conduction electrons) since it is difficult to get dehybridization in a Bloch formalism. The method of Ref. [31] uses a Wannier representation, but unfortunately such a method would be difficult to apply to a metal (and would be non-unique besides). Both myself, and Monnier and Jansen have applied an atomic-like SIC to just the f electrons in bulk solids without success, the reason again probably being the Bloch representation for the f electrons.

Another approach to take would be to simply junk the LDA. Only one practical approach exists at this time, that is the optimized effective potential method which has been extensively applied by Talman's group [33]. This method essentially extracts the Kohn-Sham potential from an interacting Hamiltonian by means of an integral equation. Talman's group originally applied this method to Hartree-Fock and discovered that the optimized effective potential [OEP] reproduced the Hartree-Fock energy and density to a high precision. Recent work has involved using a configuration-interaction Hamiltonian, and also yields good results (at considerably less effort than solving the full many-body problem). We have also applied this technique to the SIC functional discussed above with great success [29]. The problem with this method is how to extend it to solids in an unambiguous manner. The first problem is that the integral equation becomes a mess in the solid because of the required sum over the Brillouin zone, and also because of the need to construct a periodic Green's function for each state. The second problem is that there is no long-range part of the potential that can be used to fix the spurious constant which can occur because of the use of an integral equation. We have considered getting around some of these problems by using a SIC functional (so the OEP procedure only involves the SIC part of the functional, which is much smaller than the entire exchange-correlation energy). The SIC is treated in an atomic approximation inside muffin-tin spheres placed around the atomic sites (it is assumed that anything outside the spheres are conduction electrons with no SIC). This means that the OEP integral equation can be written in an atomic approximation. The problem of the spurious constant, though, is still there, and one also has the problem of which ℓ states to treat with the SIC (just the 4f?) and what occupation numbers to use in the equation (the integrated ℓ charge inside the sphere, or a set integer number?). I have tested this on $CeSn_3$ (only SIC for the 4f, integral occupation of one for the 4f, potential adjusted to match the LDA one on the sphere boundary) and did not get promising results. Certainly, though, more exploratory work needs to be done in this direction, perhaps by using a different (non-Bloch) representation.

At this point, a discussion of the multiplet problem is in order. As discussed in the first section, it is difficult to resolve multiplet structure within the LDA. One can calculate multiplet splittings using the OEP method mentioned above, but to do this requires a subtraction of total energies (the eigenvalue spectrum does not reflect the multiplet structure). The problem is exasperated in solids because in a paramagnetic system, each band possesses only spin degeneracy. It is conceivable that one could construct a functional with orbital degeneracies which would allow more than two electrons to occupy a band, but such a functional has not been constructed yet.

I would like to end this section on a rather fundamental problem, which is the nature of the mixed valent state. It is our opinion that the concept of mixed valence is probably a misnomer for cerium systems. A true mixed valent state would be a configuration-interaction mixture of two localized f configurations with different integer occupation numbers. Such a state is inconsistent with the dHvA data on $CeSn_3$ (which definitely indicates f banding). It is our opinion that in many Ce systems, one has f bands at low T, and the banding is lost at higher temperature as can be seen from the resistivity [34]. This may not be the case for Sm, Eu, and Tm systems, which may really be true mixed valent states. Within the exact density functional theory, one treats the case of fractional occupation of orbitals in a formalism that is exactly equivalent to a mixed valent notation [28]. Note that the LDA gives an incorrect description for fractional occupation cases (although SIC does). It is conceivable that one could take the coherent potential approximation [CPA] method used for random alloys to simulate a true mixed valent state. I have been told by the experts that this is difficult, but I believe that it would not be under the assumption that the f electrons are treated like core states. Note that a description in terms of fractionally occupied orbitals is not the same as a hybridized band, so a standard band calculation would not yield the desired effects. The determination of when a band description is valid versus when a localized (or fractionally occupied orbital) description is valid is a major, unresolved problem. Certainly, the exact functional description of Ref. 28, which involves a statistical ensemble of integer occupation states, needs to be explored more for solids, both in a fundamental and a computational sense.

We might also note that the CPA method mentioned above has been recently extended to treat spin entropy effects at finite temperature, and has yielded promising results on how moments disorder in Fe as the temperature is raised [23], as mentioned in the first section. This leads to some hope that such a method might show the crossover behavior mentioned above between f bands at low T and free moments at high T.

SPECULATIONS

In the previous section, some new density functional methods were explored in light of possible applications to the f electron problem. Several techniques, such as the OEP method and the CPA method, seem promising. The basic problem with applying the new density functional techniques is their proper extension to the case of a lattice. This extension needs to be thought out properly before any significant developments can be made. In particular, the use of spin degenerate bands obeying Fermi-Dirac statistics within a Bloch formalism may not be sufficient to yield the desired effects. We might remind the reader that the exotic effects one finds for the new density functional methods when one continues to fractional occupation numbers (non-analytic behavior of the total energy and discontinuities of the potential about integer occupation values, etc.) may not be resolvable in a band formalism. In particular, one might note that in the SIC formalism, the eigenvalues for unoccupied orbitals degenerate with occupied orbitals are not the same as the occupied ones! This results in the fact that no Fermi-Dirac statistics problem occurs in SIC (in LDA, one can have the total energy minimize with holes beneath the chemical potential, but not in SIC). The development of new "band" techniques, perhaps based on statisitical ensemble of localized states or use of a non-Bloch representation, clearly needs to be investigated.

In conclusion, some promising new density functional methods have been developed over the past several years which have proven to be quite successful for atoms. The extension of these methods to the lattice should yield some fundamental insights into the expression of the many-body problem in an effective one-particle format.

ACKNOWLEDGMENTS

The author would like to thank Dale Koelling and John Perdew for several years of fruitful collaboration as regards some of the ideas in this paper. This work was supported by the U.S. Dept. of Energy, Office of Basic Energy Sciences, Div. of Materials Sciences, under Contract No. W-31-109-Eng-38. The author would also like to acknowledge a grant of computer time at the Energy Research Cray X-MP at the Magnetic Fusion Energy Computing Center, by which some of these ideas have been explored recently.

REFERENCES

1. "Theory of the Inhomogeneous Electron Gas", ed. S. Lundqvist and N. March, (Plenum, New York, 1983).
2. P. W. Anderson in "Valence Fluctuations in Solids," ed. L. Falicov, W. Hanke, and M. B. Maple, (North Holland, Amsterdam, 1981), p. 451.
3. P. A. Lee, T. M. Rice, J. W. Serene, L. J. Sham, and J. W. Wilkins, Comments on Cond. Mat. Phys. 12, 99 (1986).
4. P. Hohenberg and W. Kohn, Phys. Rev. 136, B864 (1964).
5. W. Kohn and L. J. Sham, Phys. Rev. 140, A1133 (1965).
6. J. C. Slater, Phys. Rev. 81, 385 (1951).
7. J. Harris, Phys. Rev. A 29, 1648 (1984).
8. O. Gunnarsson and B. I. Lundqvist, Phys. Rev. B 13, 4274 (1976).

9. U. vonBarth and L. Hedin, J. Phys. C 5, 1629 (1972); also see Ref. 1
10. U. vonBarth, Phys. Rev. A 20, 1693 (1979).
11. J. Perdew and M. R. Norman, Phys. Rev. B 26, 5445 (1982).
12. J. Perdew, R. G. Parr, M. Levy, and J. L. Balduz, Jr., Phys. Rev. Lett. 49, 1691 (1982).
13. M. R. Norman and D. D. Koelling, to be published, J. Less-Common Metals.
14. B. I. Min, H. J. F. Jansen, T. Oguchi, and A. J. Freeman, J. Mag. Magn. Matls. 59, 277 (1986).
15. H. J. F. Jansen, A. J. Freeman, and R. Monnier, Phys. Rev. B33, 6785 (1986).
16. D. D. Koelling in "The Electronic Structure of Complex Systems," ed. P. Phariseau and W. M. Temmerman (Plenum, New York, 1984), p. 183.
17. M. S. S. Brooks, Physica 130 B, 6 (1985); M. R. Norman and D. D. Koelling, Phys. Rev. B33, 3803 (1986).
18. A. J. Arko, D. D. Koelling, and J. E. Schirber, in "Handbook on the Physics and Chemistry of Actinides," vol. 2, ed A. J. Freeman and G. H. Lander, (North Holland, Amsterdam, 1985), p. 175.
19. D. D. Koelling, Solid State Comm. 43, 247 (1982).
20. M. R. Norman and D. D. Koelling, Phys. Rev. B33 6730 (1986).
21. P. H. P. Reinders, M. Springford, P. T. Coleridge, R. Boulet, and D. Ravot, unpublished.
22. M. R. Norman and D. D. Koelling, unpublished.
23. B. L. Gyorffy, A. J. Pindor, J. Staunton, G. M. Stocks, and H. Winter, J. Phys. F 15, 1337 (1985).
24. D. C. Langreth and M. J. Mehl, Phys. Rev. B28 1809 (1983); J. P. Perdew, Phys. Rev. Lett. 55, 1665 (1985) and Phys. Rev. B33, 8822 (1986).
25. M. R. Norman and D. D. Koelling, Phys. Rev. B28, 4357 (1983).
26. M. R. Norman, unpublished.
27. J. P. Perdew and A. Zunger, Phys. Rev. B23, 5048 (1981).
28. J. P. Perdew in "Density Functional Methods in Physics," ed. R. M. Dreizler and J. da Providencia, (Plenum, New York, 1985), p. 265.
29. M. R. Norman and D. D. Koelling, Phys. Rev. B30, 5530 (1984).
30. J. G. Harrison, J. Phys. Chem. 79, 2265 (1983).
31. R. A. Heaton, J. G. Harrison, and C. C. Lin, Phys. Rev. B28, 5992 (1983).
32. M. R. Norman and J. P. Perdew, Phys. Rev. B28 2135 (1983).
33. J. D. Talman and W. F. Shadwick, Phys. Rev. A 14, 36 (1976); K. Aashamar, T. M. Luke, and J. D. Talman, J.Phys. B12, 3455 (1979).
34. M. Weger and N. F. Mott, J. Phys. C 18, L201 (1985).

NON-LOCAL CORRELATION AND POINT TRANSFORMATIONS IN DENSITY FUNCTIONAL THEORY

Eduardo V. Ludeña[1], Aníbal Sierraalta[1], Eugene S. Kryachko[2] and Antonio Hernández[3]

[1]Centro de Química, Instituto Venezolano de Investigaciones Científicas, IVIC, Apartado 21827, Caracas 1020-A, Venezuela
[2]Institute for Theoretical Physics, Kiev, 252130, USSR
[3]Universidad Simón Bolívar, Apartado 80659, Caracas 1080-A, Venezuela

I. INTRODUCTION

In spite of the impressive recent developments in computational facilities and of the extended use of computer in the treatment of quantum mechanical many body problems, the solution of Schrödinger's equation for large systems containing hundreds of electrons and nuclei, still remains an almost impossible task. Such systems, however, do not represent exceptional examples but they correspond, rather, to everyday situations arising in solid state physics, material science, catalysis, molecular biology, etc.

According to Schrodinger's formalism, an n particle system must be described by a wavefunction $\Psi(\vec{r}_1, \sigma_1, \ldots, \vec{r}_n, \sigma_n)$ which depends upon 3n spatial coordinates $\{\vec{r}_i \equiv (x_i, y_i z_1)\}$ and n spin coordinates $\{\sigma_i\}$. Thus, the difficulty in solving the problem grows at least as 3n. In fact, near-exact solutions or, more properly speaking, highly approximate solutions have only been obtained for systems containing only a few tens of electrons.

The present day importance of density functional theory |1-5| stems from the fact that it attempts to bypass the usual treatment of a quantum mechanical many-body problem based on Schrödinger's equation and to replace it by another one whose fundamental mathematical entity is the one-particle density $\rho(\vec{r}_1)$. Obviously, since $\rho(\vec{r}_1)$ depends, regardless of the size of the physical system, only upon 3 spatial coordinates, such an undertaking, if successful, would bring considerable simplifications to the many body problem.

One of the basic difficulties of density functional theory has to do with how to incorporate into a single equation for the density all physical effects arising from many-particle interactions. According to the theorem of Hohenberg and Kohn |6|, there is a one-to-one correspondence between the density $\rho(\vec{r})$ and the external potential $v(\vec{r})$. This fact opens up the possibility of writing a universal functional for the energy in terms of $\rho(\vec{r})$ alone and to cast the search for the energy minimum in the framework of a well defined procedure |7,8|. The search for this universal functional has remained, however, elusive.

In the development of density functional theory, the abstract construct known as the homogeneous electron gas has played a very important rôle both in setting the physical model upon which a more general theory could be further elaborated, as well as in determining a mathematical framework which could later be refined by means of a progressive inclusion of inhomogeneities which, of course, are present in real systems |9|. In the simplest expression of density functional theory, the introduction of the homogeneous electron gas approximation leads to the Thomas-Fermi term ($a\ \rho^{5/3}$) for the kinetic energy and also to the Dirac term ($b\ \rho^{4/3}$) for the exchange interaction energy. Corrections to this first approximation are incorporated through gradient expansions for the kinetic and potential energy densities |10-12|.

In a previous work |13,14| we have critically discussed the gradient expansion, indicating, in particular, that when asymptotic conditions are included, then the only terms which remain in this expansion are the Thomas-Fermi and the Weizsacker terms for the kinetic energy and the Dirac term for exchange |15|. But, as is well known, these terms are not sufficient to describe accurately a many electron system.

Notwithstanding these theoretical difficulties, additional terms suggested by the gradient expansion have been incorporated in order to adequately reproduce energy values of many electron systems via density functional theory |16|. However, as it has been shown by Vosko |17|, coincidence with energy values is not the most appropriate criterium for assessing the quality of an approximate functional as it may turn out that the density which is obtained by solving the corresponding Euler-Lagrange equation might give rise to unphysical features.

In any attempt to retain the basic framework of the gradient expansion, one must bear in mind that it does not lead to a proper asymptotic behavior for the energy density. An interesting way out of this difficulty has been suggested by Perdew |18| where, by means of cut-off factors these ill-behaved terms are projected out.

But, as pointed out in previous works |13,14|, the gradient expansion does not constitute the only alternative. Another approach based on a decomposition of the energy functional into local and non-local terms, and which in principle could lead to the exact representation of the energy functional, has been recently advanced |19-21|.

In the context of this approach, the Weizsacker term arises naturally as the local kinetic energy term and the Thomas-Fermi term is the first non-local correction when the correlation factor for the homogeneous electron gas is used. Thus, according to this alternative view, it becomes clear that any progress toward obtaining an accurate energy functional expressed in terms of the density must rely on a thourough analysis of the non-local contributions.

We discuss in the present work some recent results concerning the non-local kinetic energy term |22|. We show that this non-local term is a basic ingredient of density functional theory and, in particular, we indicate how this also applies to the "exact" differential equation for $\rho(r)^{1/2}$ recently advanced by Levy, Perdew and Sahni |23|. Finally, we briefly discuss the use of point transformations for the representation of the non-local correlation function |24,25|.

II. THE NATURE OF THE NON-LOCAL CORRELATION FUNCTION G(1,2)

When the reduced first order density operator $\gamma(1,2)$ is written as

$$\gamma(1,2) = \rho^{1/2}(1)\, \rho^{1/2}(2)\, G(1,2) \qquad (2\text{-}1)$$

then, the kinetic energy density becomes

$$t|\rho| = \frac{1}{8}\vec{\nabla}_1\rho(1) \cdot \vec{\nabla}_1\rho(1)/\rho(1) + \frac{1}{2}\rho(1)\,\vec{\nabla}_1\cdot\vec{\nabla}_2 G(1,2)|_{2\to 1} \qquad (2\text{-}2)$$

The first term is the well-known Weizsacker term |26| which corresponds to the local contribution to the kinetic energy |19|. The second is the non-local part. In a natural orbital approximation, the <u>exact</u> expression for $\nabla_1\nabla_2 G(1,2)|2\to 1$ is given by |15|

$$\nabla_1\nabla_2 G(1,2)|_{2\to 1} = \sum_{i<j} \lambda_i\lambda_j |X_i(1)\vec{\nabla}_1 X_j(1) - X_j(1)\vec{\nabla}_1 X_i(1)|^2/\rho^2(1) \qquad (2\text{-}3)$$

In a recent work |22|, we have analyzed this non-local term for several atoms. As a result of this analysis it was to found that this non-local term gives bell-shaped curves at precisely the intershell regions. As an illustration we present in Fig. 1, this bell-shaped curve for atomic neon which was calculated using the Hartree-Fock SCF wavefunction of Clementi and Roetti |27|. The vertical dashed line in this figure indicates the intershell radius.

The non-local correlation function G(1,2) mainly describes exchange interaction between particles. This follows from the fact that to a very good approximation the reduced second order density operator $\Gamma^2_{HF}(1,2;1,2)$ expressed in terms of Hartree-Fock functions resembles quite closely the

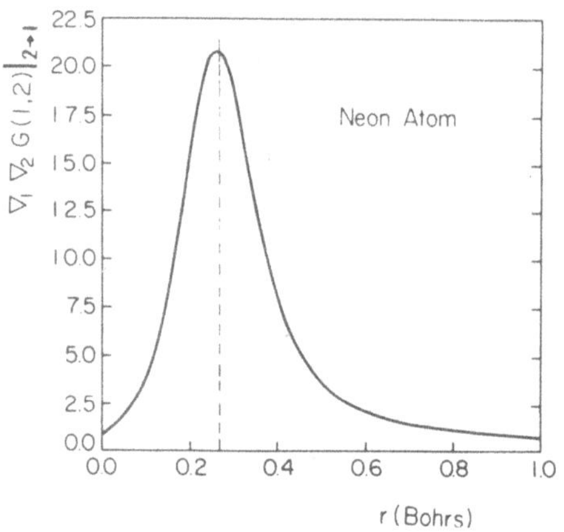

Fig. 1. Non-Local Contribution to the Kinetic energy $\nabla_1\nabla_2(1,2)_{2\to 1}$ for Atomic Neon (Arbitrary Units).

exact $\Gamma^2(1,2;1,2)$, which may always be written as

$$\Gamma^2(1,2;1,2) = \Gamma^2_{HF}(1,2;1,2)(1 + f_{CORR}(1,2;1,2)) \tag{2-4}$$

where $f_{CORR}(1,2;1,2)$ is the correlation factor. The effect of this correlation factor on the total energy is quite small as about 99% of the energy can generally be recovered from Γ^2_{HF}. But since, in the Hartree-Fock approximation

$$\Gamma^2_{HF}(1,2;1,2) = \rho_{HF}(1)\rho_{HF}(2)\left[1 - G_{HF}(1,2)G_{HF}(2,1)\right] \tag{2-5}$$

it follows that $- G_{HF}(1,2)G_{HF}(2,1)$ may be essentially regarded as a Fermi correlation factor. Bearing this in mind, and in order to assess the importance of the Coulomb correlation contribution to G(1,2), we have plotted in Fig. 2, the difference between the non-local contributions to the kinetic energy

$$\Delta\left[\vec{\nabla}_1\cdot\vec{\nabla}_2 G(1,2)|_{2\to 1}\right] = \vec{\nabla}_1\cdot\vec{\nabla}_2 G_{CI}(1,2)|_{2\to 1} - \nabla_1\nabla_2 G_{HF}(1,2)|_{2\to 1} \tag{2-6}$$

where CI denotes a highly elaborate configuration interaction wavefunction for neon computed by Bunge and Esquivel |28|, which recovers 85% of the correlation energy. The striking fact is that for neon, the correlation factor gives a negligible contribution (notice that the scale of Fig. 2 is about 1/100 of that of Fig. 1) to this non-local term. This result opens the interesting possibility, which of course must be tested in other systems, of dealing by means of different and separate expressions with exchange and Coulomb correlation effects.

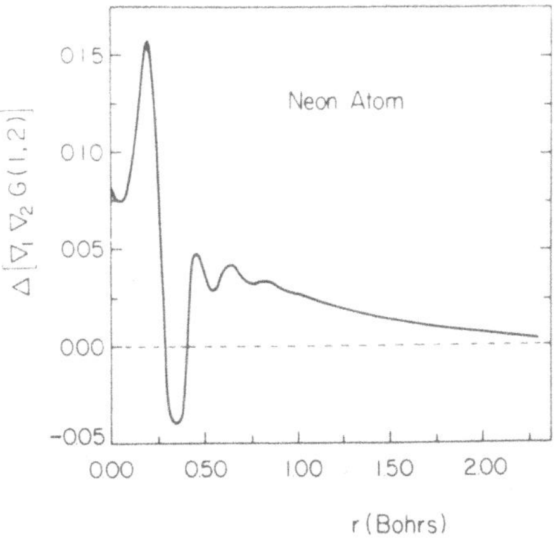

Fig. 2. Difference between the Non-Local Contributions to the Kinetic Energy for Atomic Neon Calculated with CI and Hartree-Fock Wavefunctions.

III. THE NON-LOCAL TERM IN A DIFFERENTIAL EQUATION FOR $\rho^{1/2}(\vec{r})$.

In a recent article, Levy, Perdew and Sahni |23|, have advanced the following differential equation for $\rho^{1/2}(\vec{r})$:

$$h_{eff}(\vec{r})\rho^{1/2}(\vec{r}) = \mu\rho^{1/2}(\vec{r}) \tag{3-1}$$

$$h_{eff}(\vec{r}) = -\frac{1}{2}\nabla^2 + V(\vec{r}) + v_{eff}(\vec{r}) \tag{3-2}$$

where the effective potential is given by

$$v_{eff}(\vec{r}) = \int d\vec{r}\,\tilde{\rho}\,(\vec{r}\,;\,\vec{r}_n)|\,\vec{r} - \vec{r}_n|^{-1}$$
$$+ \int d1\ldots\int d(n-1)d\sigma_n\phi^*(1\ldots(n-1);n)|H(1\ldots(n-1) - E^\circ_{(n-1)}|\phi(1\ldots(n-1);n)$$
$$+ \frac{1}{2}\int d1\ldots\int d(n-1)d\sigma_n\vec{\nabla}_n\phi^*(1\ldots(n-1);n)\,\vec{\nabla}_n\phi(1\ldots(n-1);n) \tag{3-3}$$

According to these authors, $v_{eff}(\vec{r})$ is simply a local or multiplicative operator.

In the derivation of this equation there is a minor misconception regarding the parametric dependence of the function $\phi(1\ldots(n-1);n)$ on coordinate n. This leads to the apparent absence of the non-local kinetic energy term involving $G(\vec{r}_1,\vec{r}_2)$ of Eq(2-2). For this reason, in what follows we rederive this equation in order to show the explicit occurrence of this term.

Consider the Schrödinger equation for an n-particle system

$$H(1...n)\ \Psi_0(1...n) = E_0^{(n)}\ \Psi_0(1...n) \tag{3-4}$$

The subindex zero just labels the ground state wavefunction and its corresponding energy (we assume for simplicity a non degenerate case). The reduced first order density operator for the ground state of this system is defined by

$$\rho_0(\vec{r}_n) = n \int d1...\int d(n-1)d\sigma_n \Psi_0^*(1...n)\ \Psi_0(1...n) \tag{3-5}$$

The Hamiltonian operator is

$$H(1...n) = H(n) + H(1...(n-1)) + \sum_{i=1}^{n-1} r_{in}^{-1} \tag{3-6}$$

where

$$H(n) = -\frac{1}{2}\nabla_n^2 + V(\vec{r}_n) \tag{3-7}$$

and H(1...(n-1)) is the Hamiltonian for an(n-1) particle system. Consider now the exact relationship

$$\Psi_0\ (1...n) = \rho_0^{1/2}\ (\vec{r}_n)\ \Psi_0\ (1...n)\ /\ \rho_0^{1/2}\ (\vec{r}_n)$$

$$= \rho_0^{1/2}\ (\vec{r}_n)\ \phi_0\ (1...(n-1);\ n) \tag{3-8}$$

The function ϕ_0 (1...(n-1); n) may be regarded as a conditional probability amplitude and $\rho_0^{1/2}(\vec{r}_n)$ as a marginal amplitude |35|. The notion that ϕ_0 (1...(n-1); n) depends only on coordinates 1,...,(n-1) and only "parametrically" on coordinate n, is the actual source of confusion. Actually, for a fixed $n \equiv \vec{r}_n, \sigma_n$, ϕ_0 (1...(n-1); n) depends only on coordinates 1,...,(n-1). But when n is not fixed, then ϕ_0 (1...(n-1); n) is a bona fide function of coordinates 1 through n and for this reason it must be denoted as ϕ (1...n). The confusion regarding "parameters' and actual functional "variables" has also led to difficulties when dealing with the non-adiabatic (non-Born-Oppenheimer) treatment of the motion of nuclei |35|. It is important to notice that

$$\int d1...\int d\ (n-1)\ d\ \sigma_n\ \phi_0^*\ (1...n)\ \phi_0\ (1...n)$$

$$= \int d1...\int d(n-1)\ d\ \sigma_n \frac{\Psi_0\ (1...n)}{\rho_0^{1/2}\ (\vec{r}_n)} \frac{\Psi_0\ (1'...n')}{\rho_0^{1/2}\ (\vec{r}_n')}\Bigg|\ 1'\to 1,\ ...\ n'\to n$$

$$= n\ \frac{\gamma_0(\vec{r}_n,\vec{r}_n')}{\rho_0^{1/2}(\vec{r}_n)\ \rho_0^{1/2}(\vec{r}_n')}\ \Bigg|\ \vec{r}_n'\to\vec{r}_n = n\ G\ (\vec{r}_n,\vec{r}_n')\Big|\ \vec{r}_n'\to\vec{r}_n = n \tag{3-9}$$

where Eq(2-1) has been used in order to introduce the non-local correlation function $G(\vec{r}_n,\vec{r}_n')$. It is clear from this expression that

$$\vec{\nabla}_n \cdot \vec{\nabla}_{n'} \int d1 \ldots \int d(n-1) d\sigma_n \phi_0^* (1 \ldots n)\ \phi_0(1 \ldots n) = n\vec{\nabla}_n \cdot \vec{\nabla}_{n'} G(r_n, r_n') \big|_{\vec{r}_n', \vec{r}_n} \neq 0 \tag{3-10}$$

Let us consider the expression

$$\int d1 \ldots \int d(n-1) d\sigma_n \phi_0(1 \ldots n) \,|H(1 \ldots n) - E_0^{(n-1)}|\, \phi_0(1 \ldots n)$$

$$= (E_0^n - E_0^{(n-1)}) \int d1 \ldots \int d(n-1) d\sigma_n\ \phi_0(1 \ldots n) \psi_0(1 \ldots n) \tag{3-11}$$

Using Eqs.(3-6) and (3-8) and letting $\mu = E_0^n - E_0^{(n-1)}$, we have

$$\int d1 \ldots \int d(n-1) d\sigma_n \frac{\psi_0(1 \ldots n)}{\rho^{1/2}(r_n)} \qquad H(n') \psi_0(1 \ldots n') \big|_{n' \to n}$$

$$+ \sum_{i=1}^{n-1} r_{in}^{-1}\ \psi_0(1 \ldots n) + H(1' \ldots (n-1)') \rho_0^{1/2}(r_n') \phi_0(1' \ldots n') \big|_{1' \to 1, \ldots n' \to n}$$

$$= \mu\ \rho^{1/2}(\vec{r}_n) \tag{3-12}$$

This equation may be rewritten as

$$\Bigg[\frac{1}{2n\rho(\vec{r}_n)} \left(\vec{\nabla}_n \cdot \vec{\nabla}_{n'}\ \gamma\ (\vec{r}_n, \vec{r}_n') \ \big|_{n' \to n} + V\ (\vec{r}_n)\ \rho\ (\vec{r}_n) \right)$$

$$+ \frac{2}{n} \int d\vec{r}_1\ (\ \Gamma^2\ (\vec{r}_1, \vec{r}_n; \vec{r}_1, \vec{r}_n)\ /\ \rho(\vec{r}_n)\ r_{1n}^{-1}$$

$$+ \int d_1 \ldots \int d(n-1)\ \phi_0^*\ (1 \ldots n)(H(1 \ldots (n-1)) - E_0^{n-1})\ \phi_0(1 \ldots n) \Bigg] \rho^{1/2} (\vec{r}_n)$$

$$= \mu\ \rho^{1/2}\ (\vec{r}_n) \tag{3-13}$$

The definition of the effective potential may now be easily obtained from this equation. Let us notice that the first term of this potential yields, by Eq(2-2), the Weizsacker term plus a non-local contribution to the kinetic energy. Also, $\Gamma^2\ (\vec{r}_1, \vec{r}_n; \vec{r}_1, \vec{r}_n)$ appearing in the third term is just the reduced second order density operator, which is a non-local operator. Thus, as one should expect from the nature of the quantum mechanical many body problem, an equation for $\rho^{1/2}(\vec{r}_n)$ also contains non-local terms.

IV. POINT TRANSFORMATIONS

A new and interesting approach which is particularly relevant to the analysis of the non-local contribution to the kinetic energy is based on the theory of point- (or local- scaling-) transformations. This approach has been advanced by Petkov, Stoitsov and Kryachko |24|.

The basic idea may be succintly developed as follows. Consider the transformation of a vector $\vec{r} \in R^3$ into another vector $\vec{f}(\vec{r}) \in R^3$:

$$\vec{r} \equiv (x,y,z) \xrightarrow{\mathcal{R}} \vec{f} \equiv (f_x, f_y, f_z) \equiv \vec{f}\,(\vec{r}) \tag{4-1}$$

For this transformation to exist and to have an inverse, it is sufficient that the Jacobian be different from zero:

$$D\,\{\,\vec{f}\,(\vec{r});\,\vec{r}\,\} \equiv \frac{D(f_x, f_y, fz)}{D(x,y,z)} \neq 0 \tag{4-2}$$

In particular, if we restrict this transformation to the particular type given by

$$\vec{f}\,(\vec{r}) = \frac{\vec{r}}{r}\, f\,(\vec{r}) \tag{4-3}$$

then the Jacobian becomes

$$D\,\{\,\vec{f}\,(\vec{r});\,\vec{r}\,\} = \frac{f\,(\vec{r})}{r}^{2}\; \frac{\partial f(\vec{r})}{\partial r} \tag{4-4}$$

When this transformation is applied to all coordinates $\{\vec{r}_i\}$ of an n-particle wavefunction, we obtain

$$\Psi\,(\vec{r}_1,\vec{r}_2,\ldots,\vec{r}_n) \xrightarrow{\mathcal{R}^n} \Psi\,(\vec{f}\,(\vec{r}_1), f\,(\vec{r}_2),\ldots,\vec{f}\,(\vec{r}_n)) \tag{4-5}$$

Clearly, the normalization condition on this wavefunction is preserved by this transformation:

$$\int d\vec{r}_1 \ldots \int d\vec{r}_n\, \Psi^*\,(\vec{r}_1 \ldots \vec{r}_n)\, \Psi\,(\vec{r}_1 \ldots \vec{r}_n)$$

$$= \int d\vec{f}_1 \ldots \int d\vec{f}_n\, \Psi^*\,(\vec{f}_1 \ldots \vec{f}_n)\, \Psi\,(\vec{f}_1 \ldots \vec{f}_n) = 1 \tag{4-6}$$

where

$$d\vec{f}_i = d\vec{f}\,(\vec{r}_j) = d\vec{r}_i\; D\,\{\vec{f}(\vec{r}_i);\vec{r}_i\} \tag{4-7}$$

The reduced first order density operator

$$\gamma\,(\vec{r}_1,\vec{r}_1') = n \int d\vec{r}_2 \ldots \int d\vec{r}_n\, \Psi^*\,(\vec{r}_1 \ldots \vec{r}_n)\, \Psi\,(\vec{r}_1 \ldots \vec{r}_n) \tag{4-8}$$

is transformed, upon application of Eq(4-5), into

$$\gamma(\vec{r}_1,\vec{r}_1') = D^{1/2}\,\{\vec{f}(\vec{r}_1);\vec{r}_1\}\; D^{1/2}\,\{\vec{f}(\vec{r}_1');\vec{r}_1'\}\; \gamma_f\,(\vec{f}(\vec{r}_1),\vec{f}(\vec{r}_1')) \tag{4-9}$$

where

$$\gamma_f\,(\vec{f}(\vec{r}_1)\,,\,\vec{f}(\vec{r}_1')) = n \int d\vec{f}_2 \ldots \int d\vec{f}_n\, \Psi^*\,(\vec{f}_1 \ldots \vec{f}_n)\, \Psi\,(\vec{f}_1 \ldots \vec{f}_n) \tag{4-10}$$

Thus, it follows that the density given in terms of the vector $\vec{r}$ is related to the density given in terms of the vector $\vec{f}$ by the expression

$$\rho(\vec{r}) = D\{ f(\vec{r}) ; \vec{r} \} \rho_f (\vec{f}(\vec{r})) \tag{4-11}$$

which, for the particular transformation considered in Eq(4-3), yields the non-linear differential equation

$$\rho(\vec{r}) = \frac{f(\vec{r})}{r}^2 \frac{\partial f(\vec{r})}{\partial r} \rho_f (\vec{f}(\vec{r})) \tag{4-12}$$

or equivalently,

$$\frac{\partial f^3(\vec{r})}{\partial r} = 3 r^2 \rho(\vec{r}) / \rho_f (\vec{f}(\vec{r})) \tag{4-13}$$

Let us now consider for simplicity, a spherically symmetric case and let us assume that $\rho_f(\vec{f}(\vec{r}))$ is a constant density equal to n/V, where V is the volume and n, the number of particles. Integrating Eq(4-13), we obtain

$$f(r) = \left| \frac{3V}{n} \int_0^r dr\, r^2 \rho(r) \right|^{1/3} \tag{4-14}$$

Introducing this result into Eq(4-4), it results

$$D\{ \vec{f}(\vec{r}) ; \vec{r} \} = \frac{\rho(r)}{n} \tag{4-15}$$

When this point transformation is applied to a plane wave, we get

$$V^{-1/2} e^{i\vec{k}_j \cdot \vec{r}} \xrightarrow{\mathcal{R}} V^{-1/2} D^{1/2} \{\vec{f}(\vec{r});\vec{r}\} e^{i\vec{k}_j \cdot \vec{f}(\vec{r})} \tag{4-16}$$

Using Eq(4-15), we see that the transformed orbitals are simply the equidensity orbitals discussed by Harriman |29| and others |30-34|:

$$\phi_j(\vec{r}) = \left| \frac{\rho(\vec{r})}{n} \right|^{1/2} e^{i\vec{k}_j \cdot \vec{f}(\vec{r})} \tag{4-17}$$

The details of how equidensity orbitals may be used for the evaluation of the non-local contribution to the kinetic energy, have been given elsewhere |20,33,34|. The important feature which we would like to stress here is that all these approximations are particular cases of point-transformations for the density. Thus, this theory opens a general and perhaps very fruitful avenue for the understanding of the non-local effects in density functional theory.

REFERENCES

1. J. Keller and J.L. Gázquez, eds., Density Functional Theory, Springer-Verlag Lecture Notes in Physics N°187, Berlin (1983).
2. J.P. Dahl and J. Avery, eds., Local Density Approximations in Quantum Chemistry and Solid State Physics, Plenum, New York (1984).
3. S. Lundqvist and N.H. March, eds., Theory of the Inhomogeneous Electron Gas, Plenum, New York (1983).
4. R.M. Dreizler and J. da Providencia, eds., Density Functional Methods in Physics, Plenum, New York (1985).
5. R.M. Erdahl and V.H. Smith, eds., Density Matrices and Density Functionals, Reidel, Dordrecht (in press).

6. P. Hohenberg and W. Kohn, Phys. Rev. 136:B864 (1964); W. Kohn and L.J. Sham, Phys. Rev. 140:A1133 (1965).
7. M. Levy, Proc. Natl. Acad. Sci. USA, 76:6062 (1975).
8. E.H. Lieb, Int. J. Quantum Chem. 24:243 (1983).
9. W. Kohn and P. Vashishta, in Reference 3, p. 79.
10. E.H. Lieb, Rev. Mod. Phys. 53:603 (1981).
11. N.H. March, in Reference 3, p. 1.
12. R.G. Parr, Ann. Rev. Phys. Chem. 34:631 (1983).
13. E.V. Ludeña, in Recent Progress in Many Body Theories, H. Kümmel and M.L. Ristig, eds., Springer-Verlag, Lecture Notes in Physics, N°198, Berlin (1984), p. 370.
14. E.V. Ludeña, in Condensed Matter Theories. I., F.B. Malik, ed., Plenum, New York (in press).
15. Y. Tal and R.F.W. Bader, Int. J. Quantum Chem. Symp. 12, 153 (1978).
16. D.C. Langreth and M.J. Mehl, Phys. Rev. B 28:1809 (1983).
17. S. Vosko and L.D. MacDonald, A New Generalized Gradient Expansion for Exchange Only from Atomic Exchange Energies (this issue).
18. J.P. Perdew, On the Density-Gradient Expansions for the Electronic Exchange and Correlation Energies (this issue).
19. E.V. Ludeña, J. Chem. Phys. 76:3157 (1982).
20. E.V. Ludeña, J. Chem. Phys. 79:6174 (1983).
21. E.V. Ludeña, in Reference 2, p. 287.
22. A. Sierraalta and E.V. Ludeña, Int. J. Quantum Chem. (in press).
23. M. Levy, J.P. Perdew and V. Sahni, Phys. Rev. A 30:2475 (1984).
24. I. Zh. Petkov, M.V. Stoitsov and E.S. Kryachko, Int. J. Quantum Chem. 29:149 (1986).
25. E.S. Kryachko and E.V. Ludeña, Many-Electron Energy Density Functional Theory. I. Point-Transformations and One-Electron Densities (to be published).
26. C.F. von Weizsacker, Z. Phyzik 96:431 (1935).
27. E. Clementi and C. Roetti, At. Data Nucl. Data Tables 14:177(1974).
28. A. Bunge and R.D. Esquivel (to be published).
29. J.E. Harriman, Phys. Rev. A 24:680 (1981).
30. N.H. March and W.H. Young, Proc. Phys. Soc. London 72:182 (1958).
31. W. Macke, Phys. Rev. 100:992 (1955); Ann. Phys. 17:1 (1955).
32. J.K. Percus, Int. J. Quantum Chem. 13:89 (1978).
33. G. Zumbach and K. Maschke, Phys. Rev. A 28:544 (1983); 29:1585(E) (1984).
34. S.R. Gosh and R.G. Parr, J. Chem. Phys. 82:3307 (1985).
35. O. Goscinski and V. Mujica, in Reference 5 (in press); O. Goscinski and A. Palma, Int. J. Quantum Chem. 15:197 (1979).

SYMMETRY CONSTRAINTS IN THE IONIZATION POTENTIALS AND ON THE FORMULATION OF THE HOHENBERG-KOHN-SHAM THEORY

Jaime Keller, Carlos Amador, Carmen de Teresa, and Jose A. Flores

División de Estudios de Posgrado, Facultad de Química
Universidad Nacional Autónoma de México
Ciudad Universitaria, Delegación Coyoacán
04510 México, D.F., México

ABSTRACT

Density functional-theory was given a formal structure with the Hohenberg-Kohn[1] (HK) theorems and a succesful practical procedure with the Kohn-Sham[2] (KS) scheme to construct the density function as linear combinations of squares of auxiliary functions ψ_i of symmetry i with weight c_i. The functions ψ_i obey a Schrödinger like wave equation with eigenvalue ε_i. The overall symmetry of the system has to be consistent with the "occupation" of the i-th "states". In principle the set $\{i\}$ should be complete and the total energy $E[\rho;\{c_{i,N}\}]$ has to be minimized with respect to the parameters $\{c_{i,N}\}$, then the HK-KS equations should really be written $\delta E[\rho;\{c_{i,N}\}] = 0$ and $\int \rho \, d\tau = \int \Sigma_i c_i (\psi_i)^2 \, d\tau = N$, allowing for <u>fractional</u> c_i. Also the ionization potential μ_i, (with $\delta c_i^{(i)} \approx 1$), is given by

$$-\mu_i = \Sigma_i \frac{\partial E}{\partial c_i} \delta c_i^{(i)} + \Sigma_{n>1} \frac{1}{n!} \Sigma_{j,k,\ldots} (\partial^n E/(\partial c_i \partial c_k \ldots))(\delta c_i^{(i)} \delta c_k^{(i)} \ldots)$$

with $\Sigma_j \delta c_j^{(i)} = 1$. In this way $\mu_i = -\varepsilon_i + \Delta\mu_i$ and there will be a $\mu_k \leq \mu_{i \neq k}$, corresponding to the least energy removal of one electron. Examples and results are discussed and shown. The internal symmetry of the system is an important (hidden or explicit) part of the theory and when properly considered, the results have always compared reasonable well with experiment (see Keller, Amador and de Teresa[3], also Trickey[4]).

SYMMETRY CONSTRAINED FORMULATION OF THE HOHENBERG-KOHN-SHAM THEORY.

The standard derivations of the Hohenberg-Kohn[1] theory and the Kohn-Sham[2] procedure is now well understood in relation to the ground state energies and (electronic) density many-body system. They allow the self-consistent determination of the density and, within the accuracy of the density functionals employed, the calculation of energies, energy differences, response functions and related properties. In previous papers [3,5] we have discussed both the density functionals and some ansatz for the density as well as some of the applications of the theory. In particular, we have mentioned[3] the existence, in the practical implementation of the schemes, of some inner structure or configuration either in the form of a shell structure or in the form of an assumed density of states and the corresponding occupation functions (free electron like in the Thomas-Fermi theory, for example) which define some basic features of the system. In the present paper we will further discuss this problem and illustrate it with three types of examples.

The Kohn-Sham (KS) scheme within the density functional theory, in the formal structure of the Hohenberg-Kohn (HK) theorems, provides a succesful practical procedure to construct the density function as linear combinations of squares of auxiliary functions ψ_i of symmetry i with weight c_i. The functions ψ_i obey a Schrödinger like wave equation with eigenvalue ε_i^{KS}. A further, implicit, condition is that the overall symmetry of the system has to be consistent with the assumed "occupation" of the i-th "states". In principle the set $\{i\}$ should be complete and the total energy $E[\rho;\{c_{i,N}\}]$ has to be minimized with respect to the parameters $\{c_{i,N}\}$, then the HK-KS equations should really be written

$$\delta E[\rho;\{c_{i,N}\}] = 0 \tag{1}$$

and the constraints

$$\int \rho \, d\tau = \int \Sigma_i c_i (\psi_i)^2 d\tau = N , \tag{2}$$

allowing for <u>fractional</u> c_i. Also the ionization potential μ_i of the i-th state, is given by

$$-\mu_i = \Sigma_i \frac{\partial E}{\partial c_i} \delta c_i^{(i)} + \Sigma_{n>1} \frac{1}{n!} \Sigma_{j,k,..} (\partial^n E/(\partial c_i \partial c_k \ldots))(\delta c_i^{(i)} \delta c_k^{(i)}) \tag{3}$$

with $\Sigma_j \delta c_j^{(i)} = 1$ and $\delta c_i^{(i)} \approx 1$. In this way $\mu_i = -\varepsilon_i + \Delta\mu_i$ and there will be a $\mu_k \leq \mu_{i \neq k}$, corresponding to the least energy removal of one electron. With our formulation it is clear that the internal symmetry of the system is an important (hidden or explicit) part of the theory and when properly considered, the results have always compared reasonably well with experiment (see Keller, Amador and de Teresa[3], also Trickey[4]).

It is common to approximate $\delta c_j^{(i)} \approx \delta_{ij}$, instead of optimizing the c_j for the ground state and searching for the set $\{\delta c_j^{(i)}\}$ for the least energy removal of one electron or for the largest electron affinity, but the need to optimize with respect to the set $\{c_{i,N}\}$ cannot be omitted, in fact any a priori fixing of them as $\{c^o_{i,N}\}$ should result in a Lagrange multiplier which corresponds to an additional energy producing an excited state of the system of given overall symmetry. This is not a disadvantage of the theory, on the contrary, according to eq. (1), a stationary state is obtained for each assumed $\{c^o_{i,N}\}$, and in fact the density functional theory should be useful for the study of, stationary, excited states.

This is also important for a correct formulation of the differential equation for the square root of the total density. The variation of the energy functional (M. Levy, J.P. Perdew and V. Sahni[6]), in order to do it properly, should contain additional contributions to the auxiliary potentials $\partial E/\partial \rho$, for example a kinetic energy term (the importance of the average value of the angular part of the kinetic energy has been discussed before[5])

$$V_{kin} = \Sigma_i c_i \psi_i [-\nabla_r^2 + \ell_i(\ell_i + 1)/r^2] \psi_i / \Sigma_i c_i |\psi_i|^2 - [\rho^{1/2}(-\nabla^2 \rho^{1/2})] \quad (4)$$

or an electron-hole potential term

$$V_{e-h} = \partial E_{e-h}/\partial \rho \quad (5)$$

indicating that the shell structure is being properly considered. That the mapping of the density matrix into the density imposes some necessary conditions, has recently been discussed by Ludena and Sierralta[7].

Our approach also illustrates an old question: which is the chemical potential of a several atoms system when the atoms separate from each other? The answer is that the set of symmetry coefficients will decide unambiguously from which atom and state is an electron given or taken away, in complete accordance with common experience.

In reference (3), we have introduced the idea of a vertical calculation where a standard self-consistent determination of the Kohn-Sham auxiliary functions and their Lagrange multipliers ε_k^{KS} is made, for the external potential v, using the KS auxiliary equations

$$\hat{F}_{KS} \phi_k = \varepsilon_k^{KS} \phi_k \quad (6)$$

Using χ as the electron-electron Coulomb potential and E_{xc} as the (unknown) exchange-correlation energy approximated by either a local or a non-local density functional:

$$\hat{F}_{KS} = -(1/2)\nabla^2 + v + \chi + (\delta E_{xc}/\delta\rho) \ , \quad (7)$$

this is is what one would consider a standard procedure. In (3) a "horizontal" computational scheme was also discussed. Levy, Perdew and

Sahni[6] have proposed such a scheme using only one auxiliary function: $\rho^{1/2}$, the square root of the total electronic density and as a consequence one Lagrange multiplier, μ_{LPS} which they proposed should be identified to the (least) ionization potential IP of the system. But, as we discussed in our previous paper[3], if the L-P-S scheme is followed, what one would obtain as Lagrange multiplier μ_{LPS} is not the ionization potential of the system, but the average energy per electron, even if the contrary is stated in the LPS paper. The reason for this discrepancy is that those authors identified, in the second part of their paper, some integrals with the auxiliary potential in a way which cannot be made without some symmetry constraints being introduced in the variational procedure as discussed above. It is only if $\delta\rho = \Sigma_j \delta c_j^{(i)} |\rho_j|^2$ for the lowest possible μ_i. If, on the other hand these symmetry constraints are introduced, a set of horizontal schemes is obtained, each one producing an energy eigenvalue ε_i which is an approximation to the i-th ionization potential. The minimum ionization potential of the system, that is the least energy necessary to remove one electron from the system, is obtained only as a limiting procedure.

Our equation (3) is more general than the resulting Euler equation obtained from the symmetry constrained stationary principle

$$\delta\{E[\rho;\{c_{i,N}\}] - \mu[N[\rho;\{c_{i,N}\}] - N]\} = 0 \ , \qquad (8)$$

because in eq. (3) the relaxation of the charge density of each $|\psi_j|^2$ should be included in the ionization (or electron addition) procedure, as a consequence of the set $\{c_{i,N}\}$ in eq. (8) being changed in the process.

It should be stressed here that also the most commonly employed density functionals have limitations imposed by the internal symmetry. For example, if an atom is computed with the constraint that the $\{c_{i,N}\}$ is such that an electron is missing from an inner shell, using the Kohn-Sham procedure in a stationary state, as a solution of the symmetry constrained KS equations. We calculate a symmetry constrained total energy, which can be used to obtain the energy difference ΔE between the neutral atom ground state and this excited "state", this ΔE will be a good approximation to the inner shell electron removal or electron excitation energies (of course we don't usually introduce in the energy functionals, the correlation energy between the core hole and the electrons, but in principle, such energy can be estimated at least approximately, and included in the energy functional and from its functional derivative with the density, an additional potential will improve the effective hamiltonian . But, if from the KS density of the excited state an attempt is made of computing the energy (of the ion or) of the excited atom using the Thomas-Fermi functionals, a complete failure will occur. This will happen because the Thomas-Fermi kinetic energy functional assumes a "vertical" scheme corresponding to free electrons where all the states of a momentum smaller than the Fermi momentum are occupied, and this is not, even approximately, the case for the excited atom when an inner electron has been transferred. The success of the Thomas-Fermi approximation for the kinetic energy functionals (including gradients terms) has always been in the case of ground state atoms or in the case of ions where the electron being removed or added is removed or added from the top most state.

Another important term in the determination of the ionization potentials is the so called relaxation contribution. There are some approximations which can account for the changes in the density functions ρ_i corresponding to the auxiliary KS orbitals. These changes are larger when an inner electron is removed, and as a consequence the KS auxiliary energy eigenvalues are farther from the ionization potentials in the case of core electrons. But in many calculations where only the least bound electrons are considered to be removed, the changes of the inner most shells are small, and these relaxation terms are less significant. For the electron being removed:

$$\delta\rho_i = c_i|\psi_i|^2 - (c_i + \delta c_i)|\psi_i + \delta\psi_i|^2 \approx \delta c_i|\psi_i|^2 - c_i(\psi_i^*\delta\psi_i + \delta\psi_i^*\psi_i) \qquad (9)$$

the last term representing the relaxation of the orbital.

It is convenient to split the changes in the density $\delta\rho(\underline{r})$ in two contributions

$$\delta\rho(\underline{r}) = \delta\rho^o(\underline{r}) + \delta\rho^1(\underline{r}) \qquad (10)$$

where $\delta\rho^o(\underline{r})$ is the change in density from the electron being removed itself

$$\delta\rho^o(\underline{r}) = \Sigma_i \delta c_i \rho_i(\underline{r}) \qquad (11)$$

and $\delta\rho^1(\underline{r})$ is the change in the density originated from the change in the system's potential, mainly from the electrostatic part. This second term can be obtained, self consistently in principle, from the solution of the response equations

$$\delta\rho^1(\underline{r}) = -\int d\underline{r}' R_o(\underline{r},\underline{r}';o)\,\Phi_{self}(\underline{r}') \qquad (12)$$

$$\Phi_{self}(\underline{r}) = -\int d\underline{r}'\,\delta\rho(\underline{r}')/|\underline{r} - \underline{r}'| \qquad (13)$$

where $R_o(\underline{r},\underline{r};o)$ is the response function regarding the electrons as essentially independent entities moving in the self-consistently determined potential $\Phi_{self}(\underline{r})$ which could, in principle, be obtained from the integral equation

$$\Phi_{self}(\underline{r}') = \Phi^{o}_{self}(\underline{r}') + e^2 \int d\underline{r}' d\underline{r} [R_o(\underline{r}',\underline{r}'';o)/|\underline{r}' - \underline{r}''|] \Phi_{self}(\underline{r}'') \quad (14)$$

with Φ^{o}_{self} being the Coulomb potential associated with $\delta\rho^{o}(\underline{r})$. The response function $R_o(\underline{r},\underline{r}';o)$ has a very simple form in density functional theory

$$R_o(\underline{r},\underline{r}';\omega) = \Sigma_{i\ occ}\ \psi_i^*(\underline{r})\psi_i(\underline{r}')G^V(\underline{r},\underline{r}';\varepsilon_i + \hbar\omega) + c.c. \quad (15)$$

because the Green's function $G^V(\underline{r},\underline{r}';E)$ has the following expantion

$$G^V(\underline{r},\underline{r}';E) = \Sigma_L\ Y_L^*(\hat{r})Y_L(\hat{r}')j_\ell^V(r_<\sqrt{E})h_\ell^V(r_>E)/W[j_\ell^V h_\ell^V] \quad (16)$$

where the j_ℓ^V and the h_ℓ^V are the functions regular at the origin and the one behaving as a Hankel function at infinite, respectively. The $\psi_i(\underline{r})$ are the KS auxiliary functions.

SOME NUMERICAL EXAMPLES

a) The Ionization Potentials of Atoms

In this first example we compute the ionization potential of the carbon atom from differences of total energy and from the rate of change of the energy when the 2p or the 2s electron is removed. This rate of change is computed from a finite differences procedure. It is seen that the results are reasonable for the 2p electron. But if the electron being removed is a 2s electron, poor results are obtained, from the finite differences procedure using the Thomas-Fermi (and gradients) terms for the kinetic energy and the statistical exchange approximation, the reason for this, as discussed above, is that those approximations are based on a free electron like filling up of orbitals and this cannot even approximately be the case for core electrons or inner shells.

To illustrate this failure of the free electron approximations for the energy density functional we present the results obtained with the Kohn-Sham procedure (which compare well with experiment) and with the free electron like local approximations. It is clear that the failure of the approximations to reproduce even the order of magnitude of the kinetic energy appears only when the electron being removed is an inner shell or a core electron.

Table 1. Energies in Hartree Units.

State	IP_{calc}	IP_{exp}	Electronic Configuration	$E_{kinetic}$	$E^{statistical}_{kinetic}$
2p	0.420	0.413	$1s^2 2s^2 2p^0$	36.519	36.475
2s	0.600	0.712	$1s^0 2s^2 2p^2$	13.009	5.600

Some energy calculations for the carbon atom and ion.

b) Ionic Radii

Our second example of the use of the symmetry constraints in the Kohn-Sham procedure is an application we have made to the first principles determination of ionic radii in alkaline halide crystals[8].

When two atoms, say sodium and chlorine, are brought together in the crystal, the charge transfer from the alkaline atom (Na) to the halogene atom (Cl) is enhanced by the increased interatomic electrostatic attraction, as a result a completely ionic crystal is obtained. But a calculation of the ions is not in principle a ground state calculation (we refer to the embedded ions in the crystal), nevertheless a calculation is possible for the ion in its Wigner-Seitz cell if a symmetry restriction is made for the corresponding orbital occupation. Moreover, the volume per atom is not that of the least total electronic energy but that of the minimum total free energy and, as a consequence, a finite volume per ion is to be considered. The resulting electronic pressure, that is the rate of change of the electronic energy with the Wigner-Seitz cell's volume, can be computed for each pair of ionic radii adding up to the total interatomic distance of the crystal in thermal equilibrium. The radii thus determined are within a few percent from the experimentally accepted ionic radii (the experimental values are averages over several crystal structures then they are not an exact quantity). In the case of NaCl the ionic radius of Cl^- is 3.34 a.u. (exp. 3.40 a.u.) and that of Na^+ is 1.90 a.u. (exp. 1.84 a.u.). The force constants are also obtained from this procedure.

c) Electron Affinity and Hardness.

Our last example will be a discussion of the relation between ionization potentials μ_i^-, electron affinity μ_i^+, chemical potentials ε_i and hardness χ_i, which in principle are related by the relationships

$$\mu_i^+ = \varepsilon_i + \chi_i \quad ; \quad \mu_i^- = -\varepsilon_i + \chi_i \tag{17}$$

and at the same time

$$\begin{aligned} \mu_i^+ &= E_{N+1}(n_j^\circ, n_i = n_i^\circ + 1) - E_N(n_k^\circ) \\ \mu_i^- &= E_{N-1}(n_j^\circ, n_i = n_i - 1) - E_N(n_k^\circ) \end{aligned} \tag{18}$$

or

$$\varepsilon_i = \partial E/\partial n_i \quad \text{and} \quad 2\chi_i = \partial^2 E/(\partial n_i)^2 \tag{19}$$

assuming higher order derivatives to be negligible.

A fractional occupation calculation is possible,to obtain the negative ion (F^- in our example), or the positive ion (F^+), as a continuous procedure.

In figure 1 the variation of the total energy E, the kinetic energy KE, the electron-nuclear energy EN, the electron-electron energy

EE, Coulomb part, and the exchange energy Ex, with the occupation of the $2p^n$ state, for the atomic ground state configuration $1s^2 2s^2 2p^5$ (or $n° = 5$). Energies in Hartree units.

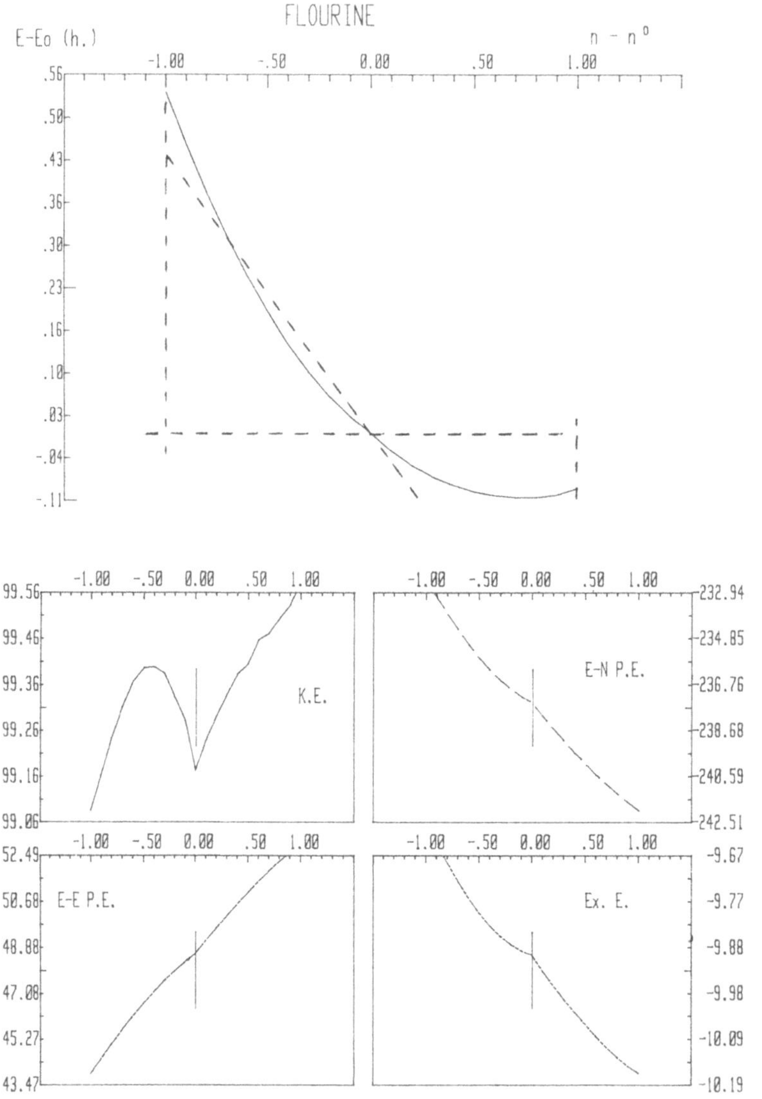

Figure 1. The variation of the total energy and its contributions, for the fluorine atom and ion, with the change in the occupation of the $2p^5$ atomic orbital. The far from linear behavior of the kinetic energy is clearly seen.

The identification made in formula (19) is not possible, mainly because of the relaxation terms given by (12), selfconsistently accounted for in our calculation. Note also that the hardness is not a small correction in formula (17).

DISCUSSION

We have analyzed both the origin of several deficiencies and of the successes of the current practice and formulation of density functional theory. We have shown that the explicit inclusion of basis sets in the Kohn-Sham procedure (the discussion of Harriman (1986)[9] is highly relevant to this point) and of the occupation numbers as symmetry constraints allow a useful formulation of a variety of problems and their solutions. We propose that a complete formulation of the theory should include at least two more features: a) the electron-hole correlation which will simultaneously provide a correct formulation of the theory of stationary excited states and of the transitions between states and b) the self consistent inclusion of the response function of the system allowing the correct description of the density relaxation processes in a variety of physical situations.

The inclusion of the electron-hole correlation solves a number of basic problems encountered in the present use of density functional theory: a) It allows the study of configurations and of configuration interaction because both Hartree-Fock level ground states and the excited configurations are then properly described. In fact the electron-hole correlation energy stabilizes the stationary states with integral occupation against Hartree-Fock level fictitious fractional occupation. The fractional occupation schemes will occupy the hole states until their eigenvalues reach the Fermi level. b) It clarifies the fundamental problem related to the physical difference between itinerant and localized states; a localized state is that where the electron-hole correlation energy in an (atomic) cell is large (and negative). A lower final total energy will correspond to the case where an electron is removed from an atomic core or from a localized atomic state, instead of the electron being removed from an extended band, where none or little electron-hole correlation energy will be recovered.

Experiments in condensed matter are consistent with single atom photoemission in the case of core electrons, and in many instances for the f and d electrons of the valence and conduction bands, in spite of the fact that this processes may break the translational symmetry of the solid.

The self consistent inclusion of the response function (eqs. 12-16) will contribute an imaginary part to the self energies of the electrons in excited states, indicating that any excited state will properly decay with probability amplitudes proportional to the imaginary parts of the self energies. Then a finite life time is expected for excited stationary states that will decay via the spontaneous emission of photons (or the corresponding gauge particles of the interactions responsible of the decay). In other words, a proper formulation of density functional theory cannot be achieved, unless the electromagnetic field is either explicitly included or at least accounted for through the self consistent use of the response function of the system for all frequencies.

In fact this last formulation includes the shell structure of the system as far as the response functions present maxima for energies corresponding to transitions to all possible states as a result of stimulated or spontaneous photon emissions or absorptions.

We have also shown that from a comparison of the Levy, Perdew and Sahni[6] equation for the square root of the density with our present approach, an additional potential should be included, in order to make certain that the Lagrange multiplier of that equation corresponds to the lowest ionization potential of the system.

REFERENCES

1. P. Hohenberg and W. Kohn, Phys. Rev. 136, B864 (1964).
2. W. Kohn and L.J. Sham, Phys. Rev. 137, A1697 (1965).
3. J. Keller, C. Amador and C. de Teresa in "Condensed Matter Theories I" p. B. Malik Ed., Plenum (1986) and J. Keller, Int. J. Quant. Chem., in press (1986).
4. S.B. Trickey, Phys. Rev. Lett. 56, 881 (1986).
5. J. Keller, C. Keller and C. Amador in "Recent Progress in Many Body Theories", p. 364, J. G. Zabolitzky, M. de Llano, M. Fortes and J.W. Clark Eds., Springer Verlag. Lecture Notes in Physics No. 142 (1981), and J. Keller and C. Amador in "Density Functional Theory", p. 269, J. Keller and J.L. Gázquez Eds., Springer Verlag. Lecture Notes in Physics No. 187 (1983).
6. M. Levy, J.P. Perdew and V. Sahni, Phys. Rev. A30, 2745 (1984).
7. E.V. Ludeña and A. Sierraalta, Phys. Rev. A32, 19 (1985).
8. X. Cruz, Thesis Facultad de Química. Universidad Nacional Autónoma de México. 1986.
9. J.E. Harriman, Phys. Rev. A34, 29 (1986).

HYPERNETTED CHAIN ANALYSES OF

DENSE PLASMALIKE MATERIALS

Hiroshi Iyetomi and Setsuo Ichimaru

Department of Physics
University of Tokyo
Bunkyo, Tokyo 113, Japan

ABSTRACT

Free energy formulas applicable to the electron-screened ion plasmas are derived in the hypernetted chain (HNC) approximation. The formulas, expressed in terms of the correlation functions, enable one to avoid the more cumbersome and less accurate calculations involving the thermodynamic integrations. The miscibility of iron atoms in the hydrogenic solar plasmas is treated as a specific example of application.

The density-functional formalism is applied to the analyses of the multiparticle correlations in ˈense plasmas, leading to an exact summation of all the bridge diagrams. With the aid of a nonlocal density-functional approximation to the direct correlation function, we derive new formulas for the bridge function, which are then used for an improvement of the HNC scheme. Consequences of various approximations are numerically examined.

1. INTRODUCTION

Various dense plasmas in nature consist of many ionic species [1,2]. Such a multi-ionic plasma creates novel issues involving the phase properties such as a limitation in the miscibility and a possible formation of an eutectic alloy [3,4]; those problems have astrophysically important consequences. A solution to such a problem depends quite sensitively on evaluation of the thermodynamic functions; a calculation with high precision is thus required. For application to the multi-ionic cases it is not only of physical interest but also of practical importance to advance accurate and workable theoretical schemes capable of describing the thermodynamic and correlational properties in dense plasmalike materials.

In this paper we present recent developments in theoretical analyses of dense plasmas on the basis of the hypernetted chain (HNC) approximation [5]. Free energy formulas applicable to the electron-screened ion plasmas are derived in the HNC approximation. The formulas, expressed in terms of the correlation functions, enable us to avoid the more cumbersome and less accurate calculations involving the thermodynamic integrations. The miscibility of iron atoms in the hydrogenic solar plasmas is treated as a specific example of application.

The density-functional formalism [6] is then applied to the analyses of the multiparticle correlations in dense plasmas, leading to an exact summation of all the bridge diagrams. With the aid of a nonlocal density-functional approximation to the direct correlation function and its modifications, we derive new formulas for the bridge function, which are then used for an improvement of the HNC scheme. The validity and accuracy of those schemes are examined through comparison with the exact Monte Carlo (MC) results.

2. THE FREE-ENERGY FORMULAS

It has been well demonstrated both empirically [1,2] and theoretically [7] that the HNC scheme offers an accurate way of describing interparticle correlations in dense Coulombic systems. The HNC equation for a classical many-particle system with a pair potential $\phi(r)$ is obtained by neglecting all the bridge-diagram contributions $B(r)$ in the following exact equation for the radial distribution function $g(r)$,

$$g(r) = \exp[-\beta\phi(r) + h(r) - c(r) + B(r)] \quad . \tag{1}$$

Here $h(r) = g(r) - 1$ and the direct correlation function $c(r)$ is related to $h(r)$ via the Ornstein-Zernike relation [5]; β is the inverse temperature in energy units.

As a unique feature of the HNC scheme, the chemical potential and hence the Helmholtz free energy can be calculated directly through integration involving the correlation functions [1,8]. For example, the excess chemical potential is given in the HNC theory as

$$\beta\mu = \frac{n}{2}\int d\vec{r}\, h(r)[h(r) - c(r)] - n\int d\vec{r}\, c(r) \quad . \tag{2}$$

The calculation of the free energy in the HNC scheme can thus be performed with an accuracy comparable to those of the internal energy and the pressure; the latter quantities involve analogous integrations. This feature of the HNC scheme has been used advantageously in the analysis of delicate thermodynamic properties of dense plasmalike materials such as phase separation in multi-ionic plasmas [9,10].

The existing derivation of the HNC free-energy formula [1,8] has not, however, anticipated a possible dependence of the interparticle potential on the thermodynamic variables such as the density and temperature. In many practical cases of application, one adopts an adiabatic approximation to the electrons, so that their screening action is taken into account by those effective interactions between ions, which depend in turn on the density and temperature of the electrons. Recently such an extension of the HNC free energy formula has been achieved for the electron-screened ionic plasmas [11]; the excess chemical potential of the screened one-ionic species (with electric charge Ze) plasma is thus obtained as

$$\begin{aligned}\beta\mu_{sc} &= \frac{n}{2}\int d\vec{r} h(r)[h(r) - c(r)] - n\tilde{c}(q=0) - n\beta\tilde{\phi}_e(q=0)\\ &+ \frac{\beta}{2}\phi^{(0)} - \frac{\beta}{6}\sum_{\vec{q}}{}' \tilde{v}(q)\left[r_s\frac{\partial}{\partial r_s}\frac{1}{\varepsilon(Q,0)}\right]_\theta S(q)\\ &- \frac{\beta}{3}\sum_{\vec{q}}{}'\tilde{v}(q)\left[\theta\frac{\partial}{\partial\theta}\frac{1}{\varepsilon(Q,0)}\right]_{r_s} S(q)\\ &- \frac{\beta}{6}\sum_{\vec{q}}{}'\tilde{v}(q)Q\frac{\partial}{\partial Q}\left[\frac{1}{\varepsilon(Q,0)}\right] S(q) \quad .\end{aligned} \tag{3}$$

Here $S(q)$ is the static structure factor, $\tilde{v}(q) = 4\pi(Ze)^2/q^2$, $\varepsilon(Q,0) \equiv \varepsilon(q/q_F, \omega = 0; r_s, \theta)$ is the electronic screening function with $q_F = (3\pi^2 n_e)^{1/3}$, $r_s \equiv (3/4\pi n_e)^{1/2} me^2/\hbar^2$, $\theta \equiv 2m/\beta\hbar^2 q_F{}^2$, $\tilde{\phi}_e(q) = \tilde{v}(q)/\varepsilon(Q,0)$,

$$\tilde{c}(q = 0) = \int d\vec{r} c(r), \quad \phi^{(0)} = \sum_{\vec{q}}{}' [\tilde{\phi}_e(q) - \tilde{v}(q)] \quad ,$$

and the $\vec{q}$ summation is carried out over those wave vectors ($\neq 0$) appropriate to periodic boundary conditions for a unit volume.

3. MISCIBILITY OF IRON ATOMS IN THE SOLAR HYDROGEN PLASMAS

As an application of the generalized HNC free - energy formula, we revisit the miscibility problem of iron atoms in hydrogen plasmas under the solar interior conditions, with a hope of shedding light on solution to the solar neutrino problem [12]. We assume the temperature and the pressure of the solar plasma in the vicinity of 1.5×10^7K and 10^5 Mbar. The relative concentration of the irons near the thermonuclear burn region is assumed to take on a value close to the cosmic abundance (2.5×10^{-5} ionic mole fraction).

The present calculations [13] have been carried out with special emphasis on the role of the screening effect arising from semiclassical electrons in the solar interior; such a semi-classical electron gas acts to screen the ion-ion interaction quite efficiently, and hence modifies the thermodynamic properties of the plasma substantially. The calculations thus improve over those of Alder, Pollock and Hansen [10] in two ways: (i) a proper account of the electronic polarization through the static screening function of the electrons, and (ii) a corresponding account of the exchange and correlation contributions to the thermodynamic functions for the electron system. Since the electrons are weakly coupled in the sun interior, the random phase approximation [2] is applicable for the description of the correlational properties of the electrons. The strong coupling effects between ions are treated accurately in the HNC scheme, as Alder et al. [10] have done.

The Gibbs free energy of mixing is expressed as the sum of the electronic, ideal-gas, and excess contributions. Qualitatively, the electronic and ideal-gas terms favor phase mixing, whereas the excess term promotes phase demixing (Fig. 1). Phase separation of the plasma mixtures takes place as a consequence of delicate balance between those physically distinct contributions.

Figure 2 shows the phase diagram for the hydrogen-iron mixture calculated in the present scheme at $P = 0.5 \times 10^5$ Mbar. The critical point for demixing takes place at $T_c \simeq 5.5 \times 10^6$K and $x_c \simeq 2.4 \times 10^{-2}$. Comparing the present results with those of Alder et al, we find an increase of T_c by 15% arising from the electronic screening effect. The increase, however, is not sufficient so as to resolve the solar neutrino dilemma through the idea [12] of a limited solubility of the iron atoms in the solar interior plasma. Figure 2 also exhibits the substantial influence of the adopted electronic equation of state exerted on the phase diagram calculations.

4. IMPROVEMENT OVER THE HNC APPROXIMATION

The calculations presented in the preceding section have been successful because the plasma density in the solar interior is not so high as to require an improvement over the HNC approximation. In many other examples of astrophysical dense plasmas, such as the interiors of Jovian planets and

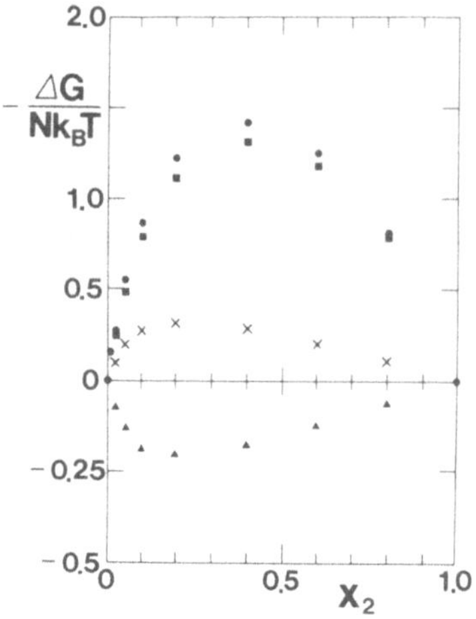

Fig. 1. Contribution to the Gibbs free energy of mixing $\Delta G/Nk_BT$ for the $H^+ - Fe^{24+}$ mixture at $P = 0.5 \times 10^5$Mbar and $T = 1.5 \times 10^7$K. The points with closed squares represent the ideal-gas contribution; closed triangles, the ionic excess contribution; crosses, the electronic contribution; closed circles, sum of all the three contributions. Note the difference in the scales above and below the abscissa.

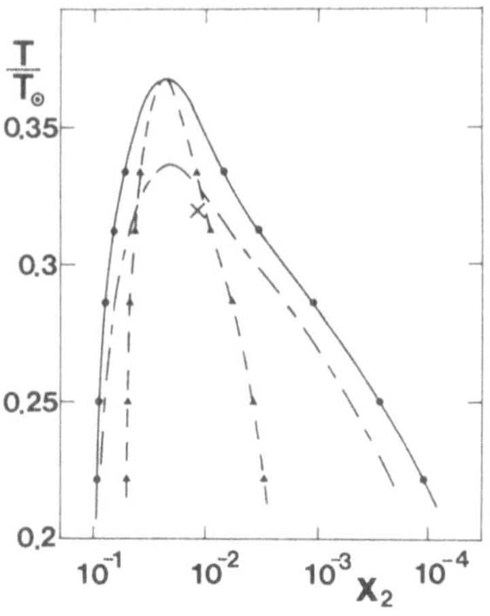

Fig. 2. Phase diagrams for the $H^+ - Fe^{24+}$ mixture with the electronic screening at $P = 0.5 \times 10^5$Mbar. The temperature T in the ordinate is normalized with $T_\odot = 1.5 \times 10^7$K, the interior temperature of the sun. The solid and dashed curves are the coexistence and spinodal curves, interpolated by the spline method with the third order polynomials; the closed circles and triangles represent the calculated points. The cross refers to the critical point obtained without the electronic screening by Alder, Pollock, and Hansen [10]. The chain curve is the coexistence curve calculated by retaining only the ideal-gas term in the equation of state for the uniform electron gas.

white dwarfs, the relevant density and temperature parameters are such that it becomes essential to develop a theoretical scheme which significantly improve over the HNC approximation. In this section we describe our recent efforts in these directions, confining ourselves to a consideration of the classical one-component plasma (OCP) consisting of ions with electric charge Ze embedded in a uniform neutralizing background of negative charges. The OCP is characterized by a single dimensionless parameter $\Gamma = \beta(Ze)^2/a$, where $a = (4\pi n/3)^{-1/3}$ is the Wigner-Seitz (ion-sphere) radius.

According to the density-functional formalism [6], the bridge function B(r), neglected in the HNC scheme, can be given in a closed form as

$$B(r) = n^2 \int_0^1 d\lambda(1 - \lambda) \int d\vec{r}_1 d\vec{r}_2 \frac{\delta c[\vec{r}_1, \vec{r}_2; n(\vec{r};\lambda)]}{\delta n(\vec{r};\lambda)} h(r_1)h(r_2) \quad . \tag{4}$$

Here $c[\vec{r}_1, \vec{r}_2; n(\vec{r})]$ is the direct correlation function in the inhomogeneous OCP with a density distribution $n(\vec{r})$, and the parameter λ measures the degree of inhomogeneity in the system, so that

$$n(\vec{r};\lambda) = n + \lambda(n(\vec{r}) - n) \quad , \tag{5}$$

with $n(\vec{r}) = ng(r)$.

The improvement of the HNC scheme is thus achieved once an appropriate evaluation of $c[\vec{r}_1,\vec{r}_2;n(\vec{r})]$ is made. To obtain a practically feasible scheme, we approximate $c[\vec{r}_1,\vec{r}_2;n(\vec{r})]$ by the direct correlation function of an equivalent homogeneous OCP with an average density n_{av},

$$c[\vec{r}_1,\vec{r}_2;n(\vec{r})] \simeq c(|\vec{r}_1 - \vec{r}_2|;n_{av}) \quad . \tag{6}$$

The nonlocal choice of the average density n_{av} is usually made in a way [6] so that

$$n_{av} = [n(\vec{r}_1) + n(\vec{r}_2)]/2 \quad . \tag{7}$$

For $c(r)$ we use the analytic formula for the HNC results given by Ng [14], which enables us to calculate the λ-integration in Eq. (4) analytically; we also evaluate $h(r)$ in Eq. (4) using the corresponding HNC results. Thus the bridge function obtained with Eqs. (6) and (7) retains the bridge contributions to all orders within those approximations; this evaluation is designated as $\tilde{B}(r)$. If we neglect the inhomogeneity of the system by setting $n(\vec{r};\lambda) = n$ in Eq. (4), the three-body contribution $\bar{B}(r)$ results.

A nonlocal treatment such as Eq. (7) may not be totally successful in describing the strong correlations in the vicinity of the origin, $r \simeq 0$, which involve the highest degree of nonlocality in $c[\vec{r}_1,\vec{r}_2;n(\vec{r})]$. Inadequacy of Eq. (7) is, in fact, revealed in the short-range behavior of $\tilde{B}(r)$, in that the values of $|\tilde{B}(0)|$ so calculated are larger, by a factor exceeding two, than the values $|B(0)_{IS}|$ estimated on the basis of the ion-sphere model for $\Gamma \gtrsim 1$; we note that $\tilde{B}(0)/B(0)_{IS} = 2.35$ at $\Gamma = 160$. The values $\bar{B}(0)$, on the other hand, are in fairly good agreement with $B(0)_{IS}$. These findings suggest that the use of a smoothed density rather than the real density in

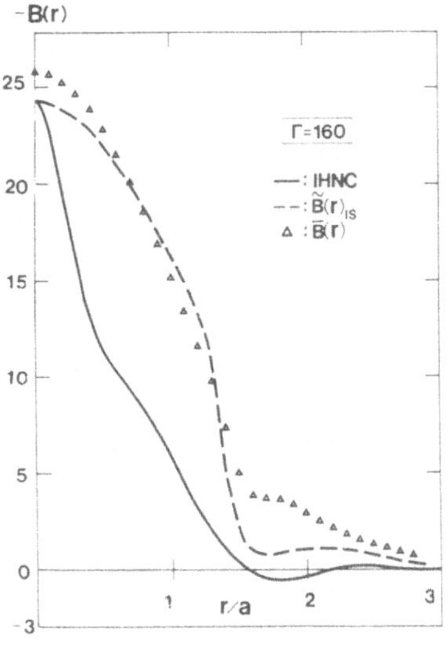

Fig. 3. Bridge function of the OCP in various schemes at $\Gamma = 160$. IHNC refers to the values in the improved HNC scheme [15]; $\tilde{B}(r)_{IS}$, those calculated from Eq. (8); $\bar{B}(r)$, the three-body bridge contribution.

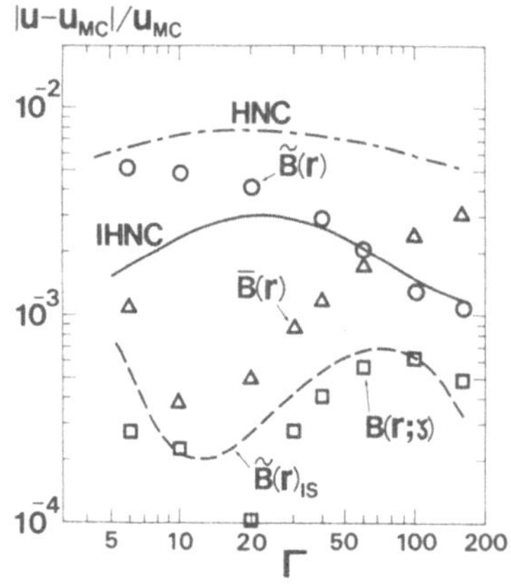

Fig. 4. Relative accuracy of the internal energy u in various schemes. u_{MC} refers to the Monte Carlo internal energy [18]; HNC, the values in the HNC scheme; $\tilde{B}(r)$, those based on Eq. (6) with Eq. (7); $B(r;\zeta)$, those based on Eq. (9). See Fig. 3 for the meaning of IHNC, $\tilde{B}(r)_{IS}$, and $\bar{B}(r)$.

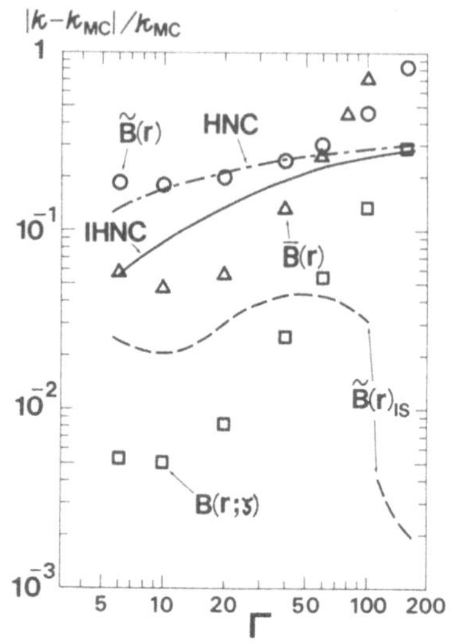

Fig. 5. Relative consistency of the compressibility sum rule in various schemes. κ_{MC} is the isothermal compressibility calculated from the MC equation of state [18]; κ, the isothermal compressibility calculated from the long-wavelength limit of the static structure factor. The meaning of the abbreviated symbols is the same as in Figs. 3 and 4.

Eq. (7) may be more appropriate for a treatment of the strong nonlocality. Such an effective density has been used recently in connection with improvement on the local density approximation in the free-energy functional [16,17].

To rectify the apparent flaw of $\tilde{B}(r)$ in the short-range domain, we tentatively renormalize $\tilde{B}(r)$ according to the ion-sphere model as

$$\tilde{B}(r)_{IS} = [B(0)_{IS}/\tilde{B}(0)]\tilde{B}(r) \quad . \tag{8}$$

We introduce also a further modified scheme in which the real density $n(\vec{r})$ is replaced in Eq. (6) by an averaged density,

$$\bar{n}(\vec{r}) = n + \zeta(n(\vec{r}) - n) \quad . \tag{9}$$

Here the parameter ζ is determined in accord with $B(0)_{IS}$ ($\zeta = 0.65$ at $\Gamma = 160$); the bridge function obtained in this scheme is referred to as $B(r;\zeta)$.

Figure 3 shows the functional behaviors of $\tilde{B}(r)_{IS}$ and $\bar{B}(r)$ at $\Gamma = 160$, together with the bridge function in the improved HNC (IHNC) scheme proposed earlier [15]; the attractive part of B(r) stemming from the lowest-order bridge-diagram contribution in the IHNC scheme is not taken into account in those newly investigated schemes.

In Fig. 4 we compare the internal energies obtained in those newly introduced schemes with the MC results [18] and those of the HNC and IHNC schemes. Figure 5 examines the internal consistency achieved in those schemes in light of the compressbility sum rule [1,2]. Such comparisons appear to indicate importance of the following aspects in achieving a

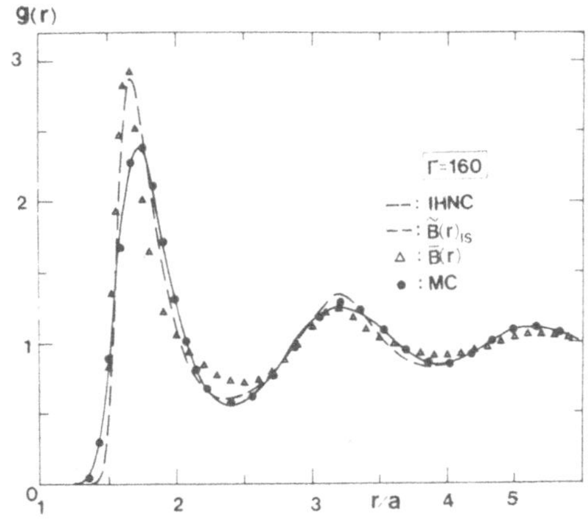

Fig. 6. Radial distribution function of the OCP at $\Gamma = 160$. MC refers to the MC data taken from Ref. 18. See Fig. 3 for the meaning of IHNC, $\tilde{B}(r)_{IS}$, and $\bar{B}(r)$.

significant improvement of the HNC approximation: (i) a renormalization of the real density according to the ion-sphere model ($\bar{B}(r)$vs. $\tilde{B}(r)_{IS}$ and $B(r;\zeta)$), and (ii) an accurate treatment of the long-ranged oscillatory behavior of g(r) in the strong coupling regime ($\tilde{B}(r)_{IS}$vs. $\bar{B}(r)$ and $B(r;\zeta)$).

As far as the thermodynamic properties of the OCP are concerned, the scheme with $\tilde{B}(r)_{IS}$ reproduces the MC results almost exactly over the whole fluid regime. Concerning the two-body correlation functions, however, there still remains a substantial room of improvement even in the scheme with $\tilde{B}(r)_{IS}$; the first peak of g(r) is overestimated appreciably as compared with the MC data (Fig. 6). Further theoretical development appears necessary before we have an accurate, first-principles evaluation of the bridge function without introduction of a free adjustable parameter.

REFERENCES

1. M. Baus and J.-P. Hansen, Phys. Rep. 59, 1 (1980).
2. S. Ichimaru, Rev. Mod. Phys. 54, 1017 (1982).
3. J.-P. Hansen, J. Phys. (Paris) Suppl. 41, C2-43 (1980).
4. D.J. Stevenson, J. Phys. (Paris) Suppl. 41, C2-53 (1980).
5. J.-P. Hansen and I.R. McDonald, Theory of Simple Liquids (Academic, New York, 1976).
6. R. Evans, Adv. Phys. 28, 143 (1979).
7. H. Iyetomi, Prog. Theor. Phys. 71, 427 (1984).
8. T. Morita, Prog. Theor. Phys. 23, 829 (1960).
9. B. Brami, J.-P. Hansen and F. Joly, Physica A 95, 505 (1979).
10. B.J. Alder, E.L. Pollock and J.-P. Hansen, Proc. Natl. Acad. Sci. USA 77, 6272 (1980).
11. H. Iyetomi and S. Ichimaru, Phys. Rev. A, 34, 433 (1986).
12. B.J. Alder and E.L. Pollock, Nature 275, 41 (1978).
13. H. Iyetomi and S. Ichimaru, Phys. Rev. A, to be published.
14. K.C. Ng, J. Chem. Phys. 61, 2680 (1974).
15. H. Iyetomi and S. Ichimaru, Phys. Rev. A 25, 2434 (1982); A 27, 3241 (1983).
16. P. Tarazona, Phys. Rev. A 31, 2672 (1985).
17. W.A. Curtin and N.W. Ashcroft, Phys. Rev. A 32, 2909 (1985).
18. W.L. Slattery, G.D. Doolen and H.E. DeWitt, Phys. Rev. A 26, 2255 (1982).

SELF CONSISTENT MODEL FOR TUNNELING ACROSS A ONE DIMENSIONAL BARRIER IN A MANY ELECTRON SYSTEM

J. W. Halley
School of Physics and Astronomy
University of Minnesota
Minneapolis, Minnesota 55455

David Dahl
Physics Department
St. Olaf College
Northfiled, Minnesota 55057

We consider a model with two metals (modeled for the present as jellium) separated by a barrier described by a potential $V(x)$. We show that the current density j can be fixed by adding a term H' to the energy which is of the form:

$$H' = -m \int_{-\infty}^{\infty} v(x) j(x) d^3 x$$

where $v(x)$ is a Lagrange multiplier function which depends on the position x. We show that $v(x)$ may be interpreted as the hydrodynamic velocity of the electrons in the junction so that $v(x)$ depends on the electron density $n(x)$ as $j = < n(x) > v(x)$. The expectation value of $j(x)$ is independent of x by current conservation. The computational part of the tunneling problem is thus reduced to the self-consistent solution of the Schroedinger equation with one additional term:

$$(H_o - mv(x)^2/2)\phi_p(x) = E_p \phi_p(x)$$

This equation is solved self-consistently with boundary conditions appropriate to waves incident from the left and right (which form an orthonormal complete set when appropriately normalized). The result shows that, within this model, a finite current can pass through the barrier with no electrostatic potential drop. We will argue that this result does not depend on the local density approximation (which is the only computational approximation in the model) but is to be expected because there is no dissipative mechanism in the model.

INTRODUCTION

Theoretical models of the tunneling of electrons from one metal normal electrode to another through a potential barrier have been studied for many years. The system is often described by a "transfer Hamiltonian model" [1] of the form $H_L + H_R + H_T$ in which the first two terms describe the two electrodes at infinite separation and the last term is intended to describe transitions of electrons between the electrodes. H_T is often treated as a perturbation. To describe the imposition of an electrostatic potential difference V across the barrier between the electrodes, one supposes that the one electron eigenstates of one of the electrodes are shifted by V relative to those of the other electrode. Such a model is the basis of the interpretation of tunneling of electrons across an oxide barrier between two normal electrodes and also of the present understanding of the tunneling electron microscope.

At least two elements are missing from such a theoretical model: In the first place there is no provision for energy dissipation from the electron system to the lattice in this model, though in a normal metal system such dissipation surely takes place when a current flows across the junction. The treatment of dissipation in quantum tunneling problems has recently received a lot of theoretical attention [2], but mainly in the context of of the tunneling associated with a macroscopic quantum variable, as in the Josephson effect, whereas we will be interested here in tunneling between normal metals in which no such macroscopic quantum degree of freedom exists. Secondly, the transfer Hamiltonian does not treat the Coulomb interactions between the electrons in a fully self-consistent manner, even if self-consistent bands are used to describe each electrode separately.

In this paper we will address the question of whether these omissions in the standard model affect the results qualitatively by carefully formulating the self-consistent problem of finding the current voltage characteristic of a junction in a simple model. We will find that if the problem is done self consistently then the electrostatic potential drop across the junction can ONLY arise from dissipation.

MODEL AND TECHNIQUE FOR FIXING THE CURRENT

For our detailed calculations we will consider two semi-infinite jellium slabs separated by a potential barrier assumed to be a function only of the coordinate x normal to the jellium faces. If the space between the jellium is empty space then the barrier arises from the fact that the uniform positive background of the jellium is absent in the space between the jellium. Such a model has been studied by J. Smith and coworkers as a preliminary model for cohesion between metals [3].

In order to calculate the current voltage characteristic in such a model, one can either fix the electrostatic potential drop across the junction and attempt to calculate the current or conversely fix the current and calculate the potential drop. The former point of view is usually taken in textbook, single particle treatments of the problem [1]. On the other hand if the problem is done in an electrostatically self-consistent way, then this approach is inconvenient and we will take the second point of view, fixing the current j and calculating the potential drop.

In the study of transport in bulk metals, the current is fixed by "shifting the fermi surface" in order to produce a current carrying state. This is easily shown to be equivalent in the case of bulk metals to fixing the current by use of a Lagrange multiplier which we will call mv so that the effective Hamiltonian becomes $H - mv\int j(x)d^3x$ where $j(x)$ is the current operator. The value of v is fixed at the end of the calculation by requiring that $< j(x) >= j$ where $< \cdots >$ means the expectation value in the current carrying state. One shows easily that this requirement leads to $j =< n > v$ where n is the electron density in the current carrying state. Thus v has the significance of the hydrodynamic velocity of the electrons in the metal.

We wish to fix the current in our model of a tunnel junction in essentially the same way. If one is to assure current conservation, however, one cannot assume that the Lagrange multiplier mv is independent of the coordinate x normal to the surface of the "metals". Instead, we introduce a Lagrange function $mv(x)$ which depends on this coordinate. Then current conservation in the form $< j(x) >= constant$ can be assured despite the fact that the density $< n > (x)$ varies and (as will be shown below) the current is still related to $v(x)$ through the relation $j =< n > (x)v(x)$. Thus we wish to find a state which minimizes the value of

$$< H > - \int mv(x) < j(x) > d^3x$$

in which $v(x)$ is to be determined by the requirement $< j(x) >= j(constant)$ at the end of the calculation.

CURRENT CARRYING STATE

Within this formulation we can attempt to find the Slater determinant

$$\Psi = \mathcal{A} \prod \psi_p$$

which minimizes the functional

$$< \Psi| \left\{ H - \int mv(x) j(x) d^3x \right\} |\Psi > .$$

In our calculations we will make the further, usual assumption that the effects of exchange can be adequately represented by a functional of the electron density

$$< n(x) >= \sum_{p,filled} |\psi_p|^2.$$

Then in the usual way, the current is

$$< j(x) >= \sum_p h/2mi \left\{ \psi_p^* \partial \psi_p / \partial x - c.c. \right\}. \tag{1}$$

Requiring stationarity of the functional $< H > - \int mv(x) < j(x) > d^3x$ then gives

$$\left\{ -h^2/2md^2/dx^2 + V_{eff}(n(x)) - (h/2i)dv/dx - (h/i)vd/dx \right\} \psi_p = E_p \psi_p$$

where in the usual way, the eigenvalues E_p have been introduced to require normalization of the ψ_p. $V_{eff}(n(x))$ takes account of classical coulomb interactions as well as exchange and makes the problem a self-consistent one.

The one electron Schroedinger equation which we have found for the one electron functions ψ_p contains two new terms which involve $v(x)$ and which are odd under reversal of the sign of x. These terms arise from the requirement that the state to be found carry current. The one electron equation is considerably simplified by introducing the new functions ϕ_p which are related to ψ_p by

$$\psi_p(x) = e^{im/h \int^x v(x')dx'} \phi_p(x). \tag{2}$$

This is a gauge transformation, corresponding to a local boost with the velocity $v(x)$. A short calculation shows that

$$\left\{ -h^2/2md^2/dx^2 + V_{eff}(n(x)) - mv(x)^2/2 \right\} \phi_p = E_p \phi_p. \tag{3}$$

The effective Hamiltonian for the ϕ_p contains only the term $-mv(x)^2/2$ arising from the requirement that the state be current carrying. This term is invariant under inversion of x. As a consequence, unless there is a spontaneous breaking of this inversion symmetry in the lowest energy current carrying state, we do not expect any net current to be contributed by the part of the wave function associated with the ϕ_p (and we do not find any such contribution in our detailed calculations described below). With this in mind it is elementary to calculate the expectation value of the current $< j(x) >$ using the expressions (1) and (2) with the result that $< j > (x) =< n > (x)v(x)$ so that $v(x)$ is the hydrodynamic velocity of the electrons as claimed above. One sees that it is possible then to think of the electrons in the junction as speeding up in the center of the junction where the density is low in order to satisfy the equation of continuity, much as a fluid speeds up in a constriction in a pipe.

To use this formulation to calculate characteristics of this current carrying state, we took the case of identical slabs of jellium on each side of the barrier. We then computed a complete, orthonormal set of eigenfunctions of the equation for ϕ_p which satisfy the following boundary conditions:

$$\phi_p^{(-)} \rightarrow \begin{cases} N_{k_R} T_{k_R}^{-1}(e^{ik_L x} + R_{k_L} e^{-ik_L x}), x \rightarrow -\infty \\ N_{k_R} e^{ik_R x}, x \rightarrow \infty \end{cases}$$

$$\phi_p^{(+)} \rightarrow \begin{cases} N_{k_L} e^{-ik_L x}, x \rightarrow -\infty \\ N_{k_L} T_{k_L}^{-1}(e^{-ik_R x} + R_{k_R} e^{-ik_R x}), x \rightarrow \infty \end{cases}$$

In more general cases, the states must include bound states of the potential if any exist and states confined to one side of the barrier if the two metals are dissimilar. Notice that determination of the states with the boundary conditions above requires the calculation of the reflection and transmission coefficients which are often stated to be interpretable in a way which gives the resistance of the barrier. We will find on the contrary that these transmission and reflection coefficients (which we have calculated in order to obtain the results given below) do not directly determine the resistance of the barrier.

The states are calculated numerically. For example, for states of type $\phi_p^{(-)}$ we start on the right with a plane wave propagating to the right and integrate the Schroedinger equation (3) to the left until we are well past the barrier on the left and then fit to the asymtotic form on the left, finally normalizing to get $N_k R$. When enough states have been calculated so that the positive background on both sides of the barrier can be neutralized by filling them, we compute the corresponding density and recompute $V_{eff}(n(x))$, continuing to self-consistency. States are filled with a weighting which includes the transverse momentum as in Lang and Kohn's calculation [4] so that in that sense the calculation is quasi-three dimensional. Standard techniques are used to solve the Poisson equation. We find that it is necessary to mix a (rather small) amount of the new potential with the old one in order to prevent oscillations and achieve convergence. The solutions converge to self-consistency in about 90 iterations. In the results shown we have used a form for $V_{eff}(n(x))$ which includes an approximate account of correlation as well as exchange (as in standard local density approximation calculations). Because the orbitals ψ_p are not members of a Slater determinant in the Kohn Sham formulation of the many electron problem, this means that, strictly speaking, our expression for the current is not quite right. We are very well convinced that the results would not be qualitatively different if, for example, we had used a Slater approximation for the exchange in $V_{eff}(n(x))$ and had not attempted to account for

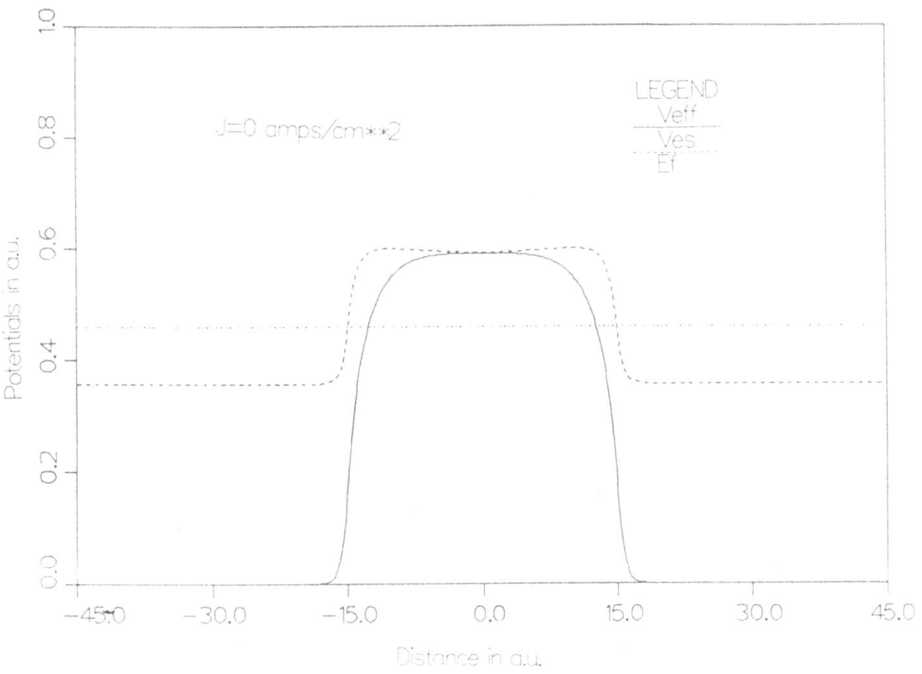

Figure 1. Self-Consistent solution to equation (3) in the case j = 0. The two jellium slabs were taken to be 30 au apart and to have electron density corresponding to $r_s = 2$. The top curve is the electrostatic potential and the lower one is Veff. Ef is the horizontal line.

correlation at all, though this remains to be checked explicitly. In the latter case, the orbitals ψ_p could be interpreted as members of a Slater determinant and the expression for the current would be exactly right.

Results are shown in Figures 1-2. The first figure shows the effective potential as well as its electrostatic part when the current is zero. Figure 2 shows the corresponding result when the current is finite. We note that both the electrostatic potential and the entire effective potential have changed shape as a consequence of the fact that we are forcing current through the barrier. On the other hand, and contrary to our initial expectations, there is no difference between the electrostatic potential far to the left and far to the right. We believe this result to be independent of many of the detailed assumptions in our calculations and we will discuss its possible significance in the next section.

DISCUSSION

In the calculation described at the end of the last section, we constructed a variational state which carried a current, which was as good an approximation to an exact eigenstate for our simple problem as many state of the art electronic structure calculations and in which the current flowed across the barrier in the absence of an electrostatic potential drop so that the barrier appeared to have zero resistance. Since this result is clearly at variance with experiments on real junctions between normal metals, we must discuss whether the fault is in the model or in the approximations

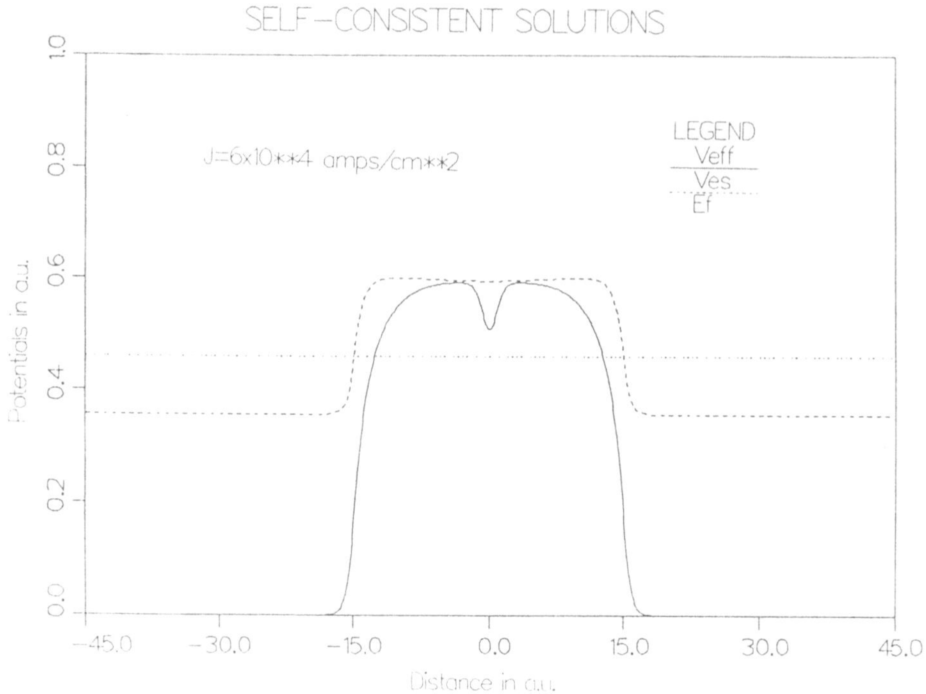

Figure 2. The same quantities shown in Figure 1, but with $j = 10^5 amps/cm^2$.

used in the calculations.

We present the following argument, which suggests to us that the problem is in the model and not in the approximations: We can apply the variational principle to the time independent Schroedinger equation without assuming any form for the many body wavefunction $\Psi(\vec{r}_1, ..., \vec{r}_N)$. (Ψ cannot, of course, be an eigenfunction of the current but we can require that $< \Psi|j|\Psi >= j$.) Then essentially the same calculations described above lead to the N-electron Schroedinger equation

$$(H - \sum_i mv(x_i)^2/2)\Phi = E\Phi$$

where

$$\Psi = exp\sum\nolimits_i \int^{x_i} mv(x_i')/hdx_i' \Phi.$$

Unless the symmetry is broken, there can again be no current associated with Φ, leading to the hydrodynamic relation $j =< n > v(x)$. But it is clear that this Schroedinger equation for Φ will lead to an electrostatic potential drop which is independent of the current because it is invariant under reversal of the direction of the current. Thus it appears that the exact solution will have the same qualitative characteristics as the approximate solution described in the preceding section. One may ask whether solutions to the equation

$$H - m \int v(x)j(x)dx\Psi = E\Psi$$

exist for all j. We have only numerical evidence on this point at present. The solutions found using the parameters shown in the figures appear numerically to be very stable.

If the fault is in the model then we may consider several possibilities: If the model is not strictly one dimensional, then the constraints of the equation of continuity become less compelling because current can leak away in the transverse directions from any finite region. This resolution appears, however, not to be consistent with what is generally thought to be the relevant experimental problem. It seems more likely that the omission of inelastic scattering leading to energy dissipation is the key ingredient of the experiments that is missing. It is clear on elementary grounds that if a current I is flowing and there is a potential drop V then heating is occuring at rate IV. While such heating could occur as a result of heating the electron gas, it is widely believed that the heat is dissipated through the lattice. Indeed, if inelastic scattering is very rapid, then electrons which, in a transfer Hamiltonian formulation, traverse the barrier once might inelastically scatter on the other side before they could traverse the barrier again and again as they would have to do in perturbation theory in order to describe a fully self-consistent solution to the problem. We think that a resolution of the apparent conflict between the present point of view and more traditional views of tunneling may possibly lie in this direction.

ACKNOWLEDGMENTS

We are grateful to H. Shore for helpful discussions of this work. Argonne National Laboratory is thanked for hospitality while this paper was written.

REFERENCES

1. J. A. Appelbaum and W. F. Brinkman, Phys. Rev. **186**, 464 (1969); Phys. Rev. B **6**, 907 (1970); C. B. Duke, G. C. Kleinman and T. E. Stakelon, Phys. Rev. B **6**, 2389 (1972); C. Caroli, R. Combescot, P. Nozieres and D. Saint James, J. Phys. C **4**, 916 (1971); **4**, 2589 (1971); T. E. Feuchtwang, Phys. Rev. B **10**, 4121 (1970); Phys. Rev. B **11**, 4135 (1974).
2. A. O. Caldeira and A. J. Leggett, Phys. Rev. Lett. **46**, 211 (1981); S. Chakravarty, Phys. Rev. Lett. **49**, 681 (1981); A. J. Bray and M. A. Moore, Phys. Rev. Lett. **49**, 1546 (1982); A. Schmid, Phys. Rev. Lett. **51**, 1506; A. O. Caldeira and A. J. Leggett, Ann. Phys. **149**, 374 (1983); S. Chakravarty and A. J. Leggett, Phys. Rev. Lett. **52**, 5 (1984); L. -D. Chang and S. Chakravarty, Phys. Rev. Lett. B **29**, 130 (1984).
3. J. Ferrante, J. R. Smith and J. H. Rose, General Motors Research Laboratories Publication GMR-4714 (1984).
4. N. D. Lang and W. Kohn, Phys. Rev. B **1**, 4555 (1970).

ELASTICITY OF CRYSTALS AND QUASICRYSTALS

Marko V. Jarić*

Lyman Laboratory of Physics
Harvard University
Cambridge, MA 02138

and

Department of Chemistry
Boston College
Chestnut Hill, MA 02167

INTRODUCTION

The last decade brought a surge of interest into density functional theories in statistical physics. In particular, the theory of solidification put forward by Ramakrishnan and Yussouff[1] was quite successful in predicting solidification parameters for hard sphere[2] (HS) and Lennard-Jones[3] (LJ) fluids. Similar theories were also used to study submonolayer phases of rare gases on graphite,[4] the glass transition,[5] and most recently the stability of icosahedral quasicrystals.[6] The accuracy of these theories can be controlled by a truncation of an expansion around a reference uniform state. The only assumptions enter through a description of this state. Consequently, Ramakrishnan[7] and several other authors suggested that a similar approach can be used for "first principle" calculations of such quantities as defect energy, liquid-solid interface,[8] elastic moduli,[9] etc.

We shall review here a density functional method for calculating elastic moduli of crystals and quasicrystals. A novelty of this calculation is in the allowance for the relaxation of the strained crystal density. Although our main interest is in quasicrystals, we shall first consider elastic moduli of HS, LJ, and (hypothetical) cobalt FCC crystals.[10] Then, we shall turn to the elastic moduli of a hypothetical cobalt icosahedral quasicrystal.[11]

---

*On leave from Department of Physics, North Dakota State University, Fargo, ND 58105. Address after Sept. 1, 1986: Center for Theoretical Physics, Texas A&M University, College Station, TX 77843.

Because of the lack of relevant experimental and theoretical results we could not make an independent check of calculated crystal moduli. However, the results which are available at different points in the phase diagrams indicate that our moduli are of correct order of magnitude. A particularly interesting result of our investigation is the discovery that the HS crystal has negative Poisson ratio at the melting point. The Poisson ratio is driven negative precisely by the above mentioned density relaxation which has purely geometric (entropic) origin in that the static free volume of a sphere opens up in the directions perpendicular to the stretching direction.

Extending calculations for crystals, we evaluated the elastic modulus tensor for icosahedral Co.[11] Our starting point was the density functional determination of icosahedral quasicrystalline structure,[6] in which the quasicrystal is represented as a cut through a six-dimensional crystal. Consequently, it was necessary to define a six-dimensional elastic modulus tensor.[12] We found that the phonon part of the tensor is positive, while the full tensor has one negative eigenvalue revealing an elastic instability and a spontaneous deformation of the icosahedral quasicrystal. If this deformation is small, it will be generated by a phonon-phason eigenstrain associate with the negative eigenvalue. The most likely (minimal) reduction of the icosahedral symmetry due to such spontaneous strain would be to D_5 or D_3 symmetry.[13] Diffraction patterns of so distorted quasicrystals reveal systematic shifts of the spots away from the ideal icosahedral positions, in good qualitative agreement with some recent experiments.[14] Therefore, the observed shifts might reflect an elastic instability rather than inhomogeneities caused by the growth process.[14]

CRYSTAL ELASTIC MODULI

Let us assume a small strain ϵ to be defined on the surface of the solid so that a point at x is moved to $x_\epsilon = (1 + \epsilon)\cdot x$ and the volume V is changed to $V_\epsilon = V/\|1 + \epsilon\|$ while the number of particles in the solid N_S and the temperature T are kept constant. Isothermal elastic modulus tensor C of a solid can now be defined by expanding the Helmholtz potential of the strained solid around the unstrained solid to second order in the strain tensor.

The density functional approach[1] allows a calculation of strained and unstrained densities [$n_\epsilon(x)$ and $n(x)$] and potentials [H_ϵ and H]. The grand potential G, which is Legendre transform of H, is the minimum of a variational potential W whose "energy" contribution is expanded around that of a reference liquid:

$$(p_L - p_S)V = G(\mu_S,V,T) - G(\mu_L,V,T) = \min_{n(x)}\Delta W \tag{1}$$

and

$$\Delta W = -N_S(\mu_S - \mu_L) - (N_S - N_L) + \int_V n(x)\ln\frac{n(x)}{n_L} \tag{2}$$

$$- \frac{1}{2}\int_V\int_V [n(x) - n_L]c_L(|x-y|)[n(y) - n_L] + \ldots,$$

where all energies are in units k_BT, μ denotes chemical potential, and subscripts S and L denotes the solid and the liquid, respectively. The direct pair correlation function of the liquid is c_L and its Fourier ransform is related to the liquid structure factor $s_L = 1/(1-c_L)$ which must be given as an input into the calculation.

An excellent approximation can be obtained for a simple solid by expanding its density into Gaussians centered at the sites of its Bravais lattice. With this Ansatz the density $n(x) \equiv n(x;n_S,a,\alpha)$ depends on the average solid density n_S, the 3x3 matrix of lattice constants a, and the symmetric 3 x 3 Gaussian-width matrix α. Consequently, the functional ΔW becomes a function of these parameters.

The free energy of a strained crystal is determined by its equilibrium density $n_\epsilon(x)$. If the crystal was a uniform continuum then $n_\epsilon(x)$ would be simply given by $n[(1+\epsilon)^{-1}\cdot x]/\|1+\epsilon\| = n[(1+\epsilon)^{-1}\cdot x;n_S,a,\alpha]/\|1+\epsilon\|$. However, this formula does not generally hold below the unit cell scale. Consequently,

$$n_\epsilon(x) = n[(1+\epsilon)^{-1}\cdot x;n_S,a_\epsilon,\alpha_\epsilon]/\|1+\epsilon\|, \tag{3}$$

where $a_\epsilon \equiv a + \Delta a$ and $\alpha_\epsilon \equiv \alpha + \Delta\alpha$ are to be determined by minimizing ΔW_ϵ given by (2) with n(x) substituted by $n_\epsilon(x)$, V by V_ϵ, and μ_S by $\mu_{S\epsilon}$. The chemical potential $\mu_{S\epsilon}$ of the strained solid is fixed by the condition that the parameter n_S should indeed minimize ΔW_ϵ.

A nonzero Δa corresponds to a change in the number of defects as the solid is strained. Due to long characteristic times for diffusion of vacancies and interstitials it is probably a good approximation to assume that in a realistic experiment this number is constant. We shall therefore take $a_\epsilon = a$. On the other hand, a nonzero $\Delta\alpha$ roughly corresponds to a change in the mean-square displacements of the atoms and it may be an mportant, previously neglected effect.[15]

It is generally not a simple task to minimize ΔW_ϵ with respect to six components of α_ϵ. However, since to lowest order $\Delta\alpha \propto \epsilon$, the elastic modulus C can be determined from

$$E_{el} = \frac{1}{2}\epsilon:C:\epsilon = \min_{\Delta\alpha}(\frac{1}{2}\epsilon:C_0:\epsilon + \Delta\alpha:C_1:\epsilon + \frac{1}{2}\Delta\alpha:C_2:\Delta\alpha), \tag{4}$$

where the fourth rank tensors C_0, C_1, and C_2 can be directly evaluated by expanding[10] ΔW_ϵ. A minimization in (4) trivially leads to the elastic modulus C.

QUASICRYSTAL ELASTIC MODULI

Density of an incommensurate crystal can be always represented as a cut through a higher dimensional periodic crystal. The physical, experimentally relevant icosahedral quasicrystal density $n^{\|}(x^{\|})$ can be viewed as a particular three-dimensional cut through a six-dimensional simple-cubic crystal density $n(x^{\|},x^{\perp})$,

$$n^{\|}(x^{\|}) = n(x^{\|},x^{\perp} = 0), \tag{5}$$

where the superscripts $\|$ and $\perp$ denote the three-dimensional physical subspace and its three-dimensional orthogonal complement, respectively. It can also be shown that minimization of (1) with respect to $n^{\|}(x^{\|})$ is formally equivalent to minimization of a six-dimensional ΔW with respect to $n(x^{\|},x^{\perp})$. The hypothetical six-dimensional and the physical liquid structure factors must be related as $c_L(Q^{\|},Q^{\perp}) = c_L^{\|}(Q^{\|})$.

The required quasicrystal minimization of ΔW was performed using the Gaussian expansion of the density in six dimensions and $c_L^{\|}$ identified with the structure factor of amorphous Co.[6] It was found that the quasicrystal is more stable than the reference liquid. The diffraction pattern determined from the calculated quasicrystal density was in good qualitative agreement with experiments on real alloys.

It can be easily verified that ΔW is invariant under displacements $x \to x + u$. The $u^{\|}$ component of u is the usual physical displacement whereas the component $u^{\perp}$ corresponds to relative displacement of the density waves. Consequently, the two displacements may be called phonon and phason displacements, respectively. Small $u^{\|}$ and $u^{\perp}$ label independent degenerate densities. Elasticity viewed as a long wavelength limit can be considered by allowing (uniform) $x^{\|}$ dependence in u: $u^{\|} = \epsilon^{\|,\|}\cdot x^{\|}$ and $u^{\perp} = \epsilon^{\perp,\|}\cdot x^{\|}$. Therefore, the same calculation as outlined for elasticity in three dimensions can be applied in six dimensions. One must only keep in mind that the only non-zero blocks of the six-dimensional ϵ are the usual physical symmetric strain $\epsilon^{\|,\|}$ and the new strain $\epsilon^{\perp,\|}$.

Consequently, the "elastic" modulus tensor C can be considered as a 15x15 matrix. Either through our direct calculation or using group theory it can be verified that for icosahedral quasicrystals C has only five independent components. The usual elastic energy $\frac{1}{2}\epsilon^{\|,\|}\cdot C^{\|,\|;\|,\|}\cdot\epsilon^{\|,\|}$ must be isotropic due to high icosahedral symmetry. Consequently, this block of C, which is the usual elastic modulus tensor, has only two independent components. The block $C^{\perp,\|;\perp,\|}$ has also only two independent components, but their tensorial structure is more complicated. Finally, the block $C^{\perp,\|;\|,\|}$ has only a single component and its tensorial structure is also nontrivial.[11] By contraction of appropriate components of C with the physical wavevector $q^{\|}$ one can obtain the phonon and phason energy.[2] All our conclusions about the number of independent components and their explicit tensorial structure agree with Ref. 12.

Diagonalization of C can be accomplished by observing that the fifteen-dimensional representation of Y spanned by ϵ decomposes into irreducible components according to: 15=5+5+4+1. Corresponding four eigenvalues are (in arbitrari units) -1724.6, 3163.1, 456.2, and 3078.9. Hence, the five-dimensional subspace associated with the first eigenvalue is unstable indicating that a "strained-icosahedral" rather than icosahedral density minimizes ΔW.

The new stable structure can be easily determined when the equilibrium distortion of the hypercubic lattice is small. In this case the "spontaneous" strain ψ can be assumed to lay in the five-dimensional eigenspace of the negative eigenvalue. ψ minimizes the Landau free energy $F(\psi) = -1724.6|\psi|^2 + \ldots$, whose Y-invariant cubic and quartic terms can be determined by expanding ΔW_ϵ. However, even before such complicated expansion and actual minimization are performed, using only group theory we can determine the most likely symmetries of the spontaneous strain.[16] These are the maximal isotropy subgroups of the five-dimensional irreducible representation of Y. In this way we obtain D_3 and D_5.[13] It turns out that the symmetries D_3 and D_5 completely fix the strain ψ except for its magnitude. Therefore, we can calculate corresponding strained densities and their diffraction patterns.

CONCLUSIONS

To summarize, we have formulted a density functional theory of elasticity. Our principal results are the negative Poisson ratio for HS solid and a martensitic instability of icosahedral Co. The first result calls for an independent verification, e.g., by Monte Carlo simulations. In order to check the accuracy of our approach it will be also necessary to extend our single point calculation for LJ solid to the entire liquid-solid transition line. The future work should, furthermore, explore potentially important contribution from the three-point correlations neglected in the lowest order expansion in (2).

The second result raises the question whether the icosahedral quasicrystalline structure is intrinsically unstable with respect to some other structure, or whether it is destabilized only at a sufficiently low temperature at which a martensitic transition would occur. This points out to the urgency of further density- functional calculations based on more realistic structure factors of two-spicies supercooled liquids. A careful experimental determination of the temperature dependence of the shifts which could discriminate quenched strains is also desirable.

ACKNOWLEDGEMENTS

This report is based on a collaboration with U. Mohanty and D. R. Nelson. I also acknowledge conversations with B. Halperin, G. Jones, T. V. Ramakrishnan and D. Turnbull. This work has been supported by the Petroleum Research Fund and NSF through the Harvard Materials Research Laboratory and grants DMR 85-14638 and CHE 85-11728.

REFERENCES

1. T. V. Ramarkrishnan and M. Yussouff, Phys. Rev. B19, 2775 (1979).
2. A. D. J. Haymet, J. Chem. Phys. 78, 4641 (1983); G. L. Jones and U. Mohanty, Mol. Phys. 54, 1241 (1985); P. Tarzona, Phys. Rev. A31, 2672 (1985).
3. C. Marshall, B. B. Laird, and A. D. J. Haymet, Chem. Phys. Lett. 122, 320 (1985); W. A. Curtain and N. W. Ashcroft, Phys. Rev. Lett. 56, 2775 (1986).
4. D. K. Fairbonet, W. F. Saam, and L. M. Sander, Phys. Rev. B26, 179 (1982); L. M. Sander and J. Hautman, Pnys. Rev. B29, 2171 (1984).
5. M. Baus and J.-L. Colot, J. Phys. C19, L135 (1986).
6. S. Sachdev and D. R. Nelson, Phys. Rev. B32, 4592 (1985).
7. T. V. Ramakrishnan, Pramana xx, xxxx (1983).
8. A. D. J. Haymet and D. W. Oxtoby, J. Chem. Phys. 74, 2559 (1981); D. W. Oxtoby and A. D. J. Haymet, J. Chem. Phys. 76, 6262 (1982).
9. M. D. Lipkin, S. A. Rice, and U. Mohanty, J. Chem. Phys. 82, 472 (1985).
10. M. V. Jarić, U. Mohanty, and D. R. Nelson, to be published.
11. M. V. Jarić and U. Mohanty, to be published.
12. D. Levine et al., Phys. Rev. Lett. 54, 1520 (1985); P. Bak, Phys. Rev. B32, 5764 (1985); P. A. Kalugin, A. Yu. Kitayĕv, and L. S. Levitov, J. Physique Lett. 46, L-601 (1985); T. C. Lubensky, S. Ramaswamy, and J. Toner, Phys. Rev. B32, 7444 (1985).
13. D_2 also corresponds to minimal reduction. However, a Landau theory analysis shows it inaccessible [O. Biham, D. Mukamel, and S. Shtrikman, (unpublished)].
14. P. A. Bancel and P. Heiney in International Workshop on Aperiodic Crystals, edited by D. Gratias and L. Michel, J. Physique Colloq. 47, C3-341 (1986); T. C. Lubensky et al. (unpublished).
15. The effects of $a_\epsilon \neq a$ and scalar $\Delta\alpha$ are independently studied by G. L. Jones (unpublished).
16. M. V. Jarić, Phys. Rev. Lett. 51, 2073 (1983).

SUPERFLUIDITY IN ^3He FILMS

Oriol T. Valls

School of Physics and Astronomy
University of Minnesota
Minneapolis, MN 55455

Zlatko Tesanovic

Department of Physics
Lyman Laboratory
Harvard University
Cambridge, MA 02138

ABSTRACT

Superfluid ^3He films in the thickness regime $k_F^{-1} \ll d < \xi$ are considered. We find that for smooth boundaries the transition temperature decreases with thickness and, contrary to what is found for s-wave pairing, exhibits only weak oscillatory behavior. Other quantities, such as Nuclear Magnetic Resonance frequency shifts, however, show discontinuous jumps. Diffusive boundary scattering has a qualitative influence in the phase diagram: for a specified degree of surface roughness there is a critical thickness d_c below which superfluidity disappears. As the thickness decreases towards d_c an intermediate region of gapless superfluidity is reached. This region appears to be experimentally accessible. Properties of the gapless state are briefly discussed.

Since its discovery fifteen years ago,[1] superfluidity in ^3He has been one of the most fascinating fields of study in condensed matter physics. Only very recently, however, have advances in experimental techniques brought about the possibility of studying two-dimensional superfluidity in ^{3}He films. Preliminary,[2] although unconfirmed, reports of superflow in such films are encouraging. The tensorial nature of the order parameter in superfluid ^{3}He hints at even greater richness and variety of two-dimensional phenomena than found in superfluid ^{4}He, its boson counterpart.[3]

Here we discuss some of the properties we expect to be major characteristics of the two-dimensional superfluid state in ^{3}He. We will consider the regime where the film thickness d is smaller than the superfluid coherence length ξ while remaining considerably larger than the interparticle separation, i.e., than the inverse Fermi wavevector k_F^{-1}. In this regime we can assume that the nature of the quasiparticle interactions in the normal state is essentially the same as in the bulk liquid. Since ξ is several hundred Å even at T = 0, while $k_F^{-1} \sim 1$ Å, the region considered is quite wide.

We will see that boundary scattering plays an important role in determining the superfluid properties in this region. If the boundaries are atomically smooth, quantum size effects become observable in several superfluid properties. These effects are due to the quantization of the quasiparticle momentum in the direction normal to the film. For purely specular scattering one can estimate that quantum size effects will be observable if the thickness d is $d < \frac{2 E_F}{\omega_o} \pi k_F^{-1}$ where E_F is the Fermi energy and ω_o the spin-fluctuation frequency.[4] This would correspond to about 40 layers of ^{3}He.

Quantum size effects can be investigated by solving the Gorkov equations for a p-wave superfluid in the appropriate geometry. Here we summarize some of the results. A more extensive discussion is given in Ref. (5). The transition temperature T_c and the gap parameter $\Delta(T = 0)$ are shown in Fig. 1 as functions of thickness, normalized to their bulk values. Both decrease when the thickness is reduced even for the case of a specularly reflecting boundary. This is due to the combined effect of zero-point motion of the quasiparticles and the angular dependence of the attractive effective interaction in the p-channel. The behavior of $T_c(d)$ in ^{3}He films is qualitatively different from that found in ordinary superconductors (s-wave pairing) which is also shown in the inset of Fig. 1. The oscillations in T_c in the latter case are due to the change in the number of "subbands" resulting from the discretization of the transverse component of the momentum. As each liquid layer fills

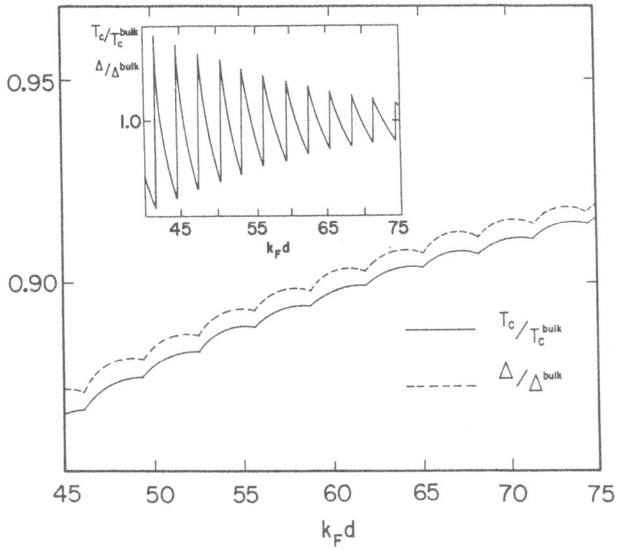

Fig. 1. The transition temperature T_c and the superfluid gap Δ of the ^{3}He film as functions of $k_F d$ ($k_F = 0.785\ A^{-1}$).

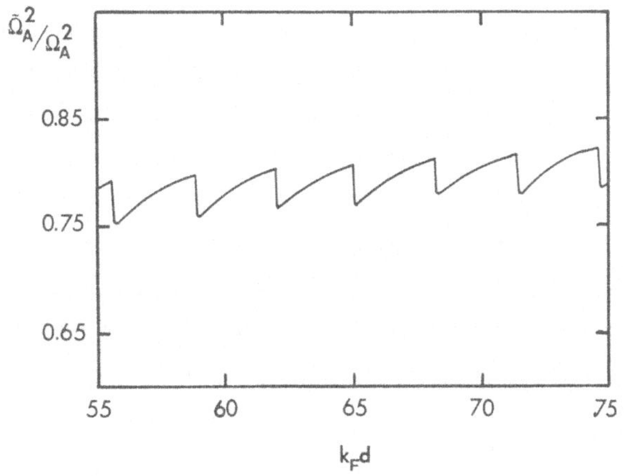

Fig. 2. The shift ^2_A in the transverse resonance frequency. Note that $\tilde{\Omega}^2_A$ (Ω^2_A) are used here without the factors $(1 - T/T_c)$ $((1 - T/T_c^{bulk}))$; they are to be multiplied by these factors in order to obtain the standard measurable frequency shifts.

up and this number changes, discontinuous jumps in T_c are caused in the s-type case. This effect has been observed experimentally in thin superconducting films.[6] In the p-wave case, the effective interaction and the gap vanish in the subband appearing at the pole of the Fermi level as one new layer is started. This results only in a discontinuity in the derivative of $T_c(d)$, rather than in $T_c(d)$ itself. The curve $T_c(d)$, therefore, appears rather featureless, and since both diffusive boundary scattering and quasiparticle interactions in ^{3}He will tend to wash out somewhat the subband structure, the transition temperature structure will be very hard, if not impossible, to observe, except for the overall downward trend. However, the shift in the transverse Nuclear Magnetic Resonance (NMR) frequency appears much more promising in this respect. Results for the A phase (which we have assumed is stable in thin films) are shown in Fig. 2, where we plot the ratio $\tilde{\Omega}_A^2/\Omega_A^2$ ($\tilde{\Omega}_A$ and Ω_A are the film and bulk shifts, respectively). These shifts are calculated using the two-dimensional generalization of Leggett's methods.[7] The jumps seen in Fig. 2 are due to the discontinuous jumps in the density of states. Their magnitude is about 5% of the overall shift and well within experimental resolution. Diffusive scattering and interactions will broaden the discontinuities but the qualitative form of $\tilde{\Omega}_A^2$ (d) will still be described by Fig. 2.

We now turn to the effect of the diffusive component of boundary scattering, induced by irregularities in the substrate. It is well-known[8] that any type of disorder is pair-breaking in the p-wave case. Therefore, it is very important to understand the qualitative and quantitative effects of boundary roughness. As we will see, these effects on the properties of superfluid ^{3}He films are overwhelmingly determined by the pair-breaking nature of random boundary scattering.

We model a film on an irregular substrate by a film between two impenetrable plates the distance between which is a function of position in the x-y plane (the plane of the film). We write $d(x,y) = d + w(x,y)$, where d is the average thickness and $w(x,y)$ is a Gaussian random function satisfying:

$$\langle w(x,y)w(x',y')\rangle = w^2\delta(x-x')\delta(y-y') \tag{1}$$

Equation (1) expresses the assumption that we are dealing with a surface rough on the atomic scale, with totally uncorrelated irregularities ("white noise" surface). The parameter w measures the rms fluctuations in thickness. While this is the simplest assumption, it does not cover some experimentally relevant cases (e.g., regular steps). The problem of quasiparticle motion in a box of variable width belongs in the general class of "random boundary condition" problems. The influence of this random boundary condition on the quasiparticle states can be represented by a set of random pseudopotentials,[9] which for $w/d^2 << 1$ take the form:

$$V_{n,m}(x,y) = \frac{\pi^2}{m^* d^3} nmw(x,y) \tag{2}$$

where $V_{n,m}(x,y)$ is the matrix element of the pseudopotential matrix between transverse momentum eigenstate of the box of thickness d. The properties of the superfluid can be extracted from the Gorkov equations, which in the presence of a rough boundary are:

$$(\omega-\xi_k-\lambda_n-\Sigma_m(\omega))G_n(k,\omega) + \Delta_n(k)F_n(k,\omega) = 1 \tag{3a}$$

$$(\omega-\xi_k-\lambda_n-\Sigma_n(-\omega))F_n^+(k,\omega) + \Delta_n^+(k)G_n(k,\omega) = 0 \tag{3b}$$

where n is the subband index. We have:

$$\lambda_n = (\pi n)^2/(2m^* d^2) \; ; \quad n = 0,1.... \tag{4}$$

and the self-energy is:

$$\Sigma_n(\omega) = \frac{\pi^4}{m^{*2} d^2} \left(\frac{\omega^2}{d^4}\right) n^2 \sum_{k,n'} n'^2 G_{n'}(k,\omega) \tag{5}$$

Σ_n contains the effects of random boundary scattering. The anomalous part of the self-energy vanishes for a "white noise" surface. Introducing:

$$\tilde{\omega}_n = \omega + i\, n^2 \Sigma_n(\omega) \tag{6}$$

the self-consistent equations for the self-energy and the gap parameter can be written as:

$$\tilde{\omega}_n = \omega + \Gamma_o \cos^2\theta_n \sum_{n'} \frac{\cos^2\theta_{n'}\ \omega_{n'}}{\cos\theta_{n'}\ [\tilde{\omega}_{n'}^2 + \Delta^2\sin^2\theta_{n'}]^{1/2}} \tag{7a}$$

$$1 = \frac{3}{2}\frac{\pi^2\lambda}{k_F d} T \sum_n \frac{\sin^2\theta_n}{\cos\theta_n\ [\tilde{\omega}_n^2 + \Delta^2\sin^2\theta_n]^{1/2}} \tag{7b}$$

where $\cos^2\theta_n \equiv (n\pi)^2/(2m^* d^2 E_F)$, λ is the bulk coupling constant, and:

$$\Gamma_o \equiv \frac{\pi^4}{2m^* d^2} \left(\frac{w^2}{d^4}\right) \left(\frac{2m^* d^2 E_F}{\pi^2}\right)^{5/2} \tag{8}$$

We can now investigate the effect of diffusive surface scattering on various physical quantities. Considering first the transition temperature, we find that the ratio of the transition temperature $T_c(d)$ in the presence of scattering to its counterpart $T_c^o(d)$ for a smooth boundary is given by:

$$\ell n \frac{T_c(d)}{T_c^o(d)} = -\langle\psi\left(\frac{1}{2} + \frac{\Gamma_o \cos^2\theta_n}{6\pi\ T_c(d)}\right)\rangle + \psi\left(\tfrac{1}{2}\right) \tag{9}$$

where ψ is the digamma function and the brackets denote the weighted average:

$$\langle\ldots\rangle \equiv \sum_n \sin^2\theta_n\ (\ldots)/\sum_n \sin^2\theta_n \tag{10}$$

From Eqs. (9)-(10) T_c can be determined as a function of d and w. The main result is that for a specified value of the surface roughness there is a critical thickness d^c below which superfluidity disappears. It is convenient to introduce the dimensionless parameter $y \equiv k_F^4 w^2$ as a measure of surface roughness. One then has that the critical thickness $d^c(y)$ is given by:

$$d^c(y) \approx 0.75\ \eta y \xi \tag{11}$$

where $\ell n \eta \equiv \langle \ln \cos^2\theta_n \rangle$. For $y = 1$ one has $k_F d^c \sim 35$. It follows that superfluidity in ^{3}He films of moderate thickness (say 10 ~ 20 layers) will not be destroyed by surface roughness unless $y \gg 1$, which would represent a very strongly disordered substrate. But surface roughness will influence the observability of quantum size effects. Surface roughness affects $\tilde{\Omega}_A$, for example, in two ways: firstly, the overall magnitude of $\tilde{\Omega}_A$ decreases, scaling down as T_c and vanishing as $d \to d_c$. Secondly, the sharp jumps in $\tilde{\Omega}_A(d)$ will be smoothed out and eventually the oscillatory character of $\tilde{\Omega}_A$ will diminish. One finds, however, that so long as $y < 1$ the size effects in $\tilde{\Omega}_A$ will be observable in the superfluid state.

Many low temperature properties of the superfluid state are determined by the character of the excitation spectrum. Thus, we consider now the dynamical quasiparticle density of states. To find this quantity we rewrite (7a) in the form:

$$u_n = \frac{\omega}{\Delta} + \rho \cos^2\theta_n \sum_{\cos\theta_{n'}} \frac{u_{n'} \cos^2\theta_{n'}}{[u_{n'}^2 + \sin^2\theta_{n'}]^{1/2}} \tag{12}$$

where $u_n \equiv \tilde{\omega}_n/\Delta$ and $\rho \equiv \Gamma_o/\Delta$. Making the ansatz $u_n = \omega/\Delta = \rho U \cos^2\theta_n$ one obtains the following equation for $U(\omega/\Delta)$:

$$U = \sum_{\cos\theta_n} \frac{(\omega/\Delta)\cos^2\theta_n + \rho U \cos^4\theta_n}{[((\omega/\Delta) + \rho U \cos^2\theta_n)^2 + \sin^2\theta_n]^{1/2}} \tag{13}$$

Once $U(\omega/\Delta)$ is known from the numerical solution of (13), the quasiparticle density of states can be found from:

$$N(\omega) = N(o)\ \mathrm{Im} \sum_{\cos\theta_n} \frac{(\omega/\Delta + \rho\ U \cos^2\theta_n)}{[\sin^2\theta_n - ((\omega/\Delta) + \rho U \cos^2\theta_n)^2]^{1/2}} \tag{14}$$

The solution $N(\omega)/N(o)$ of Eqs. (13) and (14) is plotted in Fig. 3 for various values of ρ. There are two qualitatively different types of behavior. For $\rho \leq 1.6$ the density of states vanishes at the Fermi level and $N(\omega) \sim \omega^2$ for small ω, as in the bulk A phase but with an increased coefficient of ω^2. For $\rho > 1.6$ $N(\omega)$ is finite at the Fermi surface. The system resembles, in this sense, a "dirty" superconductor. This is a "gapless" state. The low temperature thermodynamics displays temperature dependencies characteristic of the normal state but with different coefficients. Another characteristic feature is the sharply peaked structure at $\omega \sim \Delta$. This is due to the state dependence of the pair-breaking field which arises from boundary scattering, and would not

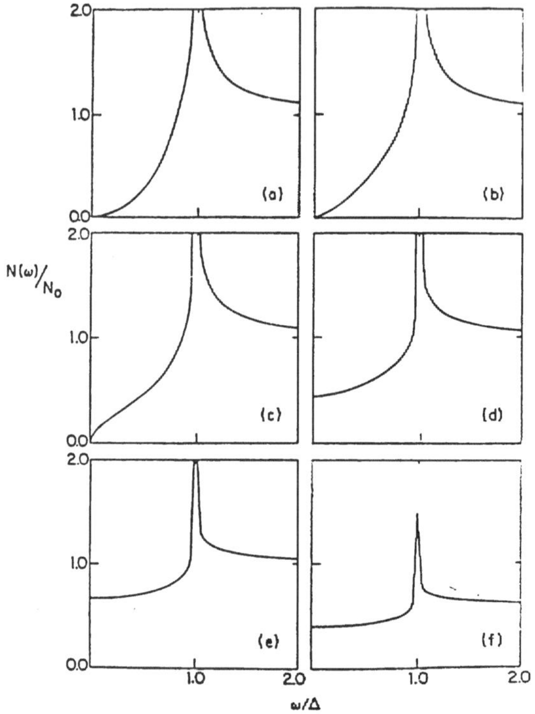

Fig. 3. The dynamical quasiparticle density of states for various values of $= \Gamma_0/\Delta$: a) $\rho = 0.0$, b) $\rho = 0.5$, c) $\rho = 1.6$, d) $\rho = 6.0$, e) $\rho = 18.0$, f) generic form of the density of states in the gapless regime exhibiting the peaked structure around $\omega/\Delta = 1$.

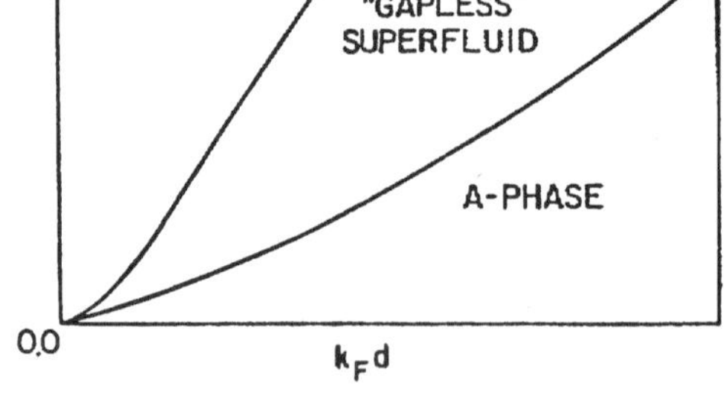

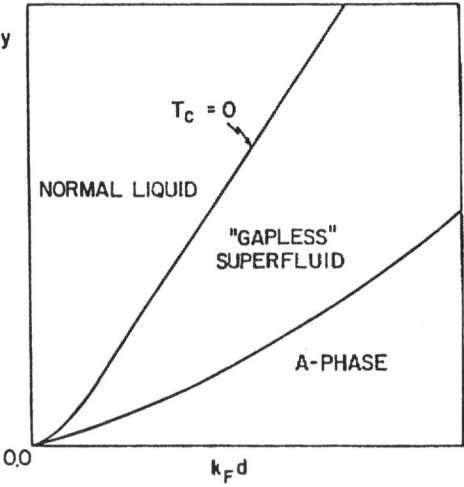

Fig. 4. The T = 0 phase diagram of superfluid ^{3}He film.

be present for uniformly distributed impurities. More details on the substrate-induced behavior can be found in Ref. (10).

The T = 0 phase diagram of the superfluid is displayed in Fig. 4. The relevant parameters are d and y. There are three different regions: (i) the A-phase-like state, (ii) the gapless state and (iii), the normal liquid, where superfluidity is destroyed by random scattering. Region (ii) is of particular interest. It is very likely that in many experimental situations with substrates of intermediate roughness one is in this region. This would make it possible to study the interplay between anisotropic superfluidity and reduced quasiparticle mean free path. By controlling the thickness one is in a position to effectively "tune" the strength of surface scattering (i.e., the parameter y). As d decreases below $d^c(y)$ the transition to the normal state occurs. In a film at very low temperatures the quasiparticle mean free path is limited by elastic surface scattering, rather than by quasiparticle collisions, as in the bulk, and this leads to modified Fermi liquid transport properties[(11)] and to weak-localization effects which impinge also[(12)] on linear response and behavior in a magnetic field.

ACKNOWLEDGEMENTS

Zlatko Tesanovic was supported in part by NSF through grants DRR 82-07431 and MPS 85-14638, and by the Harvard Materials Research Laboratory. Oriol T. Valls acknowledges support from the MEIS Center at the University of Minnesota.

REFERENCES

1. D. D. Osheroff, W.J. Gully, R. C. Richardson, and D. M. Lee, Phys. Rev. Lett. 29, 920 (1972).
2. A. Sachrajda, R. F. Harris-Lowe, J. P. Harrison, R. R. Turkington, and J. G. Daunt, Phys. Rev. Lett. 55, 1602 (1985).
3. D. Nelson, in Vol. 7 of "Phase Transitions and Critical Phenomena," C. Domb and J. L. Lebowitz, eds., Academic Press, London (1983).
4. See for example K. Levin and O. T. Valls, Phys. Rep. 98, 1 (1983).
5. Z. Tesanovic and O. T. Valls, Phys. Rev. B33, 3139 (1986).
6. B. G. Orr, H. M. Jaeger, and A. M. Goldman, Phys. Rev. Lett. 53, 2046 (1984).
7. A. J. Leggett, Rev. Mod. Phys. 47, 331 (1975).
8. R. Balian and N.R. Werthamer, Phys. Rev. 131, 1553 (1963).
9. Z. Tesanovic and M.V. Jaric, unpublished.
10. Z. Tesanovic and O.T. Valls, submitted to Phys. Rev. B.
11. P.A. Lee and T.V. Ramakrishnan, Rev. Mod. Phys. 57, 287 (1985).
12. Z. Tesanovic, unpublished.

QUANTUM MONTE CARLO AND THE EQUATION OF STATE OF LIQUID ^{3}He

Robert M. Panoff

Department of Physics, Kansas State University
Manhattan, KS 66506 USA
and
Courant Institute of Mathematical Sciences (NYU)
New York, NY 10012 USA

I. INTRODUCTION

The accurate and efficient calculation from microscopic interactions of the macroscopic properties of bulk liquid ^{3}He at zero temperature has been a perennial goal of condensed matter theorists. Seen as a proving ground for many-fermion theories and methods, ^{3}He continues to occupy the attention of many of the participants in this Workshop.[1] Significant progress has been realized over the past several years in refining inter-atomic potentials, wave functions, and calculational schema employed to attack the many-body problem.

Because of the very large computational requirements, quantum Monte Carlo methods,[2] both variational (VMC) and Green's function Monte Carlo (GFMC), have been looked upon by some as being justified only because they provide valuable and otherwise unattainable benchmarks for other computational approaches. With recent access to new supercomputers, however, we are finding that quantum Monte Carlo methods, essentially exact within some arbitrary statistical error, may now be applied in a truly cost-effective manner to go beyond ground-state energy benchmarks to yield real physical insights about the wave function and model Hamiltonians themselves.

In this paper I will briefly review the present status of Monte Carlo "technology" as it applies to the study of the ground-state properties of strongly-interacting many-fermion systems in general, and to liquid ^{3}He at zero temperature in particular. The rest of this paper is organized as follows. In Section II, variational Monte Carlo methods are reviewed and the model many-body problem to be tackled is introduced. Then I outline the domain Green's function Monte Carlo method in Section III, with mirror potentials providing a coherent framework for discussing solutions to the fermion problem. Finally, in Section IV, I will present results for the zero-temperature equation of state of ^{3}He, along with other ground-state properties derived from the many-body wave function. All of the work reported here has been carried out in close collaboration with the members of the quantum Monte Carlo group at the Courant Institute.

II. VARIATIONAL MONTE CARLO

Variational methods have been used for many years now to study the properties of quantum fluids at zero temperature. Regardless of the computational method employed to evaluate the energy expectation values, the theoretical formulation of the problem is essentially the same. For a non-relativistic system of N helium atoms which interact by means of two-body forces only, the Hamiltonian has the form

$$H = \sum_{i=1}^{N} t(i) + \sum_{1 \leq i<j \leq N} v(r_{ij}) \tag{1}$$

where $t(i) = -(h^2/2m)\nabla_i^2$ in configuration space, and m is the mass of a neutral helium atom. The state-independent atom-atom interaction potential $v(r_{ij})$ is taken to be the Aziz pair potential[3] which has proven to be an effective many-body potential for ^4He. By "effective many-body potential," I mean that the use of this pair potential alone is sufficient to reproduce many-body properties without recourse to three- or higher-body interactions.

An appropriate trial wave function ψ_T for the N particle system must be chosen. For helium systems, the usual form is a product of two- and three- (and possibly higher-) body correlations. For ^3He, a fermion system, this product wave function is also multiplied by a determinant of model states which takes into account the required antisymmetry of the wave function under interchange of particle coordinates. Ideally, correlations in the wave function should be determined by solving the variational problem

$$\frac{\delta E_v[\psi_T]}{\delta \psi_T} = 0 , \tag{2}$$

where the extremum of

$$E_V[\psi_T] = \frac{\int \psi_T^*(R) H \psi_T(R)\, dR}{\int \psi_T^*(R) \psi_T(R)\, dR} \tag{3}$$

is a minimum.

Alternatively, based on physical intuition, parameterized forms for the two- and three-body correlations, and for state-dependent correlations built into the determinant, may be adopted. The functional forms and any parameters contained therein are varied to minimize the energy. The trial wave function used in our helium work is of this form

$$\psi_T = \psi_3 \det(\exp\{i\vec{k}_i \cdot [\vec{r}_j + \sum_{l \neq j} \eta(r_{lj})\, \vec{r}_{lj}]\}), \tag{4}$$

where

$$\psi_3 = \exp[-\frac{1}{2} \sum_{i<j} \tilde{u}(r_{ij}) - \frac{\lambda_T}{4} \sum_{l} \sum_{i<j \neq l} \xi(r_{lj})\xi(r_{lj})\, \vec{r}_{li} \cdot \vec{r}_{lj}] \tag{5}$$

and

$$\tilde{u}(r) = u(r) - \lambda_T \, \xi^2(r) \, r^2 . \tag{6}$$

The two-body Jastrow pseudo-potential u(r) entering Equation 6 is taken to be an appropriately scaled solution of an Euler-Lagrange equation calculated via optimal Fermi hypernetted chain[4] (FHNC/C), or a "semi-optimized" pseudo-potential which is the solution of a Schoedinger-like equation.[5] The simpler one-parameter form of McMillan has also been used in our studies, especially for the fully-polarized phase of ^{3}He for which it seems to give slightly better variational energies. Functional forms currently used for the triplet correlation function $\xi(r)$ and the backflow correlation $\eta(r)$ are

$$\xi(r) = \exp\left[-\left(\frac{r-r_T}{w_T}\right)^2\right] \tag{7}$$

and

$$\eta(r) = \lambda_B \exp\left[-\left(\frac{r-r_B}{w_B}\right)^2\right] + \frac{\lambda'_B}{r^3} \tag{8}$$

In previous variational Monte Carlo calculations,[6-8] both $\xi(r)$ and $\eta(r)$ were multiplied by a third-degree polynomial "box" factor which had the effect of ensuring that the correlations vanished at the side of the simulation cube. This additional factor is completely superfluous, except perhaps for very small numbers of particles or very high densities, due to the effectively short range of the correlations, and serves only to scale the other parameters in the wave function. As an alternative, the pair, triplet, and backflow correlations above are instead replaced in our calculations by a sum of reflected correlations

$$f(r) = f(r) + f(2r_{side}) - 2f(r_{side}) . \tag{9}$$

In actual practice, again since the correlations which minimize the variational energy turn out to be relatively short-ranged in nature, this finite-size adaptation has no real effect on the magnitude of the correlations.

To evaluate the variational energy, the Metropolis, or $M(RT)^2$, Monte Carlo algorithm[9] is employed to carry out the required many-dimensioned integration, which avoids the need of introducing further approximation except that the integral will be done numerically. During the past two decades sufficient experience has been gained that one now has considerable confidence in this method. The energy expectation value in Equation 3 is rewritten as

$$E_v = \int \frac{\psi_T^* \psi_T}{\int \psi_T^* \psi_T dR} \; \frac{H\psi_T}{\psi_T} \, dR. \tag{10}$$

The variational energy is computed as the average of the local energy $H\psi_T/\psi_T$ evaluated at particle configurations drawn from a probability distribution proportional to $|\psi_T|^2$, as generated by the Metropolis algorithm. Besides the energy, other ground-state properties are computed as averages over these same particle configurations. If the variational energy is close to the true ground-state energy, these other properties derived from the wave function are also believed to accurately represent ground-state properties of the many-body system.

The early work by Ceperly et al.[6] showed how variational Monte Carlo could be used to sample the square of a simple antisymmetric function: a Jastrow factor times a Slater determinant of plane wave orbitals. Particles were moved one at a time. The change in such a wave function is easily calculated for both the Jastrow factor (subtracting the "old" contributions and adding the "new") and the Slater determinant (only the one row and one column corresponding to the moved particle needed to be altered). The ratio of the square of the wave function evaluated at the new and old particle positions is used as the acceptance criterion, and the inverse of the Slater matrix need be recomputed only if the proposed move is accepted. Since matrix inversion and the attendant evaluation of the determinant are expensive (computationally intensive) operations, this approach for Jastrow-type wave functions produces significant savings in computational overhead.

When Schmidt and co-workers[8] introduced triplet and backflow correlations into the trial wave function, the entire determinant had to be recomputed at each step in order to evaluate the transition probability. However, particles were still moved one at a time, since this was the most efficient way to use the serial processors available at the time. The advent of new supercomputer architectures, incorporating parallel and/or vectorized processors, has allowed us the opportunity to re-think the implementation of the Metropolis algorithm. The biggest gain in efficiency comes from moving all N particles at every Monte Carlo step resulting in a highly-vectorized code. Even though this requires a much smaller step size in order to maintain a reasonable acceptance ratio (the "mythical" 50%), sufficiently many more such moves can be accomplished this way than by moving the particles singly for a given expediture of computer time. This results in a more efficient sampling also known as a greater part of configuration space. Directed sampling, also know as "force-biased" or "smart" Monte Carlo,[6,10] may be employed to augment the size of the N-particle step.

The larger memories available in the new computers have also enabled us to extend our studies of finite-size dependence in our calculations. We have evaluated the energy expectation values for a wave function including Jastrow, triplet, and backflow correlations for systems ranging from 38 to 186 ^{3}He atoms in a periodic box, which gives us new confidence that our simulations faithfully represent the properties of the bulk system. The variational energy and the potential and kinetic energy components remain constant to within 0.1 K over the entire particle range studied.

Even though it entails some expense, which may be significantly reduced by re-weighting[6] or correlated sampling, variational Monte Carlo remains as the only reliable method to find the best parameters for a trial wave function which minimize the energy expectation value for realistic wave functions. In the course of studying the sensitivity of the variational energy to long-range correlations, we had the occasion to test the FHNC-scaling approximation,[11] which was developed by "matching" the Monte Carlo results for a given wave function at a given density. The low-order integrals over the distribution functions generated by this approximation seriously overestimate the importance of long-range correlations and underestimate the kinetic energy for the triplet and backflow parts of the wave function. As a result, at equilibrium density, the FHNC-scaling evaluation of the energy expectation value for an approximately "re-optimized" wave function is more than 0.5 K below its actual value as determined by a direct Monte Carlo calculation. The FHNC-scaling wave function, therefore, actually results in a higher variational energy than the wave function[6] upon which the method was intended to improve.

III. GREEN'S FUNCTION MONTE CARLO FOR FERMIONS

The Green's function Monte Carlo (GFMC) method[2,12-16] is based on the realization that integrating the Schroedinger equation in imaginary time is equivalent to solving a diffusion equation in real time,[17]

$$\psi(\tau+\Delta\tau) = \exp[-(H-E_T)\ \Delta\tau]\ \psi(\tau) \qquad (11)$$

and diffusion processes are easily simulated on a computer. Possible problems associated with the finite time step in Equation 11 are avoided by transforming the Schroedinger
equation to an integral equation with a shifted energy scale,

$$\psi^{n+1} = \left[\frac{E_T + E_c}{H + E_c}\right] \psi^n \qquad (12)$$

The magnitude of the trial energy, E_T, controls the growth of the configuration population and is chosen to be a good approximation to the ground-state energy; E_c is a constant added to the potential such that the spectrum of $H + E_c$ is positive.

The GFMC method is implemented by choosing a set of points {R} in configuration space and iterating the equation

$$\psi^{n+1}(R) = (E_T + E_c) \int G(R,R')\ \psi^n(R')dR', \qquad (13)$$

where G(R,R') is the Green's function for the Hamiltonian:

$$HG(R,R') = \delta(R-R'). \qquad (14)$$

In general, G(R,R') is not known analytically but may be sampled in a domain around each particle.[11,13] This requires an ancillary random walk whose expected density develops the Green's function. Use of GFMC to integrate the Schrodinger equation formally requires that the wave function entering Equation 13 be interpreted as a probability density: it must, therefore be non-negative so that the wave function may be represented as a list of configurations. For systems obeying Bose statistics, such as ^{4}He, where this condition is satisfied,the GFMC method has been applied with great success.[12,18,19]

Applying GFMC to fermion systems, most notably ^{3}He, has been an arduous task. Since the antisymmetry of the wave function, a manifestation of the Pauli exclusion principle, prevents its direct interpretation as a probabilty density, the fermion wave function must be represented as the difference of two positive densities

$$\psi_F = \psi^+ - \psi^- \qquad (15)$$

each of which separately is a solution to Equation 13, and their difference is a solution to the original Schroedinger equation for the ground-state wave function. The fermion problem in the GFMC method is essentially the exponential decay of the fermion signal in a background of symmetric noise. Consequently, the difficulty is that, even though straightforward application of GFMC, using the decomposition above, iterates towards the true fermion ground state, it may fail to do so with sufficiently small variance to extract meaningful information about the properties of the ground state.

Recently, Carlson and Kalos introduced the concept of mirror potentials[20] which results in a coherent theoretical framework for both discussing and understanding alternative solutions to the fermion problem. To the Hamiltonian of the system is added an external many-body potential that forces the ground state to be a linear combination of symmetric and antisymmetric components, of which the latter is the correct many-fermion ground state. Coupled equations for ψ^+ and ψ^- are written as

$$[H(R) + C(R)\,\psi^-(R)]\,\psi^+(R) = E\psi^+(R)$$

$$[H(R) + C(R)\,\psi^+(R)]\,\psi^-(R) = E\psi^-(R) \qquad (16)$$

The non-linearity of the problem makes it possible, in theory, to create distinct stable populations for ψ^+ and ψ^- and retain the property that the difference $\psi_F = \psi^+ - \psi^-$ satisfies the original Schroedinger equation. The additional potentials $C(R)\psi^{\pm}(r)$ are called "mirror potentials" because they act in such a way that they constrain the positive and negative random walkers to stay predominately in separate parts of coordinate space, namely those where the fermion ground state is respectively positive and negative. Random walkers whcih cross nodes tend to be reflected back to their own "proper" region of coordinate space.

The symmetric function C(R) should, in general, be chosen to be positive and finite. Its optimal form remains a difficult problem, and will be the focus of future study. However, the two most prevalent implementations of GFMC for fermions, results for which are presented in the next section, represent the two limiting choices for C(R). By choosing C(R)=0, the mirror potential method reduces to transient estimation,[9] in which ψ^+ and ψ^- separately iterate to the symmetric ground-state distribution; their difference, therefore, suffers from the problem of an exponentially decreasing signal-to-noise ratio for fixed computing time (or an exponentially increasing computational problem for fixed variance) mentioned above. Even so, with a good starting wave function--the best variational wave function, for example--transient estimation does produce a decreasing series of upper bounds which appears to converge.

The other extreme, choosing $C(R) = \infty$, recaptures the fixed-node approximation,[15,16,21] in which the random walkers which generate the probability densities are not allowed to cross the nodes of a suitably chosen trial function, again, the best available variational wave function. This approximation is equivalent to enforcing the boundary condition that $\psi \to 0$ whenever $\psi_T<0$. The fixed-node energy lies below the variational energy,[21] and is a upper bound to the true ground-state energy expectation value. In practice, it has also proven to be a close upper bound.

Since the significant computational load of the GFMC calculations for fermions is concerned with evaluating the trial wave function which is used as the importance function, a necessary ingredient for low variance,[6] both transient estimation and the fixed-node approximation are well suited for highly-vectorized supercomputer architectures. Furthermore, GFMC is ideally suited for massively parallel systems since, besides evaluating the importance function, the basic flow of the algorithm is to read a configuration R from a list, evaluate the Green's function G(R,R') which is interpreted as a transition probability for the system going from R to R', and writing the configuration back on the list a proportionate number of times. Hence, the ideal computing environment for GFMC would be a highly parallel computer, each element being a vector processor.

IV. RESULTS AND DISCUSSION

The results of our variational and Green's function Monte Carlo calculations are summarized in Table 1 and Figures 1-4. In Figure 1, I have plotted the energy per particle vs. density for our best variational wave function (dotted line) including Jastrow, triplet, and backflow correlations. Also plotted are the results from the GFMC calculations employing the fixed-node approximation (dashed line) over the same range of liquid densities, and the GFMC transient estimate result for the energy (asterisk with error bar) near the finite-density minimum. Comparing the GFMC calculations with each other and with the experimental equation of state (solid line), we conclude that the Aziz potential serves nearly as well as an effective many-body potential for ^{3}He as it does for ^{4}He.[19]

This conclusion must be qualified by the assumption that the transient estimate calculation has indeed converged to the true anti-symmetric ground state. Evidence for this assumption is given in Figure 2 which shows the energy per particle as a function of GFMC iteration, which also demonstrates the large growth of the variance in the transient estimate calculation. The upper data points in Figure 2 are from an earlier calculation which began with a variational energy of -1.95 K and included only Jastrow correlations in the importance function; in this case the growth of the variance was faster than the rate of convergence. The lower curve is our best calculation, starting with a variational energy of -2.12 K and including both Jastrow and triplet correlations in the importance function. The better starting point and the control of the variance by the better importance function seem sufficient to extract a converged energy before the variance has grown too large. Even so, the transient estimate result of -2.44 ± 0.04 K for the energy per particle is clearly the best upper bound to date.

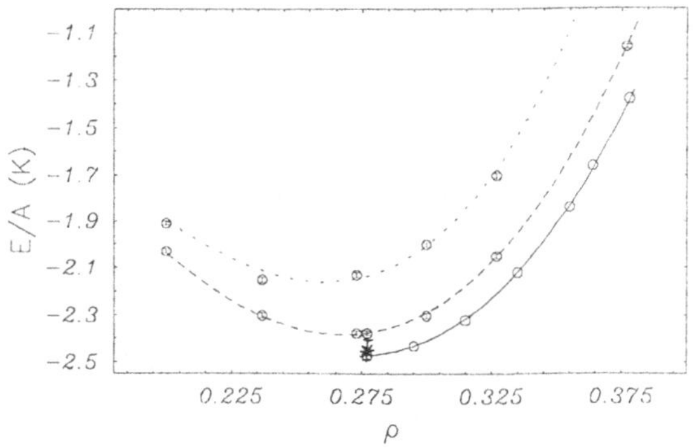

Figure 1. Equation of state of ^{3}He. Wave functions as identified in the text.

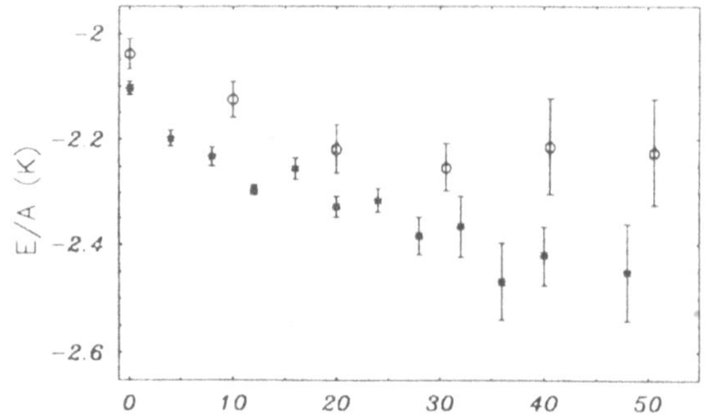

Figure 2. Energy per particle vs. GFMC iteration by transient estimation for 54 ^{3}He atoms at experimental equilibrium density.

While the absolute convergence to the lowest lying anti-symmetric state by transient estimation may still be open to question, the fixed-node calculation does converge, in fact, to the best wave function subject to the nodes of the trial function. The clear convergence of the energy is shown in Figure 3. Referring back to Figure 1, and comparing the fixed-node result with the transient estimate for the energy at the finite-density minimum, it is clear that the fixed-node approximation in GFMC yields a close upper bound to the ground-state energy expectation value. Given the near agreement of the results of these two limiting extremes of GFMC for the many-fermion problem, it is clear that a future study utilizing a more-general mirror potential is justified and likely to be fruitful.

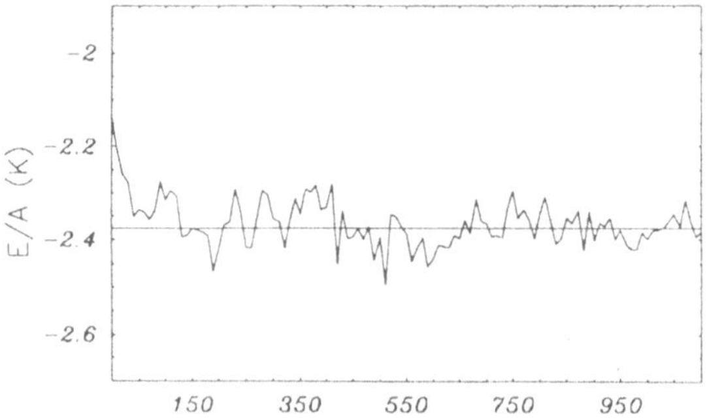

Figure 3. Energy per particle vs. GFMC iteration by fixed-node approximation for 54 ^{3}He atoms at experimental equilibrium density.

Table 1. Energies per particle for ^{3}He at equilibrium density, $\rho\sigma^3 = 0.273$, $\sigma = 2.556$ A, N = 54. All energies are in K. (Key to wave function labels: J: Jastrow correlations, T : triplet correlations, D: Slater determinant of plane wave orbitals, B: backflow correlations in the determinant.)

Wave Function	Kinetic Energy	Potential Energy	Total Energy
VMC, J·D	12.76 ± 0.06	-14.02 ± 0.05	-1.26 ± 0.03
VMC, J·T·B	12.22 ± 0.03	-14.35 ± 0.02	-2.13 ± 0.02
GFMC-FN	12.28 ± 0.04	-14.65 ± 0.03	-2.37 ± 0.01
GFMC-TE	12.40 ± 0.10	-14.84 ± 0.10	-2.44 ± 0.04
Experiment			-2.47 ± 0.01

Results for the energy per particle at equilibrium density for different wave functions are collected in Table 1. An important point to emphasize is the sensitivity of the calculations to the close cancelation of potential and kinetic energy contributions. Neither the kinetic nor potential energy changes significantly as one goes from the simple Jastrow wave function (J·D) to the wave function which also incorporates triplet and backflow correlations (J·T·B), and finally arriving at the GFMC-iterated wave functions (FN: fixed node, TE: transient estimation). The result for the kinetic energy per particle, 12.3 K, appears to be somewhat higher than the recently reported experimental value of Sokol.[22] Gaussian fits applied in the analysis of the experimental data, however, have been shown to significantly underestimate this quantity.[23]

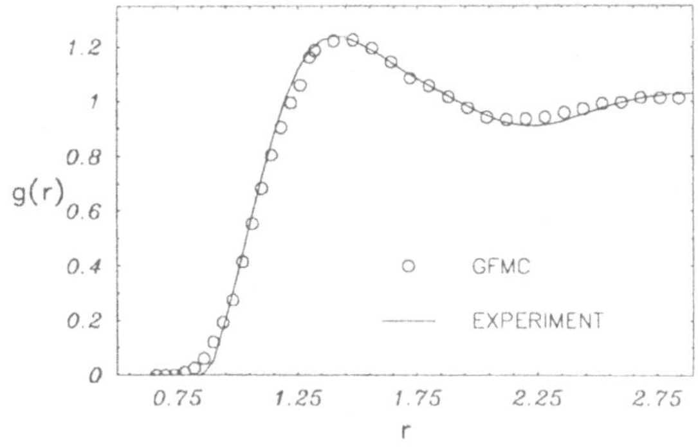

Figure 4. Comparison of calculated and experimental radial distribution functions for ^{3}He at experimental equilibrium density.

Besides showing that the Aziz potential may be used with confidence in many-body calculations for ^{3}He, our calculations indicate that we have identified the most significant ingredients of the ground-state wave function. The variational Monte Carlo curve in Figure 1 lies close to the fixed-node GFMC curve over the range of liquid densities studied. The quality of the GFMC wave function is further demonstrated by Figure 4 which compares the pair radial distribution functions, g(r), for both experiment and GFMC calculations. Nevertheless, there remains a 0.4 K difference in energy between the best trial wave function and the GFMC result. Future work will attempt to improve on both the symmetric part of the wave function and the location of the nodes of the determinant, for example, by considering a wider class of functional forms for the correlations. The success at the Jastrow level of using correlations generated by FHNC-type procedures as input to the variational Monte Carlo calculations may yet be repeated for triplet correlations.[24]

Other ground-state properties, most notably the single-particle density matrix and the momentum distribution, are also being studied, and these results will be presented elsewhere.[25] In addition, we plan to complement our study of ^{3}He by renewing our calculations on spin-polarized phases, in which we must calculate, using correlated sampling techniques, the small energy differences between systems with different degrees of polarization.

In closing, both variational and Green's function Monte Carlo will continue to be important tools to study the properties of many-fermion systems. Their more-widespread use will come about as we improve our understanding of the mirror-potential approach to the fermion problem, and as access and use of supercomputers extend to a wider community of researchers.

Acknowledgments. I would like to thank Joe Carlson, Paula Whitlock, Kevin Schmidt, Jules Moskowitz, Mike Lee, Geoffrey Chester, and Mal Kalos for their hard work, shared ideas and codes, and many stimulating discussions at the Courant Institute. This work was supported in part by the National Science Foundation under grant number DMR-8513300, by the Applied Mathematical Sciences subprogram of the Office of Energy Research, U.S. Department of Energy under contract number DE-AC02-76ER03077, and by the Devision of Nuclear Physics of the Office of High Energy and Nuclear Physics, U.S. Department of Energy under contract number DE-AC02-79ER10353.

REFERENCES

1. See "Condensed Matter Theories," Vol. 1, F. Malik, ed., (Plenum Press, New York, 1986), and present volume.
2. For good reviews of quantum Monte Carlo methods and applications, see J. Zabolitzky, in "Progress in Particle and Nuclear Physics," A. Faessler, ed., Vol. 16 (Pergamon, Oxford, 1986); and "Proceedings of the Conference on Frontiers of Quantum Monte Carlo," J. Stat. Phys. 43 (1986).
3. R.A. Aziz, V.P.S. Nain, J.S. Cerley, W.L. Taylor, and G.T. McConville, J. Chem. Phys. 70, 4330 (1979).
4. E. Krotscheck, R.A. Smith, J.W. Clark, and R.M. Panoff, Phys. Rev. B24, 6383 (1981).
5. V.R. Pandharipande and H. Bethe, Phys. Rev. C7, 1312 (1972).
6. D.M. Ceperley, G.V. Chester, and M.H. Kalos, Phys. Rev. B16, 3081 (1977).
7. M.A. Lee, K.E. Schmidt, M.H. Kalos and G.V. Chester, Phys. Rev. Lett. 46, 728 (1981).
8. K.E. Schmidt, M.A. Lee, M.H. Kalos and G.V. Chester, Phys. Rev. Lett. 47, 807 (1981).

9. N. Metropolis, A.W. Rosenbluth, M.N. Rosenbluth, A.M. Teller, and E. Teller, J. Chem. Phys. 21, 1087 (1953).
10. M. Rao and B.J. Berne, J. Chem. Phys. 77, 129 (1979).
11. E. Manousakis, V.R. Pandharipande, and Q.N. Usmani, Phys. Rev. B31, 7022 (1985).
12. M.H. Kalos, D. Levesque, and L. Verlet, Phys. Rev. A9, 2178 (1974).
13. D.M. Ceperley and M.H. Kalos, in "Monte Carlo Methods in Statistical Physics," K. Binder, ed., Topics in Current Physics, Vol. 7 (Springer, Berlin, Heidelberg, New York, 1979) Chap. 4.
14. K.E. Schmidt and M.H. Kalos, in "Applications of the Monte Carlo Method in Statistical Physics," K. Binder, ed., Topics in Current Physics, Vol. 36 (Springer, Berlin, Heidelberg, New York, 1984).
15. K.E. Schmidt and J.W. Moskowitz, J. Stat. Phys. 43, 1027 (1986).
16. J.W. Moskowitz and K.E. Schmidt, J. Chem. Phys. 85, 2868 (1986).
17. N. Metropolis and S. Ulam, J. AM. Stat. Assoc. 44, 335 (1949).
18. P.A. Whitlock, D.M. Ceperley, G.V. Chester, and M.H. Kalos, Phys. Rev. B19, 5598 (1979).
19. M.H. Kalos, M.A. Lee, P.A. Whitlock and G.V. Chester, Phys. Rev. B24, 115 (1981).
20. J. Carlson and M.H. Kalos, Phys. Rev. C32, 1735 (1985).
21. J.B. Anderson, J. Chem. Phys. 63, 1499 (1975); 65, 4121 (1976); 73, 3897 (1980).
22. P.E. Sokol, K. Sköld, D.L. Price and R. Kleb, Phys. Rev. Lett. 54, 909 (1985).
23. J. Carlson, R.M. Panoff, K.E. Schmidt, P.A. Whitlock, and M.H. Kalos, Phys. Rev. Lett. 55, 2367(C) (1985).
24. E. Krotscheck, Phys. Rev. B33, 3158 (1986).
25. P.A. Whitlock and R.M. Panoff, J. Can. Phys., "Proceedings of Banff Conference on Quantum Fluids and Solids," to be published.

TOPICS IN MULTI-COMPONENT FERMI SYSTEMS

Khandker Fazlul Quader

Physics Department
University of Illinois at Urbana-Champaign
1110 West Green Street
Urbana, Illinois 61801

Recently there has been considerable interest in multi-component Fermi systems. Some examples are spin-polarized ^{3}He, and ^{3}He-^{4}He mixtures, where the asymmetric spin populations form two-component Fermi liquids, isospin asymmetric nuclear matter, the coupled electron, proton, neutron and neutrino Fermi liquids in the astrophysical context of stellar collapse and cores of neutron stars, liquid metallic hydrogen where the interpenetrating electrons and protons form the two-component fluid, and electron-hole droplets in semiconductors. Recent experimental studies,[1,2] in particular viscosity measurements[1] in spin-polarized ^{3}He near the melting curve, have stimulated new interest. Theoretical studies in the multi-component systems include the calculations of transport properties in various systems, but under varying degrees of approximations, by Flowers and Itoh,[3] Meyerovich,[4] Mullin and Miyake,[5] and Oliva and Ashcroft.[6] The interactions in the two-component system, especially in the context of ^{3}He, have been studied by Quader and Bedell,[7] (and the references therein), Bedell and Quader,[8] Bedell and Sanchez-Castro,[9] and Hess, Pines and Quader.[10]

Here, I shall concentrate for the most part on some recent work; in particular the following:

(A) Exact, analytic calculations of multi-component transport coefficients;
(B) The scattering amplitudes;
(C) Means of obtaining the Landau interactions for a two-component system, the associated problems, and some applications.

(A) TRANSPORT IN MULTI-COMPONENT SYSTEMS

Exact calculations of the transport properties of a single component Fermi liquid at low temperatures exist.[11,12] Here I shall discuss the recent generalization by Anderson, Pethick and Quader[13] (hereafter referred to as APQ) to the multi-component case. The analytic results of APQ are exact, and expressed in terms of the results for the one-component case.

It may be recalled that the transport modes correspond to the various local conservation laws of a system. If the system is subjected to inhomogeneous static perturbations such as a temperature or a velocity gradient or gradients in the chemical potential differences, there arise in the system heat, momentum or spin flow which are limited by the quasiparticle collisions, and are proportional to the same variables which specify the local equilibrium of the system. The constants of proportionality are the transport coefficients. I shall start off by first discussing, in quite general terms, the multi-component transport, and then specialize to the two-component case.

General Case

The basic transport equation is the Landau kinetic equation

$$\frac{\partial n_{\underset{\sim}{p}i}}{\partial t} + \underset{\sim}{\nabla} n_{\underset{\sim}{p}i} \cdot \underset{\sim}{\nabla}_{\underset{\sim}{p}} \varepsilon_{\underset{\sim}{p}i} - \underset{\sim}{\nabla}_{\underset{\sim}{p}} n_{\underset{\sim}{p}i} \cdot \underset{\sim}{\nabla} \varepsilon_{\underset{\sim}{p}i} = I_i , \tag{1}$$

where $\underset{\sim}{p}$ is the momentum of a quasiparticle and ε_p is its energy, including non-equilibrium contributions. The index i refers to the species/spin of a quasiparticle. I_i is the collision integral which gives the rate at which the collisions increase the occupation of state, i, so it depends on the other quasiparticle distributions and is given by

$$I_{\underset{\sim}{p}_1 i} = -\frac{1}{V^2} \sum_{\underset{\sim}{p}_2, \underset{\sim}{p}_3, \underset{\sim}{p}_4}{}' \sum_{jk\ell} W_{ijk\ell}(1,2;3,4)\, \delta_{\underset{\sim}{p}_1+\underset{\sim}{p}_2,\ \underset{\sim}{p}_3+\underset{\sim}{p}_4}\, \delta(\varepsilon_{1i}+\varepsilon_{2j}-\varepsilon_{3k}-\varepsilon_{4\ell})$$

$$[n_{1i} n_{2j}(1 - n_{3k})(1 - n_{4\ell}) - (1 - n_{1i})(1 - n_{2j}) n_{3k} n_{4\ell}] \tag{2}$$

where the transition probability W is given by

$$\frac{2\pi}{\hbar} |\langle 3,k;\ 4,\ell|t|1,i;\ 2,j\rangle|^2 = \frac{W_{ijk\ell}}{V^2}\, \delta_{\underset{\sim}{p}_1+\underset{\sim}{p}_2,\ \underset{\sim}{p}_3+\underset{\sim}{p}_4} = W_{ij}(\theta,\phi) \tag{3}$$

where $\langle 3,k;\ 4,\ell|t|1,i;\ 2,j\rangle$ is the T-matrix for scattering of quasiparticles, see Fig. 1(a); (θ,ϕ) are the usual scattering angles, see Fig. 1(b); $p_\alpha = \alpha$. The prime in the sum is to avoid double-counting. It may be noted that the Landau kinetic equation is richer than the usual

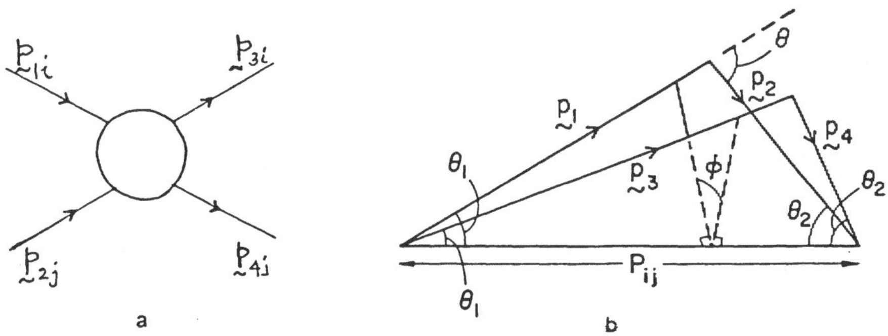

Fig. 1(a) Scattering of two quasiparticles with the convention, $p_1 = p_3 = p_i$ and $p_2 = p_4 = p_j$.

(b) The relation of θ and φ to the momenta; $\sin\theta_1/p_2 = \sin\theta_2/p_1 = \sin\theta/p_{ij}$.

Boltzmann equation for dilute gases as it includes many-body effective field contributions, and the quasiparticle velocity can depend on position and time. An analogous quantum mechanical kinetic equation exists;[14] however, in the hydrodynamic regime ($\hbar\omega \gtrsim k_BT$) relevant to this discussion, it is identical to the classical one.

I wish to make the following observations regarding the calculations: (i) Completely general scattering rates, $W_{ij}(\theta,\phi)$, and arbitrary ratios of the Fermi momentum, p_{Fi} and effective masses m_i^* are used. (ii) Since it is assumed that no non-species-conserving processes such as dipole-dipole interactions in ^{3}He, or β-decay processes in the nuclear systems are present, the number of quasiparticle of each component is conserved separately. (iii) At low temperatures, T, the quasiparticle momenta, p_α are confined near their respective Fermi surfaces; hence the Abrikosov-Khalatnikov[15] energy-momentum decoupling is possible. (iv) Since the system is almost in local thermodynamic equilibrium, one can linearize the collision term by writing

$$n_{\underset{\sim}{p}i} = n_i^o(\varepsilon_{\underset{\sim}{p}i}) - T\frac{\partial n^o_{\underset{\sim}{p}i}}{\partial \varepsilon_{\underset{\sim}{p}i}}\,\bar{\Phi}_{\underset{\sim}{p}i} \tag{4}$$

where n_i^o is the Fermi function denoting local equilibrium distribution, and $\bar{\Phi}_{\underset{\sim}{p}i}$ is a dimensionless deviation function.

With (i)-(iv), the transport equation becomes

$$-\frac{\partial n_i^o}{\partial x_1} X_i(x_1,\hat{p}_1) = I_{1i} \tag{5}$$

where $X_i(x_1,\hat{p}_1)$ is the "driver" term, and

$$I_{1i} = -\frac{m_i^* T^2}{8\pi^4 h^6} \iint dx_2 dx_3 n^o(x_1) n^o(x_2)(1 - n^o(x_3))(1 - n^o(x_1 + x_2 - x_3))$$

$$\{\sum_j m_j^{*2} \iint \frac{d\Omega}{4\pi} \frac{d\phi_2}{2\pi} \frac{W_{ij}\beta_{ij}}{1+\delta_{ij}} (\bar{\Phi}_i(p_1) + \bar{\Phi}_j(p_2) - \bar{\Phi}_i(p_3) - \bar{\Phi}_j(p_4))\}, \tag{6}$$

with $x_\alpha = (\varepsilon_\alpha - \mu_\alpha)/T$, and $\beta_{ij} = p_j/P_{ij} = p_j/(p_i^2 + p_j^2 + 2\tilde{p}_i \cdot \tilde{p}_j)^{1/2}$. For $i = j$, β_{ij} reduces to the well-known result $(2\cos(\theta/2))^{-1}$.

Next, I shall outline the general solution of Eq. (5). Since the kernel in I_{1i} is invariant under a simultaneous change of signs of all the x_α, one can consider the odd or even functions of x separately. The "driver" and the deviation function can then be generally written as

$$X_i^P(x,\hat{p}) = F_i X_Q^P(x) D_Q(\hat{p})Q \tag{7}$$

$$\bar{\Phi}_i^P(x,\hat{p}) = \gamma_i(x) D_Q(\hat{p})Q \tag{8}$$

where Q is the gradient of a local equilibrium variable, $D_Q(\hat{p})$ is an irreducible tensor quantity (e.g. spherical harmonic), $X_Q(x)$ and $\gamma_i(x)$ are functions only of the energy variable, x, F_i is a scalar function, and P denotes the oddness or evenness of the "driver".

After some variable changes,[13] the transport equation assumes the form

$$\Xi_{1i}^P(x_1) = (1 + \frac{x_1^2}{\pi^2}) \Gamma_i(x_1) - \sum_j \lambda_{ij} \int_{-\infty}^{\infty} \frac{dx_2}{\pi^2} \frac{(x_1 - x_2)}{\sinh((x_1 - x_2)/2)} \Gamma_j(x_2) \tag{9}$$

where $\Xi_{1i}^P(x)$ is related to $X_{Q,i}^P(x)$, and $\Gamma_i(x)$ to $\gamma_i(x)$. λ_{ij} is a matrix with

$$\lambda_{ii} = \frac{\sum_j (\frac{m_j^*}{m_i^*})^2 \int \frac{d\Omega}{4\pi} \frac{W_{ij}}{1+\delta_{ij}} \frac{p_j}{P_{ij}} \xi_{ij}^i(\ell,P)}{\bar{W}_i}, \tag{10}$$

$$\lambda_{ij} = \frac{\int \frac{d\Omega}{4\pi} W_{ij} \frac{\sqrt{p_i p_j}}{P_{ij}} \xi_{ij}^j(\ell,P)}{\sqrt{(\bar{W}_i \bar{W}_j)}} \quad \text{for } i \neq j, \tag{11}$$

where

$$\overline{W}_i = \sum_j \left(\frac{m_j^*}{m_i^*}\right)^2 \int \frac{d\Omega}{4\pi} \frac{W_{ij}}{1 + \delta_{ij}} \frac{P_j}{P_{ij}} , \tag{12}$$

and $\xi_{ij}^k(\ell, P)$ is given in Table 1. It is useful to see that the quasiparticle lifetime, is given by

$$\frac{1}{\tau_i(x)} = \frac{m_i^{*3} T^2}{8\pi^4 \hbar^6} \frac{x^2 + \pi^2}{2} \overline{W}_i \tag{13}$$

An orthogonal transformation diagonalizes λ_{ij} and results in a decoupled set of equations in the diagonal basis (bars on top):

$$\overline{\Xi}_i^P(x_1) = \left(1 + \frac{x_1^2}{\pi^2}\right) \overline{\Gamma}_i(x_1) - \lambda_i \int_{-\infty}^{\infty} \frac{dx_2}{\pi^2} \frac{(x_1 - x_2)}{\sinh\left((x_1 - x_2)/2\right)} \overline{\Gamma}_i(x_2) , \tag{14}$$

Equation (14) may now be solved for $\overline{\Gamma}_i(x)$ in the standard fashion.[13,14]

The transport coefficients are finally obtained by evaluating the response of the system to an applied perturbation,

$$J = T \sum_i \sum_{\underset{\sim}{p}} O_{\underset{\sim}{p}i} \frac{\partial n^o_{\underset{\sim}{p}i}}{\partial x_i} \overline{\Phi}_{\underset{\sim}{p}i} \tag{15}$$

where, similar to Eq. (7) $O_{\underset{\sim}{p}i} = G_i X_0(x_\alpha) D_0(\hat{p}_\alpha)$, and $\overline{\Phi}$ can be written in terms of $\overline{\Gamma}$.

The quantity (-J/Q) is then the desired transport coefficient; the expression appears on the top row of Table 1, N_i being the density of states for species i. The parity-dependent matrix elements, $\langle X|\nu\rangle$ are given in Ref. 13. The main point is that in (-J/Q), the sum over ν is essentially the transport coefficient for a single-component system. The sum over t picks up the eigenvalues, λ_t. For each i and k, the matrix $\tilde{S}_{tk}$ transforms from the initial basis to the eigenstates of the collision operator, while S_{it} transforms back to the initial one. Table 1 summarizes the results. As examples, thermal conductivity, κ, viscosity, η, and diffusion D are given for both the multi-component (mc) and the single-component (sc) cases, together with the corresponding "drivers" and deviation functions. It may be noted that the susceptibility at constant pressure, χ_p enters into the diffusion coefficient, D. The function $R^{\pm}(\lambda)$ are plotted in Ref. 13.

Two-Component Case

For illustrative purposes, I shall now provide some details in the two-component case. In Eq. (9), λ_{ij} is now a 2×2 matrix, and $\Xi_{1i}^P(x_1)$ and $\Gamma_i(x_1)$ are two-component vectors. The matrix

Table 1. Summary of the transport results. The first row gives the general expression. "mc" ≡ multi-component, "sc" ≡ single-component. $A_{ik} \equiv \sqrt{N_i\tau_i N_k\tau_k}$, $\tau_\kappa^{sc} = R^-(\lambda)\tau^{sc}$, $\tau_{\eta,D}^{sc} = R^+(\lambda)\tau^{SC}$ and $m = n_1 - n_2$ is the asymmetry density.

$$-\frac{J}{Q} = T\int d\Omega_p D_Q(\hat{p})D_0(\hat{p}) \sum_{ikt} \sqrt{N_i\tau_i N_k\tau_k}\, G_i S_{it} \sum_\nu \langle X_0|\nu\rangle \frac{1}{1-\frac{\lambda_t}{\alpha\nu}} \langle\nu|X_Q\rangle \tilde{S}_{tk} F_k$$

	κ	η	D
Q	$\underset{\sim}{\nabla T}$	$\frac{\partial u_x}{\partial y}$	$\underset{\sim}{\nabla}(\delta\mu)$
F_i	v_i/T	$p_i v_i/T$	$(\sigma_i - m/n)v_i$
$X_Q = X_0$	x	1	1
$D_Q = D_0$	$\hat{p}\cdot\hat{n}$	$(\hat{p}\cdot\hat{x})(\hat{p}\cdot\hat{y})$	$\hat{p}\cdot\hat{n}$
G_i	T_{v_i}	$p_i v_i$	$(\sigma_i - m/n)v_i$
$\xi_{ij}^k(\ell, P)$	$\ell = 1$, $P = -1$	$\ell = 2$, $P = +1$	$\ell = 1$, $P = +1$
$(-J/Q)_{mc}$	$\frac{\pi^2}{3} T \sum A_{ik}v_i v_k$ $S_{it}R^-(\lambda_t)\tilde{S}_{tk}$	$\frac{1}{15} \sum A_{ik}\, p_i v_i p_k v_k$ $S_{it}R^+(\lambda_t)\tilde{S}_{tk}$	$\frac{T}{3} X_P^{-1} \sum A_{ik}v_i v_k$ $(\sigma_i - \frac{m}{n})(\sigma_k - \frac{m}{n})$ $S_{it}R^+(\lambda_t)\tilde{S}_{tk}$
	↑	↑	
$(-J/Q)_{sc}$	$\frac{\pi^2}{3} TN(0)v_F^2\tau_\kappa^{sc}$	$\frac{1}{15}N(0)p_F^2 v_F^2 \tau_\eta^{sc}$	$\frac{T}{3}N(0)v_F^2 X^{-1}\tau_D^{sc}$

$$S = \begin{pmatrix} \cos \xi/2 & -\sin \xi/2 \\ \sin \xi/2 & \cos \xi/2 \end{pmatrix} \tag{16}$$

diagonalizes λ_{ij} and gives eigenvalues, $\lambda_t = \lambda_s \pm \sqrt{\lambda_a^2 + \lambda_{12}^2}$, with $\lambda_{s(a)} = (\lambda_{11} \pm \lambda_{22})/2$, A cos $\xi = \lambda_a$ and A sin $\xi = \lambda_{12}$. λ_{11}, λ_{12}, λ_{22} are given by Eqs. (10)-(12). $\xi = 0$ corresponds to a set of uncoupled Fermi liquids.

Since χ_p enters into diffusion, and will be useful for other purposes, I provide a derivation of it, which will be different from that in APQ.

A Two-Component Susceptibility, χ_p

The derivation follows along the work of Bedell,[16] who also obtained other responses in spin-polarized ^{3}He. Consider a two component system, with an asymmetry density, m due to the presence of a magnetic field $\underset{\sim}{B}_o$. Within Fermi liquid theory the change in the energy density, E is given quite generally by[14]

$$\delta E = \sum_{\underset{\sim}{p},i} \varepsilon^o_{\underset{\sim}{p}i}\, \delta n_{\underset{\sim}{p}i} + \frac{1}{2} \sum_{\underset{\sim}{p}i,\underset{\sim}{p}'j} f^{ij}_{\underset{\sim}{p}\underset{\sim}{p}'}\, \delta n_{\underset{\sim}{p}i}\, \delta n_{\underset{\sim}{p}'j}. \tag{17}$$

Here $f^{ij}_{\underset{\sim}{p}\underset{\sim}{p}'}$ is the quasiparticle interaction function to be expanded in the angle between $\underset{\sim}{p}$ and $\underset{\sim}{p}'$, i.e. $f^{ij}_{\underset{\sim}{p}\underset{\sim}{p}'} = \sum_\ell f^{ij}_\ell P_\ell(\hat{p}\cdot\hat{p}')$, and $\delta n_{pi} = n_{pi} - n^o_{pi}$, n^o_{pi} being the equilibrium distribution function in the presence of $\underset{\sim}{B}_o$. The quasiparticle energies are such that $\varepsilon^o_{F1} - \varepsilon^o_{F2} = \mu_1 - \mu_2 = 2B_o$, ε^o_{Fi} being the equilibrium Fermi energies. If the Fermi surfaces are uniformly distorted, only the $\ell = 0$ moment of $f^{ij}_{\underset{\sim}{p}\underset{\sim}{p}'}$ contributes, and to second order, the change in the distribution function is[14]

$$\delta n_{pi} = \delta\varepsilon_{Fi}\,\delta(\varepsilon^o_{Fi} - \varepsilon^o_{pi}) - \tfrac{1}{2}(\delta\varepsilon_{Fi})^2\delta'(\varepsilon^o_{Fi} - \varepsilon^o_{pi}) \tag{18}$$

with $\partial n_i/\partial\varepsilon_{Fi} = N_i(0)$. This gives

$$\delta E = \sum_i \varepsilon_{Fi}\,\delta n_i + \frac{1}{2}\sum_{i,j}\left(f^{ij}_o + \frac{1}{N_i(0)}\right)\delta n_i\,\delta n_j \tag{19}$$

The susceptibility at constant pressure, $\chi_p = (\partial B/\partial m)_p$ where $B = B_o + \delta B = (\partial E/\partial m)_n$, with δB being a perturbation on the intial field B_o. From Eq. (19) it follows that[16]

$$\left(\frac{\partial B}{\partial m}\right)_P = \frac{1}{4}\sum_{i,j}\left[\left(1 + \sigma_i\left(\frac{\partial n}{\partial m}\right)_P\delta_{ij}\right)\left(\frac{1}{N_i(0)} + \sigma_i\sigma_j f^{ij}_o\right)\right] \tag{20}$$

where $\sigma = +1(-1)$ for $i,j = 1(2)$. $(\partial n/\partial m)_P$ is obtained from the expression for pressure given by:[16] $P = P_o + \delta P = -\varepsilon + n(\partial\varepsilon/\partial n)_m + m(\partial\varepsilon/\partial m)_n$, where P_o is the equilibrium pressure. Using Eqs. (19) and (20) and setting $\delta P = 0$ (constant pressure) it is seen that

$$\left(\frac{\partial n}{\partial m}\right)_P = -\left[\frac{\sum_{i,j} \sigma_j n_i \left(\frac{1}{N_j(0)} \delta_{ij} + f_o^{ij}\right)}{\sum_{i,j} n_i \left(\frac{1}{N_i(0)} + f_o^{ij}\right)}\right] \tag{21}$$

Using Eqs. (20) and (21), χ_p is obtained:

$$[\chi_p]^{-1} = \frac{n}{2}\left[\frac{(1+N_1(0)f_o^{11})(1+N_2(0)f_o^{22}) - N_1(0)N_2(0)f_o^{12^2}}{n_1 N_2(0)(1+N_1(0)f_o^{11}) + n_2 N_1(0)(1+N_2(0)f_o^{22}) + n\, N_1(0)N_2(0)f_o^{12}}\right]. \tag{22}$$

For the case of spin diffusion in an unpolarized Fermi liquid, $n_1 = n_2 = \frac{n}{2}$, $N_1(0) = N_2(0) = \frac{1}{2} N(0)$, $f_o^{11} = f_o^{22}$, and Eq. (22) reduces to the familiar result $\chi^{-1} = (1 + F_o^a)/N(0)$, where $F_o^a = \frac{1}{2} N(0)(f_o^{11} - f_o^{12})$.

In the APQ calculations, $W_{ij}(\theta,\phi)$ have been kept completely general so the results can be used to predict the transport coefficients if $W_{ij}(\theta,\phi)$ is given, or to deduce $W_{ij}(\theta,\phi)$ from experiments.

(B) SCATTERING AMPLITUDES

The scattering probabilities, $W_{ij}(\theta,\phi)$ and the effective masses, m_i^* are determined by the many-body quasiparticle interactions in the system of interest. In some cases, such as in unpolarized ^{3}He or in ^{3}He-^{4}He mixtures, experiments are able to provide some information about the interactions. However, in multi-component cases such as asymmetric nuclear matter, or electron, neutron and proton fluids, such experimental information is mostly unavailable. In the case of spin-polarized ^{3}He the recent experiments[1,2] have been encouraging. An approach to obtaining the quasiparticle interaction would be a microscopic many-body calculation starting from the bare interactions. But, this is a difficult task even for the single-component system. Other approaches are compromises between the completely microscopic and phenomenological ways. In the nuclear cases, the measured nucleon-nucleon phase shifts can be used to obtain the scattering rates. Within Fermi liquid theory (FLT) the Landau interactions may be used to construct the scattering amplitudes in $W_{ij}(\theta,\phi)$. Though the calculation of such interactions would normally be model-dependent, they have proved to be useful in the single-component case.

For the two-component case, Quader and Bedell,[7] have obtained such scattering amplitudes. These can be used to construct $W_{ij}(\theta,\phi)$ in the s-p approximation,[14] for example. To see this, it may be noted that in the presence of a strong magnetic field, due to the anisotropy in the spin space, the quasiparticle-quasihole interaction has the form

$$f^{13;24}_{\underset{\sim}{p}\underset{\sim}{p}'} = \phi_1\delta_{13}\delta_{24}+\underset{\sim}{\phi}_2\cdot\underset{\sim}{\sigma}_{13}\delta_{24}+\underset{\sim}{\phi}_2'\cdot\underset{\sim}{\sigma}_2\delta_{13}+\phi_3\sigma^z_{13}\sigma^z_{24}+\phi_{4/2}(\sigma^+_{13}\sigma^-_{24}+\sigma^-_{13}\sigma^+_{24}) \tag{23}$$

where the ϕ's are momentum-dependent functions related to the usual $f^{\sigma\sigma'}_{\underset{\sim}{p}\underset{\sim}{p}'}$, and the σ's are spin-matrices. Thus there are four independent functions, ${}^0f^{\uparrow\uparrow}$, ${}^0f^{\downarrow\downarrow}$, ${}^0f^{\uparrow\downarrow}$, ${}^1f^{\uparrow\downarrow}$ in place of the two in the unpolarized case. The superscript 0,1 denotes the spin-projection, m_z in the p-h channel. The scattering amplitudes, ${}^{0,1}A^{\sigma\sigma'}$ are related to ${}^{0,1}f^{\sigma\sigma'}$ by coupled integral equations with the $m_z = 0$ and 1 terms decoupled from each other, see Fig. 2. Expanding $f^{\sigma\sigma'}$ and $A^{\sigma\sigma'}$ in the Landau angle, and solving the coupled equations, the scattering amplitudes can be obtained:

$${}^0A^{\sigma\sigma}_\ell = \frac{1}{D_\ell}\{f^{\sigma\sigma}_\ell[1 - f^{-\sigma-\sigma}_\ell\, \chi^{-\sigma}_\ell(q)] + f^{\sigma-\sigma}_\ell \chi^{-\sigma}_\ell(q) \tag{24}$$

$${}^0A^{\sigma-\sigma}_\ell = \frac{1}{D_\ell}\, f^{\sigma-\sigma}_\ell \tag{25}$$

$${}^1A^{\sigma-\sigma}_\ell = {}^1f^{\sigma-\sigma}_\ell/(1 - {}^1f^{\sigma-\sigma}_\ell \chi^+_\ell(q)) \tag{26}$$

where $D_\ell \equiv [1 - f^{\sigma\sigma}_\ell \chi^\sigma_\ell(q)][1 - f^{-\sigma-\sigma}_\ell \chi^{-\sigma}_\ell(q)]-(f^{\sigma-\sigma}_\ell)^2\chi^\sigma_\ell(q)\chi^{-\sigma}_\ell(q)$. $\chi^\sigma_\ell(q)$ are the usual Lindhard functions and $\chi^+_\ell(q)$ are the similar functions for p-h propagators having opposite spins; the expressions are given in Ref. 7.

Keeping up to the $\ell = 1$ moments, the $A^{\sigma\sigma'}_\ell$'s can be used to construct a $W_{\sigma\sigma'}(\theta,\phi)$. According to the convention of Sec. (A), $W_{\sigma-\sigma}(\theta,\phi)$ will include contributions both from the non-spin-flip, ${}^0A^{\sigma-\sigma}_\ell$ and the spin-flip ${}^1A^{\sigma-\sigma}$.

(C) LANDAU INTERACTIONS IN A TWO-COMPONENT SYSTEM

I shall confine the discussion to spin-polarized ^{3}He. Since spin-polarizations are expected to appreciably affect the ^{3}He properties, the changing features in $W_{\sigma\sigma'}(\theta,\phi)$, the transport coefficients, the various susceptibilities, $\chi(\Delta)$, sound speeds, field splitting of the A-phase, etc. can be studied once the polarization-dependent Landau interactions are known.

Among the approaches to obtaining the interactions is the semi-microscopic "induced" interaction model of Quader and Bedell[7,8] the paramagnon model,[17] the extension of the Gutzwiller model,[18] and the polarization potential model in the $\Delta \to 1$ limit.[19]

Here, I shall be concerned with the recent efforts of Bedell and Sanchez-Castro[9] (referred to as BS) and Hess, Pines and Quader[10] (referred to as HPQ), both of which are on a more phenomenological

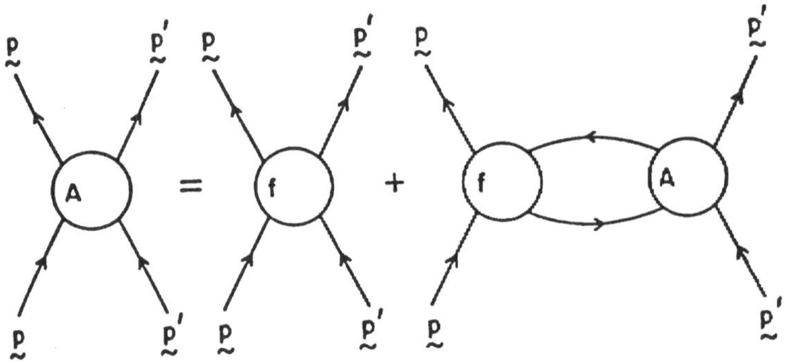

Fig. 2 Schematic representation of the integral equation for ${}^{0,1}A^{\sigma\sigma'}$ in terms of ${}^{0,1}f^{\sigma\sigma'}$.

level. Though HPQ has done a more detailed study, in essence, both BS and HPQ rely on expanding the interactions (twiddled quantities) (Sec. (B)) to some order in $\Delta = m/n$ around the unpolarized ones (untwiddled). Thus

$$ {}^{0}\tilde{f}_0^{\sigma\sigma} = {}^{0}f_0^{\sigma\sigma}\left(1 + \sum_{k=1} \alpha_k(-\sigma\Delta)^k\right) \tag{27} $$

$$ {}^{0}\tilde{f}^{\sigma-\sigma} = {}^{0}f_0^{\sigma-\sigma}\left(1 + \sum_{\text{even } k} \beta_k\Delta^k\right) \tag{28} $$

$$ {}^{0}\tilde{f}_0^{\sigma\sigma} = R^{\sigma\sigma}\, {}^{0}f_1^{\sigma\sigma}\left(1 + \sum_{k=1} \gamma_k(-\sigma\Delta)^k\right) \tag{29} $$

$$ {}^{0}\tilde{f}_1^{\sigma-\sigma} = R^{\sigma-\sigma}\, {}^{0}f_1^{\sigma-\sigma}\left(1 + \sum_{\text{even } k} \delta_k\Delta^k\right) \tag{30} $$

where $\sigma^k = \sigma$ (k odd) $= 1$(k even). $R^{\sigma\sigma'} = p_F^{\sigma}p_F^{\sigma'}/p_F^2$ appear in some of the cases of HPQ, and equals 1 in BS. Since m_σ^*'s are related by Galilean invariance to ${}^{0}\tilde{f}_1^{\sigma\sigma'}$, they can also have similar expansions. HPQ have also obtained the spin-flip interaction, ${}^{1}\tilde{f}^{\sigma-\sigma}$. In practice, the expansions are truncated at some order and the coefficients are fixed by physical arguments, sum rules (of which there are a few in the polarized case), thermodynamics, and experiments. They are then used to predict other properties.

At this point, even at the risk of sounding overly cautious, I wish to alert the reader of the following: (a) Apriori, it is unclear whether the truncation converges rapidly enough, whereby there remains a manageable number of parameters to fix. Whereas there are reasons[9] to expect a rapid convergence in the $\ell = 0$ case, the situation is more complicated for the higher ℓ's. Consequently, the results at very large polarizations may not be as justifiable as at the lower end. (b) Even at a low order of truncation, there is a rather large number of parameters. (c) Since the order of truncation, and the choice of fixing the parameters is somewhat arbitrary, the results can turn out to be artifacts of the choice.[10] The rationale is that the results justify the means, and from among the family of models will emerge some unique features about which one is able to make definite statements.

The key BS result is a maximum in $\chi_n(\Delta)$ (at constant density) and decreasing density of states, $N_\sigma(\Delta)$; hence the interpretation as "near-metamagnetism". HPQ finds that though this is a possibility, but depending on certain choices, $\chi_n(\Delta)$ can diverge at higher densities suggesting a true metamagnetic transition. Anyway, the main point is that in all the plausible HPQ models, the initial rise in $\chi_n(\Delta)$ is accompanied by a decreasing $N_\sigma(\Delta)$. This is in contradistinction to the ferromagnetic[17] and to the nearly localized[18] viewpoints, which again differ from each other.

Finally, I present some details of the HPQ work. Owing to certain physical arguments, HPQ need to employ a fewer number of parameters compared to BS. In all of their case studies, HPQ utilize the following: (i) They argue that for the fully polarized (FP) ${}^{FP}f_o^{\uparrow\uparrow}$,

$$f_0^{\uparrow\uparrow} - {}^{FP}f_o^{\uparrow\uparrow} = f_0^{\uparrow\downarrow} - f_o^{\uparrow\uparrow} \tag{31}$$

Then for the partially polarized system, with the parameter, t_1

$$f_0^{\uparrow\uparrow} - \tilde{f}_0^{\uparrow\uparrow} = (f_0^{\uparrow\downarrow} - f_0^{\uparrow\uparrow})(t_1\Delta + (1 - t_1)\Delta^2) \tag{32}$$

so that correct limiting results are obtained at $\Delta = 0$ and $\Delta = 1$. A comparison with the quadratic order expansion (Eq. (29)) of $\tilde{f}_o^{\sigma\sigma}$ eliminates one of the parameters. (ii) They apply the $\sum_\ell \tilde{A}_\ell^{\downarrow\downarrow} = 0$ sum rule to two down-spin impurities in the $\Delta \to 1$ limit to obtain another constraint. A couple of their cases are now discussed:

(I) An example of a HPQ case study is one in which the $\ell = 0$ and $\ell = 1$ interactions are truncated at the quadratic and cubic orders respectively. Magnetostriction, and the $\sum_\ell \tilde{A}_\ell^{\uparrow\uparrow} = 0$ sum rule to linear order are used to constrain the remaining linear coefficients. $(m_\downarrow^*)_{FP}$ is constrained to be 2.0 by physical arguments,[10] $R^{\sigma\sigma'} \neq 1$ and the Maxwell relation $\partial C_v/\partial B = T \frac{\partial^2 \chi}{\partial T^2}$ is used. The results (see Fig. 3) for $\chi_n(\Delta)$ and

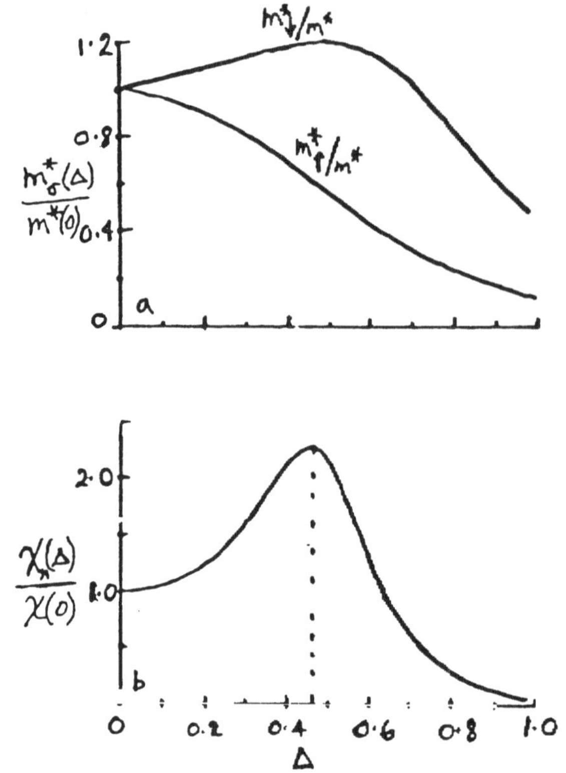

Fig. 3 (a) $m^*_\sigma(\Delta)/m^*(\Delta = 0)$ and (b) $\chi_n(\Delta)/\chi_n(\Delta = 0)$ as functions of $\Delta = m/n$ for the HPQ case I in text.

$m^*_\sigma(\Delta)$ at melting are qualitatively similar to that of BS, but they are achieved with fewer parameters.

(II) In another model, HPQ expand $m^*_\sigma(\Delta)$ (rather than $\tilde{f}_1^{\sigma\sigma'}$) to quadratic order, because the convergence is expected to be better there than in $\tilde{f}_1^{\sigma\sigma'}$, where HPQ finds that at least a cubic order expansion is needed. Here the predicted $(m^*_\downarrow)_{FP}$ is slightly above the unpolarized m^* at all densities, whereas $\chi_n(\Delta)$ diverges at higher densities at $\Delta \sim 0.5$. Thus the result here is different than that of BS.

It is hoped that out of these efforts, a coherent, comprehensive picture of the spin-polarized system will emerge. The APQ analytic results enables transport calculations for spin-polarized ^{3}He. In view of the recent viscosity measurements,[1] it is amusing to use the BS and HPQ interactions to calculate transport coefficients. This is a work in progress.[20]

I would like to thank R. Anderson, D. Hess, C. J. Pethick and D. Pines for collaborations in various parts of the work discussed. I wish to specially thank D. Hess for numerous discussions on the interactions in spin-polarized ^{3}He, and producing results at short notices.

REFERENCES

1. P. Kopietz, A. Dutta, and C. N. Archie, Stonybrook preprint (1986).
2. C. N. Archie, Stonybrook preprint (1986); G. Bonfait, et al., Phys. Rev. Lett. 53:1092 (1984); A. Dutta and C. N. Archie, Phys. Rev. Lett. 55:2949 (1985).
3. E. Flowers and N. Itoh, Ap. J. 230:847 (1979).
4. A. E. Meyerovich, J. Low Temp. Phys., 47:271 (1982); ibid, 53:487 (1983).
5. W. J. Mullin and K. Miyake, J. Low Temp. Phys. 53:313 (1983); W. J. Mullin and K. Miyake; preprint (1986).
6. J. Oliva and N. W. Ashcroft, Phys. Rev. B 35:223 (1982).
7. K. F. Quader and K. S. Bedell, J. Low Temp. Phys. 58:89 (1985).
8. K. S. Bedell and K. F. Quader, Phys. Lett. 96A:91 (1983); Phys. Rev. B 30:2894 (1984).
9. K. S. Bedell and C. Sanchez-Castro, Phys. Rev. Lett., to be published (1986).
10. D. Hess, D. Pines and K. F. Quader, in preparation (1986).
11. G. A. Brooker and J. Sykes, Phys. Rev. Lett. 21:279 (1968); J. Sykes and G. A. Brooker, Ann. Phys. 56:1 (1970).
12. H. Højgaard Jensen, H. Smith and J. W. Wilkins, Phys. Lett. 27A:532 (1968); Phys. Rev. 185:323 (1968).
13. R. Anderson, C. J. Pethick and K. F. Quader, submitted to Phys. Rev. B (1986).
14. G. Baym and C. J. Pethick in The Physics of Liquid and Solid Helium. Part II, ed. K. H. Bennemann and J. B. Ketterson (Wiley, New York, 1978) p. 116.
15. A. A. Abrikosov and Khalatnikov, Zh. Eksp. Teor. Fiz. 32:1083 (1957) [Sov. Phys. JETP 5:887 (1957)]; Rep. Prog. Phys. 22:329 (1959).
16. K. S. Bedell, Phys. Rev. Lett. 54:1400 (1985).
17. M. T. Beal-Monod and E. Daniel, Phys. Rev. B27:4467 (1983).
18. D. Vollhardt, Rev. Mod. Phys. 56:99 (1984).
19. D. Hess, D. Pines and K. F. Quader, preprint (1986).
20. D. Hess and K. F. Quader, in preparation (1986).

RAPIDLY-CONVERGENT TRUNCATION SCHEME FOR THE GROOUND STATE ENERGY OF QUANTUM FLUIDS

V.C. Aguilera-Navarro,* C. Keller, M. de Llano and M.Popovich**

Physics Department
North Dakota State University
Fafgo, North Dakota 58105

INTRODUCTION

We begin by stating our calculated equation of state for "nuclear matter", derived through 2nd order in the van der Waals perturbation scheme to be discussed later, for the spin-averaged BHK (Baker-Hind-Kahane[1)]) model nucleon-nucleon potential which is a hard-core-square-well shape rather accurately reproducing low-energy scattering and deuteron data. Figure 1 shows the energy per nucleon vs. $x \equiv k_F c$,

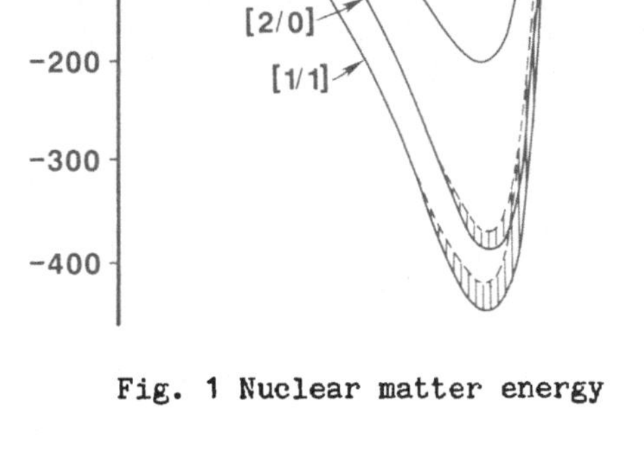

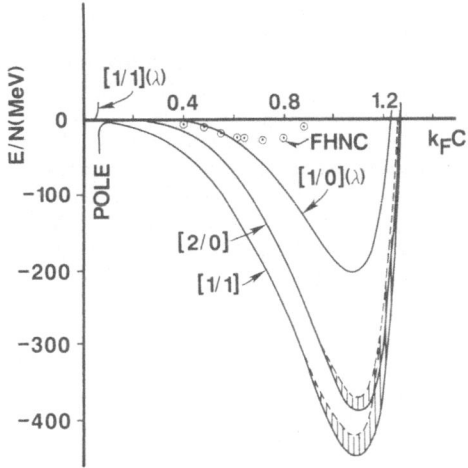

Fig. 1 Nuclear matter energy

* On leave from Instituto de Física Teórica, São Paulo, Brazil.
Work supported by FAPESP, Brazil.

** On leave from Boris Kidric Institute, Belgrade, Yugoslavia.

where c is the hard-sphere diameter and k_F the Fermi wave number. The curve labelled $[1/0](\lambda)$ refers to 1st order in the attractive potential well strength parameter λ. The full and dashed curves labelled $[2/0](\lambda)$ refer to 2nd order and bracket the uncertainty (of about 5%) in our results. Higher-order corrections are being investigated and are expected to merely lower the saturation minimum a little more. Note that even to 2nd order our method is yielding more than ten times more binding than the Jastrow-variational calculations[2] done employing Fermi hypernetted chain methods (circled dots labelled FHNC) with the same potential model. (About 50% lower energy was previously found[3] for neutron matter as compared with Jastrow-Monte-Carlo results[4] for the same NN potential utilized here). Needless to say, though the expected[5] empirical values for real nuclear matter are -17 MeV<E/N<-15MeV and 0.516<x<0.576 (for c=0.4F), nothing is known with certainty about the ground-state energy of our model potential nuclear matter as would be the case if, e.g., GFMC calculations were available for this simple interaction. The drastic order-of-magnitude discrepancy here noted would of course be clarified by such simulations or "theoretical" (computer) experiments. These, of course, are not only very expensive but also somewhat uncertain as yet for the fermion case. We now sketch the process by which we arrived at the results of Figure 1.

Many-body theory has produced few rigorous results for the microscopic description of bulk matter viewed as a system of interacting particles. Classical statistical mechanics, for example, has produced the virial expansion in powers of the number density n=N/V (N the number of particles, V the volume enclosing them) for the pressure of a fluid at given density and temperature. The expansion coefficients are multiple integrals over the (generally, pair) interactions and only up to the first seven are known for rigid spheres and up to the first five for Lennard-Jones particles.

Quantum field theory, on the other hand, has given us the well-known low density expansions for the ground-state energy of the many-fermion and many-boson systems interacting via a specific two-body potential. These expansions, however, are not power series but contain fractional powers and even logarithms of the density.

For the N-fermion system[6], for instance, each fermion pair-interaction can be characterized by low-energy scattering parameters a, R_0, $A_1(0)$, $A_0''(0)$, etc.---respectively called the S-wave scattering length and effective range, the P-wave scattering length (cubed), the(shape-dependent) S-wave second-moment-scattering-length (cubed), etc. Let the particle density be $n=N/V=\nu k_F^3/6\pi^2$, where $\hbar k_F$ is the Fermi momentum and ν the number of distinct species (2 for neutron, 4 for nuclear, etc., matter). The many-fermion ground-state energy per particle is then known to be given by the expansion(1). Here m is the particle mass and

$$\frac{E}{N} \simeq \frac{3}{5}\frac{\hbar^2 k_F^2}{2m}\Big\{1 + K_1 k_F a + K_2 (k_F a)^2 + \Big[\tfrac{1}{2}K_3\frac{R_0}{a} + K_4\frac{A_1(0)}{a^3} + K_5\Big](k_F a)^3$$

$$+ K_6 (k_F a)^4 \ln|k_F a| + \Big[\tfrac{1}{2}K_7\frac{R_0}{a} + K_8\frac{A_0''(0)}{a^3} + K_9\Big](k_F a)^4 \qquad (1)$$

$$+ o\big[(k_F a)^4\big]\Big\}$$

$K_1, K_2, \ldots, K_9$ are dimensionless coefficients given for $\nu=2$ and 4 in ref.[7], as taken from ref.[6] and corrected for errors. For either case, one has only <u>four</u> terms beyond the Fermi kinetic energy term since $K_6=0$ for $\nu=2$ while K_7, K_8 and K_9 are unknown to date for $\nu=4$. The series (1) is essentially an infinite order perturbation expansion about an ideal Fermi gas as unperturbed (reference) system.

A similar expansion exists for bosons, where only <u>three</u> coefficients are known and only the scattering parameter a enters through that order, which happens to be a log term. In the boson case it is very clear that negative a---which is the case for most physical pair-interactions---leads to trouble since the second correction term to the ideal Bose gas system involves a $a^{3/2}$ term which gives <u>complex</u> values. This can be avoided by substituting into the original low-density-ideal-gas-based perturbation expansions, the power-series expansions in terms of an appropriate <u>attractive</u> <u>interaction</u> strength parameter λ, of the scattering quantities a, R_0, $A_1(0)$, $A_0''(0), \cdots$. One thus gets the <u>rearranged</u> series

$$\frac{E}{N} \simeq A\frac{\hbar^2}{ma_0^2}x^2\sum_{i=0}^{\infty} e_i(x)\lambda^i \; ; \qquad A = \begin{cases} 2\pi & (bosons) \\ \frac{3}{10} & (fermions) \end{cases} \tag{2}$$

where $a_0 \equiv \lim_{\lambda\to 0} a$. Here $x=(na_0^3)^{1/2}$ for bosons, while $x \equiv k_F a_0$ for fermions. All terms in (2) are now real. It moreover corresponds to a perturbation scheme about the nontrivial fluid of repulsive particles (the original particles with the attractions turned off) as a reference system. Such a scheme we call a van der Waals perturbation scheme for obvious historical reasons. Any order in this perturbation scheme can be calculated without difficulty of principle; the only limitation emerges from our incomplete knowledge of the $e_i(x)$ $(i=0,1,2,\ldots)$ functions for the case of non-well-behaved pair potentials. For well-behaved ones, there are the <u>exact</u> calculations of Baker and Gammel[8] through 4th order. Their calculations refer to many fermions interacting via purely attractive square well potentials---with <u>no</u> hard-core repulsions---of unit depth and unit range R=1. Determination of the exact $e_i(x \equiv k_F R)$ envolved Monte Carlo evaluation of the multiple integrals represented by <u>one</u> Hugenholtz perturbation diagram for i=1, <u>one</u> for i=2, <u>four</u> for i=3, and <u>twenty-eight</u> for i=4. The resulting $e_i(x)$, normalized so that $e_i(0)=1$, behave as shown in Figure 2. Note that whereas $e_1(x)$ <u>increases</u> monotonically in density (as one might expect), $e_i(x)$, for i=2,3,4 <u>rapidly</u> <u>diminish</u> in density---except for the mild increase at higher densities of $e_4(x)$. Were it not for this small exception, moreover, one could assert that $e_i(x) < e_{i-1}(x)$ would be a general trend at any fixed density in the range considered. Finally, positivity, or $e_i(x)>0$, is observed for all four orders even though this behavior can be established rigorously only in 1st and 2nd order on general grounds.

In condensed matter physics one is commonly required, unfortunately, to deal with ill-behaved pair-potentials with rather singular repulsions. The infinite partial summations of ref.[6] then become mandatory. These lead to (1), and then to (2) as we have argued. Now, of course, the $e_i(x)$ naturally appear as low-density, non-power series known to as many terms as are known in the original low-density expansions: namely, 4 for fermions and 3 for bosons. Nevertheless, this is about twice of what one could recover from either ladder-perturbative or Jastrow-variational treatments. A graphical representation of our raw or truncated $e_i(x)$ series for the spin-averaged BHK potential is shown in Figure 3. The characteristic behavior observed in the exact (purely attractive square

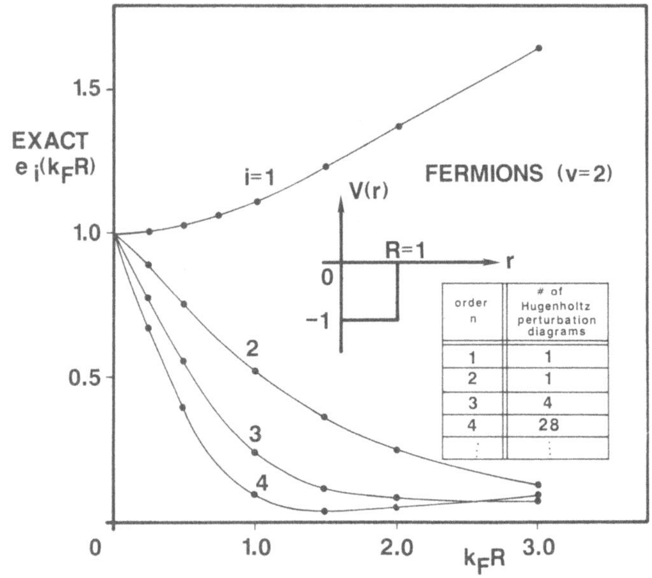

Fig. 2 Exact perturbation energies through 4th order

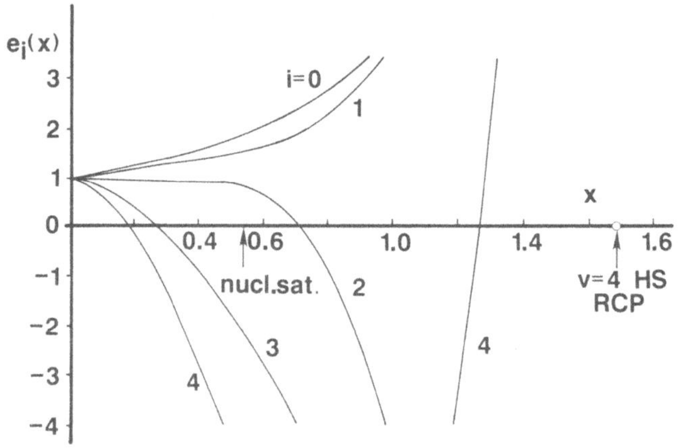

Fig. 3 Plot of Eq. (3) (below) through f_3 term

well) case of Fig. 2 is again noted for this hard-core-square-well potential but <u>only at very low densities</u>, still too low with respect to the relevant physical nuclear saturation value, marked by an arrow on Fig. 3, and which is $k_F c=0.544$ corresponding to $k_F = 1.36F^{-1}$ for $c=0.4$. The x value marked RCP refers to the random close packing density for 4-species fermion hard spheres as obtained in refs. [9,10]. One clearly needs a means of "taming" the "wild" behavior of the $e_i(x)$. To carry this out in a systematic fashion is the central task within the van der Waals perturbation scheme.

Our approach applies series analysis methods, like Padé approximants, to analytically extend the raw series $e_i(x)$ from zero density (x=0), where they are presumably exact to physical densities. We employ a new method which generalizes the Padé scheme, which we now describe briefly.

TAILING APPROXIMANTS

We present here a general method essentially inspired by, but somewhat more general than, Padé approximation[11] (both standard and generalized). It is intended to extrapolate to non-zero, physical densities the information contained in the low-density forms $e_i(x)$ for i=0,1,2... which appear in the van der Waals rearrangement for the ground state energy eq. (2).

The method is motivated by the very common situation in a physical series (regular, power, or otherwise, with, e.g., log terms) that the earlier terms of the series are numerically known to higher accuracy than the later terms in the "tail" of the expansion.

For the (4-species) many-fermion and many-boson cases to be presented here, the $e_i(x)$ have the forms

$$e_i(x) = 1 + f_1 x + f_2 x^2 + f_3 x^3 \ln x + f_4 x^3 + \ldots \qquad (fermions) \tag{3}$$

$$e_i(x) = 1 + b_1 x + b_2 x^2 \ln x^2 + b_3 x^2 + \ldots \qquad (bosons) \tag{4}$$

with f_4 and b_3 as yet unknown. In the spirit of the generalized Padé approximants discussed in ref.[10], the tailing procedure begins by expressing the least number of terms in the tails of either (3) or (4) that will allow expression as a ratio of functions N(x) and D(x) say, such that expansion about x=0 of N(x)/D(x) reproduces the initial tail expression. The next step starts with a tail expression having one more term, etc., until the whole expression---(3) or (4)---has been represented as ratios of such functions. In general then, we have the structures ($\hat{=}$ meaning "represented by"):

$$e_i(x) \hat{=} P(x) + \frac{N(x)}{D(x)} \equiv \epsilon_i(x) . \tag{5}$$

Here, P(x) is always polynomial (or zero) and the single log term in the raw series occurs <u>either</u> in N(x) or in D(x). Table 1 lists <u>some</u> of the <u>twenty-nine</u> distinct possibilities that were found by the tailing method and which could serve, potentially, to correctly extrapolate the $e_i(x)$ to non-zero x, including, hopefully, physical densities. <u>All</u> 29 forms were analyzed in detail and Table 1 is meant only as an illustration of our selection procedures. For example, the form VI from Table 1 stands for the approximant

$$\epsilon_i(x) = 1 + \frac{\alpha_1 x + \alpha_2 x^2}{1 + \beta_1 x + \beta_2 x^2 \ln x} . \tag{6}$$

Table 1

Form	# of terms in P(x)	terms in N(x)	terms in D(x)	Order 1	Order 2
I	2	$x^2, x^3\ln x$	$1, x$		
II	2	x^2	$1, x\ln x, x$		
III	2	x^2, x^3	$1, x\ln x$	p	p
IV	1	$x, x^3\ln x$	$1, x, x^2$	nmi	p
V	1	x	$1, x, x^2\ln x, x^2$		p
VI	1	x, x^2	$1, x, x^2\ln x$		
VII	1	x, x^2, x^3	$1, x^2\ln x$	p	p
VIII	0	$1, x, x^2, x^3$	$1, x^3\ln x$	p	p
IX	0	$1, x, x^3\ln x$	$1, x, x^2$	p	
X	0	$1, x$	$1, x, x^2, x^3\ln x$	p	
XI	0	$1, x, x^2$	$1, x, x^3\ln x$		
XII	0	1	$1, x, x^2, x^3\ln x, x^3$	nmi	ns
XIII	0	$1, x^3\ln x$	$1, x, x^2, x^3$	nmi	ns

p=pole; nmi=non-monotonic-increasing; ns=no solution.

Binomial expansion of the denominator here, and subsequent matching with (3), determines the constants in (6) to be

$$\alpha_1 = f_1, \quad \alpha_2 = f_2 - f_1 f_4 / f_6, \quad \beta_1 = -f_4 / f_2, \quad \beta_2 = -f_3 / f_1. \tag{7}$$

A table similar to Table 1 can be constructed for the boson case starting from (4) and yields only <u>twelve</u> distinct forms.

NUCLEAR MATTER

The $e_i(x)$ series appearing in (2) have been derived[10] through the computer algebraic manipulation scheme known as MACSYMA. They are valid for any central two-body potential. The coefficients in these $e_i(x)$ are thus explicitly known in terms of the λ-expansions coefficients for the a, R_o, $A_1(0)$, ... parameters in eq. (1). These latter expansions have been computed[13] for several realistic two-body potentials of interest like the Aziz, Lennard-Jones, Kolos-Wolniewicz, etc. potential functions.

For the special case of hard-core-square-well potentials--- the BHK for fermions and the Burkhardt[15] for bosons---we may use the expansions of ref.[7] For 4-species-BHK-fermions we have

$$\frac{E}{N} = \frac{3}{10}\frac{\hbar^2}{mc^2} x^2 \Big\{ 1 + 1.061033x + 0.556600x^2 + 1.300600x^3 + 1.408600x^4 \ln x + \dots$$

$$- 1.111524x \left[1 + 1.049199x + 1.584802x^2 + 5.310462x^3 \ln x + \dots \right] \lambda -$$

$$- 0.444620x \left[1 - 0.324720x - 5.359754x^2 - 15.551611x^3 \ln x + \dots \right] \lambda^2 - \tag{8}$$

$$-0.179971x\left[1-1.66636x-7.892368x^2+0.0734071x^3\ln x+\ldots\right]\lambda^3$$
$$\left.-0.0729279x\left[1-3.003423x-6.323303x^2+27.10563x^3\ln x+\ldots\right]\lambda^4+\ldots\right\}.$$

Note the reduction by a factor of approximately 2.5 from one order to the next, at small x. We shall use the "averaged-BHK" λ-value as defined in ref.[2] This is simply

$$\lambda=\frac{1}{2}\left(\lambda_{sing}+\lambda_{trip}\right)=\frac{1}{2}\left(2.368705+3.084251\right)=2.726478, \tag{9}$$

where either λ_{sing} or λ_{trip} stand for $mv_0(R-c)^2/\hbar^2$ with v_0 and R the well depth and range, respectively, and c the hard core diameter. The original, truncated (non-power) series in the square brackets of (8) multiplying λ^i are the $e_i(x)$ alluded to before and graphed in Fig. 3. For $e_0(x)$ we use the representation $\epsilon_0(x)$ of ref.[9]

To "tame" these $e_i(x)$ (i=1,2,3,4) we impose the physical constraint at random close packing density $x=x_B$ which requires that their corresponding <u>approximants</u> $\epsilon_i(x)$ must conform to

$$\epsilon_1(x_B)=const \quad ; \qquad \epsilon_i(x_B)=0 \qquad (i=2,3,4) \tag{10}$$

as detailed in ref.[10][12] This amounts to noting that at the ultimate fluid density of the hard-sphere system the energy per particle must be <u>linear in</u> λ since

$$\frac{E}{N}\xrightarrow[x\to x_B]{}\infty-\frac{1}{2}v_0\left[\frac{4}{3}\pi R^3 n_B-1\right], \tag{11}$$

where the "∞" term arises from the hard-sphere energy divergence at x_B and the square parenthesis is just the number of particle centers found within the attractive-well range as the Bernal density $n_B=2x_B^3/3\pi^2c^3$ is approached. Thus, (11) determines the non-zero constant in (10) and the constraints (10) will in turn determine, for each i, the unknown coefficient f_4 in the corresponding $e_i(x)$ of (3).

Not all 29 possible forms for either $\epsilon_1(x)$ or $\epsilon_2(x)$ are eligible approximants, however. Table 1 lists the properties of 12 of them on the last two columns. All cases marked with a "p" develop a pole in the physical interval $0<x<x_B$, and are hence discarded. Also rejected are forms like XII or XIII for first-order which <u>decrease</u> in x. Figure 4 illustrates these as well as other forms from Table 1, and compares the truncated form $e_1(x)$ (dashed curve). Only forms I, II, V, VI, and XI remain which are still acceptable. Of these I and XI are selected since, <u>being distinct forms analytically</u>, they are very close to each other numerically over the whole physical range, as Figure 5 illustrates. Note that the difference between the two is <u>less than 2%</u> of the maximum value of $\epsilon_1(x)$ which is $\epsilon_1(x_B)=3.011$. In both Figs. 4 and 5 the open circle marks the Bernal density value associated with $x_B=1.536$ for the 4-species fermion hard sphere system.

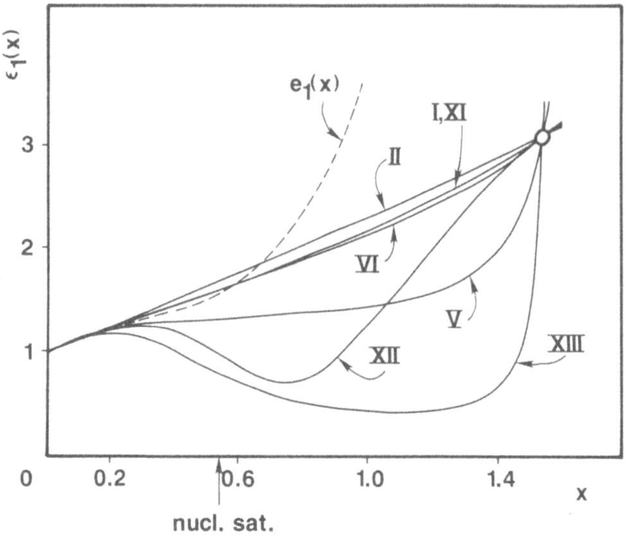

Fig. 4 First order energy approximants.

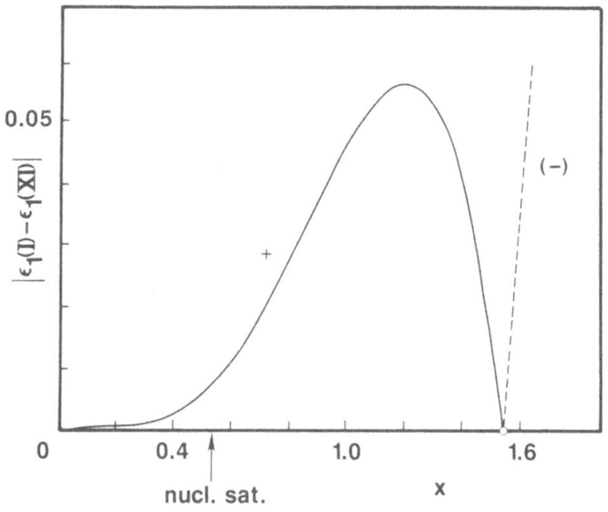

Figure 5. Difference between forms I and XI, Table 1, for first order energies.

For second order, a glance at Table 1 shows that only forms I, II, VI, IX, X and XI are acceptable. These are all graphed in Figure 6 and compared with the truncated series $e_2(x)$ (dashed curve). Forms I and VI were selected for being very similar in value---within 0.2% (at x=0.544) of the maximum value of $\epsilon_2(x)$ which is $\epsilon_2(0)=1$. We retain both forms, however, since their difference can increase to about 7% at higher densities. With form I (form VI) we get the dashed (full) energy vs k_Fc curves in Figure 1; the shaded area represents the uncertainty of the method, or about 10% of maximum binding. As mentioned before, the striking result is the order-of-magnitude larger binding, for the averaged-BHK model potential, with respect to FHNC variational results.[2)]

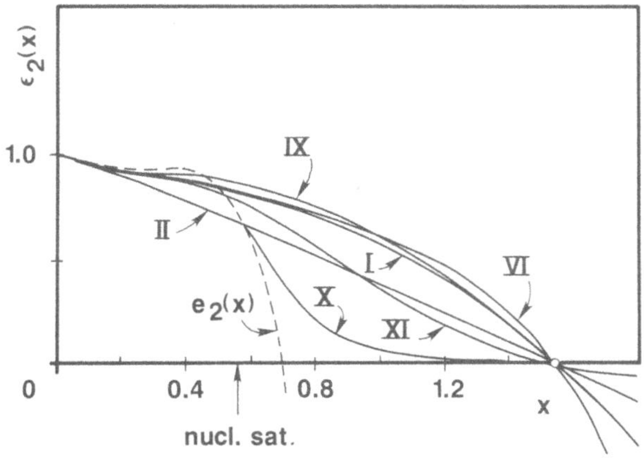

Fig. 6. Accepted approximants for second order energies

LIQUID HELIUM FOUR

The same general method has been applied to liquid helium four starting from the low-density expansion[14)] for the ground state energy per particle of a many-boson system. The Burkhardt[15)] hard-core-square-well pair-potential was used. Its attraction v_0, range R and hard-sphere diameter c parameters were determined by him so as to make the potential "phase-equivalent" to the Lennard-Jones helium-helium interaction for which, significantly, there are GFMC calculations[16)] which virtually coincide with the experimental energy vs density curve for the liquid helium-four ground state.

We only sketch matters here since results through third order were reported at the last Workshop,[12)] and we are here basically extending them through fifth order to acertain stability and convergence. The starting point is the rearrangement (2) with the constant $A=2\pi$ and $x=(nc^3)^{1/2}$ with n the particle density. The $e_i(x)$ have the form (4) with the coefficients b_1 and b_2 given, for any central potential, in ref.[10)] in terms of the λ-expansion coefficients of the S-wave scattering length a. Again, b_3 is unknown for all order i. For zeroeth order it has been determined for hard spheres in ref.[17)] through a global fit to GFMC data.[18)] This gave a value of $x_B=0.7082$. For order i=1,2,... we can again impose the constraints (10) at this value of x_B in order to fix this coefficient b_3 for each order and each generalized Padé form to the truncated expression (4). The tailing procedure discussed before now yields precisely twelve distinct forms. These are described in Table 2, in entirely the same general manner as with Table 1 for 4 species fermions. The abbreviations p, nmi and ns have the same meanings as in Table 1. In addition, we have imposed two new criteria, both present in the exact (coreless) results of Figure 2, and which, more importantly, have thus far been corroborated in the 4-species fermion problem (with a hard core) through fourth order. These two conditions are: i) positivity in orders three and higher and ii) monotonicity in order, for fixed density, namely that $\epsilon_i(x) > \epsilon_{i+1}(x)$. (We found that without these two additional requirements the expected convergence in the final equation-of-state results within the van der Waals scheme was completely

Table 2

FORM	#of terms in P(x)	terms in N(x)	terms in D(x)	Order 1	2	3	4	5
I	1	$x, x^2 \ln x^2$	$1, x$	p	p	p	vp	p
II	1	x, x^2	$1, x\ln x^2$		vp		p	p
III	1	x	$1, x\ln x^2, x$				vp	p
IV	0	$1, x$	$1, x, x^2\ln x^2$	p	p	p	vp	vp
V	0	$1, x, x^2\ln x^2$	$1, x$	p	p	p	vp	p
VI	0	1	$1, x, x^2\ln x^2, x^2$	nmi	ns	ns	ns	ns
VII	0	$1, x, x^2\ln x^2$	$1, x^2$	p	ns	ns	ns	ns
VIII	0	$1, x^2\ln x^2, x^2$	$1, x$	p	vmo	vmo	vp	vp
IX	0	$1, x, x^2$	$1, x^2\ln x^2$	nmi	p	p	vp	vp
X	0	$1, x$	$1, x^2\ln x^2, x^2$	vp	ns	ns	ns	ns
XI	0	$1, x^2\ln x^2$	$1, x, x^2$	p	ns	ns	ns	ns
XII	0	$1, x^2$	$1, x, x^2\ln x^2$	nmi	p	p		

destroyed). Thus, in Table 2 "vp" and "vmo" respectively mean violation of (i) and (ii) above. This all leads to only forms II and III as acceptable for order one; III for order two; II and III for order three; and XII for orders four and five.

In first order forms II and III differ by at most 0.0075; this is only 0.25% of the maximum value of $\epsilon_1(x)$ which is $\epsilon_1(x_B)=2.95542$ because of constraint (10). Thus we retain form II and discard III. In order two only one form survives, namely III. In order three forms II and III are still both acceptable and so we retain both of them; as before, this will determine the uncertainty of the final result. In orders four and five only form XII has survived. These forms are graphed in Figure 7, and are all unique save for third order where the two pertinent forms are shown in full (III) and dashed (II) curves. It is both interesting and instructive to compare Figures 2 and 7.

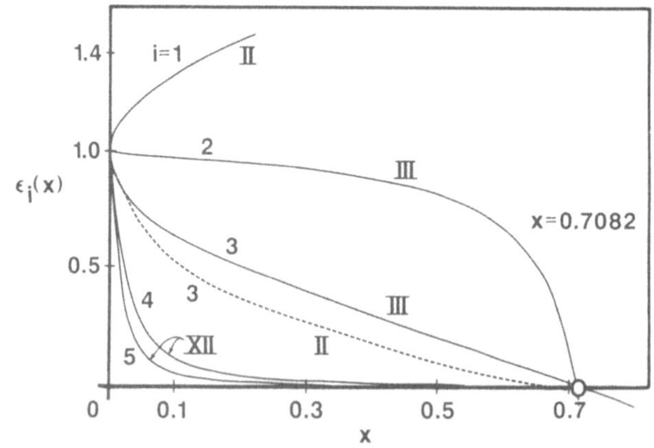

Fig. 7. Boson perturbation energies

Finally, the Padé approximants $[L/M](\lambda)$ to the λ-power-series (2), with $e_i(x)$ replaced by the $\epsilon_i(x)$ approximants selected as just described, were constructed for L,M=0,1,...,5, with $1 \leq L+M \leq 5$. The resulting energy-per-particle vs density curves are qualitatively similar to those of Figure 1 and will not be presented. We do present, however, in Figure 8 an illustration of the rapidity of convergence to a single ground state saturation energy E^{sat}_{theor}, in units of the empirical[19] liquid helium four value of $E^{sat}_{exp} = -7.14$ K/atom, as one goes from third through fifth order, using the form III for $\epsilon_3(x)$. With form II for $\epsilon_3(x)$ one gets about 1 K less binding per particle, out of the total of about 15 K/particle predicted with form III. This then is our uncertainty. Very swift convergence is observed in Figure 8 to about 2.1 times the empirical binding. Even faster is the convergence to an equilibrium density of about 1.5 times the experimental value. These discrepancies with (laboratory) experiments are of course not necessarily significant

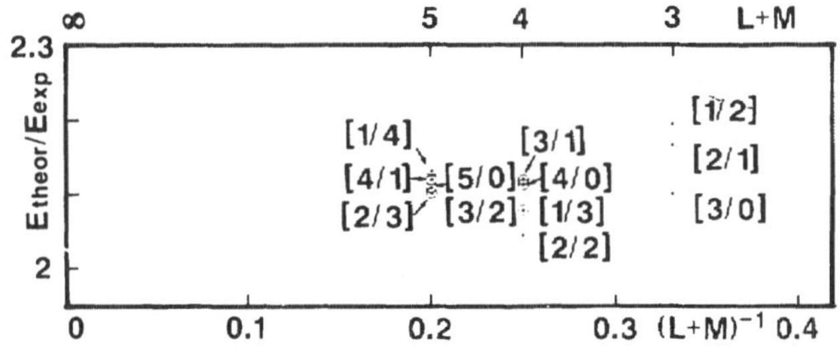

Figure 8. Calculated to empirical liquid ^{4}He binding energy

because real liquid helium four may not be adequately described by the two-body potential model used here, and there are doubtlessly present many-body forces in the real liquid.

CONCLUSIONS

We have presented the results of detailed generalized Padé analyses of log-bearing low-density series for the perturbation corrections to the ground state energy per particle of a many-fermion as well as a many-boson system. The perturbation scheme is not one about the ideal gas state but about the highly nontrivial hard sphere fluid. The fermion system is nuclear matter with the nucleons interacting pairwise via the Baker-Hind-Kahane (BHK) hard-core-square-well potential while the boson system is liquid helium four with the atoms interacting pairwise with a hard-core-square-well potential which is phase-equivalent to the corresponding Lennard-Jones potential function.

The set of generalized Padé approximants constructed is sufficient to obtain reasonable representations of the fermion problem through second order and the boson problem through fifth order. By reasonable we mean with general behavior characteristic of the exact calculations through fourth order of coreless fermions interacting via a purely attractive square well.

These representations are then employed to set up the perturbation calculation of the ground state energies of both systems. About ten times more binding is obtained in the nuclear matter, through second order, than FHNC variational calculations. For the helium four case fifth order is seen to be more than enough to get a stabilized, converged result.

The crucial question is: to what extent one can rely on the present truncation scheme to give accurate information about the many-body Schrodinger ground-state energy for a specific(two-body) hamiltonian? This question will definitely be answered when further (computer) experiments (viz., GFMC) are done, even if only for pure hard-sphere (fermion) systems.

REFERENCES

1. G.A. Baker, Jr., M.F. Hind & J. Kahane, Phys. Rev. C2: 841(1970)

2. J.W. Clark, M.T. Johnson, P.M. Lam & J.G. Zabolitzky, Nucl. Phys. A283:253(1977)

3. V.C. Aguilera-Navarro et al, "Constructive Methods for the Ground State Energy of Neutron Matter" (to be published)

4. K.E. Schmidt, priv. comm.

5. B.D. Day, Rev. Mod. Phys. 50:495(1978)

6. G.A. Baker, Jr. Rev. Mod. Phys. 43:471(1971)

7. G.A. Baker, Jr., et al, Phys. Rev. A26:3575 (1982)

8. G.A. Baker, Jr. et al, Phys. Rev. 132, 1373(1963); and ref.[6)]

9. G.A. Baker, Jr. et al, Ann. Phys. (N.Y.) 153:283(1984)

10. M. Fortes et al, "Van der Waals Perturbation Theory for Boson and Fermion Ground State Matter" (to be published)

11. G.A. Baker, Jr. & P. Graves-Morris, Pade Approximants, Vols. 13 and 14 of Encycl. of Math. and its Appls., ed. by G. C. Rota (Addison-Wesley, Reading, MA, 1981

12. M. Fortes et al, Cond. Matter Theory. (IX Workshop) ed. by F. B. Malik (Plenum Press, 1986)

13. L.P. Benofy et al, Phys. Rev. A 33:3749(1986)

14. N.M. Hugenholtz & D. Pines, Phys. Rev. 116:489(1959)

15. T.W. Burkhardt, Ann. Phys. (N.Y.) 47:516(1968)

16. M.H. Kalos et al, Phys. Rev. B 24:115(1981)

17. G.A. Baker, Jr. et al, Phys. Rev. B 24:6304 (1981)

18. M.H. Kalos et al, Phys. Rev. A 9:2178(1974)

19. P.R. Roach et al, Phys. Rev. A 2:543(1970)

BOSON-MIXTURES AT NON-ZERO TEMPERATURES

K. E. Kürten and M.L. Ristig

Institut für Theoretische Physik
Universität zu Köln
D-5000 Köln 41, W.-Germany

Experimental and theoretical studies of quantum fluid mixtures are of actual interest and can provide important contributions for an understanding of fundamental macroscopic quantum many-body effects. Liquid ^{4}He provides the simplest theoretical model of a binary boson mixture by assuming that a fraction of ^{4}He atoms my be distinguished (say by 'coloring' them) from the rest. As a representative sample of more complex binary quantum fluid mixtures we may list mixtures of spin-polarized hydrogen and tritium ($H^{\downarrow}$-$T^{\downarrow}$) /1/ or of isotopic helium atoms (^{6}He-^{4}He) /2/, solutions of tritium in helium ($T^{\downarrow}$-^{4}He) /3/ or the classic ^{3}He-^{4}He) system /4/.

One of the most interesting questions concerns the critical behaviour for the first order phase transitions of such systems. A suitable generalization of the variational density-matrix theory recently developed for one-component homogeneous boson fluids /5/ offers an adequate approach for addressing this question and for examining, more generally, the phase diagrams of quantum mixtures on the basis of an ab-initio procedure.

Here, we outline briefly a realization of this formal program for homogeneous binary boson mixtures at finite temperatures based on the hypernetted-chain approximation and on the separability approximation for the entropy /5/. We emphasize, however, that the variational formalism may be developed on a more advanced level since the approximations adopted at present can be systematically improved.

The density-matrix approach permits an optimization of the partial structure functions $S_{11}(k)$, $S_{22}(k)$ for components i=1,2 and of the mixed structure function $S_{12}(k)$ characterizing the binary mixture and of the associated elementary excitation energies $\varepsilon_1(k)$ and $\varepsilon_2(k)$ for the first sound mode and the concentration mode, respectively. The procedure is based on the Gibbs-Delbrück-Molière minimum principle for the trial Helmholtz free energy at temperature $T=k_B\beta^{-1}$,

$$F = (E_o+U_o)-TS_o. \tag{1}$$

Adopting the separability assumption the entropy S_o and the internal energy functional (E_o+U_o) are explicitely given by

$$S_o = (2\pi)^{-3} V \int d\underline{k} \; \mathrm{Tr}\{(1+\underline{n})\ln(1+\underline{n}) - \underline{n} \ln \underline{n}\}, \tag{2}$$

$$E_o = \varrho^{-1} V \{\varrho_1^2 E_{11} + 2\varrho_1\varrho_2 E_{12} + \varrho_2^2 E_{22}\}, \tag{3}$$

$$2U_o = -(2\pi)^{-3} V \int d\underline{k} \; \mathrm{Tr} \; \underline{S} \; \underline{\nu}^*, \tag{4}$$

the system being characterized by the partial densities ϱ_1, ϱ_2, the total density ϱ and confined to volume V. The structure matrix $\underline{S}$ represents the elements $S_{ij}(k)$ and matrix $\underline{n}$ is defined by $\underline{n}=(\exp\beta\underline{\varepsilon}-1)^{-1}$ with matrix $\underline{\varepsilon}$ having the eigenvalues $\varepsilon_1(k)$ and $\varepsilon_2(k)$. The energy components E_{ij} are the integrals

$$2E_{ij} = \varrho \int v_{ij}^*(r) g_{ij}(r) d\underline{r} \tag{5}$$

which involve the partial radial distribution functions $g_{ij}(r)$ and the effective potentials $v_{ij}^*(r)$ explicitely defined in Ref. /6/. The trace (4) is generated from the elementary excitations present at non-zero temperatures via the potential matrix

$$\underline{\nu}^* = -2\underline{S}^{-1}\underline{\varepsilon}_o^{\frac{1}{2}} \underline{n}(1+\underline{n})\underline{\varepsilon}_o^{\frac{1}{2}} \underline{S}^{-1} \tag{6}$$

where the (diagonal) matrix $\underline{\varepsilon}_o$ represents the kinetic energy elements $\varepsilon_{oi}(k)=\hbar^2k^2/2m_i$ for particles of species i=1,2.

The associated Euler-Lagrange equations which determine the optimal matrices $\underline{S}$ and $\underline{\varepsilon}$ and their optimal eigenvalues may be cast into the form of a generalized Feynman matrix equation

$$\underline{\varepsilon} \; \mathrm{tgh} \; \frac{\beta}{2}\underline{\varepsilon} = \underline{\varepsilon}_o^{\frac{1}{2}} \underline{S}^{-1} \underline{\varepsilon}_o^{\frac{1}{2}} \tag{7}$$

and a generalized Bogoliubov matrix equation

$$\underline{\varepsilon}^2 = \underline{\varepsilon}_o^{\frac{1}{2}} (\underline{\varepsilon}_o + 2\underline{\nu})\underline{\varepsilon}_o^{\frac{1}{2}} \; . \tag{8}$$

Eq. (8) contains a self-consistent potential matrix $\underline{\nu}$ which is determined by the structure of the medium itself. In hypernetted-chain approximation this structural relation reads

$$\nu_{ij}(r) = v_{ij}g_{ij} + (g_{ij}-1)(w_{ij}+\nu_{ij}^*) + \frac{\hbar^2}{2}\left(\frac{1}{m_i}+\frac{1}{m_j}\right)(\underline{\nabla}\sqrt{g_{ij}})^2 \; . \tag{9}$$

Here, $v_{ij}(r)$ is the bare interaction potential between a particle of species i and a particle of species j. Functions $\nu_{ij}(r)$ and $\nu_{ij}^*(r)$ are the Fourier inverse of the elements $\nu_{ij}(k)$ and $\nu_{ij}^*(k)$ of the matrix $\underline{\nu}$ and of $\underline{\nu}^*$, respectively. Quantities $w_{ij}(r)$ are the Fourier inverse of the elements $w_{ij}(k)$ of a matrix $\underline{w}$ which represents the induced potential defined by eq. (45) of Ref. /5/ generalized in a straight-forward fashion to binary mixtures.

Eqs. (7)-(9) constitute a coupled set of equations for the optimal partial structure functions $S_{ij}(k)$ and the elementary excitation energies

$\varepsilon_i(k)$. These energies depend linearly on the wave number, $\varepsilon_i(k)=\hbar c_i k$, as $k\to 0$. The velocities c_1 and c_2 are functions of temperature and density ρ. If one or both of them vanish the system becomes unstable. In particular, $c_2=0$ signals an instability against fluctuations of the concentration of the admixed component and thus phase separation sets in.

The generalized density matrix formalism may be employed for a numerical study of boson mixtures such as listed above. Here, we present some of our results on the simplest example of a binary isotopic mixture: the two components consist of ^{4}He-atoms, i.e., $m_1=m_2=m_{He}$, but the components are assumed to be distinguishable (by some additional property). In this case the eqs. (7)-(9) simplify somewhat since the matrices $\underline{S}$, $\underline{\varepsilon}$ and $\underline{\gamma}$ may be simultaneously diagonalized. This transformation may be achieved at given wave number k and concentration $x=\rho_2/\rho$ by a unitary matrix of the form

$$U = \begin{pmatrix} \sqrt{1-x} & -\sqrt{x} \\ \sqrt{x} & \sqrt{1-x} \end{pmatrix} \quad (10)$$

generating the eigenvalues $\varepsilon_i(k)$ and $S_i(k)$ for the matrix $\underline{\varepsilon}$ and $\underline{S}$, respectively. Corresponding to the interpretation of $\varepsilon_1(k)$ and $\varepsilon_2(k)$ as the excitation energies of the first sound mode and the concentration mode the structure functions $S_1(k)$ and $S_2(k)$ describe, respectively, the correlations of the total density fluctuations $(\rho_{\underline{k}}^{(1)}+\rho_{\underline{k}}^{(2)})$ and of the difference $(x\rho_{\underline{k}}^{(1)}-(1-x)\rho_{\underline{k}}^{(2)})$, $\rho_{\underline{k}}^{(i)}=\sum_{\ell}^{N_i}\exp i\underline{k}\cdot\underline{r}_{\ell}^{(i)}$ being the density fluctuation operator for particles of species i=1,2.

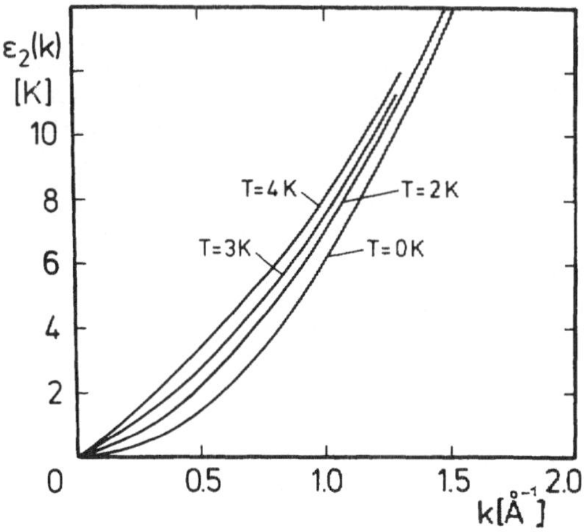

Fig. 1: Elementary excitation energies ε_2 for the concentration mode in ^{4}He at various temperatures and saturation density ρ = 0.02185 Å^{-3} as function of wave number k.

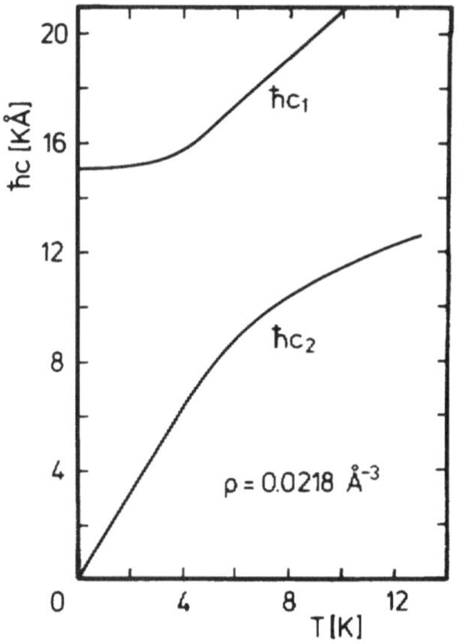

Fig. 2: The velocities of first sound $\hbar c_1$ and of the concentration mode $\hbar c_2$ at density $\rho=0.02185$ Å^{-3} as functions of temperature T. The results are independent of concentration x within the numerical accuracy of the calculation.

Within numerical accuracy the calculated energies $\varepsilon_i(k)$ and structure functions $S_i(k)$ - as well as the corresponding distribution functions $g_i(r)$ - are found to be independent of concentration x. Moreover, $\varepsilon_1(k)$ and $S_1(k)$ agree numerically very well with the single-component results for ^{4}He reported in Ref. /5/. Our numerical results on the energies $\varepsilon_2(k)$ are shown in Fig. 1 at total density $\rho=0.02185$ Å^{-3} and at various temperatures. Fig. 2 depicts the interesting behaviour of the velocity c_2 describing the propagation of the concentration mode compared with c_1 for the ordinary sound. The concentration mode becomes soft only at vanishing temperature indicating that the phase separation line in the x-T diagram is at T=0 as physically expected. For temperatures $T<4$K the velocity c_2 is extremely well represented by a linear relation $\hbar c_2=l_o T$ with a characteristic length $l_o \simeq 1.54$ Å. Consequently, the structure function $S_2(k)$ exhibits a singularity $S_2(0)=\hbar^2(m_{He} l_o^2 T)^{-1}$ at k=0 as the temperature vanishes. Fig. 3 represents our results on the spatial correlation function $g_2(r)$. At small T the function is of long range becoming short-ranged with increasing temperature. Its amplitude increases from zero at T=0K to a maximum of about 0.12 at T~2K decreasing for higher temperatures.

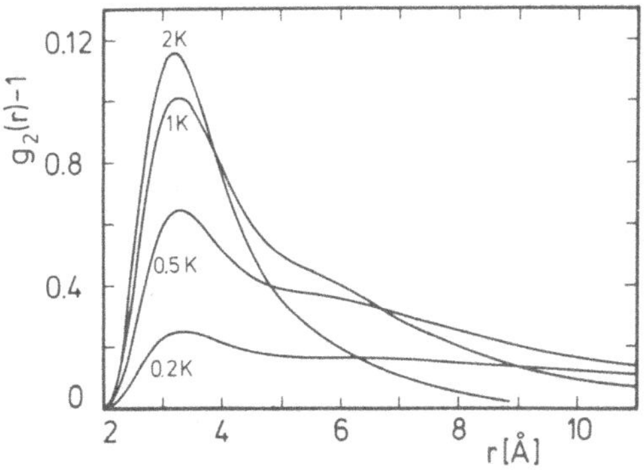

Fig. 3: The radial distribution function $g_2(r)$ at various temperatures T representing the correlations between the fluctuations of concentration.

References

/1/ M.D. Miller and L.H. Nosanow, Phys. Rev. B15, 4375 (1977)
/2/ L.H. Nosanow, J. Low Temp. Phys. 23, 605 (1976)
/3/ K.E. Kürten and M.L. Ristig, Phys. Rev. B31, 1346 (1985)
/4/ G. Baym, in: The helium liquids, ed. by J.G.M. Armitage and I.E. Farquar, Academic press (1975)
/5/ G. Senger, M.L. Ristig, K.E. Kürten and C.E. Campbell, Phys. Rev. B33, 7562 (1986)
/6/ M.L. Ristig, S. Fantoni and K.E. Kürten, Z. Phys. B51, 1 (1983)

ELEMENTARY EXCITATIONS IN TWO DIMENSIONAL ELECTRON GAS ARRAYS *

P. Hawrylak, J.-W. Wu, and J. J. Quinn

Brown University

Providence, RI 02912 USA

ABSTRACT

Two examples of elementary excitations in layered electron gas systems are considered, magneto-excitons of a system with integral Landau level filling and plasmons of a quasiperiodic system. For the former case, the simplest elementary excitations are obtained by promoting an electron from the highest occupied Landau level to the lowest unoccupied one. The single particle picture doesn't adequately describe these excitations because the resulting electron and hole interact to form a magneto-exciton. The energies of the singlet and triplet magneto-excitons can be calculated in perturbation theory if the Coulomb energy is small compared to the cyclotron energy. For a periodic array of two dimensional electron gas layers, the magneto-exciton dispersion broadens into a band with Bloch wavenumber k_z describing the relative phase of the excitation amplitude on neighboring layers. A qualitative comparison of the elementary excitations of these simple systems with the magneto-roton excitations and Laughlin quasiparticles in the quantum Hall effect is made. The second example, plasmons of a quasiperiodic layered electron gas system, can be studied using a transfer matrix method. The band structure found is quite similar to the electron energy bands of a quasiperiodic potential. As the size of the quasiperiodic array tends to infinity, an infinite number of very narrow bands is found, which display typical self-similar Cantor set structure. The behavior of the total band width as function of size, which is a measure of localized vs. extended behavior of the plasmons, is discussed. The global scaling properties of the spectrum and the infrared resonant absorption are studied.

I. Two dimensional electron gas layers in a strong magnetic field

The behavior of a two dimensional electron gas in the presence of a quantizing magnetic field is a difficult mathematical problem. The reason for this is that in the absence of many body interactions the single particle eigenstates are highly degenerate, and no suitable perturbation expansion exists. The situation is considerably simpler if the Landau levels involved are either completely filled or completely empty. In that case the ratio of the Coulomb energy e^2/ℓ, where $\ell = (\hbar c/eB)^{1/2}$ is the magnetic length, to the cyclotron energy $h w_c$ can play the role of the small parameter necessary to a well defined perturbation theory. The low lying excitations consist of singlet or triplet electron-hole pairs.

Because of their interaction with one another, the electron and hole form a bound state usually referred to as a magneto-exciton. The binding energy of the magneto-excitons has been studied by a number of authors [1]. For the most thorough analysis the reader is referred to the work of Kallin and Halperin [1].

For the case in which only the lowest Landau, lowest spin level is occupied, the energy of the n = 1 magneto-exciton in the x-y plane is given by (to first order in perturbation theory)

$$E(K) = \hbar w_c + 1/2\left(\frac{\pi}{2}\right)^{1/2}(e^2/\ell)\{1 - \exp(-K^2\ell^2/4)(1 + 1/2K^2\ell^2)$$
$$I_o(K^2\ell^2/4) - 1/2K^2\ell^2 I_1(K^2\ell^2/4) - (2/\pi)^{1/2}K\ell\exp(-\exp(-K^2\ell^2/4)]\}.$$

Here I_o and I_1 are modified Bessel functions. The momentum of the exciton's center of mass is $\hbar\vec{K}$, and $\ell^2\vec{K}\times\hat{z}$ is the expectation value of the particle hole separation. This energy is plotted as a function of $K\ell$ as the dashed line in Fig. 1a. For $K\ell \to \infty$, the binding of the exciton goes to zero. The difference between E(K) and hw_c is completely attributable to the exchange energies of the particle and hole, each acting as independent quasiparticles. As $K\ell \to 0$, $E(K) \to \hbar w_c$, indicating that the exciton binding energy exactly cancels the exchange energy as required by Kohn's theorem [2].

It is clear that the energy of a single magneto-exciton is independent of the position of the center of mass. If two magneto-excitons are present however, they will interact with one another. For large separations of their centers of mass they interact via dipole-dipole coupling since they each have dipole moments proportional to the momenta of their centers of mass. For separations of the order of their size, the interaction is more complex and exchange between excitons becomes important. As far as the authors are aware no one has investigated bi-exciton or multi-exciton complexes in these systems although they have been studied for a related problem, that of a two-dimensional semimetal containing electron and hole bands. When the system is totally isotropic the Fermi level lies at the midpoint of the band overlap and all Landau levels are either completely filled or completely empty. The magneto-exciton formed by promoting an electron from the highest filled electronic level in one band to the lowest empty electronic level in the other has been rather thoroughly studied in the literature [3]. The question of exciton complexes and the possibility of phase transitions to new ground states have also been investigated.

For a periodic array of two-dimensional electron gas layers, the 2-D magneto-excitons on different layers interact with one another. The coupled modes form bands with Bloch wavenumber k_z which describes the relative phase of the excitations on neighboring layers [4]. For a system with layer spacing "a", the dispersion relation (Eq.1) is modified by having the structure factor $S(K, k_z) = \sinh Ka\,[\cosh Ka - \cos k_z a]^{-1}$ multiply the last term in the square bracket, i.e. the term $-(2/\pi)^{1/2}\,K\ell\,\exp(-K^2\ell^2/4)$. For $Ka>>1$, the structure factor approaches unity and every layer supports an independent magneto-exciton. For smaller values of Ka the magneto-exciton (or magneto plasmon) band has a finite band width as shown by the solid lines in Fig. 1a for the case $a = 10\ell$. In three dimensional electron gas problems, these modes are called cyclotron waves or magneto-plasmons.

In the quantum Hall [5] effect fractionally charged quasiparticles and quasiholes are the elementary excitations. Laughlin [6] first proposed the incompressible liquid ground state and described the fractionally charged excitations. The gap for creating a quasiparticle-quasihole pair, unlike the cyclotron energy of the filled Landau level problem, is itself dependent on the electron-electron correlations and thus proportional to e^2/ℓ. For the purpose of picturing the elementary excitation of the Laughlin state the magneto-exciton problem is a qualitatively useful analogy. At a fixed value of the applied magnetic field and of the electron concentration, the elementary excitations are created in pairs. Because they are charged, they interact to form magneto-excitons (or, in analogy with He^4, magneto-rotons). The magneto-exciton spectrum has been determined by Girvin et al. [7] using the Laughlin ground state. The dispersion relation is very similar to that of the magneto-exciton of a filled Landau level. We may surmise that at very large values of the quasiparticle separation the magneto-exciton energy should be that of an independent quasielectron and quasihole (as first proposed by Laughlin). For smaller values of the separation, the quasiparticle and quasihole form a bound state, the magneto-roton. The possibility of bi-excitons and multi-exciton complexes seems worth investigation, particulary for this system where the quasiparticle energies can be of the same order as the binding energies. The analogy with the filled Landau level magneto-exciton problem might give some useful insight, however the complexity of the mathematical problem when no small parameter is present makes it impossible to use the same approach to calculating energy values.

II. Plasmons in quasiperiodic arrays of two-dimensional electron gas layers

There has been a lot of interest in electronic and phonon properties of one-dimensional quasiperiodic systems [1]. Recently, Das Sarma, Kobayashi and Prange [2] and Hawrylak and Quinn [3] pointed out that plasmons in artificially structured materials exhibit many interesting phenomena such as localization transition and critical states in a similar way that electrons do. We will discuss here a particular system which exhibits only critical behavior.

Plasmons in periodic semiconductor superlattices, such as modulation doped GaAs-GaAlAs, are now well understood [4]. Plasmon can propagate along superlattice direction, and the allowed plasmon frequencies form bands characterized by a Bloch index k. The case of a quasiperiodic superlattice [3] is more challenging because the Bloch theorem is not applicable. Recently, a quasiperiodic semiconductor superlattice has been grown by Merlin et al. [5]. It consists of two building blocks A and B having thicknesses a and b respectively arranged in the Fibonacci sequence. Each block is composed of GaAs and GaAlAs layers. If the GaAlAs region is doped with donors, a layer of quasi two-dimensional electrons can be produced in every block A and B. Such a system can be thought of as an array of electron gas layers, separated by distances a or b, arranged in a quasiperiodic (Fibonacci) sequence. Because electrons do not tunnel between the layers, the problem of collective charge density excitations is essentially that of obtaining a self-consistent solution of the Poisson equation for the induced potential (charge density) on every layer. This problem of solving a quasiperiodic Poisson equation is similar to the problem of solving a quasiperiodic Schrodinger equation, which has been studied in detail [6, 7]. It is now well established that the spectrum of a quasiperiodic Schrodinger operator is a Cantor set having pure point components (localized states), components with absolutely continuous measure (extended states) and singular components with critical (chaotic) states. We find a similar behaviour for the plasmon spectrum. The case of a continuous incommensurate modulation of electron density in equally spaced layers has been studied by Das Sarma et al. [2]. Using a duality transformation these authors demonstrated the existence of "mobility edge" in the plasmon spectrum. For the strength of the Coulomb interaction between the layers smaller than a certain critical value, all plasmon states become localized, while for Coulomb interaction above the critical value all plasmon states are extended. Critical states exist precisely at the mobility edge. Any arbitrary small deviation from the critical point, as is bound to happen in experiment, will produce extended or localized states. Our main interest is in critical states which are the least understood at present. We will show that the system studied here is entirely "critical", i.e., it never admits neither extended nor localized states, and is therefore ideally suited for the study of critical states.

Let us consider F_m two-dimensional electron gas (2DEG) layers, where

F_m is a Fibonacci number, i.e., F_m satisfies a recursion relation $F_{m+1} = F_m + F_{m-1}$ with $F_o = 1$, $F_1 = 1$. Let ℓ label the layers and z_ℓ be the position of layer ℓ along the superlattice axis. The distance between layer ℓ and $\ell + 1$ is called d_ℓ, and d_ℓ can be either a or b. The set $\{d_\ell\}$ of a's and b's is generated using the Fibonacci rule F[a,b] = [ab,a] [7]. In practice we start with a single 2DEG layer at z = 0 followed by a semiconductor layer of thickness a. In the second step, according to the Fibonacci rule, a 2DEG layer followed by semiconductor layer of thickness b, is added. This gives a string {a,b}. The third generation produces the string {a,b,a} by replacing the element a with two elements a and b and the element b with the element a in the string {a,b}. We continue this process m times. In the m-th generation there are F_m elements of the string. This includes F_{m-1} elements a and F_{m-2} elements b. Note that the ratio of the number of elements a to the number of elements b approaches the "golden mean" value $\tau = (1 + \sqrt{5})/2$.

The potential ϕ, induced on every layer by a perturbation with frequency ω and wave vector q parallel to the layers satisfies integral equation.

$$\phi(\ell) = \sum_{\ell'} v_q \Pi(q,\omega) V(\ell,\ell')\phi(\ell') \tag{1}$$

where $v_q = 2\pi e^2/\varepsilon q$, Π is a polarizability of a 2D electron gas and $V(\ell,\ell') = \exp\{-q|z_\ell - z_{\ell'}|\}$. In the long wavelength limit the product $v_q\,\Pi$, defined as X, can be written as $(\omega_q/\omega)^2$, where $\omega_q^2 = 2\pi e^2 nq/m^*$ is the two-dimensional plasma frequency. Here n is electron density, ε is the background dielectric constant, and m^* is the electron mass. We can rewrite Eq. 1 in the form of the tight binding "Hamiltonian"

$$\bar{\omega}^{-2}\phi(\ell) = \sum_{\ell'\neq\ell} V(\ell,\ell')\phi(\ell') + V(\ell,\ell)\phi(\ell) \tag{2}$$

where $\bar{\omega}^{-2}$ plays the role of the eigenvalue, $V(\ell,\ell')$ are the hopping matrix elements, and the potential on the site $V(\ell,\ell)$ is equal 1. The solution to the eigenvalue problem, Eq. 2, gives the plasma modes of the system.

Much of the progress in studying the quasiperiodic Schrodinger operator has been achieved by the transfer matrix method. Eq. 2 cannot be cast in this form due to the long range hopping matrix elements $V(\ell,\ell')$, unless only nearest neighbour hopping is included. Progress is made by another approach [3]. We write the solution to the Poisson equation $\left(\frac{d^2}{dz^2} - q^2\right)\phi(z) = 0$ in the region between layer ℓ and $\ell + 1$ as

$$\phi_\ell(z) = A_\ell e^{-q(z-z_\ell)} + B_\ell e^{q(z-z_\ell)};\ \phi(\ell) = \phi(z_\ell). \tag{3}$$

Standard electromagentic boundary conditions across the electron layer allow us to connect (A_ℓ,B_ℓ) with $(A_{\ell+1}, B_{\ell+1})$ via a transfer matrix T

$$\begin{pmatrix} A_{\ell+1} \\ B_{\ell+1} \end{pmatrix} = T_\ell \begin{pmatrix} A_\ell \\ B_\ell \end{pmatrix};\ T_\ell = \begin{Bmatrix} (1+X)e^{-qd_\ell}; & Xe^{qd_\ell} \\ -Xe^{-qd_\ell} & ;\ (1-X)e^{qd_\ell} \end{Bmatrix} \tag{4}$$

where susceptibility $X = v_q\Pi$ is the relevant variable. Matrix T is a 2 x 2 matrix with a unit determinant. Note that $\{d_\ell\}$ is a Fibonacci sequence so the string of matrices $\{T_\ell\}$ is also a Fibonacci sequence of matrices T_a and T_b $\{\ldots T_aT_bT_aT_bT_aT_aT_bT_a\}$ where $T_{a(b)}$ are matrices T with d = a(b). Eq. 4 is conveniently studied by the rational approximation method. A rational approximation "m" to a Fibonacci sequence consists of a periodic sequence of unit cells containing F_m matrices T obtained in "m"-th generation of Fibonacci sequence. The spectrum consists of F_m bands and F_{m-1} gaps. The bands consist of those values of X for which the trace of the transfer matrix across the unit cell is between -2 and +2. Following [3] we define matrix $M_m = \prod_{i=1}^{F_m} T_i$. Matrix M_{m+1} is a product of two previous Fibonacci products of matrices T, i.e., $M_{m+1} = M_{m-1}M_m$. Defining $x_m = (1/2)\mathrm{Tr}(M_m)$ we have a recursion relation for x_m [3,6]

$$x_{m+1} = 2x_m x_{m-1} - x_{m-2} \tag{5}$$

The starting conditions are $x_1 = (1/2)\mathrm{Tr}(T_a) = \cosh(qa) - X\sinh(qa)$, $x_0 = (1/2)\mathrm{Tr}(T_b) = \cosh(qb) - X\sinh(qb)$, and $x_{-1} = \cosh(q(a-b))$. The value of x_{-1} has been determined by requiring that Eq. 5 with x_1, x_0, as defined above, gives $x_2 = (1/2)\mathrm{Tr}(T_bT_a)$. The recursion relation (Eq. 5) has an important invariant equal to

$$\lambda^2 = -1 + x_m^2 + x_{m-1}^2 + x_{m-2}^2 - 2x_m x_{m-1} x_{m-2} \tag{6}$$

This quantity remains constant at every step of the recursive formula Eq. 6. Note that here λ^2 is a function of X. The three-dimensional map given by Eq. 4 and Eq. 5 has been studied by Kohmoto and Oono [6] for the case of λ^2 = const. Fixed point analysis yields escaping, periodic and chaotic orbits.

We now turn to the plasmon spectrum obtained using Eq. 5. We have set her b = 1, a = τ, and q = 1 (b is the unit of length). In Fig. 1 we plot the bands of X for various rational approximations "m" to the Fibonacci sequence, with F_m being the size of the unit cell. When $x_m < 1$, X is allowed, otherwise X is forbidden. This band structure is very similar to that obtained by Hofstadter [7], Kohmoto et al. [6], and Ostlund et al. [7]. As $m \to \infty$ we see an infinite number of very narrow bands which have a typical selfsimilar Cantor set structure. An important quantity here is the total width B of allowed values of X. As in [6] we find that B scales with a size of the unit cell as $B \sim F_m^{-\delta}$. The scaling index δ can be identified as a diffusion constant of the map [6]. The dependence of ln(B) on the $\ln(F_m)$ is shown in Fig. 2. The inset in Fig. 2 shows schematically the dependence of ln(B) on $\ln(F_m)$ for extended, localized and critical states [6]. The spectrum of plasmons is clearly critical, with exponent δ depending on q. The measure of the criticality here is more subtle however. We see that there are 2 large gaps and 3 bands. Even though the bands are selfsimilar, their measure B scales differently with the number of bands. The dependence of ln(B) for bands I (low X) and III (high X) on the ln(F) is shown in Fig. 2. Clearly B_3 scales to zero faster than B_1. In sense, the low X (high frequency) part of the spectrum corresponds to more extended states than the high X (low frequency) part of the spectrum. This dependence is due to the dependence of λ^2 on X.

Global scaling properties of the plasmon spectrum can be analyzed using the ideas proposed recently for dynamical systems by Halsey et al. [8] and applied to a quasiperiodic tight binding Hamiltonian by Tang and Kohmoto [6]. Let $\{S_m\}$ be the set of F_m bands of X. We define the measure of each band "i" in the set of $\{S_m\}$ as $p_i = 1/F_m$. Let the width of the i-th band be w_i. Then the scaling index α tells us how the measure scales with the band width i.e. $p_i = (1/F_m) = w_i^{\alpha}$. The set of bands with same scaling exponent α has a fractal dimension f(α) i.e. the number of elements N in this set is given by $N(\alpha) = w_i^{-f(\alpha)}$. The function f(α) is obtained by introducing a partition function Γ as

$$\Gamma m(r,s,\{S_m\}) = \sum_{i=1}^{F_m} (F_m)^{-r} w_i^{-s} \tag{7}$$

The condition

$$\Gamma(r,s) = \lim_{m \to \infty} \Gamma m(r,s,\{S_m\}) = 1 \tag{8}$$

uniquely determines the function S(r). Then the scaling indices α and the function f(α) are given by

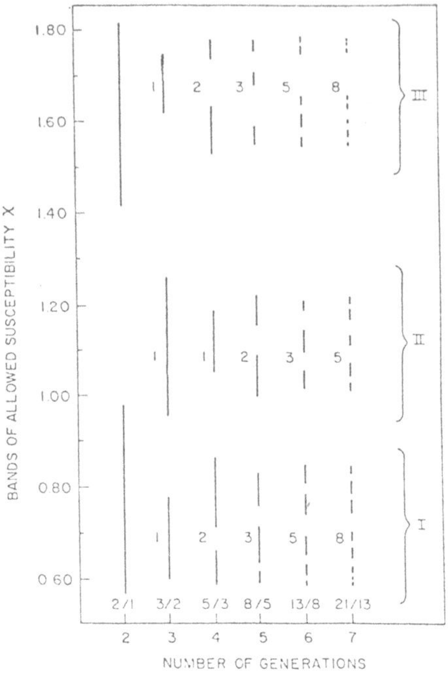

Fig. 1 The allowed values of susceptibility X for $qb \sim 1$. The band structures for m = 2, 3, ..., 7 are shown. Note that $X^{-1} \sim \bar{\omega}^2$. Roman numbers denote 3 major bands and greek numbers enumerate bands for each "m".

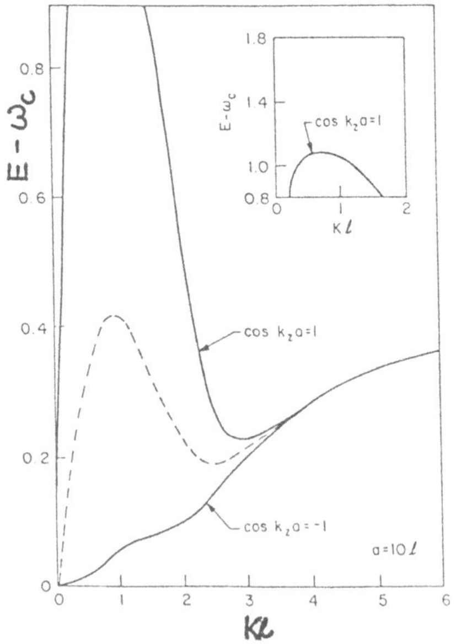

Fig. 1a A plot of $E(K) - hw_c$ vs. $K\ell$. The dashed line is the result for a single layer. The edges of the magneto-exciton band for the case $a = 10\ell$ are indicated by the solid line. The insert shows the upper band edge on a reduced scale.

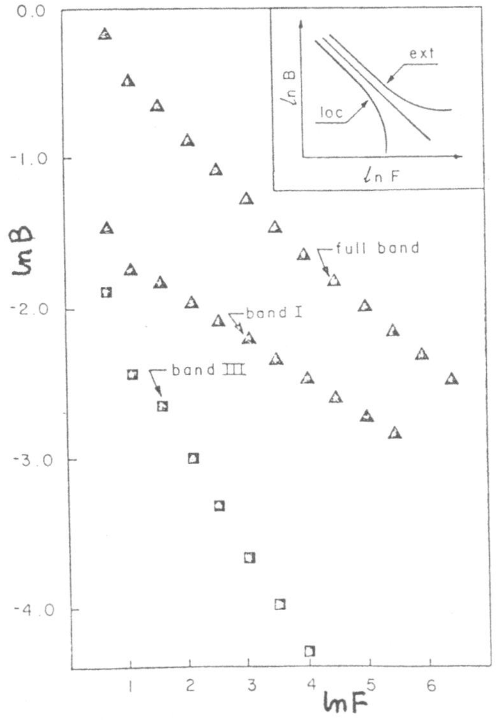

Fig. 2 The dependence of lnB on lnF for full band and bands I and III of the spectrum of Fig. 1. Here B is the sum of allowed susceptibility values and F is the size of the unit cell. $F \to \infty$ represents quasiperiodic system. The inset shows the typical dependence of lnB on lnF for localized, extended and critical spectra.

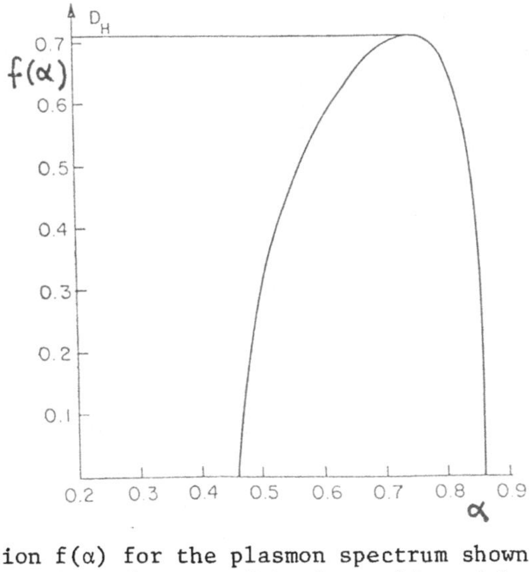

Fig. 3 The function $f(\alpha)$ for the plasmon spectrum shown in Fig. 1a. Here f is a fractal dimension of a set of bands scaling with the scaling index α.

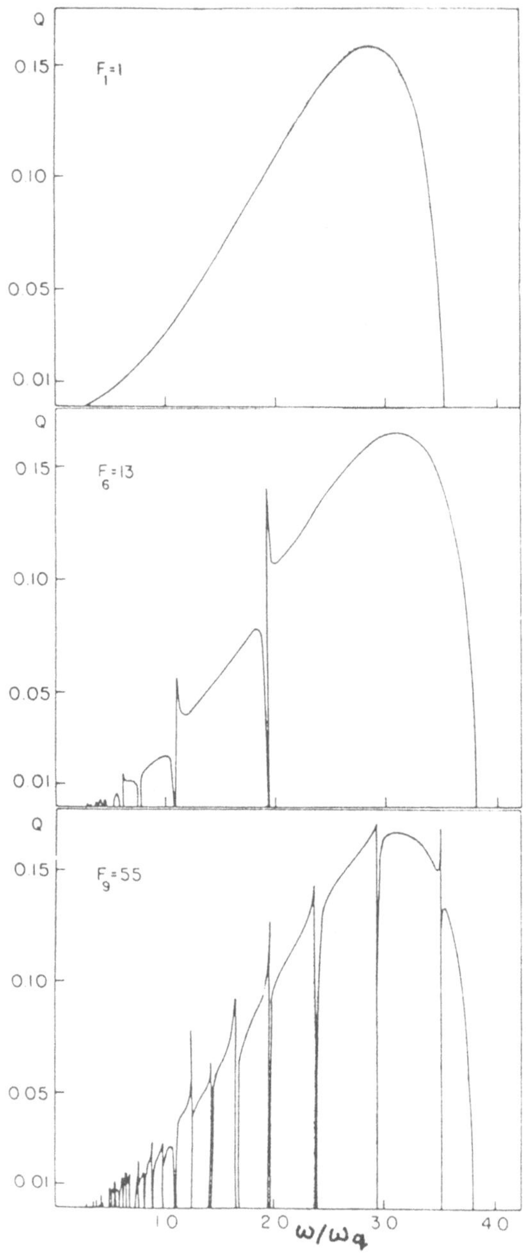

Fig. 4 The evolution of the plasmon infrared absorption spectrum $Q(\omega)$ with the Fibonacci generation number "m". The cases with m = 1, m = 6 and m = 9 are shown. Here F_m denotes the number of layers in a unit cell and qb = 0.1.

$$\alpha(r) = \frac{ds}{dr};\ f(r) = r\alpha(r) - s(r) \tag{9}$$

The function $f(\alpha)$ for the plasmon spectrum i.e. the fractal dimension of the set of bands of X scaling with the same scaling index α is shown in Fig. 3 for qb = 1.0 and for the finite number of bands $F_{15} = 987$. There is a finite interval of the scaling indices α, $0.459 < \alpha < 0.861$. The most probably scaling exponent $\alpha = 0.745$ corresponds to the maximum of $f(\alpha) = 0.709$, which is the Hausdorff dimension of the set. Our results for the plasmon spectrum are similar to those of Tang et al. [6] for the quasiperiodic tight binding Hamiltonian.

We now turn to the possibilities of experimental observation of the plasmon bands in quasiperiodic superlattices. The Raman scattering has been discussed in Ref. 5 and as already mentioned, it probes the local density of plasmon states in the momentum space and not the total band structure. The appropriate technique is the infrared absorption of electromagnetic waves in superlattices, achieved by the grating coupler on the surface of the superlattice. The electric field of the wave passing through the diffraction grating can be represented by a Fourier series in $\cos q_s x$, where $q_s = 2\pi s/L$, L is the period of the diffraction grating and s = 0, 1, 2, ... The harmonics with s > 1 correspond to the electric field which attenuates exponentially with increasing distance from the diffraction-grating plane. These evanescent waves interact resonantly with plasma oscillations of the superlattice if the frequency of the incident radiation is in the plasmon band. The theory of resonant infrared absorption in periodic superlattices has also been formulated by Krasheninnikov and Chaplik [9]. The total work Q performed by an external field $\phi_0 e^{iqx-qz} e^{i\omega t}$ on the system per unit time is

$$Q(4\pi b/\varepsilon_0 |\phi_0|^2 \omega_q) = -\mathrm{Im}\{ \frac{\omega}{\omega_q} \cdot qb \cdot X \cdot \sum_\ell \phi_{tot}(\ell) e^{-qz_\ell} \} \tag{10}$$

Here $\phi_{tot}(\ell)$ is the total potential on layer "ℓ". For a periodic, semi-infinite system with a unit cell obtained in the m-th generation of a Fibonacci sequence, the total potential can be obtained by a transfer matrix method with appropriate boundary conditions. The full details of the calculations will be published elsewhere [10]. The absorption bands for $F_1 = 1$, $F_6 = 13$ and $F_9 = 55$ layers per unit cell are shown in Fig. 4 where the value of qb = 0.1 has been chosen, and it can be achieved experimentally. The absorption band for a system of equally spaced layers ($F_1 = 1$) is in complete agreement with calculation of Krasheninnikov et al. [9], who used the Wiener-Hopf method. Absorption occurs in the frequency interval corresponding to the plasmon band. As the number of bands increases the power spectrum becomes more and more fragmented as the bands begin to form a Cantor set. Clearly high frequency bands give higher absorption than than the low frequency bands. Note also what seems to be a set of divergencies or spikes in some of the bands. These are actually smooth shoulders but we have no physical explanation for their origin. In experiments all sharp features would be rounded off by a broadening due to the finite electronic mobility.

The infrared resonant absorption spectrum could provide experimental verification of the interesting properties of plasmon bands in quasiperiodic semiconductor superlattices.

Footnotes and References

Supported by the U. S. Army Research Office, Durham.

Part I

1. Yu. A. Bychkov, S. V. Iordanskii, and G. M. Elianshberg, Sov. Phys. JETP Lett. 33, 143 (1981); C. Kallin and B. I. Halperin, Phys. Rev.

B30, 5655 (1984); G. F. Giuliani and J. J. Quinn, Phys. Rev. B31, 6228 (1985).
2. W. Kohn, Phys. Rev. 123, 1242 (1961).
3. I. V. Lerner and Yu. E. Lozovick, Sov. Phys. JETP 51, 588 (1980) and 55, 691 (1982); Yu. A. Bychkov and E. I. Rashba, Solid State Comm. 48, 399 (1983).
4. J.-W. Wu and J. J. Quinn, to be published.
5. D. C. Tsui, H. L. Stormer, and A. C. Gossard, Phys. Rev. Lett. 48, 1559 (1982).
6. R. B. Laughlin, Phys. Rev. Lett. 50, 1395 (1983).
7. S. M. Girvin, A. H. MacDonald, and P. M. Platzman, Phys. Rev. Lett. 54, 581 (1985).

Part II

1. For a review see B. Simon, Adv. Appl. Math 3, 463 (1982) and J. B. Sokoloff, Phys. Rep 126, 189 (1985).
2. S. Das Sarma, A. Kobayashi, and R. E. Prange, Phys. Rev. Lett. 56, 1280 (1986).
3. P. Hawrylak and J. J. Quinn, Phys. Rev. Lett. (In press).
4. A. C. Tselis and J. J. Quinn, Phys. Rev. B 29, 3318 (1984).
5. R. Merlin, K. Bajema, and R. Clarke, F.-Y. Juang and P. K. Bjattacharya, Phys. Rev. Lett. 55, 1768 (1985).
6. M.Kohmoto, L. P. Kadanoff, C. Tang, Phys. Rev. Lett. 50, 1870 (1983); M. Kohmoto, Phys. Rev. Lett. 51, 1198 (1983); M. Kohmoto and Y. Oono, Phys. Lett. 102A, 145 (1984); C. Tang and M. Kohmoto, to be published.
7. S. Ostlund, R. Pandit, D. Rand, H. J. Schellnhuber and E. D. Siggia, Phys. Rev. Lett. 50, 1873 (1983); S. Ostund and R. Pandit, Phys. Rev. B29, 1394 (1984); D. R. Hofstadter, Phys. Rev. B 14 2239 (1976).
8. T. C. Halsey, M. Jensen, L. P. Kadanoff, I. Procaccia, B. Schraiman, Phys. Rev. B 33, 1141 (1986).
9. M. V. Krasheninnikov and A. V. Chaplik, Sov. Phys. JETP 57, 624 (1983).
10. P. Hawrylak and J. J. Quinn, to be published.

QUARK CLUSTER MODEL FOR HIGH ENERGY REACTIONS WITH NUCLEI

James P. Vary and Avaroth Harindranath

Physics Department
Iowa State University
Ames, IA 50011

This paper reviews the consequences of multi-quark correlations in nuclei which arise through the particular cluster decomposition of the nuclear ground state specified by the quark cluster model (QCM). The critical radius of a three quark cluster, R_c, controls the formation probability of larger clusters. The value of $R_c = 0.50$ fm is fixed by fits to deep inelastic lepton nucleus scattering data. The QCM prediction for the ratio of nuclear structure functions in the $x > 1$ domain is discussed as a critical test of the model. We also present an extended discussion of the application of the QCM to the nuclear Drell-Yan process.

1. MODEL ASSUMPTIONS

The quark cluster model (QCM) was proposed[1-5] to explain the deep inelastic electron-^{3}He scattering (DIS) results from SLAC[6]. In the QCM for any application one assumes the nucleus at all times is organized into color singlet clusters. We label the clusters by their leading Fock space component in the infinite momentum frame[7] as three-quark (3-q), six-quark (6-q) etc. clusters. Larger clusters are assumed to form by the overlap of smaller clusters. Specifically a 3-q cluster is assumed to have a critical radius, R_c such that clusters of 6, 9, etc., quarks are defined by the number of 3-q clusters joined by center of mass separations $d \leq 2R_c$. The overlap probabilities, and coordinate and momentum space probability distributions for these clusters are obtained in the rest frame from conventional nuclear wavefunctions.

We have assumed a sharp transition radius, R_c, for two reasons. First, it leads to a simple orthonormal cluster decomposition of a many-body wavefunction. Second, the limited data available can determine at most one parameter of the model. One may entertain the idea of relaxing this restriction as additional data become available or as quantitative results from non-perturbative calculations of QCD emerge. There are established methods to generate an orthonormal cluster decomposition with a transition function of finite spatial extent so that one may generalize the QCM in a straightforward way. Since we view the QCM as a model which allows us to write down the quark distributions in nuclei we expect it will be ultimately superceded by direct calculations within non-perturbative QCD. In the meantime we expect the QCM to form a basis for the uniform interpretation of a variety of high energy lepton-nucleus and hadron-nucleus interactions. In this context we are able to summarize all our efforts to date with the claim that all the data we have examined are consistent with $R_c = 0.50 \pm 0.05$ fm which is the same value determined by the initial fits[1] to DIS data on ^{3}He.

To apply the QCM to DIS and to the Drell-Yan (DY) process[8,9] we make two additional assumptions. First, as is customary in parton phenomenology, we assume that the participating quark (or antiquark) is quasifree. That is, we ignore initial and final state interactions. In the case of DY there has been much discussion of this approximation.[10] Second, since the cluster from which the participating quark originates is, by the definition of cluster configurations, spatially isolated from the remaining clusters, we assume the cluster is also quasifree. Consistent with this second assumption we assign to an i-quark cluster a mass equal to $\frac{i}{3}$ times the nucleon mass. This last assumption is also equivalent to neglecting quark exchange processes between clusters. Encouraging support for this approximation is obtained from the demonstration by Frankfurt and Strikman[11] that the leading exchange graph correction to a 3-q cluster contribution to the European Muon Collaboration (EMC) effect[12] virtually disappears for $x > 0.3$.

2. ^{3}He DATA AND THE DETERMINATION OF R_c

The earliest effort[1] obtained a best fit to DIS data from SLAC on ^{3}He[6]. Subsequent analyses[2–5] with improved nuclear wavefunctions yield no major differences from the first fits.

We will now summarize the model details for DIS on ^{3}He. To conserve on time and space this will be presented in a simplified version which is nevertheless valid at high Q^2. For proper accounting of $O(\frac{Q^2}{M^2})$ effects and thresholds see references[1–4].

The variables employed are the 4-momentum transfer of the photon squared (Q^2), the lab energy loss of the lepton (ν), the mass of the nucleon (m), the baryon number of the nucleus (A) and the Bjorken $x = \frac{Q^2}{2m\nu}$ which has the range $0 \leq x \leq A$. Then the measured DIS cross section multiplied by ν and divided by the Mott cross section gives the nuclear structure function $\nu W_2(\nu, Q^2)$ if the data are restricted to sufficiently small lab scattering angles of the lepton. The QCM gives

$$\nu W_2(\nu, Q^2) = \sum_{\text{quarks } j} e_j^2 \frac{x}{A} P_j(x)$$

where e_j is the charge on quark j and $P_j(x)$ is the probability that quark j carries fraction $\frac{x}{A}$ of the total nuclear 4-momentum P in the infinite momentum frame. We take weighted averages of up and down quark distributions to obtain a nucleus dependent $P(x)$:

$$P(x) = \sum_{\text{clusters } i} \tilde{p}_i \mathcal{P}_i(x)$$

where $\tilde{p}_i$ is the probability the quark is obtained from an i-q cluster and $\mathcal{P}_i$ is the x-distribution of quarks from an i-q cluster in the nucleus. The quantities $\tilde{p}_i$ depend sensitively on R_c and are obtained by overlap integrals calculated with wavefunctions appropriate for each nucleus.[13] The distributions $\mathcal{P}_i(x)$ are given by

$$\mathcal{P}_i(x) = \int_0^A dy \int_0^{\frac{i}{3}} du \; n_{\frac{q}{i}}(u) N_{\frac{i}{A}}(y) \; \delta\left(\frac{u}{\frac{i}{3}}\frac{y}{A} - \frac{x}{A}\right)$$

which is the convolution of the probability $n_{\frac{q}{i}}(u)$ that a quark carries momentum fraction $\frac{u}{i/3}$ of the cluster's momentum with the probability $N_{\frac{i}{A}}(y)$ the cluster carries momentum fraction $\frac{y}{A}$ of the total nuclear momentum P. The delta function selects those probability products which give a quark the required momentum $\frac{x}{A}P$.

For 3-q clusters the distributions from best fits to the data and from QCD evolution equations are employed.[14] Since the $\tilde{p}_i$ decrease rapidly for increasing i (when R_c is taken in the range of 0.50 fm) we truncate the cluster sum at the 9-q cluster term when applying the model to heavier nuclei. Then, using $\bar{n}_{\frac{q}{i}}(\frac{3u}{i}) = \frac{i}{3} n_{\frac{q}{i}}(u)$, the quark distributions for 6-q and 9-q clusters are taken from counting rules,[15,16] Regge behaviour and QCD evolution to be

$$\bar{n}_{\frac{q}{6}}(v) = (B[0.5, 11 + \alpha\bar{s}])^{-1} v^{-\frac{1}{2}} (1-v)^{10+\alpha\bar{s}}$$

$$\bar{n}_{\frac{q}{9}}(v) = (B[0.5, 16])^{-1} v^{-\frac{1}{2}} (1-v)^{15}$$

where $B[a,b]$ is Euler's beta function, $\bar{s} = ln[\alpha_s(Q_0^2)/\alpha_s(Q^2)]$, Q_0^2 =1.8 GeV$^2/c^2$, α=2.4 and we have neglected the QCD evolution of the 9-q cluster distribution. For our initial applications[9] to the DY process, however, we will employ the quark distributions for 6-q clusters given by Ref. 16 and we neglect the small contributions from 9-q clusters.

The role played by Fermi motion in the analyses performed to date has been minor compared to the change in quark distributions from 3-q clusters to 6-q clusters. Hence it is sufficient to adopt a simplified treatment of Fermi motion with

$$N_{\frac{i}{A}}(y) = \frac{1}{\sqrt{2\pi}\sigma} \exp -[\frac{(y-\frac{i}{3})^2}{2\sigma^2}]$$

for clusters with $i < 3A$ and $\delta(y - \frac{i}{3})$ for $i = 3A$. A simple estimate for the A-dependence of σ is obtained by using a smooth parametrization of the Fermi momentum k_F calculated in the Hartree-Fock approximation over a range of nuclei[17]. This simple approximation consists of

$$k_F = \frac{1.16A^{\frac{1}{3}}}{\langle r^2\rangle^{\frac{1}{2}}}$$

and

$$\sigma = (1/5)^{\frac{1}{2}} \frac{k_F}{m}$$

since the deviation in the calculated mass rms radius of a nucleus $< r^2 >^{\frac{1}{2}}$ from systematics is seen to be correlated with the deviation in the calculated rms momentum of a single nucleon.[17] In the initial application to DY we have neglected the contributions of Fermi motion.

In order to fit the existing ^{3}He data[6] we incorporate the contributions of the nucleon quasielastic peak. These contributions have been carefully evaluated by the Hannover group[18] and we employ their results. Hence the final form of the nuclear structure function is

$$\nu W_2^{\text{total}}(\nu, Q^2) = \nu W_2 + \tilde{p}_3 \nu W_2^{q-el}(\nu, Q^2)$$

We fit the ^{3}He data sets for $7.26 \leq E \leq 14.70$ GeV since they span $0.8 \leq Q^2 \leq 4$ GeV$^2/c^2$ and obtain $R_c = 0.50 \pm 0.05$ fm. This value of R_c implies $(\tilde{p}_3, \tilde{p}_6, \tilde{p}_9) = (0.88, 0.11, 0.01)$ respectively for the quark cluster probabilities in ^{3}He when semi-realistic wavefunctions of ^{3}He are used in the overlap calculations. Some fits are shown in Refs. 1–4. Note especially that the ^{3}He data span $0 \leq x \leq 3$ and the most sensitive region for 6-q admixtures is $1 \leq x \leq 2$. Similarly, the most sensitive region to 9-q admixtures is $2 \leq x \leq 3$. These are the regions where traditional nuclear physics models produce far too small a cross section to explain the data. It is the strong sensitivity to multi-quark cluster admixtures that yields the small uncertainty in R_c.

We also note here that, as described in Refs. 2, 3, and 16, the QCM gives an adequate description of the EMC effect provided one accounts for the difference in the average nuclear densities of a heavy nucleus compared to deuterium.

3. TESTING OF QCM WITH $x > 1$ DATA

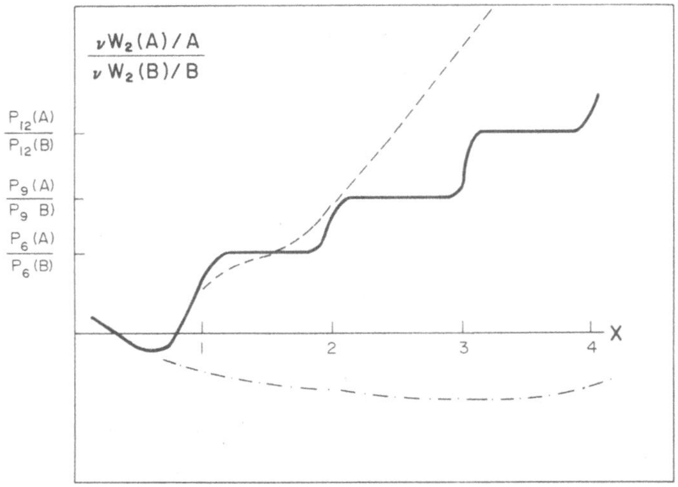

Figure 1. Characteristic behaviour of the ratio of nuclear structure functions per nucleon for different models over a wide kinematic range of x. The QCM gives the solid curve. The dahed curve is due to the model of Ref. 19. The dashed-dot curve approximates the predictions of Refs. 20 and 21.

This brings us to the question of how may we best determine the correct model among the many proposed for the quark structure of nuclei. Experiments in the $x > 1$ region should be decisive.[2,5] Fig. 1 presents a few predictions for an extended range of x for the characteristic behaviour of the ratio of structure functions of a heavy nucleus A to a light nucleus B with $B > 4$. The QCM predicts a sequence of steps in a stair case where the height of a step in the region $n - 1 < x < n$ with $n > 1$ is the ratio of $3n$-q cluster probabilities of the heavier to the lighter nucleus. By contrast rather smooth behaviour is predicted so far by other models. The dashed curve is the type of behaviour expected from the relativistic wavefunction model of Frankfurt and Strikman[19] where short range correlations give rise to a shoulder in the $1 < x < 2$ region. Another relativistic wavefunction treatment by Garsevanishvili and Menteshashvili[20] and the color dielectric model of Pirner and collaborators[21] predict behaviour indicated by the dash-dot curve. In the color dielectric model the quarks at very high Q^2 are free to move essentially throughout the volume of the entire nucleus. This naturally leads to softer momentum distributions of quarks in larger nuclei. Exactly what values of Q^2 for which the color dielectric model is expected to be valid must yet be specified. If the color dielectric picture is valid at high Q^2 the QCM will still be valid if we introduce a Q^2dependence for R_c. Then, as Q^2 increases R_c will increase in a manner predicted by the color dielectric model. In this case the steps in Fig. 1 for $x > 1$ will drop with increasing Q^2 and will eventually fall below unity until the curve reaches the smooth prediction shown for the color dielectric model. This union of the QCM with the color dielectric model produces a result which contrasts the work of Refs. 22 and 23 where the 6-q cluster probability always rises with Q^2 and the probability in a heavy nucleus never equals that of a lighter nucleus. Thus the QCD evolution of the model in Refs. 22 and 23 will produce a curve for Fig. 1 which always remains above unity for $x > 1$.

Clearly, the wide range of behaviour predicted and, in particular, the dramatic signature of the QCM motivates experiments in the $x > 1$ region.

Of course at $x = A$ the nucleus recoils intact and the experiment obtains the elastic form factor. We refer the reader to Ref. 24 for an extended presentation of the application of the QCM to the elastic nuclear form factors. Particular emphasis has been placed on the longstanding mystery of the $A = 3$ form factor and considerable success has been achieved.[24]

4. NUCLEAR DRELL-YAN PROCESS WITHIN THE QCM

There has been much recent interest in the possibility that the nuclear Drell-Yan (DY) process[8] would either distinguish between models successful in explaining the DIS experiments or would further refine their ingredients.[9,25]

In the hadron-hadron center of momentum frame we denote the total energy by $\sqrt{s}$. For hadrons A and B the 4-momenta are $P_A = (\frac{\sqrt{s}}{2}, 0, 0, \frac{\sqrt{s}}{2})$ and $P_B = (\frac{\sqrt{s}}{2}, 0, 0, -\frac{\sqrt{s}}{2})$. Let $x_1(x_2)$ denote the fraction of longitudinal momentum carried by quark 1(2) in hadron $A(B)$. Then the longitudinal momentum of the lepton pair with invariant mass M is given by

$$P_L = p_1 + p_2 = (x_1 - x_2)\frac{\sqrt{s}}{2} .$$

The kinematical variable $\tau \equiv x_1 x_2$ becomes $\frac{M^2}{s}$ since we are consistently neglecting the transverse momentum of the lepton pair. Then $P_L = (x_1 - \frac{M^2}{s x_1})\frac{\sqrt{s}}{2}$, yielding

$P_L^{\max} = (1 - \frac{M^2}{s})\frac{\sqrt{s}}{2}$. We also employ

$$x_F = \frac{P_L}{P_L^{\max}} = \frac{x_1 - x_2}{1 - \tau} .$$

Experiments measure laboratory quantities sufficient to determine M, P_L and the lepton pair transverse momentum p_T. We consider only p_T-integrated cross sections.

According to the naive DY model[8] the differential cross section for the process $AB \to \mu^+\mu^- X$ is given by

$$\frac{d\sigma}{dM^2} = \frac{4\pi\alpha^2}{9M^2} \sum e_a^2 \int dx_1 dx_2 F_a(x_1, x_2)\delta(M^2 - x_1 x_2 s),$$

where

$$F_a(x_1, x_2) = q_a^A(x_1)\bar{q}_a^B(x_2) + \bar{q}_a^A(x_1)q_a^B(x_2) .$$

Here the summation is over the flavor index a. Further, q_a^A is the quark distribution of flavor a in hadron A and $\bar{q}_a^B$ is the antiquark distribution of flavor a in hadron B. Thus

$$\frac{d^2\sigma}{dx_1 dx_2} = \frac{4\pi\alpha^2}{9sx_1x_2} \sum e_a^2 F_a(x_1, x_2) .$$

Data is sometimes presented after transforming to the variables x_F and M yielding,

$$\frac{d^2\sigma}{dM^2 dx_F} = \frac{4\pi\alpha^2}{9M^4}(1 - \tau) \sum e_a^2 \frac{x_1 x_2}{x_1 + x_2} F_a(x_1, x_2) .$$

Due to the constraints of space and time we will only discuss the nucleon-nucleus and pion-nucleus DY process here. For a more complete presentation including detailed comparisons with available data, a discussion of the K factor, the A-dependence, and nucleus-nucleus DY processes the reader is directed to Ref. 9.

The expression for the DY cross section depends on a product of quark momentum distribution functions as opposed to the linear dependence appearing in DIS cross section. By focusing on selected values of projectile x_1, DY can provide new information regarding target distribution functions. Projectile valence terms are dominant for $x_1 > 0.3$ and projectile sea distributions are dominant for small values of x_1. Thus at large x_1 DY measures the antiquark distributions of the target nucleus. Due to the assumptions of QCM, the valence quarks carry a smaller fraction of the total momentum in a 6-q cluster than in a 3-q cluster. If we assume that gluons carry the same momentum fractions in all clusters then a certain enhancement of the sea is required to conserve the total momentum.[16] We adopt this assumption for the present work.

In Fig. 2 we present the ratio of DY cross sections for Fe and D as a function of x_2 for two characteristic values of x_1. Small values of x_1 yield a ratio of cross section

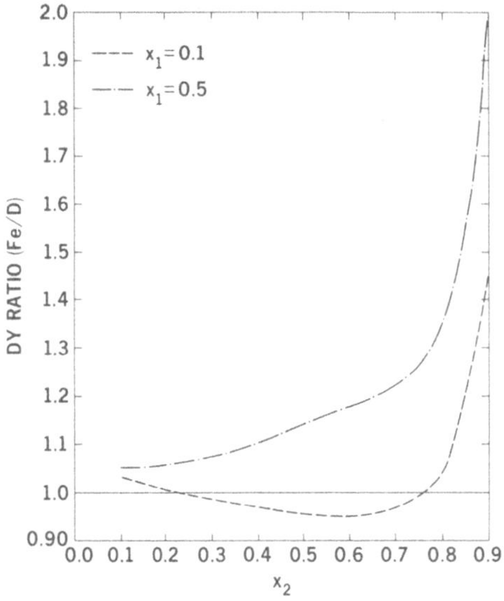

Figure 2. QCM prediction for the ratio of proton-nucleus DY cross sections for Fe and D as a function of x_2. The two different choices of x_1 indicate where the ratio is sensitive to different ingredients of the QCM. For x_1=0.1(0.5) the ratio is dominated by valence (sea) quark distributions of the target.

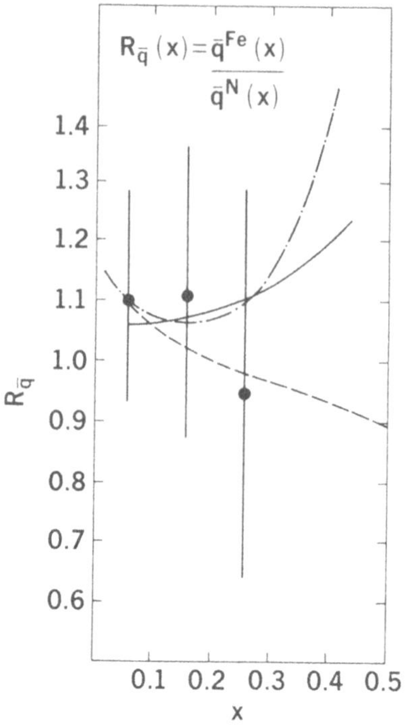

Figure 3. Predictions of QCM (solid), pion exchange model[29] (dot-dash), and the rescaling model[22,23,30] (dashed) are compared with the data of Abramovicz et al.[28], for the ratio of antiquark distributions in Fe and D. The latter two predictions are taken from Ref. 26.

similar to the ratio of valence quark contributions to the DIS cross section.[16,27] Large values of x_1 yield a ratio of the sea quark contributions which display the enhancement arising from the assumed gluon behaviour. Clearly, if data can be obtained at different values of x_1 it would be possible to seperately test the valence and the sea distributions within the QCM.

The ratio of the Fe and the nucleon antiquark distributions has been measured in a deep inelastic neutrino scattering experiment.[28] In Fig. 3 we compare the QCM result for this ratio with the data. For comparison we present the ratio of antiquark distributions obtained from the pion exchange model[29] and the rescaling model[22,23,30] as summarized in Ref. 26. The existing experimental error bars make it impossible to draw any definite conclusions.

In the pion-nucleus DY process we restrict our considerations to $x_1 > 0.4$ and therefore neglect sea quark distribution in pion. Hence for pion- nucleus the ratio of DY cross sections closely resemble the DIS cross section ratio. We have plotted the pion-nucleus DY cross section ratio for Fe and D as a function of x_F for constant M in Fig. 4. Results are shown with and without six quark clusters. Here, $\sqrt{s} = 20$ GeV. For this value of c. m. energy and $M = 4$ GeV, x_F varying from 0.0 to 0.8 corresponds to x_2 varying from 0.19 to 0.04. Thus, as Berger[26] has pointed out, measurements of the ratio of cross sections in this kinematical domain is of great interest in the light of differences between EMC[12] and SLAC[31] data on DIS from nuclear targets.

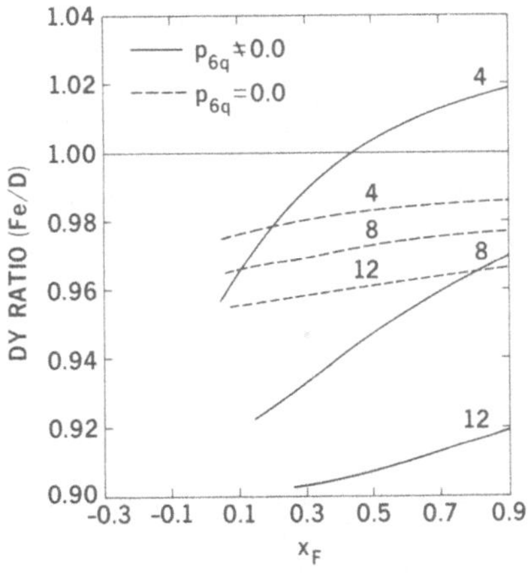

Figure 4. Predictions for the ratio of pion-nucleus DY cross sections with and without six quark clusters for Fe and D. The ratios are given as a function of x_F for three different values of M at $\sqrt{s} = 20.0$ GeV.

DY measurements of the sea quark (and hence the antiquark) distributions are as fundamental as the DIS measurements. Future DY experiments could provide precise determination of these ratios for $x < 0.4$ and would serve to eliminate some models of the EMC effect and to refine others. In the QCM these DY measurements could serve to fix what has, to this stage, been assumed for the gluon and sea quark distributions.

ACKNOWLEDGEMENTS

This work was supported in part by the U.S Department of Energy under contract No. DE-AC02-82ER40068, Division of High Energy and Nuclear Physics.

REFERENCES

1. H. J. Pirner and J. P. Vary, Phys. Rev. Lett. **46**, 1376 (1981).
2. J. P. Vary, Nucl. Phys. **A418**,195c (1984); J. P. Vary, in *Hadron Substructure in Nuclear Physics*, W.-Y. Hwang and M. H. Macfarlane, eds., AIP Conf. Proc. No. 110 (New York) 1984.
3. H. J. Pirner, *International Rev. of Nucl. Phys., Vol. II* (Singapore) 1984.
4. J. P. Vary and H. J. Pirner, *Recent Progress in Many-Body Theories*, eds.H. Kummel and M. L. Rustig (Springer-Verlag, Heidelberg, 1984) Lecture Notes in Physics **198**, p. 1.
5. J. P. Vary, *Proceedings of the VII International Seminar on High Energy Physics Problems, Multiquark Interactions, and Quantum Chromodynamics*, Dubna, 1984.
6. D. Day *et al.*, Phys. Rev. Lett. **43**,1143 (1979).
7. For a recent review of rules for the application of perturbative QCD to nuclear systems see S. J. Brodsky, *Short-Distance Phenomena in Nuclear Physics*, D. H. Boal and R. M. Woloshyn, eds., Plenum Publishing Corporation, 1983, p. 141 and references therein.
8. S. D. Drell and T. -M. Yan, Phys. Rev. Lett. **25**, 316 (1970) and Ann. Phys. **66**, 578 (1971). For extensive reviews that cover higher order QCD effects see R. Stroynowski, Phys. Rep. **71**, 1 (1981) and I. R. Kenyon, Rep. Prog. Phys. **45**, 1261 (1982).
9. A. Harindranath and J. P. Vary, Phys. Rev. **D34** 3378 (1986).
10. A. H. Mueller, in *Proceedings of the Drell-Yan Workshop*, Fermi Lab,1982; S. J. Brodsky, in *Progress in Physics, Vol .8*, A. Jaffe,G. Parisi and D. Ruelle, eds., Birkhauser (Boston) 1983, p. 1.
11. L. L. Frankfurt and M. I. Strikman, *Proceedings of the VII International Seminar on High Energy Physics Problems*, Dubna, 1984 and Moscow preprint, 1984.
12. J. J. Aubert *et al.*, Phys. Lett. **123B**,123 (1983).
13. For a review of these overlap probabilities and the method of extrapolation to heavier nuclei, see M. Sato, S. A. Coon, H. J. Pirner and J. P. Vary, Phys. Rev. C**33**, 1062 (1986).
14. A. J. Buras and K. J. F. Gaemers, Nucl. Phys. **B132**, 249 (1978).
15. D. Sivers, S. J. Brodsky and R. Blankenbeckler, Phys. Rep. **23C**, 1 (1976), and references therein.
16. C. E. Carlson and T. J. Havens, Phys. Rev. Lett. **51**, 261 (1983).
17. M. Sandel, J. P. Vary and S. I. A. Garpman, Phys. Rev. C**20**, 744 (1979).
18. H. Meir-Hajduk, Ch. Hajduk, P. U. Sauer and W. Theis, Nucl. Phys. **A395**, 332 (1983).
19. L. L. Frankfurt and M. I. Strikman, Phys. Rep. **76**, 215 (1981).
20. V. Garsevanishvili and Z. Menteshashvili, JINR, E2-84-314, Dubna (1984).
21. G. Chanfray, O. Nachtmann and H. J. Pirner, Phys. Lett. **147B**, 249 (1984).
22. F. E. Close, R. G. Roberts and G. G. Ross, Phys. Lett. **129B**, 346 (1983).
23. R. Jaffe, F. E. Close, R. G. Roberts and G. G. Ross, Phys. Lett. **134B**, 449 (1984).
24. J. P. Vary, S. A. Coon, H. J. Pirner, in *Few Body Problems in Physics*, Vol. II, ed. B. Zeitnitz (North-Holland, Amsterdam,1984), p. 683; *Proc. of Int. Conf. on Nuclear Physics*, ed. R. A. Ricci and P. Blasi (Tipografia Compositori, Bologna, 1983), p. 320; *Hadronic Probes and Nuclear Interactions* (Arizona State University) Proceedings of the Conference on Hadronic Probes and Nuclear Structure, AIP Conf. Proc. No. 133, ed. J. R. Comfort, W. R. Gibbs and B. G. Ritchie (AIP, New York, 1985).

25. R. P. Bickerstaff, M. C. Birse and G. A. Miller, Phys. Rev. Lett.**53**, 2532 (1984); R. P. Bickerstaff, M. C. Birse and G. A. Miller, Phys. Rev. D**33**, 3228 (1986) ; Y. Gabellini, J. L. Meunier and G. Plaut, Z. Phys. **C28**, 123 (1985); N. P. Zotov, V. A. Saleev and V. A. Tsarev, JETP Lett. **40**, 965 (1985).
26. E. L. Berger, Nucl. Phys. **B267**, 231 (1986).
27. H. J. Pirner in *Particle and Nuclear Physics*, edited by A. Faessler (Pergamon, Oxford, 1985), p. 361; J. P. Vary, Nucl. Phys. **A418**,195c, (1984).
28. H. A. Abramovicz *et al.*, Z. Phys. **C25**, 29 (1984).
29. E. L. Berger, F. Coester, and R. B. Wiringa, Phys. Rev. D**29**, 398 (1984); E. L. Berger and F. Coester, Phys. Rev. D**32**, 1071 (1985).
30. F. E. Close *et al.*, Phys. Rev. D**31**, 1004 (1985).
31. A. Bodek *et al.*, Phys. Rev. Lett. **50**, 1431 (1983) and **51**, 534 (1983); R. G. Arnold *et al.*, Phys. Rev. Lett. **52**, 727 (1984).

MULTIPAIR EXCITATIONS AND DYNAMIC RESPONSE OF THE METALLIC ELECTRON GAS IN TWO AND THREE DIMENSIONS

D. Neilson*, F. Green*,
D. Pines+, and J. Szymański**

* School of Physics, University of New South Wales
P.O. Box 1, Kensington, Sydney, N.S.W. 2033, Australia

+ Department of Physics, University of Illinois at
Urbana-Champaign, 1110 W. Green Street, Urbana,
Illinois 61801, U.S.A.

** Telecom Australia Research Laboratories
770 Blackburn Road, Clayton, Victoria 3168, Australia

We demonstrate, using a quantitative microscopic formalism, that multipair excitations significantly affect the dynamic response of the electron gas even for densities as high as $r_s = 2$ in the bulk. For the two-dimensional electron gas in GaAs/GaAlAs heterostructures we find that the effects of multipair excitations are even more pronounced, indicating that the standard single particle description of 2D electron systems should be used with caution. We discuss the essential role that multipair contributions at high frequency play in maintaining the frequency moment sum rules for the dynamic response of the electron gas.

I. INTRODUCTION

Recently a new microscopic theory was proposed by Green, Neilson, and Szymański (GNS)[1,2] for the electron gas at metallic densities. The theory uses an infinite subset of perturbation terms from the ground-state correlation energy. This infinite subset of terms includes all two-particle scattering correlations within the electron gas. The terms are chosen so that the theory exactly satisfies the conservation laws[3] while reproducing the known behaviour of the electron gas in the limits of large and small momentum transfer, in the metallic density range. Conservation, together with correct behaviour at these limits, then ensures the theory's accuracy over the entire range of momentum transfer.

The present authors have proved elsewhere[4] that the GNS model exactly satisfies the third frequency moment sum rule in addition to all of the lower-order rules, notably the f-sum rule.[1] We have used the GNS model to quantitatively study the high-frequency response of the metallic electron gas at large momentum transfers.[4] We showed that the characteristic high-frequency contributions to the resulting dynamic Structure Factor constitute a sensitive measure of multipair correlations for the electron gas in the bulk.

In this paper we extend our analysis to the case of the quasi-two-dimensional electron gas found in gallium arsenide/gallium aluminium arsenide heterostructures. We demonstrate that the multipair effects found to be significant in the 3D case are even more pronounced in the case of a quasi-2D electron gas. This has important implications for the applicability of the single particle picture in the study of the correlated electron gas in two dimensions. In Section II we review the ideas underlying our analysis of multipair contributions to the dynamic response. In Sec. III we summarize the properties of GaAs/GaAlAs heterostructures. We present our results in Sec. IV and our conclusions in Sec. V.

II. SINGLE PAIR AND MULTIPAIR EFFECTS

We first review some basic concepts and definitions from our previous work[1,2,4] on multipair effects in the 3D electron gas. We adopt a precise definition of single pair and multipair contributions to the polarization function $\chi(\vec{q},\omega)$ as follows. Single pair contributions to $\chi(\vec{q},\omega)$ are all those terms which involve at any instant in time only a single electron-hole pair excitation. All other contributions to $\chi(\vec{q},\omega)$ are then multipair.

With this definition one can easily isolate all the single pair contributions from the complete GNS set of terms presented in detail in Ref. 2. Out of this large number of terms only the limited set involving a single polarization bubble, with the electron propagator and the hole propagator repeatedly scattering off each other via their unscreened Coulomb interaction, constitute single pair contributions to $\chi(\vec{q},\omega)$; the electron and hole propagators are themselves renormalized to infinite order with (unscreened) Hartree-Fock self-energy insertions. All other terms belonging to the GNS set involve the simultaneous excitation of two or more electron-hole pairs. If, for example, one screens the bare Coulomb interactions the result immediately becomes a multipair term. The complete set of single pair terms corresponds precisely to the infinite-order static Hartree-Fock summation originally evaluated by Dharma-Wardana and Taylor.[5]

All other terms in the GNS set contain multi-particle effects. It is convenient to consider these effects within three distinct categories: (a) self-energies, (b) electron-hole vertex corrections, and (c) t-matrix.

(a) Self-energies. The renormalized single particle propagator $G(\vec{p},p^0)$ with self-energy insertions $\Sigma(\vec{p},p^0)$ is given by the solution of the usual Dyson equation, and the corresponding proper polarization function is

$$\chi^{se}(\vec{q},\omega) = 2\sum_{\vec{k}}\int_{-\infty}^{\infty}\frac{dk^0}{2\pi i}\; G(\vec{k},k^0)\; G(\vec{q}+\vec{k},\; \omega+k^0)$$

$$= 2\sum_{\vec{k}}\int_{-\infty}^{E_F} d\omega'\; A_{\vec{k}}(\omega')\int_{E_F}^{\infty} d\omega''\; A_{\vec{q}+\vec{k}}(\omega'')\;\frac{2(\omega''-\omega')}{\omega^2-(\omega''-\omega'-i\eta)^2}\;, \quad (1)$$

where $A_{\vec{k}}(\omega)$ is the spectral density for the propagator $G(\vec{k},\omega)$ and E_F is the renormalized Fermi energy. The complete GNS set of self-energy insertions $\Sigma(\vec{p},p^0)$, which includes both single pair and multipair contributions, consists of the familiar Random Phase Approximation (RPA) screened self-energies together with the self-energies associated with t-matrix ladders of bare particle-particle interactions plus their corresponding exchange terms.

Equation (1) contains both single pair and multipair contributions. We can also write down an expression for $\chi_{sp}^{se}(\vec{q},\omega)$ for which the renormalized propagators include only single pair self-energy insertions. From our previous discussion, these are the static Hartree-Fock self-energy insertions, $\Sigma_{HF}(\vec{k})$. The single pair contribution is then

$$\chi_{sp}^{se}(\vec{q},\omega) = 2 \sum_{\substack{|\vec{k}|<k_F \\ |\vec{k}+\vec{q}|>k_F}} \frac{2(\varepsilon_{HF}(\vec{q}+\vec{k}) - \varepsilon_{HF}(\vec{k}))}{\omega^2 - (\varepsilon_{HF}(\vec{q}+\vec{k}) - \varepsilon_{HF}(\vec{k}) - i\eta)^2} \qquad (2)$$

with $\varepsilon_{HF}(\vec{k}) = k^2/2m + \Sigma_{HF}(\vec{k})$, and with k_F as the Fermi momentum.

(b) Electron-hole Vertex Corrections. We refer here to scattering of an electron and a hole within a single polarization bubble. In this case the multipair effects arise from the dynamic screening of the interaction. To separate out the multipair contribution, we evaluate the total contribution to $\chi(\vec{q},\omega)$ to lowest order in the screened electron-hole interaction and compare this with the corresponding single pair contribution $\chi_{sp}^{vc}(\vec{q},\omega)$ evaluated to the same order. We therefore define the proper polarization function

$$\chi^{vc}(\vec{q},\omega) = \int\frac{d^4k}{(2\pi)^4}\int\frac{d^4p}{(2\pi)^4} G(k)G(k+q)V_{RPA}(k-p)G(p)G(p+q) \qquad (3)$$

where $V_{RPA}(k - p)$ is the dynamically screened RPA interaction. $\chi_{sp}^{vc}(\vec{q},\omega)$ is defined analogously, but with the bare interparticle interaction $V(k - p)$ replacing $V_{RPA}(k - p)$.

(c) t-Matrix. The remaining multipair contributions of significance in the momentum-transfer region we are examining, all fall into the general category of t-matrix type scattering terms between a pair of electron-hole polarization bubbles. These terms are purely multipair, so that in this case there is no comparison of total versus single pair contributions. The structure of these terms is specified fully in Ref. 4.

III GaAs/GaAlAs HETEROSTRUCTURE

We now summarize the physical properties of a gallium arsenide/gallium aluminium arsenide heterostructure (see Figure 1). This type of interface represents a sharp compositional discontinuity in an otherwise homogeneous semiconductor lattice. The formation of space charges in the neighbourhood of the discontinuity results in the confinement of electrons within a thin layer at the interface. The electrons are, however, highly mobile along the plane of the interface.

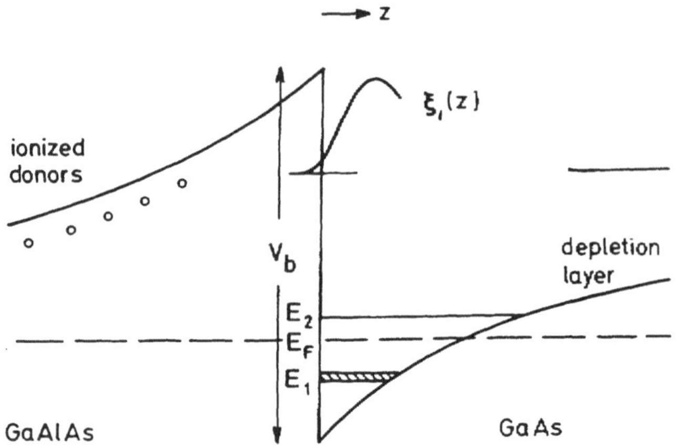

Fig. 1. Confining potential well for the quasi-2D electron gas in a GaAs/GaAlAs heterostructure. $\xi_1(z)$ is the self-consistent envelope function for electrons in the lowest subband E_1.(*) Subbands E_2 and higher are assumed to be unoccupied.

(*) Calculated with local exchange/correlation potential included (Ref.6).

The electrostatic interaction between two electrons in the lowest subband is given by

$$V(\vec{R} - \vec{R}') = \int_{-\infty}^{\infty} dz \int_{-\infty}^{\infty} dz' \; \xi_1^2(z) \frac{e^2}{\kappa|\vec{r} - \vec{r}'|} \xi_1^2(z') \quad , \tag{4}$$

where z is the space coordinate transverse to the layer, $\vec{R}$ is a vector parallel to the layer with $\vec{r} = (\vec{R},z)$ being the complete 3D space vector, and κ is the dielectric constant. The function $\xi_1(z)$ is the self-consistent solution to the transverse wavefunction for electrons in the first subband (see Ref. 6). We can write the Fourier transform of the effective 2D interaction of Eq. (4) as

$$V(q) = \frac{2\pi e^2}{\kappa q} F(q) \quad . \tag{5}$$

The Form Factor F(q) is calculated from the numerical wavefunction $\xi_1(z)$, and can be parametrized:

$$F(q) = \frac{1 + x_1(q/b) + x_2(q/b)^2}{(1 + x_3(q/b))^3} \quad . \tag{6}$$

The inverse length b is determined by the characteristics of the heterostructure substrates:[7]

$$b^3 = \frac{48\pi e^2 m^*}{\kappa\hbar^2} (N_d + \frac{11}{32} N_s) \quad , \tag{7}$$

where N_d is the number of ionized acceptors per cm^2 in the depletion layer, N_s is the number density of the 2D electron gas (cm^{-2}), and m* is the effective mass for the electrons.

The separation parameter r_s for electrons in this 2D geometry is given in terms of k_F and the effective Bohr radius a* by $r_s = \sqrt{2}/(k_F a^*)$.

Our calculation for the GaAs/GaAlAs heterostructure adopts the following material parameters: effective mass $m^* = .07m_e$ and dielectric constant $\kappa = 13$ (m* and κ are assumed to be the same in both GaAs and GaAlAs); acceptor concentration $3X10^{14}$ cm^{-3} in GaAs, resulting in $N_d = 8X10^{10}$ cm^{-2}; barrier height at the interface $V_b = .3eV$. N_s-dependent parameters are shown in the Table:

Table 1

N_s (cm^{-2})	r_s	E_F (meV)	E_1 (meV)	E_2 (meV)	b ($\mathring{A}^{-1}$)	x_1	x_2	x_3
$2.0X10^{11}$	1.28	37.4	30.5	51.8	.0284	.624	.152	.686
$0.5X10^{11}$	2.57	22.1	20.4	38.2	.0247	.580	.138	.661

IV. RESULTS

In Figure 2 we show our calculation of multipair contributions to the dynamic Structure Factor $S(\vec{q},\omega)$ for the quasi-2D electron gas in a GaAs/GaAlAs heterostructure, defined as usual by

$$S(\vec{q},\omega) = -\frac{1}{\pi} \text{Im}\, \chi(\vec{q},\omega)$$

$$= -\frac{1}{\pi} \cdot \frac{\text{Im}\, \chi^{sc}(\vec{q},\omega)}{|1 - V(\vec{q})\chi^{sc}(\vec{q},\omega)|^2} \quad , \tag{8}$$

where $\chi^{sc}(\vec{q},\omega)$ is the proper polarization function for the system. In Figure 3 we show for comparison our previous results for the 3D metallic electron gas. We remark that Fig. 3 includes all single pair corrections

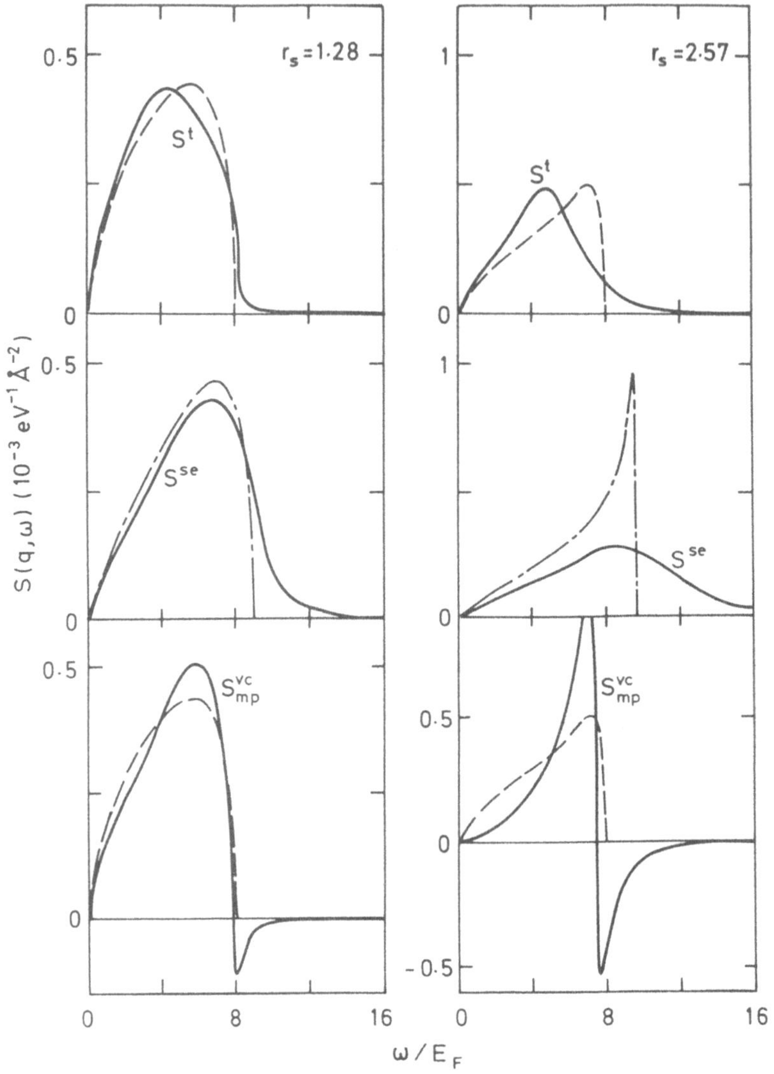

Fig. 2. Distinct multipair contributions to $S(\vec{q},\omega)$ for the quasi-2D electron gas, compared with RPA (dashed line), for momentum transfer $q = 2k_F$. Also shown is the static Hartree-Fock self-energy contribution S^{se}_{sp} (dot-dashed line).

S^t : t-matrix .

S^{se} : total GNS self-energy .

S^{vc}_{mp} : multipair vertex correction .

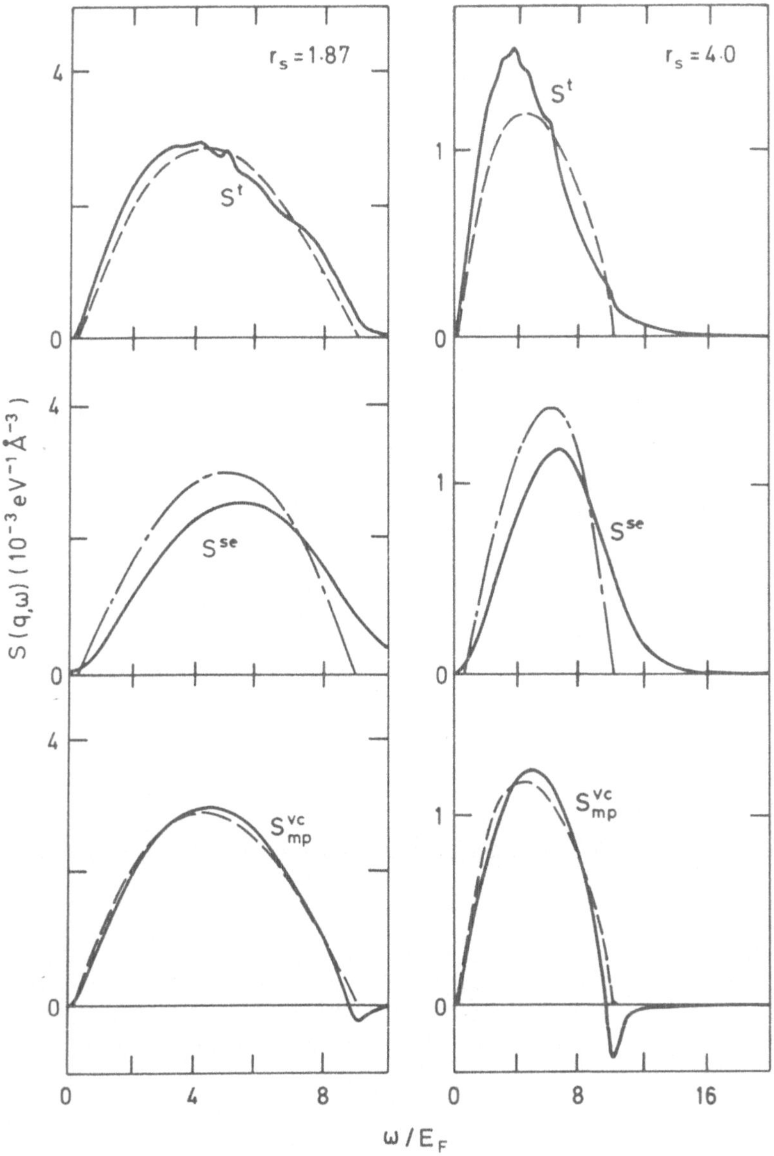

Fig. 3. Distinct multipair contributions to $S(\vec{q},\omega)$ for the 3D electron gas, compared with RPA plus total static Hartree-Fock S_{sp} (dashed line), for momentum transfer $q = 2.1k_F$. Also shown is the static Hartree-Fock self-energy contribution (dot-dashed line).

to the RPA coming from the infinite-order static Hartree-Fock series, while Fig. 2 does not; the corresponding all-order corrections for the quasi-2D electron gas will be discussed elsewhere.[8]

In comparing Figs. 2 and 3, we stress two important features:

(i) The multipair contributions for the quasi-2D electron gas are significantly larger than for the bulk electron gas at corresponding values of r_s. The 2D effects depend more sensitively on the precise value of r_s than do their counterparts in 3D. We should also emphasize that our reported results are sensitive to the width of the quasi-2D layer. If we reduce the width of the layer from $\sim$ 100 Å, we find that the discrepancy between the results for 3D and those for 2D increases dramatically. This is associated with the hardening of the interparticle potential as strict two-dimensionality is approached (see Eq. (5)).

(ii) There are systematic high-frequency cancellations among the tails of the self-energy, electron-hole vertex, and t-matrix scattering contributions to $S(\vec{q},\omega)$ in the quasi-2D electron gas. These cancellations follow a pattern closely similar to that found in our 3D analysis.[4] In the present case we find the following high-frequency asymptotic behaviour: $S^{se}_{mp} \sim \omega^{-4}$ and $S^{vc}_{mp} \sim -\omega^{-4}$, with leading-order cancellation between S^{se}_{mp} and S^{vc}_{mp}; for the t-matrix terms we have $S^{t} \sim -q^2\omega^{-5}$, with leading-order cancellation between S^{t} and $(S^{se}_{mp} + S^{vc}_{mp})$. Finally, the overall form of the high-frequency tail for $S(\vec{q},\omega)$ in the quasi-2D case is

$$S(\vec{q},\omega) \underset{\omega \to \infty}{\sim} \frac{11}{32}\pi \left(\frac{x_2 b}{(x_3)^3}\right)^2 N_s^2 e^4 q^4 \omega^{-6} \quad . \qquad (9)$$

V. SUMMARY

The main points that we wish to emphasize in this paper are:

(1) Multipair excitations have a significant effect on the dynamic response of the electron gas. This is of course expected in the low-density limit, but we have demonstrated that in 3D these effects are significant even at the highest metallic densities. Our current results indicate that multipair excitations have an even more crucial role to play in determining the response of the correlated 2D electron gas in actual GaAs/GaAlAs heterostructures. This is an important point to bear in mind for practical studies of heterostructures, where it has been the usual practice to treat electron correlation effects within a single particle picture, modified by linear RPA screening.

(2) A distinguishing feature of multipair correlations is the appearance of a high-frequency tail in $S(\vec{q},\omega)$ outside the single pair excitation range. This tail has a uniquely defined asymptotic fall-off. The form of this fall-off is determined by delicate but highly specific cancellations among quite different multipair excitations: self-energy terms, electron-hole vertex corrections, and t-matrix terms. The pattern of high-frequency cancellations among these different modes of excitation is imposed on the GNS model by the conservation equations.[4]

ACKNOWLEDGEMENT

Part of this work was carried out at the University of Nottingham (U.K.) and one of us (D.N.) would like to thank Professor L. J. Challis and the Low Dimensional Structures Group for their hospitality. Support from the U.K. S.E.R.C. is gratefully acknowledged.

REFERENCES

1. F. Green, D. Neilson, and J. Szymański, Phys. Rev. B31, 2779 (1985); Phys. Rev. B31, 5837 (1985).
2. F. Green, D. Neilson, and J. Szymański, Phys. Rev. B31, 2796 (1985).
3. G. Baym and L. P. Kadanoff, Phys. Rev. 124, 287 (1961).
4. F. Green, D. Neilson, D. Pines, and J. Szymański, in *Recent Progress in Many Body Theories*, R. A. Smith and P. J. Siemens editors, Springer, Berlin (1986), and to be published.
5. M. W. C. Dharma-Wardana and R. Taylor, J. Phys. F10, 2217 (1980).
6. F. Stern and S. Das Sarma, Phys. Rev. B30, 840 (1984).
7. F. Stern and W. E. Howard, Phys. Rev. 163, 816 (1967).
8. F. Green, D. Neilson, J. Szymański, and R. Taylor, unpublished.

THE RESPONSE FUNCTION OF THE HARD-SPHERE FERMI GAS

E. Mavrommatis,* R. Davé, and J. W. Clark

McDonnell Center for the Space Sciences
and Department of Physics
Washington University, St. Louis, MO 63130, USA

INTRODUCTION

This paper is devoted to a microscopic study of the elementary excitations of a hard-sphere Fermi gas which are produced by a weakly interacting probe. We apply linear-response theory in a correlated basis and calculate the density-density response function $\Pi(q,\omega)$ and corresponding dynamic structure factor $S(q,\omega)$, as functions of the transferred momentum $\hbar q$ and energy $\hbar\omega$. Systems with single-particle level degeneracy $\nu = 4$ and $\nu = 2$ are considered, in correspondence with symmetrical nuclear matter and pure neutron matter. The results are referred to a hard-core radius of $c = 0.4\ \text{fm}^{-1}$, again with nuclear systems in mind.

In spite of the fact that the hard-sphere Fermi gas is one of the simplest of all strongly-interacting quantum many-body systems, there is remarkably little quantitative information on its dynamical response. This alone provides adequate motivation for the present study. Surely this model problem should serve as a testing ground for theories of elementary excitations and attendant approximations. If such theories can reach agreement on the hard-sphere problem, they can be used with more confidence on problems involving realistic interactions, and the roles played by attractive and repulsive forces can be better separated and understood.

Czyż and Gottfried[1] examined the response function of the hard-sphere Fermi gas within Green's function perturbation theory, through second order in $k_F c$. Their calculations focused on the region of relatively large q and ω where effects of dynamical correlations can be readily distinguished from effects of statistical correlations (and, for a finite nucleus, effects of finite size).

Here we shall employ an approach which is in some ways complementary to that of Ref. 1. The longer-range goal is a calculation of the linear dynamical response of realistic neutron matter and nuclear matter, within the correlated random-phase approximation[2–5] (CRPA). The latter approximation extends the usual RPA method to strongly-interacting Fermi systems. While treating explicitly only 1p-1h excitations, the correlated RPA scheme includes in an average way important dispersive and polarization

*On leave of absence from the Depatment of Physics, University of Athens, Panepistimiopolis, Athens 15771, Greece.

effects of the medium along with the geometrical correlations induced by the strong repulsions. Here we take a preliminary step toward the larger goal by evaluating the dynamic structure function for a hard-sphere potential in a simplified approximation to CRPA proposed by Krotscheck.[5] In this approximation, the particle-hole force U is assumed to be local in the sense that it is allowed to depend only on the momentum transfer in the direct particle-hole channel. The particular local choice made by Krotscheck (called here LCRPA) has the convenient feature that it requires, as input to U, only quantities which are routinely generated in a Jastrow-Fermi-hypernetted-chain treatment of the ground state. The LCRPA scheme has been applied with some success to the description of elementary excitations, and more especially the self-energy and effective mass, in both unpolarized[6] and polarized[7] liquid ^{3}He as well as the electron gas[8].

The correlated RPA theory may be given a structure essentially identical to that of ordinary RPA, and the derivation of formulas for $\Pi(q,\omega)$ and $S(q,\omega)$ within the full theory and within its local version proceeds in the standard manner. However, the new theory is distinguished from ordinary RPA by the fact that the particle-hole interaction is a truly microscopic quantity, obtained by dressing the bare interaction as prescribed in the method of correlated basis functions (CBF).

OUTLINE OF CORRELATED RPA THEORY

As is well known, the ordinary random-phase approximation may be extracted as the small-amplitude limit of time-dependent Hartree Fock theory. To adapt that derivation to strongly-coupled systems requires simply a systematic replacement[3] of all energy eigenstates $|\Phi_m\rangle$ of the noninteracting Fermi system by the corresponding correlated-basis states

$$|\Psi_m\rangle = F\,|\Phi_m\rangle I_{mm}^{-1/2} \quad , \quad I_{mm} = \langle\Phi_m|F^\dagger F|\Phi_m\rangle \quad , \tag{1}$$

where F is a suitable static correlation operator, e.g. of Feenberg or Jastrow form. One arrives at the following set of supermatrix equations in place of the usual RPA equations:

$$\begin{bmatrix} \mathrm{A} & \mathrm{B} \\ \mathrm{B}^* & \mathrm{A}^* \end{bmatrix} \begin{bmatrix} \mathrm{x} \\ \mathrm{y} \end{bmatrix} = \hbar\omega \begin{bmatrix} \mathrm{M} & 0 \\ 0 & -\mathrm{M}^* \end{bmatrix} \begin{bmatrix} \mathrm{x} \\ \mathrm{y} \end{bmatrix} \quad . \tag{2}$$

Here, x and y are column matrixes and A, B, and M are square matrices whose elements carry particle-hole labels, e.g., $\mathrm{x} = (x_{ph})$ and $\mathrm{A} = (A_{ph;p'h'})$. The solutions of these equations yield approximate excitation energies $\hbar\omega$ together with amplitudes x^*_{ph}, y^*_{ph} for finding a given particle-hole pair present in or absent from the corresponding excited states. The matrices A and B (respectively Hermitian and symmetric) are now constructed in terms of the CBF effective-interaction vertex $V(12)$, along with the CBF single-particle energies $e(p)$ and $e(h)$ assigned to particles and holes.[5] The matrix M is expressible in terms of the CBF nonorthogonality vertex $N(12)$. Explicitly,

$$\begin{aligned} A_{ph;p'h'} &= [e(p) - e(h)]\delta_{pp'}\delta_{hh'} + \langle ph'|V(12)|hp'\rangle_a \quad , \\ B_{ph;h'p'} &= \langle pp'|V(12)|hh'\rangle_a \quad , \\ M_{ph;p'h'} &= \delta_{pp'}\delta_{hh'} + \langle ph'|N(12)|hp'\rangle_a \quad . \end{aligned} \tag{3}$$

The effective interaction $V(12)$ is in turn determined by the operators $W(12)$ and $N(12)$ and the single-particle energies $e(k)$. These CBF ingredients may be evaluated rather accurately by Fermi-hypernetted-chain procedures, in the case of a Jastrow correlation factor (adopted here in specific calculations). For further details, see Refs. 2-5.

The CRPA equations (2) can be solved, with considerable effort, by standard diagonalization techniques on a suitable mesh. The nonorthogonality of the correlated basis, which is responsible for the appearance of a nontrivial metric matrix M, introduces a cumbersome energy dependence which is not present in ordinary RPA. Fortunately, most of this energy dependence can be transformed away be recasting the theory in terms of a particle-hole-irreducible effective particle-hole interaction.[4] Briefly, the reformulation runs as follows. First one introduces a "correlation supermatrix"

$$\mathrm{C} = \begin{bmatrix} (C_{ph;p'h'}) & (C_{ph;h'p'}) \\ (C_{hp;p'h'}) & (C_{hp;h'p'}) \end{bmatrix}$$

$$= \begin{bmatrix} (\langle ph'|N(12)|hp'\rangle_a) & (\langle pp'|N(12)|hh'\rangle_a) \\ (\langle hh'|N(12)|pp'\rangle_a) & (\langle hp'|N(12)|ph'\rangle_a) \end{bmatrix} \tag{4}$$

and a corresponding "interaction supermatrix" W in which the vertex $N(12)$ is replaced by $W(12)$. The particle-hole-irreducible components of these matrices, denoted respectively by X and X′, are then extracted via the relations

$$\mathrm{C} = \mathrm{X} + \frac{1}{2}\mathrm{CX} \ , \qquad \mathrm{W} = (1 + \frac{1}{2}\mathrm{C})\mathrm{X}'(1 + \frac{1}{2}\mathrm{C}) \ . \tag{5}$$

It can be checked that neither X nor X′ so determined contains any diagrams which can be visually identified as being particle-hole reducible (i.e., separable into two disjoint parts by cutting a single pair of particle-hole lines). (For example, no chain diagrams appear in X.) Setting

$$\Omega = \begin{bmatrix} ([e(p)-e(h)-\hbar\omega-i\eta]\delta_{pp'}\delta_{hh'}) & 0 \\ 0 & ([e(p)-e(h)+\hbar\omega+i\eta]\delta_{pp'}\delta_{hh'}) \end{bmatrix} , \tag{6}$$

we may then form a supermatrix

$$\mathrm{U}(\omega) = \mathrm{X}' - \frac{1}{4}\mathrm{X}\Omega\mathrm{X} \tag{7}$$

which plays the role of a particle-hole-irreducible particle-hole interaction. In terms of Ω and $\mathrm{U}(\omega)$, the CRPA equations (2) may be rewritten as

$$[\Omega + \mathrm{U}(\omega)] \begin{bmatrix} \hat{\mathrm{x}} \\ \hat{\mathrm{y}} \end{bmatrix} = 0 \ , \tag{8}$$

where $\begin{bmatrix} \hat{\mathrm{x}} \\ \hat{\mathrm{y}} \end{bmatrix} = (1 + \frac{1}{2}\mathrm{C}) \begin{bmatrix} \mathrm{x} \\ \mathrm{y} \end{bmatrix}$. Apart from a very minor residual energy dependence of U (henceforth ignored), these equations are formally identical with those of ordinary RPA and may be attacked by quite conventional procedures. Nevertheless, their detailed solution is still a computationally demanding task. Accordingly, we have chosen instead to implement a simple approximation scheme which allows ready comparison with semiphenomenological approaches, notably the polarization-potential description.[9]

LOCAL VERSION OF CORRELATED RPA (LCRPA)

For our exploratory calculations we adopt a *local* approximation to the correlated RPA scheme, meaning that the matrix elements of $N(12)$ and $W(12)$ in the particle-hole channel and indeed the corresponding matrix elements forming U of Eq. (7), are converted to functions only of the momentum transfer $q = |\mathbf{p} - \mathbf{h}|$ in the direct particle-hole channel. There are several ways in which locality can be imposed on the particle-hole force of correlated RPA theory. The most compelling is that devised by Krotscheck, based on the requirements that the approximation to the $N(12)$ matrix elements preserve the relation of these matrix elements to the static structure function and that the approximation to the $W(12)$ matrix elements preserve their analogous role in the optimization condition for the Jastrow two-body correlation function $f(r)$. Within the Krotscheck approximation, the particle-hole interaction matrix is specified by

$$U_{ij} = A^{-1} U(q) \delta(\mathbf{p} + \mathbf{p}' - \mathbf{h} - \mathbf{h}') \quad , \tag{9}$$

where i (respectively j) can stand for pair labels ph or hp (respectively $p'h'$ or $h'p'$), A is the number of particles in the normalization volume, and the local particle-hole interaction

$$U(q) = \Delta(q) S^{-2}(q) + \frac{\hbar^2 q^2}{4m} [S^{-2}(q) - S_F^{-2}(q)] \tag{10}$$

is built simply out of the static structure functions $S(q)$ and $S_F(q)$ of the interacting and noninteracting systems, respectively, and the Fourier transform $\Delta(q)$ of the functional derivative of the Jastrow ground-state energy with respect to $\ln f^2(r)$. For optimal Jastrow correlations, the term in Δ obviously drops out, and $U(q)$ is determined entirely by the static structure function, which is generated automatically in solving the FHNC/C equations[10] for the Jastrow ground state.

With such a local particle-hole force, one can invoke standard formulas[11] to obtain the corresponding approximate density-density response function,

$$\Pi(q,\omega) = \frac{\Pi_o(q,\omega)}{1 - U(q)\Pi_o(q,\omega)} \quad , \tag{11}$$

and therewith the LCRPA approximation for the dynamic structure function $S(q,\omega) = -\pi^{-1}\mathrm{Im}\Pi(q,\omega)$. The response function $\Pi_o(q,\omega)$ is the particle-hole propagator of the free Fermi gas, i.e., the Lindhard function. The form of (11) tells us that the LCRPA must be a rather drastic approximation in some respects. In particular, we note that the approximation to the response function contains the response function of the noninteracting system as a factor. This means that dynamical correlation effects in regions of the (q,ω) plane where $\Pi_o(q,\omega)$ vanishes are not accessible in the LCRPA description, which accordingly fails *qualitatively* in this respect. It is in fact just such effects which were examined by Czyż and Gottfried,[1] so we shall not be able to make any useful comparison with their work. We hasten to add that although the (q,ω) domain corresponding to incoherent particle-hole excitations is the same in LRPA as in the free system, the RPA denominator in expression (11) introduces nontrivial correlation effects in that region, and in some cases gives rise to a collective mode (zero sound) outside the single-pair continuum. LCRPA will doubtless suffer, at a *quantitative* level, from the *static* nature of the effective interaction $U(q)$ entering (11); for example, dynamic screening is known to be important in the electron gas at metallic densities.[12] Moreover, one does not expect the momentum dependence of the self-energy to be adequately reproduced within this scheme, particularly in the very delicate case of unpolarized liquid ^{3}He (cf. Refs. 6). In spite of its deficiencies, LCRPA offers a simple and straightforward microscopic underpinning for phenomenological theories of comparable structure, such as the polarization-potential model.[9]

RESULTS AND DISCUSSION

Our LCRPA calculation of $\Pi(q,\omega)$ and $S(q,\omega)$ for the hard-sphere Fermi gas is based on a static structure factor $S(q)$ derived from a Jastrow trial ground state with one-parameter correlation function

$$f(r) = \begin{cases} 0 \ , & \text{if } r \leq c \ , \\ 1 - e^{-\mu(r-c)} \ , & \text{if } r > c \ . \end{cases} \tag{12}$$

The parameter μ is determined at each chosen Fermi wave number k_F so as to minimize the Jastrow energy expectation value in FHNC/C approximation.[13] This correlation function gives results for the Jastrow energy almost identical to those obtained in FHNC/0 approximation using the solution of an Euler-Lagrange equation for $f(r)$ derived from a two-body cluster approximation to the energy expectation value.[14] We shall make the tentative assumption that our $f(r)$ adequately simulates the solution of the exact Euler equation based on the full Jastrow energy expectation value and therefore set $\Delta(q)$ equal to zero in (10). This assumption should be checked by estimating the functional derivative of the energy for the assumed $f(r)$; however, a preferable next step, in progress, is repetition of our calculation for a correlation function satisfying an Euler equation within the FHNC/C scheme.[10]

The static structure function $S(q)$ corresponding to the choice (12) has been computed directly from the relevant structural formula of FHNC/C theory, namely Eq. (A22) of Ref. 10. Accordingly, it has the correct (namely, *linear*) asymptotic dependence on q at long wavelengths.

Numerical results on dynamical response have been obtained for the hard-sphere system with level degeneracy $\nu = 4$ at the k_F value 1.5 fm^{-1} (i.e., somewhat above the equilibrium density of symmetrical nuclear matter, which corresponds to $k_F = 1.35$ fm^{-1}); and for the $\nu = 2$ system at $k_F = 0.5$, 1.0, 1.5, and 2.5 fm^{-1} (i.e., bracketing the density range which would be of most interest for neutron stars). For both level degeneracies we considered a range of momentum (wave number) transfer q from 0 to about 3 fm^{-1}.

Let us first examine the results for the $\nu = 4$ system. The corresponding effective particle-hole potential $U(q)$ of Eq. (10) is displayed in Fig. 1. Note that $U(q)$ contains a factor of the density ρ in its definition according to (9) and so has dimensions of energy. The system does not sustain a zero-sound mode for this potential at the corresponding Fermi wave number. In Fig. 2 we provide a comparison of $S(q,\omega)$ and $S_F(q,\omega)$ at three values of q, namely 0.061, 0.491, and 1.472 fm^{-1}. Relative to the noninteracting Fermi gas, the strength is shifted to higher energies; this feature becomes more prominent as the momentum transfer increases. We also call attention to the sharp peak appearing at the upper end of the allowed range of energy transfer. The behavior of $S(q,\omega)$ as a function of q for $\hbar\omega$ fixed at 10, 50, and 90 MeV is shown in Figure 3.

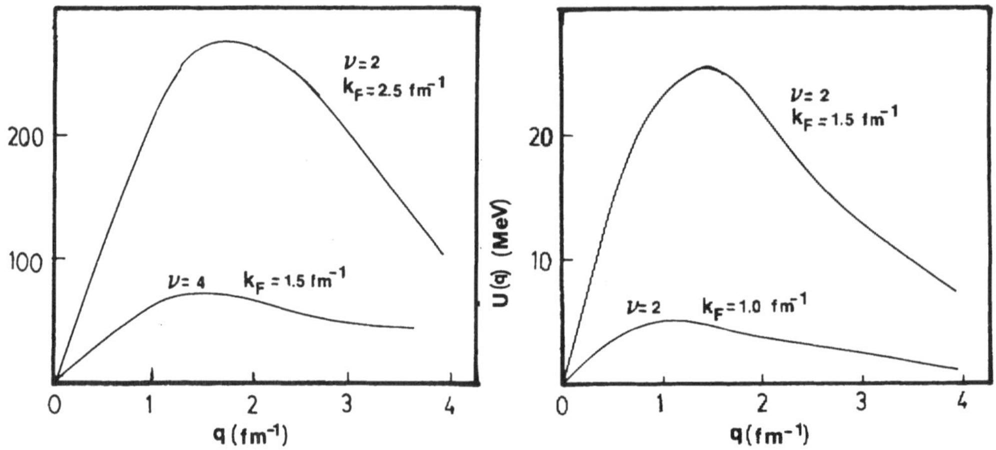

Fig. 1. Effective particle-hole interaction $U(q)$ as a function of wave number (momentum) q for the hard-sphere Fermi gas with level degeneracy $\nu = 4$ at $k_F = 1.5$ fm^{-1} and with level degeneracy $\nu = 2$ at $k_F = 1.0$, 1.5, and 2.5 fm^{-1}.

Now consider the results for the $\nu = 2$ system, where exchange effects are more important. The particle-hole interaction used in our calculation is plotted in Fig. 1 for the cases $k_F = 1.0$, 1.5, and 2.5 fm^{-1}. Only at the highest density of the four considered for $\nu = 2$, i.e., at $k_F = 2.5$ fm^{-1}, was a zero-sound mode present; this mode does not emerge from the particle-hole continuum until a q value of around 0.49 fm^{-1} is reached, and survives until about $q = 2.2$ fm^{-1}. Illustrative results for the energy dependence of the dynamic structure function for the four k_F choices are displayed in Fig. 4, at $q = 0.982$ fm^{-1}. The location and strength of the zero-sound mode occurring for the $k_F = 2.5$ fm^{-1} case are indicated. The corresponding zero-sound dispersion relation is plotted in Fig. 5, along with the upper bound of the particle-hole continuum.

As a check on our calculation, we have tested for deviations from the ω^0 and ω^1 sum rules,[15]

$$\int_0^\infty S(q,\omega)\, d(\hbar\omega) = S(q) \quad , \tag{13}$$

$$\int_0^\infty S(q,\omega)\omega\, d\omega = \frac{q^2}{2m} \quad , \tag{14}$$

for our approximate $S(q,\omega)$, as a function of q for each ν and k_F considered. These sum rules are well satisfied. In the $\nu = 2$ case at $k_F = 2.5$ fm^{-1}, it is of course necessary to include the strength of the zero-sound mode.

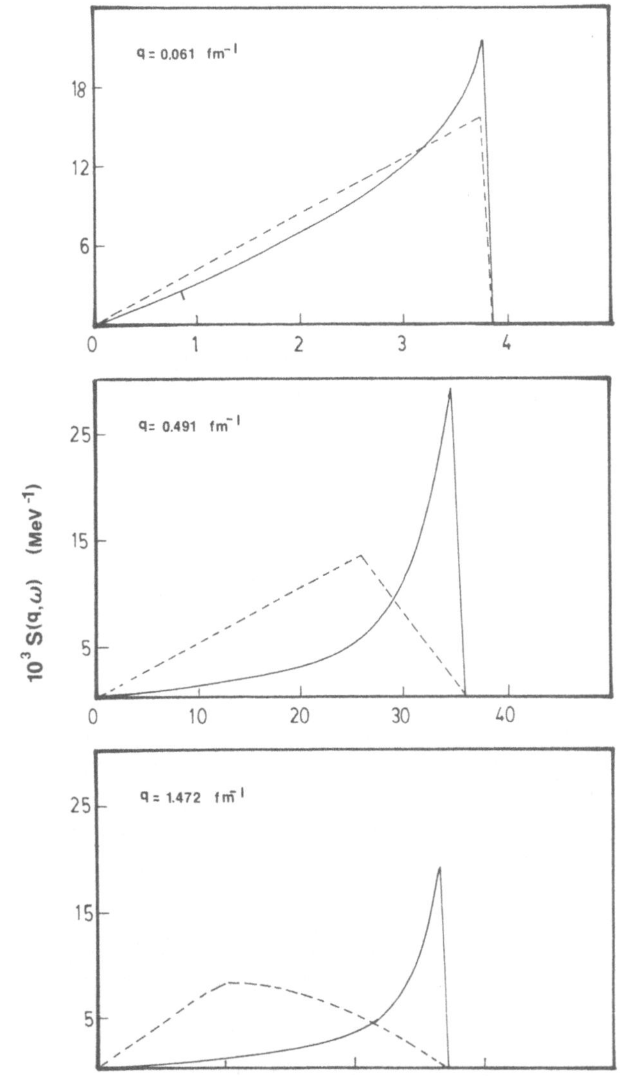

Fig. 2. Dynamic structure factor $S(q,\omega)$ as a function of energy transfer $\hbar\omega$ for the $\nu = 4$ system at $k_F = 1.5$ fm^{-1} for fixed momentum transfers $q = 0.061$, 0.491, and 1.472 fm^{-1} (solid curves), with corresponding results for the free system (dashed curves).

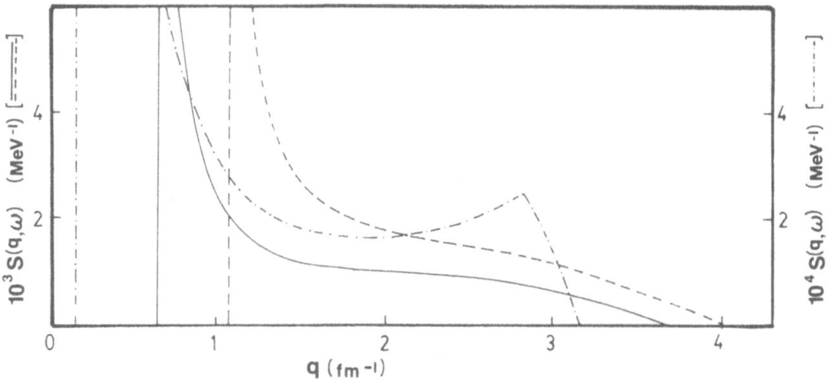

Fig. 3. Dynamic structure factor $S(q,\omega)$ as a function of momentum transfer q for the $\nu = 4$ system at $k_F = 1.5$ fm^{-1} for the following fixed energy transfers: $\hbar\omega = 10$, 50, and 90 MeV (dot-dashed, solid, and dashed curves, respectively). Respective maximum values obtained are $10^4 S(0.184, 10/\hbar) = 124.02$, $10^3 S(0.675, 50/\hbar) = 32.211$, and $10^3 S(1.104, 90/\hbar) = 20.068$ MeV^{-1}.

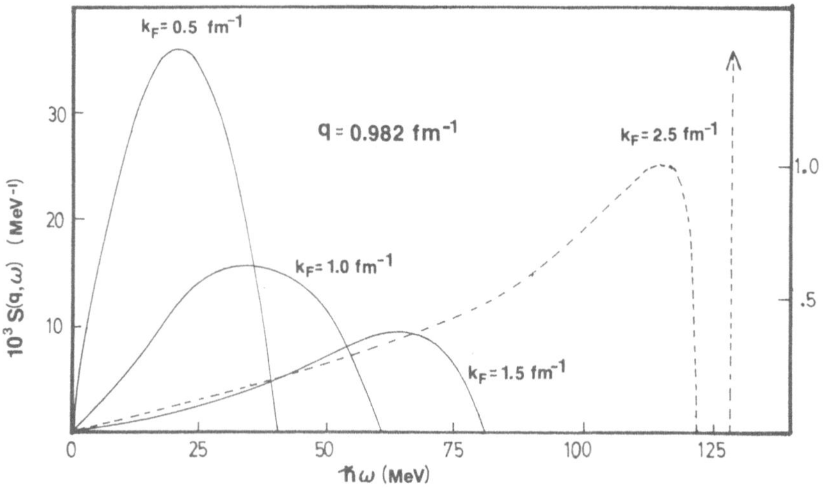

Fig. 4. Dynamic structure factor as a function of energy transfer $\hbar\omega$ for the $\nu = 2$ system at $k_F = 0.5$, 1.0, 1.5, and 2.5 fm^{-1} for fixed momentum transfer $q = 0.982$ fm^{-1}. Dashed traces refer to 2.5 fm^{-1}. Scale of solid traces [dashed traces] is on left [right] vertical axis. The zero-sound contribution to $S(q,\omega)$ at $k_F = 2.5$ fm^{-1} is $[0.124\ \delta(\hbar\omega - \hbar\omega_{zs})]$ MeV^{-1}.

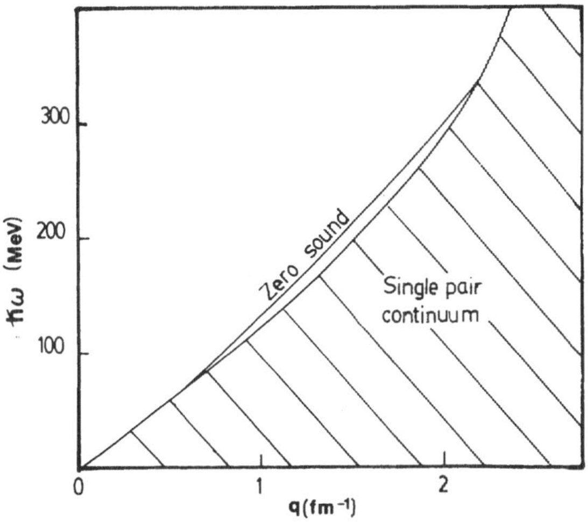

Fig. 5. Zero-sound dispersion curve for the $\nu = 2$ system at $k_F = 2.5\ \mathrm{fm}^{-1}$. The upper boundary of the continuum of single-pair excitations is also shown.

In summary, the results we have obtained for the dynamical response of the hard-sphere Fermi system within the local correlated random-phase approximation appear (for the most part) to be qualitatively reasonable and may be of semi-quantitative value in some ranges of density, momentum transfer, and energy transfer. Since other relevant calculations are lacking, it is difficult to judge the accuracy of the present attempt. Perhaps the most evident flaw arises from the nonoptimality of the two-body correlation function adopted here: Setting $\Delta(q) = 0$, the particle-hole interaction $U(q)$ vanishes at $q = 0$ (cf. Fig. 1), whereas an optimal $f(r)$ would give $U(q) \neq 0$. This deficiency will be removed in calculations presently under way. However, in the above calculations, it has the consequence of suppressing the zero-sound mode, which does not surface until rather high densities are reached. A more essential limitation of the present study, which may or may not introduce significant error, is its reliance on a local particle-hole force. We plan to investigate the validity of the local approximation to CRPA in some representative cases by solution[16] of the exact CRPA equations for the "full" effective irreducible particle-hole interaction as supplied from FHNC-FHNC' theory.[10] We must stress in conclusion that the absence of explicit multipair excitations from the description and the unrealistic confinement of the nontrivial dynamical response to certain regions of the (q,ω) plane are due to CRPA itself and can only be corrected by extension of that theory.

Following the lead of Refs. 4-8, the response function calculated here can be used to evaluate the self-energy and estimate the effective mass for the hard-sphere system.

ACKNOWLEDGMENTS

One of us (E. M.) acknowledges the hospitality extended to her during her leave of absence at the Department of Physics, Washington University. This work was supported in part by the U. S. National Aeronautics and Space Administration under NASA Innovative Grant NAGW-122 to Washington University, and in part by the Condensed Matter Theory Program of the Division of Materials Research of the U. S. National Science Foundation under Grant No. DMR-8519077. The FHNC code employed was developed from one due to M. F. Flynn. We thank A. Fabrocini, M. F. Flynn, K. Quader, and especially E. Krotscheck for helpful discussions.

REFERENCES

1. W. Czyż and K. Gottfried, Ann. of Phys. **21**, 47 (1963).
2. J. W. Clark, Springer Lecture Notes in Physics **138**, 184 (1981).
3. J. M. C. Chen, J. W. Clark, and D. G. Sandler, Z. Physik **A305**, 223 (1982).
4. E. Krotscheck, Phys. Rev. A **26**, 3536 (1982).
5. J. W. Clark and E. Krotscheck, Springer Lecture Notes in Physics **198**, 127 (1984).
6. B. L. Friman and E. Krotscheck, Phys. Rev. Lett. **49**, 1705 (1982); E. Krotscheck, in *Quantum Fluids and Solids, Sanibel, Florida, 1983*, Proceedings of the Symposium on Quantum Fluids and Solids, ed. E. D. Adams and G. G. Ihas (AIP, New York, 1983), p. 132.
7. E. Krotscheck, J. W. Clark, and A. D. Jackson, Phys. Rev. B **28**, 5088 (1983).
8. E. Krotscheck, Ann. of Phys. **155**, 1 (1984).
9. C. H. Aldrich II and D. Pines, J. Low Temp. Phys. **32**, 689 (1978); D. Pines, Springer Lecture Notes in Physics **198**, 113 (1984), and references cited therein.
10. E. Krotscheck, R. A. Smith, J. W. Clark, and R. M. Panoff, Phys. Rev. B **24**, 6383 (1981).
11. G. E. Brown, *Many Body Problems* (North Holland, Amsterdam, 1969).
12. F. Green, D. N. Lowy, and J. Szymanski, Phys. Rev. Lett. **48**, 638 (1982).
13. M. F. Flynn, private communication.
14. S. Rosati, in *The Meson Theory of Nuclear Forces and Nuclear Matter*, ed. D. Schütte, K. Holinde, and K. Bleuler (B. I. Wissenschaftsverlag, Mannheim, 1980).
15. D. Pines and P. Nozieres, *Theory of Quantum Liquids* (Benjamin, New York, 1966).
16. N. H. Kwong, Ph.D. Thesis, California Institute of Technology (1982), unpublished.

DEGENERATELY-DOPED SEMICONDUCTORS IN STRONG MAGNETIC FIELDS

B.I. Halperin

Department of Physics
Harvard University, Cambridge, MA 02138

INTRODUCTION

The properties of a disordered, strongly interacting electron system, in a strong external magnetic field, are not very well understood at the present time. Thus there is strong incentive to carry out experiments on a variety of materials, to explore the behavior under different conditions.

Two systems which have been studied extensively, in high magnetic fields and low temperatures, are the narrow band-gap semiconductors InSb and $Hg_{1-x}Cd_xTe$. Metal-insulator transitions have been reported in both of these systems.[1-7] Most recently, studies of degenerately-doped n-type Ge, have observed the occurrence of a metal-insulator transition in that system as well.[8]

In the present article, I shall briefly discuss some reasons why germanium is an interesting system to study, and I shall review some of the theoretical considerations that may apply to the field-induced metal-insulator transition in various doped-semiconductor systems. Particular emphasis will be given to the uniaxial spin-density-wave state proposed by Celli and Mermin in 1965,[9] and to the remarkable transport properties that should be observed if such a state can be realized.

PROPERTIES OF n-TYPE GE

Degenerately-doped n-type germanium is an interesting system to study in strong magnetic fields for a number of reasons. The material properties of Ge are well understood, and the magnetic quantum limit, in which all electrons are in the lowest Landau level, can be achieved at relatively low fields $H \sim 5$ Tesla. By contrast, we estimate that the field necessary for magnetic freezeout is much larger.[10] Germanium differs from previously studied semiconductors ($Hg_{1-x}Cd_xTe$ and InSb)[1-7] in a number of important respects: 1) Electrons in Ge are highly anisotropic. The conduction band consists of four equivalent valleys which are ellipsoids of revolution about the <111> crystal axes: the longitudinal effective mass along the <111> axes is $m_\ell = 1.64\, m_0$, and the transverse effective mass is $m_t = 0.082\, m_0$. For a field $\vec{H}$ along a particular <111> axis, electrons in the corresponding valley have a cyclotron mass $m_c = m_t$, while electrons in the other three valleys have cyclotron mass $m_c \cong 1.7\, m_t$ which is still small compared to m_0. 2) The field H_q required to reach the quantum limit can be varied without changing the electron density, by using uniaxial <111> stress to lift the four-fold valley degeneracy, and thus increase the Fermi energy. For $\vec{H}$ along the <111> stress axis, H_q is maximum in the "fully stressed" case, where all electrons are in one valley oriented along the stress axis. The value of H_q is computed to be 5.6 Tesla in this case, for a carrier concentration $n = 6 \times 10^{17}$ cm^{-3}. 3) Unlike work[1-5] on $Hg_{1-x}Cd_xTe$, *both spin states remain occupied* in n-type Ge at fields $H \sim 10$ Tesla, due to the relatively small g-factor for electrons[11]: $g = 1.6$. This property favors spin- over charge-density waves, and can have important consequences for the orientation of the wavevectors for these instabilities, as discussed below.

THEORETICAL CONSIDERATIONS

Strong magnetic fields can induce a metal-insulator transition in degenerately doped semiconductors at low temperatures, by a process usually termed magnetic freezeout.[12] The field compresses the wavefunctions of electrons bound to donors,[13] which reduces the overlap between wavefunctions and leads to the formation of localized electronic states. This phenomenon has been known and experimentally studied[14,15] for many years in narrow bandgap semiconductors such as InSb. If the electronic states are localized, then low-temperature electrical conduction will occur by phonon-induced hopping.[16] One would then expect that the ratio $\rho_{\parallel}/\rho_{\perp}$ should be a relatively weak function of temperature and magnetic field, but should gradually *decrease* with increasing magnetic field $\vec{H}$, as the wavefunctions become relatively more compressed in the direction perpendicular to $\vec{H}$.

Strong magnetic fields can also act to induce collective electronic phenomena, however. For theoretical model systems[9,17-20] with a uniform positive background, sufficiently high fields and low temperatures induce the electrons to form charge- or spin-density waves. For a system with one type of carrier in the strong field limit in which only one spin state is occupied, a multiple CDW state with wavevector-components both parallel and perpendicular to $\vec{H}$ is favored[17,18] in order to minimize the electrostatic self-energy. This state, termed a Wigner crystal in work[1-3] on $Hg_{1-x}Cd_xTe$, is expected to be insulating in both directions, and is thus difficult to distinguish experimentally from magnetic freezeout.[21]

SPIN-DENSITY-WAVE THEORY

In 1965 Celli and Mermin,[9] using a screened Hartree-Fock approximation, concluded that the ground state in the case where both spins are populated would contain a single spin-density wave with wavevector $\vec{Q}$ parallel to $\vec{H}$. Celli and Mermin discussed only a model with isotropic effective mass, but we expect that similar results should obtain in the anisotropic case. Thus the predictions of their theory may be compared with experiments on fully stressed Ge samples, where one electron-valley is occupied.

As long as the magnetic field is parallel to a symmetry axis of the electron pocket, we expect that the wavevector $\vec{Q}$ of the spin-density wave will point along the symmetry axis, and its magnitude will be just equal to $k_F(\uparrow) + k_F(\downarrow)$, where $k_F(\sigma)$ is the Fermi wavevector for electrons of spin σ, defined for the motion parallel to the magnetic field in the absence of the spin-density wave. (We take this direction to be the $\hat{z}$-direction.) With this choice of $\vec{Q}$, all the electron states are full whose energy lies below the position of the energy gap which is created by the spin-density wave, while all the electron states above the gap are empty. For any other choice of Q_z, there will be filled states above the gap or empty states below the gap, a situation which would be quite unfavorable for the total energy.

When $\vec{Q}$ takes the value specified above, the period λ of the spin-density wave is related to the field strength and carrier density, (in cgs units) by

$$\lambda = \frac{eH}{\pi n \hbar c} \tag{1}$$

For a fully stressed Ge sample, at $n = 6 \times 10^{17}$ cm^{-3} and $H = 15$ T, this gives $\lambda = 120$ Å

ANISOTROPIC TRANSPORT

The transport properties of the spin-density wave state should be quite remarkable. Because there is an energy gap at the Fermi energy, the diagonal elements of the conductivity tensor should vanish in the limit $T \rightarrow 0$, while the off-diagonal Hall conductance is nonzero and given by its classical value. ($\sigma_{xx} = \sigma_{yy} = \sigma_{zz} = 0$, $\sigma_{xy} = -\sigma_{yx} = nec/H$.) Inverting this tensor, one finds that the system is an insulator for electric fields parallel to H ($\rho_{\parallel} \rightarrow \infty$), but a perfect conductor in the perpendicular direction ($\rho_{\perp} \rightarrow 0$) for $T \rightarrow 0$. These properties remain true in the presence of a weak impurity potential, as long as: (1) the topological integrity of the spin-density wave is maintained, (2) there are only localized states at the Fermi level, and (3) there remains a band of extended states below the Fermi energy, which is able to carry the Hall current as required.[22,23]

We may remark that each period of the spin-density wave is a layer which contains two electrons per quantum of magnetic flux, and its transport properties are just what one finds in the quantized Hall effect. If the spin-density-wave period is given by Eq. (1), then the product of the number of layers per unit length ($1/\lambda$) with the quantized Hall conductance per layer ($2e^2/h$) is equal to the classical Hall conductance, $\sigma_{xy} = nec/H$. If there is pinning of the spin-density-wave period due to impurities or defects, we might expect the wavelength to deviate slightly from Eq. (1), and we might expect some hysteresis in the wavelength when the magnetic field strength is varied at low temperatures. This should lead to a hysteresis in the Hall effect at low temperatures, which may be observable.

THE UNSTRESSED CASE

The case of unstressed or partially stressed germanium with $\vec{H}$ parallel to <111> is more complicated than the fully stressed case because several inequivalent electron pockets can be occupied in the unstressed case. At sufficiently large fields, however, we expect that the <111> pocket aligned parallel to $\vec{H}$ will be emptied out, because of the larger diamagnetic energy in that pocket, and the electrons will be distributed equally among the three remaining pockets. One possible ground state in this situation would have *charge*-density waves for each pocket with $\vec{Q} \parallel \vec{H}$. If these CDW's are 120° out of phase with each other, there will be no charge density oscillation at the wavevector $\vec{Q}$, so that the unfavorable term in the Coulomb self-energy will not be present. The relation between the period of the CDW or SDW and the magnetic field strength and carrier density [Eq. (1)] must be modified to read $\lambda = 3eH/(\pi n \hbar c)$, when there are three valleys contributing.

An alternate possibility would be to have separate spin-density waves in each of the three occupied valleys. In each valley, the wavevector $\vec{Q}$ of the spin-density wave would be tilted away from the direction of the magnetic field towards the <111> axis corresponding to the valley. If one adds up the conductivity tensors of the three occupied valleys, however, one finds once again that the only nonzero components are $\sigma_{yx} = -\sigma_{xy}$.

For $\vec{H}$ parallel to <100> the four valleys have equivalent energies. In the absense of some transition which breaks this symmetry, the four valleys should then be equally occupied. Again, there are a variety of SDW and CDW states to be considered.

IMPURITY EFFECTS

A major difference between actual materials and the situation considered above is that the random impurity potential in a semiconductor is necessarily strong. For semimetals or metals such as graphite or $NbSe_3$, where the *electronic coupling* is weak, impurities are known to suppress the transition temperature T_c of CDW instabilities.[19,24] However the electronic system in Ge doped to donor concentrations near the metal-insulator transition is necessarily in the strong electron-electron coupling limit where quantitative theory breaks down.

It is certainly possible that the impurity potential will overwhelm the collective effects in the strong electronic-coupling case as well as the weak. However, it also seems possible that some remnant of a spin- or charge-density-wave will persist under favorable conditions.

As one possibility, suppose that impurities destroy the long-range order and even the topological order of the spin-density wave state, but still permit the formation of coherent domains that are large compared to the spin-density-wave period. Suppose also that within a domain, the impurity potential is too weak to localize the electrons, so that there is a *local* conductivity tensor with $\sigma_{xy} \neq 0$, but $\sigma_{xx} = \sigma_{yy} = \sigma_{zz} = 0$. Then the macroscopic resistivity will be determined by the nature of the defects which destroy the topological order.

As a particular example, we may suppose that the domains are separated from each other by a system of high-resistance walls. Then electrical transport perpendicular to $\vec{H}$ would be limited by the conductance of the walls. There would then be a resistance anisotropy that depends on the size of the domains, as well as the properties of the walls.

Another possibility is that the spin-density-wave structure is disordered by a collection of dislocation lines. It can be shown that an isolated dislocation line in the spin-density structure will give rise to a group of extended electron states near the Fermi energy, analogous to the edge states in the quantized Hall effect.[23,25] Electrons in these extended states will flow in a unique direction along the dislocation, and the extended states will not be localized by a weak impurity potential.

Consider a pair of dislocations of opposite sign which traverse the sample from one face to another. Under equilibrium conditions, the currents along the two dislocations will cancel each other, but this will not be the case when there is a voltage difference between the two faces. Specifically one finds that there will be a net electric current along the dislocation pair of magnitude $\lambda\sigma_{xy} V$, where V is the voltage difference between the two faces, λ is the period of the spin-density-wave structure, and σ_{xy} is the bulk Hall conductivity in the absense of dislocations. If one now imagines a set of dislocations following random-walk paths across the sample in all directions, one is led to a macroscopic conductivity tensor whose diagonal elements σ_{xx}, σ_{yy} and σ_{zz} are finite at $T = 0$. Thus the resistance would be finite for current parallel to $\vec{H}$ and there would be a nonzero dissipation for current flows perpendicular to $\vec{H}$. The magnitude of the diagonal conductivity elements would be proportional to the density of dislocation-lines, and would become comparable to the Hall conductivity σ_{xy} if the distance between dislocations and the persistence length (i.e., the length of a straight portion of a dislocation) are comparable to the period λ.

SYMMETRY CONSIDERATIONS

Analysis of electrons in a uniform positive background suggests that the ground state in that case involves a spin-density wave in which the oscillatory spin-density is linearly polarized along an arbitrary direction in the $x-y$ plane.[9,23] Of course, there is also a uniform magnetization in the z-direction, parallel to the applied magnetic field. There will also be an induced charge-density oscillation with wavevector twice that of the spin-density wave.

If spin-orbit effects are ignored, the Hamiltonian of the system is invariant under a rotation of all spins about the z-axis, so that the spin-density wave polarization represents a continuous broken symmetry additional to its broken translational symmetry. If impurities are present and the translational periodicity is destroyed, there could still be a broken spin-symmetry provided that there continues to be a (nonperiodic) oscillatory spin component along a single axis in the $x-y$ plane. The usual hydrodynamic considerations predict that there will be *spin-wave modes* associated with fluctuations in the direction of this axis, and we expect that the frequency of the spin-wave mode will vanish linearly with wavevector k, for $k \rightarrow 0$, in the absence of spin-orbit interactions.[26]

The existence of a broken symmetry in the groundstate of the system would imply that there is a well-defined phase transition, as a function of temperature or magnetic field strength, where the broken symmetry first occurs. However, when spin-orbit interactions are taken into account, the presence of impurities leads to random magnetic fields in the $x-y$ plane, which will destroy the symmetry of the Hamiltonian under spin rotations mentioned above, and will cause the direction of the oscillatory spin polarization to vary from one place to another in the sample.[27] In this case the ground state does not have a true broken symmetry, and there need not be a sharp phase transition as the temperature is lowered or the magnetic field increased. Similarly, if there were a charge-density wave (rather than a spin-density wave), which was predominantly along one axis but whose long-range order was destroyed by the impurities, there would be no broken symmetry in the ground state, and there need not be a sharp phase transition in this case.

ACKNOWLEDGMENTS

I have benefitted from many helpful discussions with R.M. Westervelt, M. Burns, G.A. Thomas, H. Fertig, Z. Tesanovic, and P.F. Hopkins. Helpful comments from P.A. Lee, H. Fukuyama, and M. Rasolt are also gratefully acknowledged. This work was supported in part by NSF grant DMR85-14638.

REFERENCES

1. T.F. Rosenbaum, S.B. Field, D.A. Nelson, and P.B. Littlewood, *Phys. Rev. Lett.* 54:241 (1985).
2. S.B. Field, D.H. Reich, B.S. Shivaran, T.F. Rosenbaum, D.A. Nelson, and P.B. Littlewood, *Phys. Rev.* B33:5082 (1986).
3. G. Nimtz, B. Schlicht, E. Tyssen, R. Dornhaus, and L.D. Hass, *Solid State Commun.* 32:669 (1979).
4. M. Shayegan, V.J. Goldman, H.D. Drew, D.A. Nelson, and P.M. Tredrow, *Phys. Rev.* B32:6952 (1986).
5. V. Goldman, H.D. Drew, M. Shayegan, and D.A. Nelson, *Phys. Rev. Lett.* 56:968 (1986).
6. G. DeVos, F. Herlach, *in*: "Application of High Magnetic Fields in Semiconductor Physics," G. Landwehr, ed., Springer-Verlag, Berlin (1983).
7. J.P. Stadler, G. Nimtz, B. Schlicht, and G. Remenyi, *Solid State Commun.* 52:67 (1984).
8. M.J. Burns, P.F. Hopkins, B.I. Halperin, G.A. Thomas, and R.M. Westervelt, to be published. Preliminary data presented at this workshop showed a large anisotropy in the electrical resistivity. Upon further investigation, however, it appears that the anisotropy was a result of strong planar striations in the dopant concentration in the germanium crystal from which the samples were cut.

9. V. Celli and N.D. Mermin, *Phys. Rev.* 140:A839 (1965).

10. A standard estimate for the impurity density n_c at the metal-insulator transition is (Ref. 16) $n_c r_\perp^2 r_\parallel = 1/64$, where $r_\perp$ and $r_\parallel$ are the "radii" of an isolated impurity perpendicular and parallel to the symmetry axis. Using a trial wavefunction as in Yafet *et al.* (Ref. 13), we estimate that the field necessary for freezeout in the fully stressed case is $H \cong 25$ Tesla.

11. G. Feher, D.K. Wilson, and E.A. Gere, *Phys. Rev. Lett.* 3:25 (1959).

12. See for example, K. Seeger, "Semiconductor Physics," 2nd ed., Springer-Verlag, Berlin (1982).

13. Y. Yafet, R.W. Keyes, and E.N. Adams, *J. Phys. Chem. Solids* 1:137 (1956).

14. R.W. Keyes and R.J. Sladek, *J. Phys. Chem. Solids* 1:143 (1956).

15. Yu. G. Arapov, A.B. Davydov, M.L. Zvereva, V.I. Stafeev, and I.M. Tsidil'kovskii, *Fiz. Tekh. Poluprovedn.* 17:1392 (1983); [*Sov. Phys. Semicond.* 17:885 (1983)].

16. N.F. Mott and E.A. Davis, "Electronic Processes in Non-crystalline Materials," Clarendon Press, Oxford (1979); B.I. Shklovskii and A.L. Efros, "Electronic Properties of Doped Semiconductors," Springer-Verlag, Berlin (1984).

17. H. Fukuyama, *Solid State Commun.* 26:783 (1978).

18. W.G. Kleppmann and R.J. Elliot, *J. Phys.* C8:2729 (1975).

19. D. Yoshioka and H. Fukuyama, *J. Phys. Soc. Jpn.* 50:725 (1981).

20. See also: H. Schulz and M. Youssoff, *Z. Physik* 267:41 (1974); R. Gerhardts and P. Schlottmann, *Z. Physik* B34:349 (1979); P. Schlottmann, *Z. Physik* B34:363 (1979); and references therein.

21. Data on small bandgap semiconductors such as $Hg_{1-x}Cd_xTe$ have been given several alternate explanations. See Refs. 1–7; also M. Ya. Azbel', *Solid State Commun.* 54:127 (1985); M. Ya. Azbel' and O. Entin-Wolhman, *Phys. Rev.* B32:562 (1985).

22. See M. Ya. Azbel', *Phys. Rev.* B26:3430 (1982); H.L. Störmer, J.P. Eisenstein, A.C. Gossard, W. Wiegmann and K. Baldwin, *Phys. Rev. Lett.* 56:85 (1986).

23. Further details will be given elsewhere.

24. Y. Iye, P.M. Tedrow, G. Timp, M. Shayegan, M.S. Dresselhaus, A. Furukawa, and S. Tanuma, *Phys. Rev.* B25:5478 (1982); G. Timp, M.D. Dresselhaus, T.C. Chieu, G. Dresselhaus, and Y. Iye, *Phys. Rev.* B28:7393 (1983).

25. B.I. Halperin, *Phys. Rev.* B25: 2185 (1982).

26. B.I. Halperin and P.C. Hohenberg, *Phys. Rev.* 177:952 (1969).

27. *Cf.*, Y. Imry and S.-K. Ma, *Phys. Rev. Lett.* 35:1399 (1975).

UNIVERSAL CONDUCTANCE FLUCTUATIONS IN DISORDERED METALS

Patrick A. Lee

Department of Physics
Massachusetts Institute of Technology
Cambridge, MA 02139

Recently it has been found both experimentally[1-3] and theoretically[4-8] that the electrical conductance in disordered metals exhibits fluctuations as the magnetic field B or the chemical potential μ is varied. The RMS magnitude of these fluctuations for a given sample is the same as that from sample to sample and is of the order $\delta G \approx e^2/h$, independent of the degree of disorder, the sample size, spatial dimensions, provided the temperature is low enough that the inelastic scattering length L_{in}, is larger than any of the sample dimensions[5,7]. $L_{in}=(D\tau_{in})^{1/2}$ is the distance an electron diffuses during the inelastic scattering time τ_{in} if D is the diffusion constant. It is to be emphasized that such fluctuations are not time-dependent noise. Instead, the conductance $G(\mu,B)$ is a deterministic, albeit fluctuating function of its arguments, for a given realization of the impurity configuration.

More recently the question of the sensitivity of the conductance of a given metal to a small change in the impurity configuration is addressed[9,10]. It is found that in d dimensions, the motion of a single impurity by a distance large compared with k_F^{-1} causes the conductance to change on the average by the amount δG_1, where

$$G_1^2 \approx (e^2/h)^2 \ (\Omega/N_i \ell^d) \ (L/\ell)^{2-d}$$

where N_i/Ω is the number of impurities per unit volume, ℓ is the mean free path and L is the sample dimension. In particular, for thin films of thickness $t > \ell$, $\delta G_1 \approx (e^2/h)(k_F\ell)^{-1}(\ell/t)^{1/2}$, independent of sample size L. This remarkable sensitivity helps explain the size of 1/f noise due to defect motion in disordered metals. We also predict a novel 1/f noise at low temperature in metallic glasses due to two level tunnelling centers.

References

1. C.P. Umbach, S. Washburn, R.B. Laibowitz, and R.A. Webb, Phys. B30:4048 (1984).
2. G. Blonder, Bull. Am. Phys. Soc. 29:535 (1984).
3. J. Licini, D. Bishop, M. Kastner, and J. Melngailis, Phys. Rev. Lett. 55:1987 (1985).
4. A.D. Stone, Phys. Rev. Lett. 54:2692 (1985).
5. P.A. Lee and A.D. Stone, Phys. Rev. Lett. 55:1622 (1985).

6. P.A. Lee, A.D. Stone, and H. Fukuyama, preprint.
7. B.L. Altshuler, Pis'ma Zh. Eksp. Teor. Fiz. 41:530 (1985); [JETP Lett. 41:648 (1985)].
8. B.C. Altshuler and D.E. Khmelnitskii, Pis'ma Zh. Eksp. Teor. Fiz. 42:291 (1985); [JETP Lett. 42:359 (1986)].
9. S.C.Feng, P.A. Lee, and A.D. Stone, Phys. Rev. Lett. 56:1960 (1986), and Erratum 56:2772 (1986).
10. B.L. Altshuler and B. Spivak, JETP Lett. 42:447 (1986).

A SCALING THEORY OF LOCALIZATION AND SUPERCONDUCTIVITY

M. Ma

Physics Department
University of Cincinnati
Cincinnati, Ohio 45221

Abstract

Scaling theories incorporating localization, electron-electron interactions and superconductivity consistently are required to understand the destruction of superconductivity in disordered systems. I discuss the ingredients that go into such theories and what difficulties are encountered. I present results for systems with spin-orbit scattering. En route, I review briefly the dilemma of the metal-insulator transition problem.

Introduction

I want to discuss here attempts to understand the destruction of superconductivity by disorder. The situation closely parallels that of the metal-insulator transition problem. It is not my intention here to show how the scaling equations are derived, and the interested reader is referred to the appropriate references.

The traditional belief that superconductivity is insensitive to normal disorder is inconsistent with the experimental observations of superconductivity destruction with increasing disorder. The popular (and most likely correct) view is that this is due to localization and/or disordered enhanced electron-electron correlations. While approaches[1] which treat the disorder perturtatively (in $1/k_f\ell$ or $1/E_F\tau$, where ℓ is the mean free path and τ the elastic scattering time), have been successful in explaining the initial drop in T_c (especially in thin films), they cannot address the eventual breakdown of superconductivity. Some kind of scaling approach is needed to, for example, answer questions like why some systems undergo a superconductor-insulator transition (SIT) and others a superconductor-metal-insulator transition (SMIT) at $T = 0$.

One might ask what features need to be included in the scaling theory. Some theories have taken non-interacting scaling theory for localization and studied the ramification of this with respect to superconductivity.[2-5] While these can be illuminating, the complete scaling theory must include the interplay between localization, coulomb correlations, and electron-phonon interactions (this point is reinforced by the importance of correlation effects in the metal insulator transition problem). I will discuss here some attempts in that direction. Nevertheless, I will restrict myself to assuming that the electron-phonon coupling can be modeled by an effective bare attractive interaction between electrons, which is independent of the degree of disorder. Furthermore, the superconductivity instability will be considered only within mean field theory. The true critical phenomena of such systems can only be understood if thermal[3,5] and quantum[6] fluctuations are included.

For non-superconducting materials (e.g. doped semi-conductors), a scaling theory for the metal-insulator transition problem has been partially achieved, by using the fact that 2 is the lower critical dimension, i.e., the metallic state is unstable against any finite disorder in 2D at T = 0. Thus in $2+\epsilon$ dimensions, the critical disorder is of $O(\epsilon)$ and is small for $\epsilon \ll 1$, thus allowing a pertubative renormalization group (RG) approach. For superconducting systems, the superconducting state is stable against weak disorder, since the localization-interaction instability is cutoff at T_c. This is a rather obvious point but is sometimes missed. Thus a pertubative RG can only give information concerning low T_c systems. Nevertheless, within the spirit of ϵ-expansion one can hopefully say something meaningful.

Before we proceed let us first recall the perturbative processes that can lead to potential singularties in 2D[7]. The underlying physics is charge conservation, which means that charge fluctuations are diffusive. This then implies the "particle-hole" propagator (Fig. 1) must necessarily have the diffusive form.

$$L_d^0\ (\vec{q},\omega) = \frac{1}{2\pi N_0 \tau}\ \frac{1}{-i|\omega| + D_0 q^2} \tag{1}$$

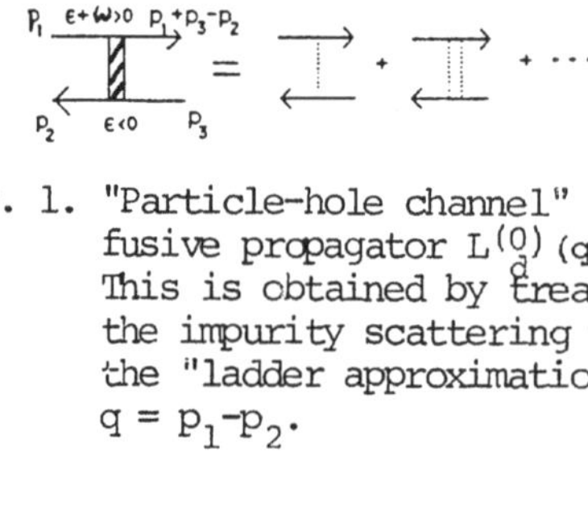

Fig. 1. "Particle-hole channel" diffusive propagator $L_d^{(0)}(q,\Omega)$. This is obtained by treating the impurity scattering with the "ladder approximation." $q = p_1 - p_2$.

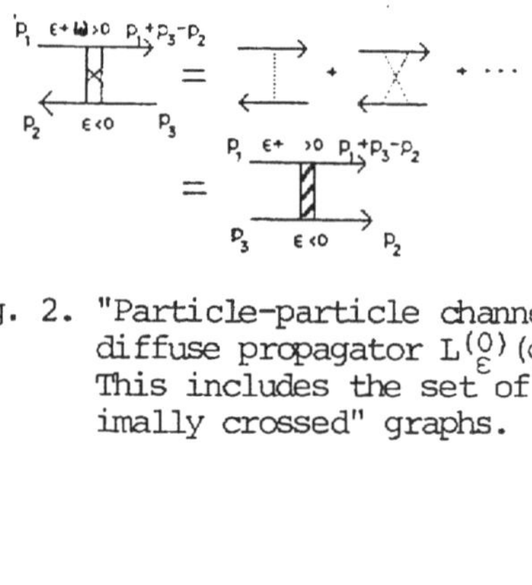

Fig. 2. "Particle-particle channel" diffuse propagator $L_c^{(0)}(q,\Omega)$ This includes the set of "maximally crossed" graphs.

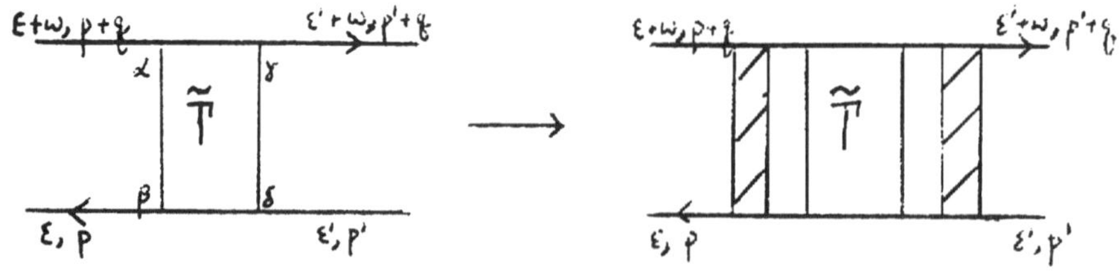

Fig. 3

Here $D_0 = V_F^2\tau/d$ is the "bare" diffusion constant.

i) If time reversal invariance is obeyed, the "particle-particle" propagator (Fig. 2) must also have the same form.

$$L_c(\vec{q}, \omega) = \frac{1}{2\pi N_o\tau}\,\frac{1}{-i|\omega| + D_0 q^2}\,, \quad \vec{q} = \vec{p}_1 + \vec{p}_3 \tag{2}$$

The pole at $\omega = 0$ and $\vec{p}_1 = -\vec{p}_3$ signifies coherent back-scattering and gives rise to the weak-localization logarithmic correction to the conductivity in 2 dimensions.

ii) As pointed out by Altshuler et. al.,[8] interactions between electrons, when coupled with their diffusive motion, can also result in logarithmic corrections. This can be understood by noting that any interaction process in the particle-hole channel must be dressed appropriately by diffusion ladders (Fig. 3, α, β, γ are spin indices), provided $(\epsilon+\omega)\epsilon<0$ and $(\epsilon'+\omega)\epsilon'<0$.

Thus, even if $\tilde{\Gamma}$ is small, the effective amplitude diverges for small q and ω. The physics is most transparent if we decompose $\tilde{\Gamma}$ into singlet and triplet amplitudes $\tilde{\Gamma}_s$ and $\tilde{\Gamma}_t$. These are then dressed by singlet (change) and triplet (spin) diffusive propagators respectively. While charge fluctuations are always diffusive, spin fluctuations are so only if spin is conserved.

iii) The same situation holds in time-reversal invariant systems for interaction processes in the particle-particle channel. We denote the amplitude in this channel by $\tilde{\Gamma}_c$ (Fig. 4).

In order to study the destruction of superconductivity, a temperature scaling approach will be used. All frequency variables introduced above should be replaced by the appropriate matsurbara frequencies. High matsurbara frequency processes can then be integrated out perturbatively. Works on the metal-insulator transition problem (particularly those of Finkel'stein)[9] have taught us that in general the only small parameter is the dimensionless resistance ($t = 1/(2\pi)^2 N^0 D$) and the $\tilde{\Gamma}$'s, on which we have been vague so far, are static amplitudes obtained from infinite resummations in the interaction.[9,10] Because interaction processes affect the single-particle density of states, it was thought initially that this will play a vital role. Subsequently it was realized that there are much cancellations and all density of states effects can be absorbed into a redefinition of the interactions[9-11]: $\tilde{\Gamma}\to\Gamma = N_R^2/N_o^2 . \tilde{\Gamma}$. Instead, the frequency renormalization is determined by a parameter z, s.t. under

Fig. 4

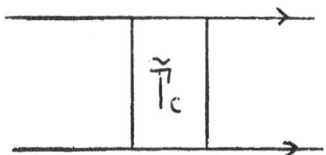

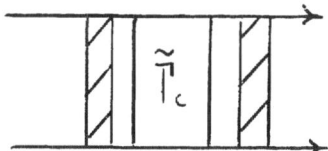

renormalization $|\omega_n| \to z|\omega_n|$ in eqs. (1) and (2). For coulomb interaction, charge conservation implies the constraint [9,10]

$$z - 2\,\Gamma_s = 0.$$

The behavior of the system under renormalization strongly depends on which processes are soft (diverges for small q and ω) at low temperature. This will depend on whether spin is conserved and time-reversal invariance obeyed. We can think of this as determining the university classes (but see previous comment on thermal and quantum fluctuations). We begin with the most general case of potential scattering only.

Potential Scattering

For repulsive interactions, Finkel'stein[9] has derived the group equations in $2+\epsilon$ dimensions.

$$dt/dy = -\tfrac{\epsilon}{2}t + t^2\left[3 - 3\,\frac{z+\Gamma_t}{\Gamma_t}\ln(1+\Gamma_t/z) - \Gamma_c/z + 1 + 1)\right] \tag{4a}$$

$$d\,\Gamma_t/dy = t\left[z/2 + \Gamma_t/2 + \Gamma_c + 4(\Gamma_t)(\Gamma_c)/z + 2(\Gamma_t)^2\,\Gamma_c/z^2 + 2(\Gamma_t)^2/z\right] \tag{4b}$$

$$dz/dy = t\left[-z/2 + 3\,\Gamma_t/2 + \Gamma_c\right] \tag{4c}$$

$$d\,\Gamma_c/dy = t\,(z + 3\,\Gamma_t)/2 - (\Gamma_c)^2/z \tag{4d}$$

where $y = -\ln zT$. These equations are correct to lowest order in t and Γ_c, but to all orders in z and Γ_t. 1+1 in (4a) comes from localization and the universal singlet ($=2\Gamma_s/z$) contributions.

We would like to use these equations, with Γ_c negative to study superconductivity. The last term of (4d) is in fact the Cooper ladder summation, with 1/z coming from renormalization of the particle-particle kernel. In the absence of disorder, Γ_c will scale to $-\infty$ at the BCS transition temperature T_c. This will be affected by the factor 1/z and by the first term, which tries to make Γ_c less negative. The 4 equations then allow us to study the interplay between superconductivity, localization and interaction. But there are a few problems. One should be wary that we need to consider Γ_c becoming large, while the equations require Γ_c small, in particular, superconductivity fluctuations are not included in the resistance equation (4a). Secondly Γ_c consists of two parts. An attractive contribution from the electron-phonon coupling, and a repulsive part from the renormalization of the screened coulomb interaction from E_F to ω_D, the Debye frequency. The RG equations start to apply at τ^{-1}. If $\tau^{-1}<\omega_D$, then $\Gamma_c(\tau^{-1})<0$ if $\Gamma_c(\omega_D)$ is. But if $\tau^{-1}>\omega_D$, then we have to scale the RG equations down to ω_D with a repulsive Γ_c and it may be that an attractive interaction is never achieved.[2] We will concern ourselves only with the more interesting situation of superconductivity destruction even when Γ_c $(\min\,(\tau^{-1},\omega_D))<0$. Finally, while the part of Γ_c due to electron-phonon coupling can have no diffusive vertex correction at long wavelength[12], the repulsive part does, and it is possible that a combined Γ_c is nonsensical. Fortunately, the RG equations to this order do not contain such corrections for Γ_c.

Even within the above limitations we cannot say much concrete for the case of potential scattering only. The difficulty here is largely just that in the metal-insulator transition problem. There it is found that the low-temperature state[9,10] is a well-behaved metal if $t_o < t_c$, of order ϵ and depends on z_o and Γ_{to}. But for $t_o = t_c$, z, Γ_t, and Γ_t/z blow up as some power of T as $T \to 0$, while t remains finite. For $t_o > t_c$, the behavior is similar, but z etc. diverge with a different power. The implications of this are not totally clear. It has been shown[10] that the magnetic suscriptibility

$$\chi = N_o z(1+\Gamma_t/z) \quad (5)$$

Thus χ also diverges at $T \to 0$, suggesting a magnetic instability. Since t is finite, one might consider a paramagnetic metal-ferromagnetic metal or metallic spin glass transition[13]. This leaves unanswered the question of what happens to the metal-insulator transition. A new idea[14] is that this is in fact the metal-insulator transition, and $t_o = t_c$ describes the onset of quasiparticle localization ($D_o = D/Z$[15]). The behavior for $t_o > t_c$ (t, hence D apparently still finite) is just due to the RG equations being no longer applicable. It leaves unexplained the diverging "Stoner enhancement" Γ_t/z.

Returning to the superconductivity problem, if the pure transition temperature T_{co} is small, then T_c decreases to zero for some $t_o < t_c$. Increasing disorder past that leaves us essentially left with the metal-insulator transition problem. Thus the transition is SM?T, with ? denoting the uncertainty discussed above. If T_{co} is large, so that T_c remains finite at t_c, then we are left with S??T.

Spin-Orbit Scattering

As in the metal-insulator transition problem, well-behaved scaling equations are obtained if part or all of the triplet fluctuations are suppressed. This can be because of Zeeman splitting caused by a strong magnetic field[9,11] or a lack of spin conservation due to spin-flip scattering[9,11] or spin-orbit scattering. In the first two cases, time-reversal invariance is also broken, and the superconductivity problem is decoupled from the localization-interaction problem. The spin-orbit scattering case on the other hand involves full interplay between localization, interaction, and superconductivity.

For positive Γ_c, the group equations have been derived by Castelleni et. al.[16]

$$dt/dy = -\frac{\epsilon}{2}\, t + t^2\left(\tfrac{1}{2} - \tfrac{1}{2}\gamma\right) \quad , \quad (5a)$$

$$dt/dy = t\left(\tfrac{1}{2} + \tfrac{1}{2}\gamma - \tfrac{1}{2}\,\gamma^2\right) - \gamma^2 \quad . \quad (5b)$$

where $\gamma = \Gamma_c/z$. The differences between equations (5) and (4) are the suppresion of triplet contributions in both particle-hole and particle-particle channels, and the coefficient $1/2\ t^2$ here, which is due to contributions from the singlet and from antilocalization. These equations are valid for $zT < \tau_{so}^{-1}$, the spin-orbit scattering rate. Thus, in addition to the previous warnings concerning setting Γ_c negative for superconductivity, equation (4) must be used to scale the system from T^{-1} to τ_{so}^{-1}. In fact, if $\tau_{so}^{-1} \ll T^{-1}$, the divergences discussed before may occur at some $zT > \tau_{so}^{-1}$ and equation (5) may never apply. We will assume $\tau_{so}^{-1} \sim \tau^{-1}$, i.e., strong spin-orbit scattering systems, where this is not an issue.

The RG flows are shown in Fig. 5[17].

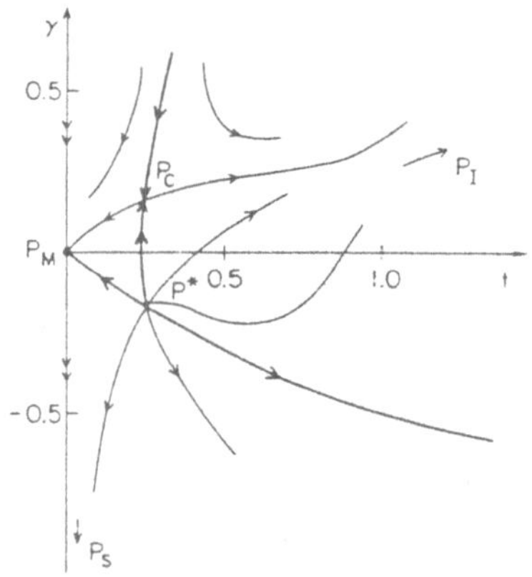

Fig. 5. Scaling trajectories in the t-γ plane. Phase boundaries are shown with thick lines. Figure is schematic.

The behavior of the system at low T depends on whether it scales to the insulating fixed point P_I at ($t = \infty$, $\gamma > 0$), the metallic fixed point P_M at (0, 0), or the superconducting fixed point P_S at (0, $-\infty$). There is also an insulating-superconducting fixed point at (∞, $-\infty$), but we believe it to be an artifact of the lowest order in Γ_c approximation. Inclusion of superconducting fluctuations in the resistance equation should bring t to zero. There is a totally unstable fixed point P* at (ϵ, $-(\epsilon/2)^{1/2}$). For $|\gamma_o|<(\epsilon/2)^{1/2}$, the system scales towards P_S for small t_o, towards P_M for intermediate t_o, and towards P_I for large t_o, i.e., an SMIT with increasing disorder. Note that $\gamma>0$ at P_I, and the interaction between localized states is repulsive within this approximation. For $|\gamma_o|>(\epsilon/2)^{1/2}$, the intermediate metallic phase is no longer accessible, and the transition is an SIT. Since the above information are obtained by analyzing flows in the vacinity of P* and P_M, they should be valid in $2+\epsilon$ dimensions in spite of the lowest order in t and Γ_c approximation. The conclusion here that small Γ_c systems undergo SMIT while large Γ_c systems undergo SIT agrees with simple-minded expectation and not with the idea that the system will somehow renormalize itself so that superconductivity is always destroyed at the "mobility edge".

In order to extend this to d=3, we set $\epsilon\approx 1$. The value of γ at P* is ~-0.7. As this is physically unlikely (remember we are in a regime where any unsystematic density of states effects are washed out by the disorder), we conclude systems with strong spin-orbit scattering should undergo SMIT.

We can also say something about the mean field T_c. Close to a SM boundary, the RG flow is initially parallel to P^*-P_M. T_c is then determined by the extrapolated intersect on the γ-axis, and so

$$T_c \sim \exp\left(-k/(n-n_c)\right) . \tag{6}$$

where n_c is (say) the critical density, and k a non-universal constant. This mean field behavior can be charged by fluctuations. Close to a SI boundary, T_c is dominated by the initial scaling towards large t. If equation (5a) is used, one obtains

$$T_c \sim \exp(-k'/(n-n_c)) \ , \quad (7a)$$

k' non-universal; while an exponential type behavior for t would give

$$T_c \sim (\ln|\ln n-n_c|)^{\delta} \ , \quad (7b)$$

δ model dependent. These will change if there is a non-perturbative (finite t) fixed point on the SI boundary. Although the above statements on T_c are derived from the present RG flows, they are in fact general behavior of T_c close to a SMT or SIT[18].

Conclusion

I have tried to explain why a scaling theory is necessary to understand the superconductivity-localization problem. I pointed out the similarities and differences between doing this here and in the metal-insulator transition problem. In the case of strong spin-orbit scattering, I think one can actually make some concrete statements. In fact, studying this particular case allows us to make general predictions concerning T_c near an SMT or an SIT.

Acknowledgement

I thank C. Castellani, C. DiCastro, E. Fradkin, H. Fukuyama, and P. A. Lee for discussions and close collaborations. I apologize to the authors and the readers for any references omitted by negligence.

REFERENCES

1. S. Maekawa and H. Fukuyama, Physica 107B, 123 (1981); J. Phys. Soc. Jpn. 51, 1380 (1982); H. Fukuyama, Physics 126B, 306 (1984).

2. P. W. Anderson, K. A. Muttalib, and T. V. Ramakrishnan, Phys. Rev. B 21, 117 (1983).

3. A. Kapitulnik and G. Kotliar, Phys. Rev. Lett. 54, 473 (1985).

4. M. Ma and P. A. Lee, Phys. Rev. B 32, 5658 (1985).

5. L. N. Bulaevskii and M. V. Sadovskii, J. Low. Temp. Phys. 59, 89 (1985).

6. M. Ma, B. I. Halperin, and P. A. Lee, Phys. Rev. B 34, 3136 (1986).

7. See P. A. Lee and T. V. Ramakrishnan, Rev. Mod. Phys. 57, 287 (1985) for a more complete review.

8. B. L. Altshuler, A. G. Aronov, and P. A. Lee, Phys. Rev. Lett. 44, 1288 (1980).

9. A. M. Finkel'stein, Z. Phys. B 56, 189 (1984); JETP 57, 97 (1983).

10. C. Castellani, C. DiCastro, P. A. Lee, M. Ma, S. Sorella, and E. Tabet, Phys. Rev. B 33, 6169 (1986).

11. C. Castellani, C. DiCastro, P. A. Lee, M. Ma, Phys. Rev. B 30, 1596 (1984).

12. A. Schmid, Z. Phys. 259, 421 (1973).

13. H. Fukuyama, J. Phys. Soc. Jpn. 54, 2092 (1985).

14. C. Castellani, G. Kotliar, P. A. Lee, unpublished.

15. W. L. McMillan, Phys. Rev. B 31, 2750 (1985).

16. C. Castellani, C. DiCastro, G. Forgacs, and S. Sorella, Sol. St. Comm. 52, 261 (1984).

17. M. Ma and E. Fradkin, Phys. Rev. Lett., 56, 1416 (1986).

18. B. Feuston and M. Ma, unpublished.

ELECTRON LOCALIZATION

IN ONE-DIMENSIONAL INCOMMENSURATE POTENTIALS

Marshall Luban

Ames Laboratory-USDOE and Department of Physics
Iowa State University, Ames, Iowa 50011

A one-dimensional potential V(x) consisting of two periodic components whose periods are incommensurate can give rise to localized electron states. Conventional numerical methods to calculate the localized eigenstates, either of the Schrodinger equation or the tight-binding models (TBM), are plagued by a numerical instability originating in a second solution which diverges for $|x|\to\infty$. We present a practical numerical algorithm for an arbitrary TBM which is free of this difficulty and which provides the localized eigenstates and their energies to ultra-high precision. The underlying method incorporates a generalization of a classic theorem of Pincherele for the existence of minimal solutions to three-term recurrence relations. We present numerical results for the Aubry model for $\Delta>1$, where $\Delta=V_I/(2V_H)$ is a dimensionless coupling constant, V_I is the strength of the incommensurate component of the one-electron potential and V_H is the nearest-neighbor hopping matrix element. We propose that the well-known transition from localized to extended eigenstates which occurs as Δ is reduced towards unity is accompanied by an incipient infinite degeneracy of the localized states. Our numerical results for the energy difference of any pair of nearly-degenerate localized states is well described by a power law, $(\Delta-1)^Y$, but with a non-universal exponent Y.

I. INTRODUCTION

The physics of independent electrons in a one-dimensional almost-periodic potential is a subject which has attracted a great deal of interest in the past six or seven years. Although there has been considerable progress,[1,2] there remain many intriguing open questions and interesting new phenomena yet to be discovered. Our motivation for studying such problems is to a considerable extent provided by the successful fabrication of layered materials with many exotic properties.[3,4]

To set the stage, consider the one-particle Schrodinger equation

$$\psi''(x)=(2m/h^2)[V(x)-W]\psi(x), \quad V(x)=V_P(x)+V_I(x), \tag{1}$$

where V_P is a periodic function with the period $\underline{a}$ of an underlying lattice and V_I is a second periodic function with period $\underline{b}$. If one chooses a/b to be a rational number, say M/N, the compound potential V is still periodic but with a new, larger period Na and the methods for dealing with such potentials are covered in standard textbooks of solid-state physics. In this case all of the physical eigenstates of (1) are extended, i.e., bounded yet non-normalizable states. Suppose instead that a/b is an irrational number, which insures that V is non-periodic; in fact, V exhibits the properties of an almost-periodic (or quasi-periodic) function. There is then a new feature in the physics of electrons in potentials of this type. In particular, after some thought one can convince oneself that, in addition to extended states, for sufficiently strong "incommensurate component" V_I the compound potential will possess localized eigenstates, which are defined as normalizable solutions of (1), akin to familiar bound states. To understand this point, refer to Fig. 1, where the dotted line corresponds to a bona fide quantum energy eigenvalue. Classically the particle would be localized; it would simply oscillate back and forth within one of the three wells shown, wherever it was initially placed, say the left-most well. Quantum-mechanically, the particle can be expected to tunnel to the right, to the nearby well. Yet we would not expect that the particle will tunnel to the right-most well, in part because of the intervening large hump in V but more importantly because the form of the potential in the right-most well differs so radically from that of the wells to its left. The right-most well can also be expected to localize our quantum particle, but not for the current choice of the energy.

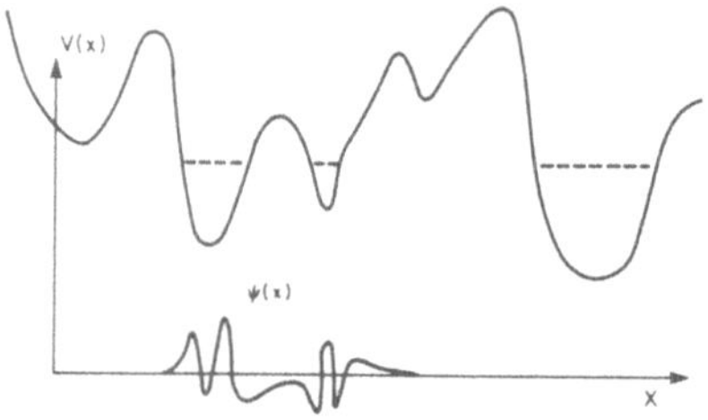

Fig. 1. One dimensional potential V(x) and a schematic localized eigenfunction $\psi(x)$ corresponding to the energy indicated by the dashed line.

An analytic calculation of the form of the localized eigenstates of (1) and the corresponding energy eigenvalues seems out of the question, except perhaps if we diligently hunt for a suitably stripped-down choice of V. On the other hand, even if we are prepared to resort to numerical methods for solving (1), we are up against a very serious obstacle generic to localized solutions which originates in the fact that if F and G denote two linearly independent solutions, their Wronskian, $F(x)G'(x)-F'(x)G(x)$ is independent of x. Thus, if F denotes a localized eigenstate, a solution which necessarily vanishes for $|x|\to\infty$, it follows that $|G|$ must necessarily diverge to infinity. Any numerical procedure will necessarily ignite the growing solution, because of inevitable round-off, and it is the growing solution which will in fact dominate the results.

The tight-binding approximation of (1) would seem to provide us with a more tractable form. Instead of a second-order differential equation we are faced with a linear second-order difference equation of the form

$$f_{n+1} + f_{n-1} = b_n f_n, \quad b_n = (W-\varepsilon_n)/V_H \ . \tag{2}$$

To pass from (1) to (2) one constrains ψ to be expressible as a linear combination of the form $\psi(x)=\Sigma_n f_n\phi(x-na)$ based on a single Wannier orbital ϕ, $V_H=\int dx\,\phi(x)H_P(x)\phi(x+a)$ is the matrix element for hopping from one site to a nearest-neighbor site, $\varepsilon_n=\int dx\,\phi(x)V_I(x+na)\phi(x)\simeq V_I(na)$ is the local site energy, and $H_P = T + V_P$ with T the kinetic energy operator. Overlap matrix elements involving more distant neighbors are ignored and we set our energy scale so that $\int dx\,\phi(x) H_P(x)\phi(x) = 0$. If f_n and g_n denote two linearly independent solutions of (2), the analogue of the Wronskian relation is $f_n g_{n+1} - f_{n+1} g_n = c$, independent of n. Again we encounter the serious obstacle that if f_n is a localized solution of (2), any other linearly independent solution of (2) will necessarily diverge in magnitude for $|n| \to \infty$. In particular, round-off will introduce this growing solution and inevitably mutilate our numerical results for the localized solution we are seeking. If we were prepared to limit our attention to extended solutions of (2) there would be no numerical instability problem to cope with, but we are specifically interested in the occurrence of localized states in incommensurate potentials.

In the following I will demonstrate a numerical procedure[5,6] for straightforwardly obtaining the localized eigenstates and eigenvalues of (2) for an arbitrary choice of ε_n. This method is completely free of the numerical instability described above, and it yields results of ultra-high accuracy. This will be illustrated for the Aubry model,[7] where ε_n is chosen as $\varepsilon_n = \varepsilon_0 \cos(2\pi q n)$ with q irrational, the most popular model to date of the form (2). The importance of a successful numerical technique lies in what new clues, new phenomena, new ideas can be provided which cannot otherwise be generated by analytical methods at this time. Of particular interest in the Aubry model is the occurrence of an apparently sharp transition from extended to localized eigenstates as the strength of the incommensurate component, conveniently measured by a dimensionless coupling parameter $\Delta = \varepsilon_0/(2V_H)$, is set equal to unity. The numerical results we obtain provide an enormous amount of quantitative information concerning the behavior of the localized eigenstates as Δ is allowed to decrease towards unity. In addition we will provide a detailed mechanism for the transition from localized to extended states, in terms of an incipient infinite degeneracy of the localized eigenstates as they smoothly evolve into extended states. The resulting ideas are easily grasped and the mathematical details are quite straightforward once we exploit a beautiful theorem from analytic number theory due to N. B. Slater.[8]

II. NUMERICAL ALGORITHM

The method to be described in this Section is based on a classic theorem in the theory of continued fractions due to S. Pincherele.[9] Here I will only provide a statement of the method; details of the derivation are given elsewhere.[6] With b_n defined in (2) we generate the two sequences r_n and s_n according to the rules

$$r_n = (b_n - r_{n+1})^{-1}, \quad s_n = (b_{-n} - s_{n+1})^{-1}, \quad (n = N, N-1, \ldots, 1). \qquad (3)$$

The quantities r_n and s_n are related to the site amplitudes in (2) by the relations $r_n = f_n/f_{n-1}$, $s_n = f_{-n}/f_{-(n-1)}$, and n is a positive integer. Note that (3) is a backward iteration scheme. Now one could expect that the values of r_1 and s_1 depend on the starting (seed) values for r_{N+1} and s_{N+1}. However, it turns out that the interesting case is where r_1 and s_1 are independent of the seed values, to the accuracy of the computing system (say quadruple precision on a VAX system). (A more precise statement is that r_1 and s_1 are said to be independent of the initial seeds if unique results are obtained in the limit $N \to \infty$ when any value is used for r_{N+1} and s_{N+1}.) In practice the value N=500 (1000) is usually sufficient if one employs double (quadruple) precision arithmetic. In essence Pincherele's theorem states that the independence to seed values occurs if and only if, for a given choice of W, Eq.(2) possesses a pair of linearly independent solutions, f_n and g_n, such that f_n/g_n decreases to zero as n approaches either of $+\infty$ or $-\infty$. A localized solution decreases to zero for both $+\infty$ and $-\infty$, and this will occur for a tiny subset of those energies W for which independence to seed values occurs. It also follows from Pincherele's theorem that r_1 and s_1 will be dependent on seed values if (2) admits extended solutions. If r_1 and s_1 are independent of seed values then we can define the unique function $\Phi(W)=b_0(W)-r_1(W)-s_1(W)$. The key point is that a localized solution of (2) situated in the vicinity of site n=0 occurs for those energies W for which $\Phi(W)$ has a zero. If Φ exists but is non-zero then (2) fails to possess a localized solution; that is, all solutions of (2) will diverge for at least one of the limits $n\to\pm\infty$. Finally, if Φ fails to exist, because of the dependence of r_1 and s_1 to seed values, (2) possesses extended states for such values of W.

The requisite computer programs are very simple and the computations are readily performed, typically by automatic scanning, varying W for a given choice of the model form for ε_n. The accuracy of the results for the energy eigenvalues are limited only by the accuracy of the computer you employ and/or your patience. In practice, if you want accurate values of the site amplitudes f_n for a localized eigenstate for of order several thousand lattice sites it is necessary to determine the energy eigenvalue to of order thirty digits. The site amplitudes are obtained using the product formulas $f_n=f_0r_1r_2\cdots r_n$ and $f_{-n}=f_0s_1s_2\ldots s_n$ with n a positive integer. Of course one must be careful to choose N somewhat larger than the largest chosen value of n. Calculations in quadruple precision are very slow and so if one only wants the eigenvalue spectrum and/or the values of f_n for say $|n|<500$ one can get by using double precision. In an energy interval where Φ exists this function is a monotonic function of W with pairs of successive zeros interlaced by a simple pole. The form of Φ as one approaches the edge of a band of extended states (where Φ no longer exists) is very interesting, but this is discussed elsewhere.[6] Instead, we now display a small selection of our results for the localized eigenstates of the Aubry model, in particular just above the critical line.

III. THE AUBRY MODEL

As remarked in the Introduction, this model is defined by the choice $\varepsilon_n=\varepsilon_0\cos(2\pi qn)$. We rewrite (2) in dimensionless form as

$$f_{n+1}+f_{n-1}=2[E-\Delta\cdot\cos(2\pi qn)]f_n \qquad (4)$$

where $E = W/(2V_H)$, and as before $\Delta = \varepsilon_0/(2V_H)$. Additionally, one could include an arbitrary phase angle in the argument of the cosine, but this will not be done here. Aubry[7] has employed a self-dual property implied by the form of ε_n to argue that, independent of q, $\Delta=1$ is a critical line in the EΔ-plane, whereby all physical eigenstates of (4) are extended if $\Delta<1$ and localized if $\Delta>1$. In addition, he has given arguments for an exponential decay of the localized eigenstates, of the form $f_n=F_n/\Delta^n$ with F_n remaining bounded and nonzero for $|n|\to\infty$. These states are thus characterized by a localization length $\xi=1/\ln\Delta$, independent of energy and q and diverging as Δ is decreased towards the critical line. In an effort to shore up Aubry's essentially heurisitic arguments[10] a number of important rigorous existence theorems have been derived.[2] Nevertheless, it is fair to say that a satisfactory proof of either of Aubry's two claims has yet to be given, even though most people are inclined to believe that the claims are generally correct. Using our numerical algorithm we have accumulated an overwhelming amount of evidence supporting Aubry's two claims for the choices of q we have studied. In addition, we have been able to establish a host of detailed quantitative properties of the localized eigenstates. In this regard, all of the numerical results have been obtained together with James H. Luscombe of the Ames Laboratory at Iowa State University. We also have a specific proposal for the mechanism of the transition from localized to extended states as Δ is decreased towards unity, and this is presented in the following Section.

The characteristics of the localized eigenstates depend very heavily on the choice of the parameter q in (4). In particular, if q is a quadratic algebraic irrational (an irrational number satisfying a quadratic algebraic equation with integer coefficients) such as the Golden Mean $q=(\sqrt{5}-1)/2 = 0.6180339...$, the characteristics of the states are very different from those when q is chosen to be a transcendental number. (Recall that a transcendental number is one which cannot be expressible as a root of a polynomial with rational coefficients.) To clarify this point, recall that Euclid's algorithm provides us with an optimal sequence of rational numbers which converge to the value of any given irrational number q. The full sequence of rational approximants is known for only a handful of transcendental numbers, and from among these we will consider the class discovered by Euler, where $q=\tanh(1/K)$ with K any nonzero integer. For the case K=2, the first elements of the Euclidean sequence are 1/2, 6/13, 61/132, 860/1861, 15541/33630, 342762/741721,... . The denominators are of particular interest to us. These are called the "quasi-periods" of q; for example, if we evaluate the term $\cos(2\pi qn)$ at sites n, n+132, n+264, n+396,..., we find almost-periodic behavior. The rapid growth of the denominators in this case enables us to speak loosely in terms of a potential which appears locally to be periodic as we traverse the lattice with a current quasi-period, e.g., 132, which ultimately is superseded by a larger quasi-period, (1861 in this example) after a relatively large number (14) of cycles having the smaller quasi-period. By contrast, in the case of the Golden Mean the corresponding sequence obtained using Euclid's algorithm is given by 1/2, 2/3, 3/5, 5/8, 8/13, 13/21, 21/34, 34/55,... (the numerators and denominators are the famous Fibonacci numbers). Of particular importance to us is the fact that the operative notion of quasi-periods does not apply for the Golden Mean, or for any other quadratic irrational for that matter, because any given quasi-period is so rapidly superseded that the potential does not appear locally periodic. As sloppy as the notion might sound, the choice of a quadratic irrational for q yields a poten- tial which is somewhat akin to that of a random potential, whereas if q is chosen to be a transcendental number, as we have seen, the potential is closer to that of a periodic potential. The distribution of quasi-periods for the potential is a qualitative distinction which can be expected to have a profound influence on the wave function. After these background remarks we examine some numerical results for these choices of q.

In Fig. 2a we provide a semilog plot of the probability $|f_n|^2$ for finding the electron at each of the lattice sites in the range $0<n<500$ for the choice of parameters $q=\tanh(1/2)$ and $\Delta=1.05$. The first 28 digits of the energy eigenvalue are listed in the caption. Note the oscillations upon the oscillations. A period of 13 is clearly discernible as is the larger period of 132. With the help of a magnifying glass you might also be able to see period 2 oscillations. Note that the peak of each of the period 132 oscillations drops monotonically as we proceed to the right. In Fig. 2b we provide the probabilities for sites 1500 to 2000. Note that the peak at around site 1861 is larger than at the previous peak site, n=1729. In fact, the probability at site 1861, of order 10^{-77} (!), is about six or seven orders of magnitude larger than we would expect based on the monotonic drop from peak to peak for the earlier sites. This is a clear manifestation of the quasi-period 1861. Recall that the next quasi-period is 33630 and so we can forego the honor of continuing the calculations out to such distant sites. In fact, if we recall Aubry's claim for the n-dependence of f_n and crudely estimate $|f_n|^2$ by Δ^{-2n}, for $\Delta=1.05$ and n=1861 we obtain 1.36×10^{-79} which is close to the value given on the graph. For site n=33,630 the estimate for the probability is about 10^{-1425}. To obtain accurate data out to site n=2000 it is absolutely mandatory to perform the calculations to quadruple accuracy. Calculating an energy eigenvalue to better than 28 digits is a matter of necessity rather than fetish! In any event, this example illustrates the type of precision that can be obtained for the localized eigenstates using our

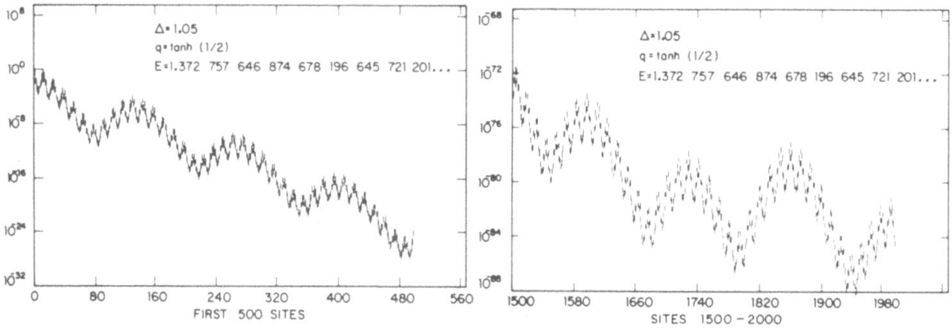

Fig. 2. Semi-log plot of the site probabilities for the Aubry model with parameters q=tanh(1/2) and Δ=1.05. (a) sites n=0 to n=500; (b) sites n=1500 to n=2000. As explained in the text the data reveals the existence of the first four quasi-periods 2, 13, 132, 1861 associated with the present choice of q.

numerical algorithm. The results clearly reflect the presence and the role of the quasi-periods. Figure 2 also provides strong support for Aubry's proposed localization length. Actually, a far more convincing supporting argument is provided by a semilog plot of the quantity $|f_n|^2 \Delta^{2n}$ versus lattice site. If Aubry's claim is correct the results should be essentially independent of Δ, and for large $|n|$ should remain bounded while not decreasing to zero. We have observed precisely such behavior for a wide variety of choices of Δ, for the present choices of q as well as numerous others.

In Fig. 3 we present a semilog plot of the site probabilities for the case where q equals the Golden Mean and Δ=1.1 and for the energy eigenvalue listed in the caption. The absence of the oscillations that we see in Fig. 2 is quite striking. In fact, apart from what superficially looks like superimposed small-amplitude noise, the data points virtually fall on a straight line, i.e., $|f_n|^2$ decays exponentially for increasing $|n|$. The slope of this straight line is consistent with Aubry's expression for the localization length. The "noisy", oscillation-free character of the data is in accord with our earlier remarks concerning the expected behavior for q chosen as a quadratic irrational. A more detailed

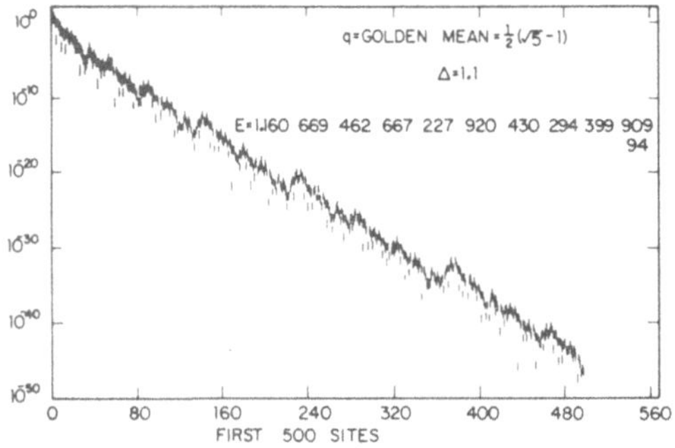

Fig. 3. Semi-log plot of the site probabilities for the Aubry model with parameters q=(√5 - 1)/2 and Δ=1.1. The plot can be visualized as a straight line with superimposed "noise" of small amplitude. The slope of the straight line is in accord with Aubry's expression for the localization length.

examination of the data in the present case reveals the fact that the quantity $\ln(\Delta^{2n}|f_n|^2)$ displays very striking behavior which can be replicated by a compositional algorithm very similar to that which generates the Fibonacci numbers. A lengthy report on this issue is given elsewhere.[11] Ostlund and Pandit[12] have discussed certain aspects of scaling and self-similarity of the localized eigenstates of the Aubry model for q given by the Golden Mean, but for a different choice of phase factor.

In summary, in this Section we have displayed some numerical results for localized eigenstates of the Aubry model obtained using our numerical algorithm. We have seen that the number-theoretic properties of the parameter q play an important role. Furthermore, the results are in full agreement with Aubry's claims for irrational[10] q that there is a sharp transition between extended and localized states which occurs for $\Delta=1$ and that the localized eigenstates exhibit exponential decay with a localization length given by $\xi=1/\ln\Delta$ that is independent of energy.

IV. INFINITE DEGENERACY

In this Section we want to come to grips with the details of the transition from localized to extended states as Δ is decreased towards unity. In particular, we will propose a specific mechanism for this transition.

For an irrational value of q the site energy $\Delta\cdot\cos(2\pi qn)$ is a nonperiodic function as we increment the discrete (integer) site label n. Imagine employing a site M as the origin of our lattice. The Aubry equation, (4), can be written as $f_{n+1}+f_{n-1}=2\{E-\Delta\cdot\cos[2\pi q(M+n)]\}f_n$ where f_n is the probability amplitude for finding the electron at the site n steps to the right of site M. By suitably choosing an integer M we can arrange matters so that for <u>all</u> integers n the quantities $\cos(2\pi qn)$ and $\cos[2\pi q(M+n)]$ will <u>differ</u> by less than some prescribed value. Indeed, this is the defining characteristic of an almost periodic function and the set {M} are called translation numbers. The physical implication of this property is the following. Suppose we have found an eigenstate localized in the vicinity of the origin of our infinite linear lattice having an energy E_0. In view of the above remarks we must expect that there exists a very similar looking eigenstate which is localized in the vicinity of site M with energy, E_M, <u>very nearly</u> equal to E_0. These two eigenstates certainly cannot be strictly degenerate for we have just seen that (4) is almost but not quite invariant under translation. In fact, there is an <u>infinite</u> number of integers M which will meet the above criterion (and we <u>will even</u> give the algorithm for listing all of them!). And so we should think of a countable infinity of nearly degenerate eigenstates localized along the lattice about these special sites M, each associated with the single eigenvalue E_0. Note that by specifying the set {M} we are in essence able to provide the <u>spatially-dependent density of localized states</u> for a prescribed energy window and prescribed segment of the <u>lattice</u>. An estimate of the size of the energy differences E_0-E_M is easily provided by noting that if we use the Aubry equation, (4), choosing our origin first at site 0 and then at site M, the shift in phase angle, η_M, whose magnitude, being the lesser of either {Mq} or 1-{Mq}, is a very small quantity. (Recall the notation that if x is a real number, [x] and {x} denote the integer and fractional parts of x, respectively.) That is, think of the effect upon the eigenvalue and eigenstate if we first employed (4) as it stands and then when qn is replaced by $qn+\eta$, where η is a very small phase angle. A little thought will convince you that the energy shift brought about by choosing η non-zero should be even in that quantity. Thus we must expect that E_0-E_M is of order $(\eta_M)^2$. We have quantified this simple argument by developing a form of renormalized perturbation theory with η serving as the small perturbation.[11] The results we obtain for E_0-E_M using this perturbation theory to second order are in excellent agreement with those we obtain by directly calculating E_0 and E_M using our numerical algorithm. To be more specific, the second order perturbation result takes the form $E_0-E_M=C\cdot(\eta_M)^2$, where $C(\Delta,E_0)$ is a quantity which exclusively utilizes the site amplitudes for the eigenfunction localized about site 0. Of special importance, C is independent of M. Thus if C is calculated for a given choice of Δ and E_0 one immediately knows the energy shift E_0-E_M for all of the countable infinity of special sites M. Thus far we have been successful in calculating C to high accuracy if Δ is chosen larger than about 1.01. The origin of our difficulties if Δ is chosen closer to the critical line is the slow convergence of several infinite series, which contain the site amplitudes, appearing in our expression for C. For Δ approaching unity the localization grows very large, i.e., the site amplitudes decrease very slowly to zero, and thus the very slow rate of convergence.

The set of sites M discussed in the previous paragraph can be calculated very simply. Suppose we require that Mq differ in magnitude from an integer by less than a small prescibed positive quantity η. The determination of these special integers M is provided by the following theorem of analytic number theory first discovered in 1950 by N.B. Slater.[8] For given values of q and η there exists a unique pair of integers μ_1 and μ_2, determined by an algorithm to be described momentarily. The requisite values of M are to be found exclusively among the set of linear combinations of μ_1 and μ_2 with integer coefficients. That is, we need only scan the set $k_1\mu_1+k_2\mu_2$, with k_1 and k_2 integers, testing whether the constraint set by η is met. The values of μ_1 and μ_2 are obtained as follows. Define the sequence of numbers a_k and b_k by the algorithm, $a_0=0$, $a_1=1$, $b_0=1$, $b_1=\{q\}$, $a_{k+1}=a_{k-1}+[b_{k-1}/b_k]a_k$ and $b_{k+1}=\{b_{k-1}/b_k\}b_k$. Finally, let m be the least integer with $b_m<|\eta|$ and let m′ be the least integer with $b_{m-1}<|\eta|+m'b_m$. After these long-winded preliminaries, we have $\mu_1=a_m$ and $\mu_2=a_{m-1}+m'a_m$. To illustrate the use of the algorithm, let us return to the case q=tanh(1/2) and suppose $\eta=5\times10^{-5}$. The recipe gives $\mu_1=1861$ and $\mu_2=31769$. Thus the values of M for which Mq will differ in magnitude from an integer by an amount not exceeding 5×10^{-5} are to be found among the set of integers of the form $1861k_1+31769k_2$. Note that 1861 is a quasiperiod for this choice of q whereas 31769 is not. (However, the next quasi-period is the sum of these two integers.) If we restrict M to lie in the range $0<M<10^5$ we quickly find that there are only eight acceptable values of M. These are given by 1861, 31769, 33630, 35491, 65399, 67260, 69121, 99029. Note that each of these integers is a linear combination of μ_1 and μ_2 with integer coefficients. For the next five segments each containing 10^5 lattice sites one finds an additional 9, 10, 12, 12, 10 special sites. Once again, the physical significance of these special sites is that there exists a localized state in the vicinity of each of them which differs only slightly in form from that in the vicinity of site 0 and whose energy E_M differs from E_0 by an amount of order $\eta^2=2.5\times10^{-9}$. All of these predictions have been confirmed by direct calculation using our numerical algorithm.

The most interesting regime is where Δ is decreased towards the critical line. We believe that the transition from localized to extended states as Δ is decreased towards unity is accompanied by an incipient infinite degeneracy between the nearly degenerate states based on the special sites M. In particular, we propose that $C(\Delta,E_0)\to0$ for $\Delta\to1+$. Stated differently, the occurrence of extended states for $\Delta<1$ is a manifestation of an infinite degeneracy of states which had existed as nearly degenerate localized states just above the critical line. This idea is confirmed by our numerical calculations. We find that for all choices of q the value of $C(\Delta,E_0)$ decreases towards zero as we decrease the value of Δ towards unity. Generally we find that C is rather well described by a power-law of the form $C\propto(\Delta-1)^Y$. However, the critical exponent Y is definitely non-universal. In particular Y depends upon both q and E_0. This is somewhat discouraging; if the exponent were universal one could consider making a concerted effort to calculate it, but instead we are up against a vast collection of values. In a way this problem makes one think of critical point phenomena as a far simpler topic of theoretical physics!

There are, however, trends in the value of Υ which are associated with the physics of localization in incommensurate potentials. We find that the values of Υ are far smaller in the case of q chosen as the Golden Mean than they are for q=tanh(1/2). Moreover, the latter are smaller still than what they are for the case q=tanh(1/5), which features as its quasi-periods 5, 76, 1905, 66751, 3005700, 165380251,... . We believe that this trend for Υ for these few choices of q has a straightforward physical explanation. Comparing the cases q_1=tanh(1/2) and $q_2=(\sqrt{5}-1)/2$, the inequality $\Upsilon_1 \gg \Upsilon_2$ implies that the localized states are more nearly degenerate in the former case than in the latter. This is consistent with our earlier remark that the site energies in the case $q=q_1$ are somewhat reminiscent of a periodic potential (for which extended states occur for all Δ) then for $q=q_2$ which is somewhat reminiscent of a random potential. Finally, if $q=q_3$=tanh(1/5) the ratio of successive quasiperiods is even greater than that for $q=q_1$ and not surprisingly $\Upsilon_3>\Upsilon_1$.

V. SUMMARY AND OUTLOOK

In this work we have described a simple numerical algorithm which can provide the form of localized eigenstates and their energies for tight-binding models of the form (2), to arbitrarily great precision, free of the numerical instability discussed in Sec. I. We have provided a small sample of our results for the Aubry model. The number-theoretic characteristics of the parameter q, and especially the distribution of the quasi-periods, be it a quadratic irrational or a transcendental number, dramatically affect the form of the eigenstates. With the aid of N. B. Slater's theorem we can identify an infinite subset of sites of the lattice about which there exist nearly degenerate localized eigenstates for $\Delta>1$. Furthermore, we have proposed that the transition from localized to extended states, as Δ is decreased towards unity, is to be thought of in terms of an incipient infinite degeneracy among these states. A quantitative description of the onset of this degeneracy suggests a power-law dependence on $\Delta-1$, with a non-universal critical exponent.

There remain a large number of open issues but we will here raise only three. First, in the case of the Aubry model is there any chance for progress concerning the non-universal critical exponent Υ? Second, what can be learned from a study of other tight-binding models which do not possess the self-dual property of the Aubry model? With the aid of the numerical algorithm described in this work, the road is open for a systematic study of any given model. Finally, can one develop a reliable numerical method for obtaining the localized eigenstates of the Schrodinger (differential) equation for an electron in an incommensurate potential? After all, the real motivation for studying a tight-binding version of the Schrodinger equation is because of our inability to overcome our analytical and numerical handicaps. In any event, with the progress[3,4] of recent years in developing new artificial structures, to keep pace with developments in the laboratory we will have no choice but to search hard for solutions to our present theoretical inadequacies.

ACKNOWLEDGEMENTS

It is a great pleasure to acknowledge the enormous contribution of James H. Luscombe (Ames Laboratory) to each and every part of the subject material presented here, let alone his many helpful suggestions which greatly improved the manuscript.

Ames Laboratory is operated for the U.S. Department of Energy by Iowa State University under contract no. W-7405-Eng-82. This work was supported by the Director for Energy Research, Office of Basic Energy Sciences.

REFERENCES AND FOOTNOTES

1. For a recent review of the physics literature see J. B. Sokoloff, Phys. Reports 126, 189 (1985).
2. A useful although somewhat outdated guide to the mathematical literature of one-particle Schrodinger equations with random and almost-periodic potentials can be found in B. Simon, Adv. Appl. Math. 3, 463 (1982).
3. See, for example, R. Merlin, K. Bajema, R. Clarke, F.-Y. Juang, and P. K. Bhattacharya, Phys. Rev. Lett. 55, 1768 (1985) who have reported the successful fabrication of quasi-periodic GaAs-AlAs heterostructures.
4. See the review article by C. M. Falco and I. K. Schuller in Synthetic Modulated Structures, edited by L. L. Chang and B. C. Giessen, (Academic, New York, 1985) devoted to the construction and properties of layered metallic structures with layer thicknesses in the 10-100A range.
5. An early version of this procedure is given in M. Luban and B. N. Harmon, Solid State Commun. 51, 199 (1984).
6. A more detailed derivation is given in "Localized Eigenstates of One-Dimensional Tight-Binding Models: A New Algorithm", M. Luban and J. H. Luscombe, (to be published.)
7. S. Aubry and G. Andre, Ann. Israel Phys. Soc. 3, 133 (1980).
8. N. B. Slater, Proc. Cambridge Philos. Soc. 46, 525 (1950).
9. S. Pincherele, Giorn. Mat. Battaglini 32, 209 (1894). A more accessible reference is R. B. Jones and W. J. Thron, Continued Fractions: Analytic Theory and Applications, in Encyclopedia of Mathematics, edited by G-C. Rota, (Addison-Wesley, London, 1980), Vol. 11, Sec. 5.3 and App. B.
10. Aubry's duality arguments are appropriately termed "heuristic arguments" and definitely do not apply for all irrational q. The duality argument can be given for all irrational q, yet B. Simon and J. Avron [Bull. Amer. Math. Soc., 6, 81 (1982)] have given a rigorous proof that the states for $\Delta > 1$ are not localized if q is chosen from a very restricted class of irrational numbers. The irrational numbers considered in this work are not of this class and for these we find that Aubry's claims are fulfilled.
11. M. Luban and J. H. Luscombe, (to be published).
12. S. Ostlund and R. Pandit, Phys. Rev. B29, 1394 (1984).

TRANSPORT IN METAL ALLOYS AND RESISTIVITY SATURATION

James C. Swihart

Department of Physics
Indiana University
Bloomington, IN 47405

William H. Butler

Oak Ridge National Laboratory
Oak Ridge, TN 37831

INTRODUCTION

We consider here two problems in the electrical resistivity of metals and alloys. These consist of the residual resistivity of random substitutional alloys as a funtion of composition and the phenomenon of resistivity saturation in metals and alloys. Much has been written on both topics.[1,2] We present a unified treatment that covers both phenomena with one model. This model was first presented by Butler[3] and is based on the Korringa-Kohn-Rostocker band structure method blended with the Coherent-Potential-Approximation (KKR-CPA).[4]

Random substituional alloys (or simply random alloys) can readily be produced by metallurgists from many different metals, and they consist of the following. Consider a binary random alloy made up of metal A in concentration (atomic fraction) c_A and metal B in concentration $c_B=1-c_A$. We can visualize the structure of the alloy by starting with pure metal A on a perfect lattice and then replacing A atoms one at a time by B atoms at random sites until we have the proper concentrations. The probability of finding an A atom at any site is c_A independent of which atoms occupy neighboring sites. That is, there are no correlations between the types of atoms on different sites.

If one could not distinguish between the A and B atoms in such an alloy, we would have a perfect crystal with complete translational invariance, and electrons in Bloch states would not be scattered by the lattice. For such a case, the zero-temperature electrical resistance would vanish. However the electrons can distinguish between A and B atoms and are thus scattered by the lattice even at zero temperature. Thus there is a nonzero residual resistivity in random alloys that depends on concentration. According to some early ideas of Nordheim[5] based on weak scattering and the Born approximation, the residual resistivity in random alloys due to the disorder should be proportional to $c_A(1-c_A) = c_B(1-c_B)$. We shall see that such a dependance is valid for some alloy systems but breaks down for others.

KKR-CPA

The KKR-CPA method[4] has been found to describe quite accurately a number of physical properties of random alloys if the potentials are determined self-consistently for each concentration considered.[6,7] It is a one-electron effective medium theory and uses nonoverlapping spherical muffin-tin potentials and the local-density approximation. The only input into a calculation are the atomic numbers of the constituents, the lattice structure, lattice constants, and concentrations. It is even possible to calculate the lattice constants,[7] but for the calculations we report here the lattice constant is an input parameter set equal to the experimental value for each concentration.

In the CPA, one treats the random alloy as a one-component metal on a perfect lattice (the effective medium) with a complex t-matrix for the scattering of an electron by a single atom. The effective medium is determined self-consistently by solving the CPA equation

$$c_A\,\tau^{coo(A)} + c_B\,\tau^{coo(B)} = \tau^{coo} \qquad (1)$$

The τ's are scattering-path operators[8] (SPO) related to the one-particle Green's function by[9]

$$G(\vec{r},\vec{r}\,',E) = \frac{2m}{\hbar^2}\sum_{L,L'} \tau^{mn}_{LL'}\,\phi^m_L(\vec{r}_m,E)\,\phi^n_{L'}(\vec{r}\,'_n,E) + \text{terms that are real.} \qquad (2)$$

We need only the imaginary part of G for this paper; so we can forget about the real terms in Eq. (2). Here $L=(\ell,\mu)$ and L' are the angular momentum components and m and n are the unit cells in real space into which $\vec{r}$ and $\vec{r}\,'$ respectively fall. $\phi^m_L(\vec{r}_m,E)$ is the real solution to the Schroedinger equation of angular momentum L for the muffin-tin potential at site m and enery E which is regular at the origin with $\vec{r}_m \equiv \vec{r} - \vec{R}_m$ where $\vec{R}_m$ is the position of the center of cell m.

The SPO τ^{mn} is the propagator for an electron from site n to site m with all possible multiple scatterings off of other atoms in between. In Eq. (1) τ^{coo} is the SPO from site zero to site zero for the effective medium (fictitious atoms of type c). $\tau^{coo(A)}$ is for the same structure but with an A atom at the origin in the effective medium (all other atoms of type c), while $\tau^{coo(B)}$ is the same but with a B atom at the origin. Thus Eq. (1) is a natural condition for the effective medium. It states that the propagator for an electron going from the origin (occupied by an A atom) back to the origin times the probability c_A that there is an A atom at the origin plus the propagator (from zero to zero) for a B atom at the origin times the probability c_B that there is a B atom at the origin is just the zero-zero propagator for an effective medium atom at the origin. In all three cases,

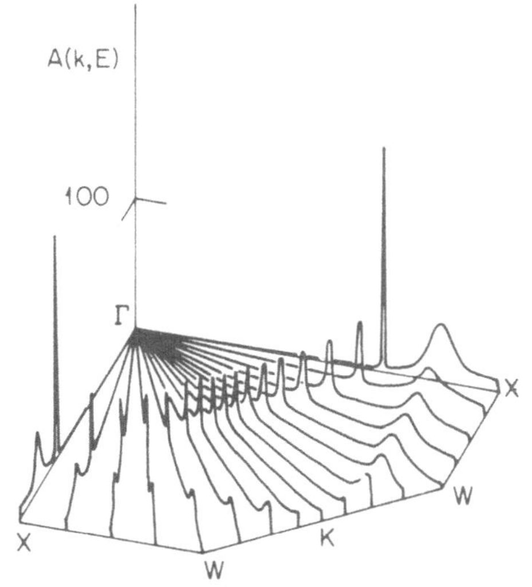

Fig. 1. Calculated Fermi-energy Bloch spectral function for $Ag_{0.2}Pd_{0.8}$ as a function of wave vector in the (100) plane. Note the sharp "s-p"-band and the broad d-band at the Fermi surface. This figure is taken from Ref. 1.

the atom at the origin is embedded in the effective medium.

Once the scattering-path operator is determined, the one-particle Greens function can be calculated from Eq. (2). (It is more convenient to take the Fourier transform to obtain G(k,E)). Of particular interest is the Bloch spectral function $A_B(\vec{k},E) = -\frac{1}{\pi}\sum_m \text{Im}\, G(\vec{k}+\vec{G}_m,E)$ from which one can obtain the density of states and the density of electrons. Fig. 1 is a plot from Ref. 1 of the spectral function $A(\vec{k},E = E_F)$ as a function of $\vec{k}$ for various directions in the first Brillouin zone from $Ag_{20}Pd_{80}$. AgPd alloys form an fcc lattice for all concentrations.

If there were no disorder, the spectral function would be zero except at the k values for which $\epsilon_{\vec{k}}$ of the band structure equals E of the spectral function. For these values of k, there would be a delta-function peak. The lattice disorder in the random alloy broadens the delta function peaks. Thus the spectral function is a convenient way of representing the band structure of the alloy in which the energy bands are smeared.

Note that there are two self-consistency conditions that must be satisfied in the KKR-CPA. The first is the CPA equation (1) which must be satisfied for particular potentials at the A and B atoms. Once the τ's (and hence the Greens functions) are determined, one finds the electron densities at an A or B site embedded in the effective medium. By means of the local-density approximation the potential at an A atom (or a B atom) in the effective medium is determined. This must be the same potential used in solving the CPA equation. Thus in calculations one has two loops, one which determines the local potential--electron density and the other which solves the CPA equation, both by successive approximations. In this way, the charge transfer from A to B atoms (and even the redistribution of charge within an atom) for a given concentration is taken into account.

RESIDUAL RESISTIVITY IN RANDOM ALLOYS

The formalism for calculating the electrical resistivity of random alloys was reported in Ref. 3, and we shall briefly review it here. The model is based on the Kubo formula[10] which in the one-electron approximation is

$$\sigma_{\mu\nu}(E) \simeq \sum_{\alpha}\sum_{\alpha'} \langle\alpha|j_\mu|\alpha'\rangle\langle\alpha'|j_\nu|\alpha\rangle\delta(E-\epsilon_\alpha)\delta(E-\epsilon_{\alpha'})$$
$$\simeq \int d\vec{r} \int d\vec{r}\,' \left[\nabla_\mu \ \mathrm{Im}\ G(\vec{r},\vec{r}\,')\right]\left[\nabla'_\nu \ \mathrm{Im}\ G(\vec{r}\,',\vec{r})\right] \tag{3}$$

where $|\alpha\rangle$ and $|\alpha'\rangle$ are one-electron states of energy ϵ_α and $\epsilon_{\alpha'}$ respectively.

The integrals over space are replaced by integrals over individual cells together with sums over the different cells. The expression (2) is used for G and then the conductivity tensor becomes a product of two current (or momentum) matrix elements at cells m and n and of τ^{mn} and τ^{nm}. This product is summed over all cells m and n. The result is configuration dependent because τ^{mn} and τ^{nm} are configuration dependent, and also because the current matrix elements at cells m and n depend on which atoms are present at m and n.

What is required is a configuration average of the conductivity. The configuration average is quite different for terms m=n in the double sum over cells than for terms with m≠n. In the former set of terms, the m=n site can have an A atom with probability c_A or a B atom with probability c_B. In the latter set of terms, the probability of finding an A or a B atom at site m is c_A or c_B respectively independent of the species at site n (the assumption that the alloy is random). It is natural then to break the conductivity into two parts

$$\sigma_{\mu\nu} = \sigma^{(0)}_{\mu\nu} + \sigma^{(1)}_{\mu\nu} \tag{4}$$

where $\sigma^{(0)}_{\mu\nu}$ consists of the terms with m=n and $\sigma^{(1)}_{\mu\nu}$ of the terms with m≠n.

In carrying out the configuration average, we use the CPA condition of Eq. (1) to eliminate many terms. The reader who is interested in the details should see Ref. 3. $\sigma^{(0)}_{\mu\nu}$ has the form

$$\sigma^{(0)}_{\mu\nu} \simeq \sum_{\alpha} c^\alpha \tau^{coo} \tilde{J}^\alpha_\mu \tau^{coo} J^\alpha_\nu \tag{5}$$

where the sum is over the constituent atoms, each of concentration c^α. $\tilde{J}$ is closely related to J, the momentum matrix element in the cell at the origin. Use has been made of the fact that in the effective medium in the CPA, there is a periodicity with the lattice and thus the evaluation of Eq. (5) at any cell is equal to the evaluation at the cell at the origin.

The $\sigma^{(1)}$ term is more complicated and becomes in the CPA

$$\sigma^{(1)}_{\mu\nu} \simeq \sum_{\alpha} \sum_{\beta} c^{\alpha} c^{\beta} \tilde{J}_{\mu}^{\alpha} \left[1-\chi w\right]^{-1} \chi \tilde{J}_{\nu}^{\beta} \tag{6}$$

The deviation of $\left[1-\chi w\right]^{-1}$ from one gives the vertex correction. Here

$$\chi = \sum_{n\neq 0} \tau^{con} \tau^{cno} \tag{7}$$

and

$$w = \sum_{\alpha} c^{\alpha} x^{\alpha} x^{\alpha} \tag{8}$$

with

$$x^{\alpha} = \left[1-\Delta m^{\alpha} \tau^{coo}\right]^{-1} \Delta m^{\alpha} \tag{9}$$

and

$$\Delta m^{\alpha} = \left[(t^{c})^{-1}-(t^{\alpha})^{-1}\right] \tag{10}$$

where t^{c} is the t-matrix for one of the coherent scatterers (one of the fictitious effective medium atoms) and t^{α} is the t-matrix for the αth type atom (A or B).

We have developed an extensive system of computer programs to carry out calculations of the residual conductivity as a function of concentration for actual alloy systems. The computer programs are built on programs developed by G.M. Stocks and co-workers to determine the KKR-CPA one-electron spectral function and electronic density of states. Our calculations use as input muffin-tin potentials determined self consistently (within local density approximation) for the particular binary alloy system and concentration we are considering. The potentials are provided by G.M. Stocks and co-workers.

Two important elements in the calculation of the conductivity that needed to be added to the programs of Stocks are the determination of the square of the scattering-path operator τ (i.e. the χ of Eq. (7)) and the calculation of the matrix elements of the current operator on both the A- and B-atom sites. The latter calculation is straightforward but tedious. The calculation of the necessary τ^2, however, is not so simple.

The square of the q-dependent SPO's must be determined and then integrated over the entire reduced Brillouin zone. We carry out the calculations with most quantities expressed in terms of angular-momentum components, and we have confined ourselves in completed calculations up to now to ℓ values less than or equal to 2. Thus we consider nine different $L(\equiv\ell,m)$ values. At first sight it might appear that there are 9^4 different $\tau_{L_1L_2}\tau_{L_3L_4}$ that must be calculated. However, because of selection rules for the matrix elements of the current operators, we must have $\ell_4=\ell_1\pm1$ and $\ell_3=\ell_2\pm1$. This limits the number of required τ^2 to $36^2=1296$, still too many to be integrated over the Brillouin zone in a practical calculation. We have been able to carry out actual calculations of the conductivity only because we recognized that, on the basis of the point symmetry within the CPA of the alloys we are considering, there are still fewer nonvanishing and independent τ^2 than 36^2. We have shown that for cubic symmetry (which includes all of the systems we have considered so far) there are no more than 114 independent τ^2. We have proved this together with identifying which 114 different τ^2 we need calculate by a combination of group theory and the carrying out of examples of point operations by computer.

We have completed calculations on the residual resistivity as a function of concentration for AgPd, CuZn, CuGa, and CuGe together with a calculation for one concentration of NiMo. These results have been reported in a recent Physical Review Letter.[11]

We first chose the AgPd alloy because the electrical conductivity had already been calculated for this system.[1] The earlier calculation obtained quite good agreement with experiment, at least over the palladium rich part of the concentration range, by assuming a relaxation-time approximation, by assuming that the energy bands are well-defined, and by neglecting scattering-in terms. The assumption of well-defined bands is reasonable for AgPd because the calculated electron spectral function as a function of k at the Fermi energy has a sharp peak at $k=k_F$ as can be seen in Fig. 1. In the

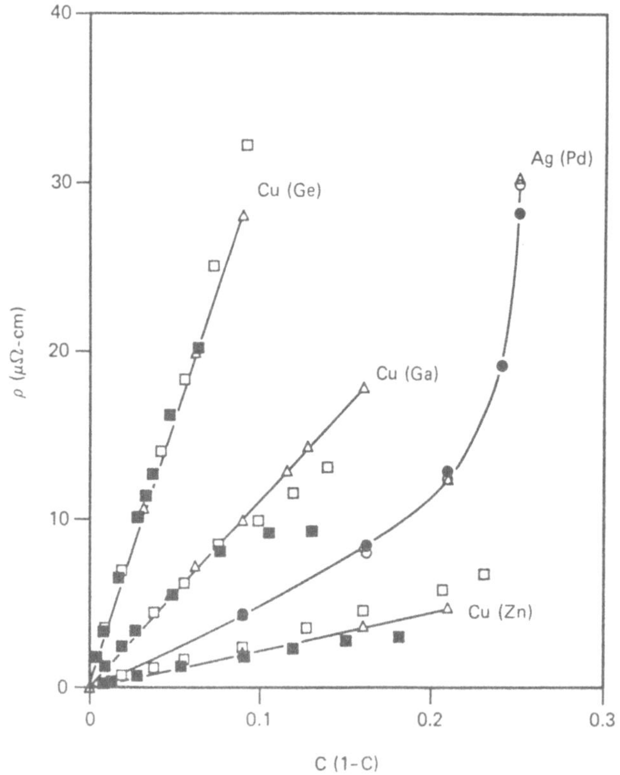

Fig. 2. Calculated and experimental residual resistivities of Cu(Zn), Cu(Ga), Cu(Ge), and Ag(Pd) alloys. Open triangles are calculated values. The open symbols [squares for the copper alloys and circles for Ag(Pd)] are for cold-worked samples, while the closed symbols are for annealed samples. This figure is taken from Ref. 11.

silver-rich region of the alloy, the earlier calculation[1] gave results for the resistivity that are up to twice as large as experiment. The authors attributed this difference to the neglect of scattering-in terms. Thus it was of interest to see whether our calculation in which the scattering-in terms were not neglected would give better agreement with the experimental data.

Fig. 2 is a plot taken from Ref. 11 of residual resistivity of several alloys as a function of c(1-c) taken from Ref. 11. On such a plot, we obtain a straight line if Nordheim's rule holds. Looking first at the AgPd data which is plotted for Pd concentrations from zero to 0.5 (at which point c(1-c)=0.25), we see a large deviation from Nordheim's rule. The open triangles are our calculated points. At concentrations of 0.1, 0.2, and 0.3 the calculated points are obscured by the measured points,[12-14] open circles for annealed samples and closed circles for cold-worked samples. The solid line is drawn through the calculated points as a guide to the eye. We have excellent agreement with experiment for the residual resistivity of AgPd in the silver-rich region of concentration. For this concentration range, vertex corrections are important as was predicted in Ref. 1. We agree with the calculated results of Ref. 1 (but not with experiment) if we neglect vertex corrections in our calculations. The same holds for the thermopower, i.e. we have good agreement with experiment for $c_{Pd}<0.5$ if we include vertex corrections, but this agreement disappears if we neglect vertex corrections.

In contrast to the results discussed in the previous paragraph, we found for silver concentrations less than 50% the calculated resistivities are higher than experiment and not as good as that of the earlier calculation.[1] We did determine that in this concentration range, vertex corrections are not very important. We attribute our calculated conductivity being too low to the fact that our calculations up to now have been limited to angular momentum $\ell_{max}=2$. This should be sufficient for s-p alloys with no partially filled d-bands (such as AgPd for Ag concentrations greater than 50%), but we probably need to go to $\ell_{max}=3$ for alloys with partially filled d-bands (such as AgPd with Ag concentrations less than 50%). The reason is that we need to calculate the current matrix elements for all of the electrons at the Fermi surface. If there are d-electrons at the Fermi surface, then they can connect through the current matrix elements (with ℓ going to $\ell\pm1$) to p and to f-states. But with $\ell_{max}=2$, we do not consider f-states and hence cut out some of the contributions to the conductivity.

We have nearly completed the development of the computer programs to handle $\ell_{max}=3$. This is not a trivial undertaking since among other things we now need to integrate 495 independent τ^2 over the Brillouin zone, up from 114 for $\ell_{max}=2$. Also several other matrices that are diagonal for $\ell_{max}=2$ have off-diagonal elements for $\ell_{max}=3$. In the meantime we have investigated three different copper alloy systems which do not have partially filled d-bands. These are CuZn (α-brass), CuGe, and CuGa, all in the α-phase (copper rich concentrations). As we see in Fig. 2, we get excellent agreement with experiment in all three cases.[11] The experimental results[15-19] are denoted by open squares for cold worked samples and closed squares for annealed samples. A straight line was drawn through the calculated points showing for these cases of relatively weak scattering that Nordheim's rule is predicted by the theory. Comparing with experiment we see that the results with cold-worked samples tend to be above the

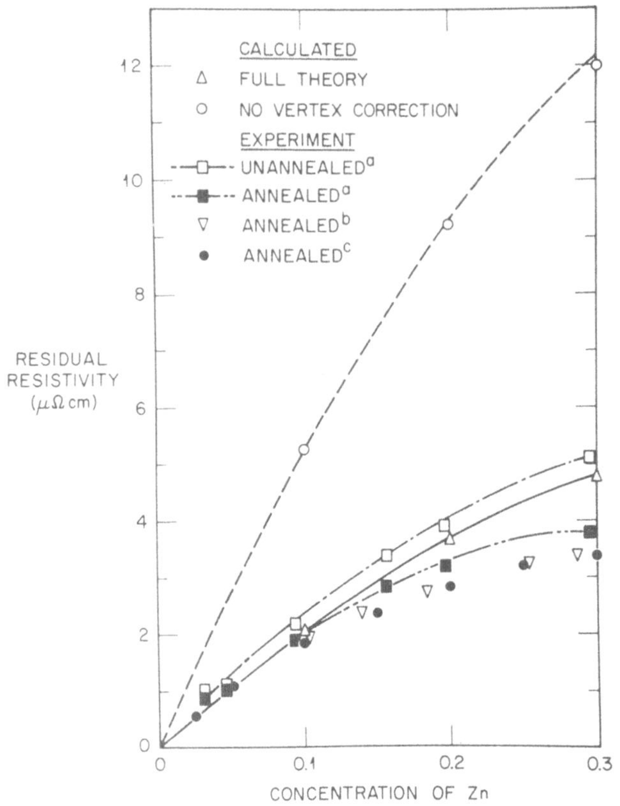

Fig. 3. Calculated and experimental residual resistivity of Cu(Zn) as a function of concentration. The open triangles are our calculations with vertex corrections while the open circles are without vertex corrections. The open squares are measurements on cold-worked samples,[15] while the closed squares,[15] inverted triangles,[16] and closed circles[17] are measurements on annealed samples.

calculated results while the annealed-sample results tend to be below the calculations. An exception is for CuGa at the higher Ga concentrations. The deviations here from our calculated results and from Nordheim's relation may be due to the fact that these concentrations are close to the boundary of another phase. Cold-worked samples should be the closest to having random ordering, which is the situation assumed by the theory.[3] On the other hand these samples have additional scattering centers such as dislocations. Thus we would expect the cold-worked samples to have resistivities higher than our calculated results. On the other hand, annealed samples should have fewer dislocations and other defects. However they may have some short-range chemical order which would tend to reduce the resistivity below our calculated results.

The main differences between CuZn, CuGa, and CuGe is that for equal concentrations, gallium scatters more strongly than zinc and germanium more strongly than gallium. This is in qualitative agreement with Linde's "law" which states that the scattering rate for different impurities in the same host should vary as $(\Delta Z)^2$ where ΔZ is the difference in the number of valence electrons of the host and impurity atoms.[20] It may be seen both from the calculations and from the experimental data, however, that the resistivity increases substantially faster than predicted by Linde. The observed ratios of the resistivity per impurity atom of Zn, Ga, and Ge in Cu are approximately 1:5:14 compared to 1:4:9 which would be predicted by Linde's "law."

Fig. 3 gives the residual resistivity of α-brass. The calculations are for three concentrations with the open triangles being the results when vertex corrections are included (the same calculations shown in Fig. 2) while the open circles are with vertex corrections neglected. We see that for this system (unlike Pd-rich AgPd) vertex corrections are extremely important and that without vertex corrections we do not come close to the experimental results. The open squares are experiments[15] with cold-worked wires which fall above our calculations as is to be expected, while the closed squares,[15] inverted triangles,[16] and closed circles[17] are experimental results with annealed wires and these fall below our calculations, again as expected.

Another random alloy system of interest is NiMo. The self-consistent potentials have been determined in the KKR-CPA scheme[21] for $Ni_{0.8}Mo_{0.2}$ and the Bloch spectral-density function is quite smooth as a function of k at the Fermi energy without the sharp peaks found in weaker-scattering systems such as AgPd. This is shown in Fig. 4. Preliminary calculations of the residual electrical resistivity[22] based on the quasiparticle lifetime approximation give a resistivity of the order of twice the measured value[23]

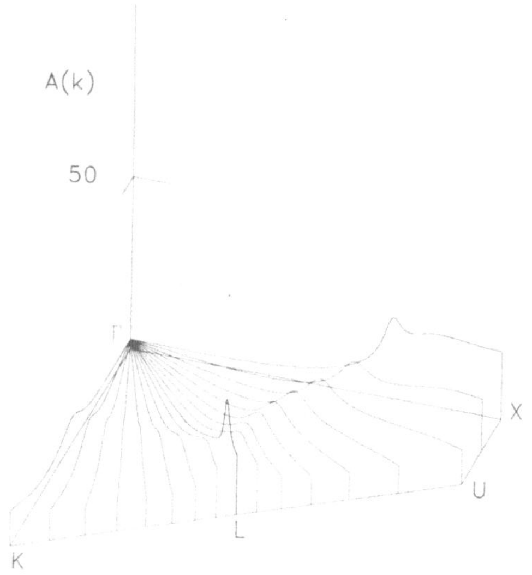

Fig. 4. Calculated Fermi-energy Bloch spectral function for $Ni_{0.8}Mo_{0.2}$ as a function of wave-vector in the (110) plane. Note the lack of sharp structure such as in Fig. 1. This plot was kindly provided by D. M. Nicholson and G. M. Stocks.

of about 120 μΩ cm while our more sophisticated calculation,[11] even with $\ell_{max}=2$, is quite close to the experimental value.

RESISTIVITY SATURATION

Resistivity saturation was first predicted by Mott[24] in which he argued that when the mean-free path becomes of the order of the interatomic spacing, the resistivity should stop increasing with increasing disorder. Mott's conjecture leads to a saturation resistivity of about 500 μΩ cm. Experimentally this saturation shows up[2] when the disorder is increased either by increasing temperature or by increasing defects by for example radiation damage. The experimentally observed saturation resistivity[2] is somewhat lower than that of Mott, being closer to 150 μΩ cm.

We have made a preliminary study of the possibility of saturation in Butler's theory[3] by looking at a very simple model. The model consists of 50% A atoms with spherical square-well muffin-tin potential +V and 50% B atoms with square-well muffin-tin potential -V. We have studied the effect of gradually increasing the potential difference between the A and B atoms, that is of increasing V. For small differences (2V) in the two potentials, the first Born approximation should be valid and the resistivity should be proportional to the square of the potential difference. Fig. 5 shows the results of our calculated resistivity as a function of V^2. We see that for small V, we have a straight line showing the Born approximation does indeed hold. However for larger V, the curve bends over and saturates at a resistivity of about 150 μΩ cm which is the same order of magnitude as experimental saturation.

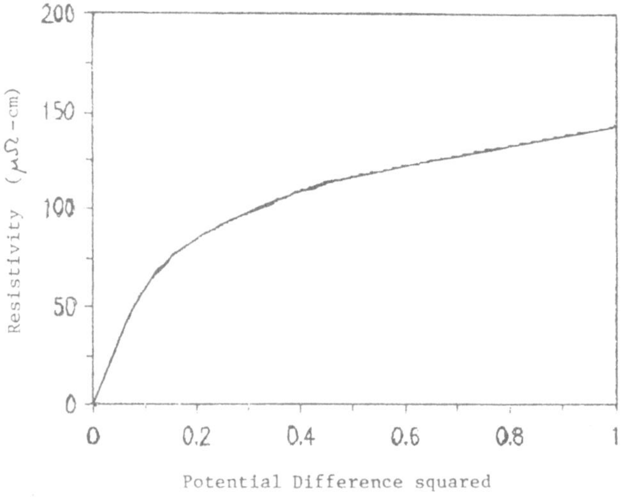

Fig. 5. Resistivity of square-well potential model of a binary alloy as a function of the square of the potential difference 2V. See the text for a description of the model.

On looking at the contributions to the resistivity as the potential V is increased, we have found the following. For low values of V, the conductivity $\sigma^{(0)}$ of Eq. (4) is very much smaller than $\sigma^{(1)}$ so that σ is determined almost entirely by $\sigma^{(1)}$. However as V is increased, $\sigma^{(1)}$ decreases as expected, but the change in $\sigma^{(0)}$ is much less than the change in $\sigma^{(1)}$. In fact for this particular model $\sigma^{(0)}$ actually increased with increasing V. (For another model, we found that $\sigma^{(0)}$ decreased with increasing disorder, but we still obtained saturation.) Thus Butler's theory[3] seems to have built into it a type of parallel resistor model. Parallel resistor models have been used before but based on different concepts to explain resistivity saturation.[25]

ACKNOWLEDGMENTS

We wish to thank D.M. Nicholson, G.M. Stocks, and R.C. Ward for help in carrying out the calculations reported here. We are also indebted to them for valuable comments and discussions. The Indiana University part of this work was supported by National Science Foundation Grant No. DMR 81-17013 while the Oak Ridge National Laboratory part was sponsored by the Division of Materials Sciences, U.S. Department of Energy under Contract No. DE-AC05-84OR21400. One of us (JCS) wishes to thank the Metals and Ceramics Division of Oak Ridge National Laboratory for its hospitality and support as a visitor while much of this work was carried out.

REFERENCES

1. W. H. Butler and G. M. Stocks, Phys. Rev. B29:4217 (1984), and the references given there.
2. See P. W. Anderson, in "Highlights of Condensed Matter Theory: Enrico Fermi Summer School," F. Bassani, F. Fumi, and M. P. Tosi, eds., North Holland, Amsterdam (1985), p. 767 for an interesting discussion of resistivity saturation. Further references are given there.
3. W. H. Butler, Phys. Rev. B31:3260 (1985).
4. G. M. Stocks, W. M. Temmerman, and B. L. Gyorffy, Phys. Rev. Lett. 41:339 (1978).
5. L. Nordheim, Ann. Phys. 9:664 (1931).
6. F. J. Pinski, J. Staunton, B. L. Gyorffy, D. D. Johnson, and G. M. Stocks, Phys. Rev. Lett. 56:2096 (1986).
7. D. D. Johnson, D. M. Nicholson, F. J. Pinski, B. L. Gyorffy, and G. M. Stocks, Phys. Rev. Lett. 56:2088 (1986).
8. B. L. Gyorffy and M. J. Stott in "Band Structure spectroscopy of Metals and Alloys," D. J. Fabian and L. M. Watson, eds., Academic Press, New York (1972).
9. J. S. Faulkner and G. M. Stocks, Phys. Rev. B21:3222 (1980).
10. R. Kubo, J. Phys. Soc. Jpn. 12:570 (1957).
11. J. C. Swihart, W. H. Butler, G. M. Stocks, D. M. Nicholson, and R. C. Ward, Phys. Rev. Lett. 57:1181 (1986).
12. W. R. G. Kemp, P. G. Klemens, A. K. Sreedhar, and G. K. White, Proc. Roy. Soc. London, Ser. A233:480 (1956).
13. B. R. Coles and J. C. Taylor, Proc. Roy. Soc. London, Ser. A267:139 (1962).
14. R. W. Westerlund and M. E. Nicholson, Acta Metall. 14:569 (1966); W.-K. Chen and M. E. Nicholson, Acta Metall. 12:687 (1964); W. H. Arts and A. S. Houston-MacMillan, Acta Metall. 5:525 (1957). These authors

give the change in resistivity as a function of the deformation. This information allows one to separate approximately the change in resistivity due to the deformation-induced randomization of the atoms and the increase in resistivity due to dislocation and other defects induced by the deformation.
15. H. A. Fairbank, Phys. Rev. 66:274 (1944).
16. W. G. Henry and P. A. Schroeder, Can. J. Phys. 41:1076 (1963).
17. C. Y. Ho, M. W. Ackerman, K. Y. Wu, T. N. Havill, R. H. Bogaard, R. A. Matula, S. G. Oh, and H. M. James, J. Phys. Chem. Ref. Data 12:183 (1983).
18. R. S. Crisp, W. G. Henry, and P. A. Schroeder, Philos. Mag. 10:553 (1964).
19. W. Koester and H.-P. Rave, Z. Metallk. 55:750 (1964). The room temperature results reported here were adjusted to low temperature by the assumption that the temperature-dependent resistivity as a function of concentration reported in Ref. 12 was not affected by cold work.
20. J. O. Linde, Ann. Phys. 15:219 (1936).
21. D. M. Nicholson and G. M. Stocks, private communication.
22. D. M. Nicholson, private communication.
23. T. S. Lei, K. Vasudevan, and E. E. Stansbury, in "High Temperature Ordered Intermetallic Alloys," C. C. Koch, C. T. Liu, and N. S. Stoloff, eds., Materials Research Society Symposia Proceedings Vol. 39, Materials Research Society, Pittsburgh (1985). p. 164.
24. N. F. Mott and E. A. Davis, "Electronic Processes in Non-Crystalline Materials," Oxford University Press, Oxford (1971).
25. P. B. Allen, in "Superconductivity in d- and f-Band Metals," H. Suh. and M. B. Maple eds., Academic Press, New York (1980), p. 291.

BALLISTIC DEPOSITION ON SURFACES

P. Ramanlal and L.M. Sander

Department of Physics
The University of Michigan
Ann Arbor, MI 48109, USA

INTRODUCTION

In recent years considerable interest has developed in the formation of random structures under non-equilibrium conditions. Much of our understanding of the geometry of these structures and its relationship to their formation mechanism has come from the study of simple models by means of both computer simulation and theoretical methods. One of the most fundamental of these models is the ballistic aggregation model in which particles are added to a growing structure via linear (ballistic) trajectories. Other simple models which have been studied intensively include the Eden[1] model, diffusion limited aggregation[2] (DLA) and diffusion limited cluster-cluster aggregation.[3,4]

In the ballistic aggregation model particles are added one at a time to a growing cluster or aggregate of particles using linear trajectories which have randomly selected positions and directions. In the most extensively studied version of this model the particles are spherical. This model leads to clusters with a porous structure which has sometimes been described in terms of a fractal dimensionality[5] (D) which is smaller than the Euclidean dimensionality (d) of the space in which the simulation is carried out. However, based on recent larger scale computer simulations[6,7] and theoretical arguments[8] a consensus that the fractal dimensionality of ballistic aggregates is equal to their Euclidean dimensionality has now developed. This means that the internal structure of ballistic aggregates is uniform on all but short length scales.

The ballistic deposition model can be simplified considerably by confining the particles to sites on a square (or cubic) lattice. The particles then follow trajectories which are normal to the surface and stick permanently when they reach an unoccupied site which is adjacent to an occupied site on the surface. This model generates porous structures which are uniform on all but short length scales. However, the active zone[9] (sites at which further growth can occur) exhibits non-trivial scaling behavior. The active zone for two dimensional square lattice ballistic deposition was first investigated by Family and Vicsek[10] using deposits grown to a mean height $\overline{h}$ on strips of width ℓ with periodic boundary conditions in the lateral direction. Family and Vicsek measured the "width of the active zone" (ξ) defined by:

$$\xi^2 = \overline{(h_i-\overline{h})^2} \ , \quad (1)$$

where h_i is the height of the active zone at the position of the ith site on the original surface and $\overline{h}$ is the mean height of the active zone. They found that the dependence of ξ on $\overline{h}$ and ℓ could be expressed in terms of the scaling relationship

$$\xi \sim \ell^{\alpha} f(\overline{h}/\ell^{\gamma}) \ , \quad (2)$$

where the scaling function f(x) has the properties f(x) = constant for large x and $f(x) \sim x^{\nu}$ ($\nu=\alpha/\gamma$) for small x. This means that for small heights ($\overline{h}<<\ell$) $\xi \sim \overline{h}^{\nu}$ and for large heights ($\overline{h}>>\ell^{\gamma}$) $\xi \sim \ell^{\alpha}$. From their numerical results Family and Vicsek found that $\alpha = 0.42 \pm 0.03$ and $\nu = 0.30 \pm 0.02$.

It has recently been shown that the width of the active zone[9] for the 2d Eden model with strip geometry can be described by the same scaling form (Eq. (2)) as the ballistic deposition model with scaling exponents (α and γ) approximately equal for both models.[10-11] Recently Kardar, Parisi and Zhang[12] have proposed a model for the evolution of the active zone for both the Eden growth and ballistic deposition processes. A Langevin equation is proposed for the local growth of the active zone which leads to the result $\alpha=1/2$ and $\gamma=1.5$ in 2d which is consistent with available computer simulation results for both models if the possibility of significant corrections to scaling in the computer simulation results are taken into account. Since the active zone shows non-trivial scaling, it is natural to inquire whether the interface is itself a fractal curve. In fact, it is not.

For if the active zone were a self similar fractal we would expect to find step sizes in the height of the active zone ($\delta h_i=|h_i-h_{i+1}|$) which could take on arbitrarily large values (for sufficiently large values of $\overline{h}$ and ℓ). Instead we find numerically that the distribution of step sizes (δh) is an exponentially decaying function of δh so that large step sizes are extremely improbable. In fact we shall show the step size to be bounded. This means that the active zone of ballistic aggregates should be described in terms of self-affine rather than self-similar fractal geometry.[5,13,14]

We have also investigated a 2d model in which all the δh values are restricted to a constant value of 1.0. In this model we start with a "surface" which has a height of 0 at odd values of i (the index describing the position along the surface) and 1 at even values of i (like a castle wall). Surface sites are then selected randomly and the height is incremented by <u>two</u> lattice units if and only if the height of both of the next nearest neighbors is greater than that of the randomly selected site (i.e., if $h_{i+1} - h_i = 1$ and $h_{i-1} - h_i = 1$). The extension to 3d is straightforward. The active zone for this model exhibits the scaling behavior given in Eq. (2) with numerical values for the exponents α and γ very close to those predicted by Kardar et al.[12] In this paper we show how this modified ballistic deposition model can be mapped onto a spin exchange model and derive values for the exponents α and γ which are in good agreement with the theoretical results of Kardar et al. and with our simulations. In contrast, in 3d our analytic results from the spin exchange model do not agree with simulations or with Ref. 12. (However, Ref. 12 does not agree with our simulations either! The situation in 3d is, at the moment, very puzzling).

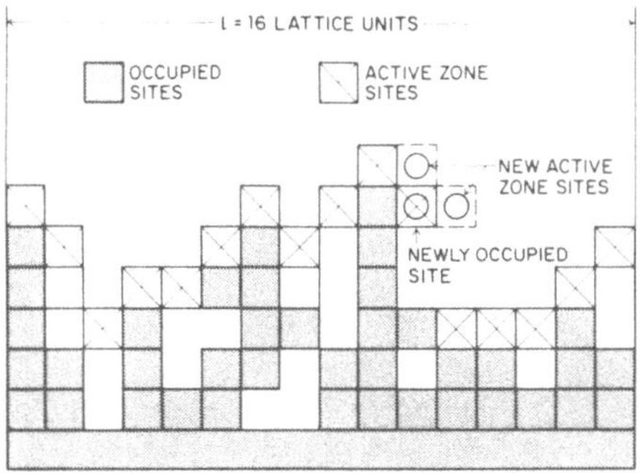

Fig. 1. Ballistic Aggregation

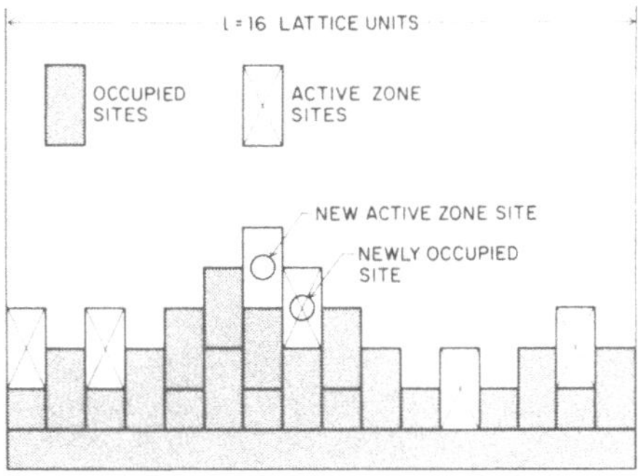

Fig. 2. The Castle-wall Model

Simulation

Fig. 1 shows an early stage in a 2d simulation of ballistic deposition. In this particular model, the "adjacent" unoccupied sites are those which are nearest neighbors to an occupied site of the active zone. This model will be referred to as the nearest neighbor on NN model.

Fig. 2 shows the "single step" ballistic deposition model. In this model two sites are added at a randomly selected active site, defined such that the height of the deposit is greater at the two neighboring positions (i.e., $h_{i+i} > h_i$ and $h_{i-i} > h_i$). The simulation starts off with sites of odd index (i) having a height of 0 and those of even index having a height of 1. In this model, the height of its nearest neighbors by exactly one lattice unit $|h_i - h_{i+1}| = 1$. In 3d we have the same condition with its four next nearest neighbors.

Scaling of the Interface for 2d Square Lattice Simultions

The exponent ν was first obtained using very wide strips (2^{18} sites long) on which particles were deposited until a mean height ($\bar{h}$) of 5,000 lattice units was reached.

For active zones with a mean height in the range $50 < \bar{h} < 500$ lattice units we find that the exponent ν has an effective value of 0.331 ± 0.006 and for active zone heights in the range $500 < \bar{h} < 5000$ we find $\nu = 0.308 \pm 0.011$. The range of uncertainty given here and elsewhere in this paper represent the 95% confidence limit. The value obtained for ν is in good agreement with the result obtained by Family and Vicsek[10] (0.30 ± 0.2) from smaller scale simulations.

To obtain an estimate for the value of the exponent α, Eq. (3), deposits have been grown on strips of width ℓ for $\ell < 512$. Deposits have been grown to a height much greater than ℓ^{γ} (a conservative estimate of 5/3 for the value of γ was used) and the dependence of the width of the active zone, ξ, on ℓ was determined only for deposit heights greater than

$20\ell^{\gamma}$. The dependence of ξ on ℓ can be expressed as $\xi \sim \ell^{\alpha}$ where the exponent α has a value of about 0.47. This value is much closer to the theoretical value[12] of 0.5 than the result $\alpha = 0.42 \pm 0.03$ obtained by Family and Vicsek.

A quantity which is of interest in understanding the nature of the surface of ballistic aggregates is the distribution of step heights, δh, in the active zone, where $\delta h_i = |h_i - h_{i+1}|$. We consider the sum of the step heights (ΔH) defined by

$$\Delta H = \sum_{i=1}^{\ell} \delta h_i \quad , \tag{3}$$

where the length of the interface in an L' metric is $\Delta H + \ell$, from simulations with a strip width (ℓ) of 2^{18}. We find that ΔH first increases rapidly with increasing deposit thickness ($\overline{h}$) but soon approaches a limiting value. Least squares fitting straight lines to the dependence of $\ln(\Delta H)$ on $\ln(\overline{h})$ for deposit heights in the range $500 < \overline{h} < 5000$ lattice units for five simulations gives a value for the effective exponent η, defined by

$$\Delta H \sim \overline{h}^{\eta}, \tag{4}$$

of 0.00278 ± 0.00024 indicating that ΔH becomes independent of $\overline{h}$. For deposit heights in the range 2500-5000 lattice units $\Delta H/\ell$ has the value of 1.13600 ± 0.00005. Thus the length of the interface does not grow with $\overline{h}$ and is proportional to ℓ.

We have also determined the distribution of step sizes at various stages during the deposition process, and find that the dependence of $N(\delta h, \overline{h})$ on δh can be expressed as:

$$N(\delta h, \overline{h}) \simeq A e^{-k\delta h} \quad , \tag{5}$$

where the constant k decreases with increasing $\overline{h}$ but the distribution retains its exponentially decaying form and the decay constant (k) approaches a constant value of about 0.39. This indicates that large steps are (exponentially) improbable and that the surface of the ballistic aggregate is not a self-similar fractal.

Discussion of 2d Simulation Results

The foregoing results on step size distribution are rather unexpected. As we have remarked, there are a very small number of large steps. The surface cannot be a self-similar fractal. In fact, we can prove the stronger result: The asymptotic value of the length of the surface is proportional to ℓ.

The proof of this bound on the length of the interface is simple, if, as proved for ballistic aggregation,[8] the density of the aggregate does not indefinitely decrease. The essence of the argument is that if a particle is deposited on the edge of a step of height δh, then the mean height of the interface $\overline{h}$ increases by $\delta h/\ell$ where ℓ is the width of the sample. Thus if n is the number of particles deposited per unit of substrate and ρ is the density of the deposit we have:

$$\frac{1}{\rho} = \frac{\partial \overline{h}}{\partial n} \sim \langle \delta h \rangle \quad , \tag{6}$$

where $\langle \delta h \rangle = \Delta H/\ell$ is a uniform average over the interface. Of course,

$\ell(\langle\delta h\rangle+1)$ is the length of the surface, as above. A rigorous analysis[15] gives for the d=2 NN model:

$$\frac{1}{\rho} - 1 < \langle\delta h\rangle < \frac{2}{\rho} \quad , \tag{7}$$

but the argument should hold in any space dimension.

One can also present a simple argument for the distribution of step heights δh as follows. Qualitatively each step can do two things: it can grow in height by accreting a particle on top of itself, and it can advance in the direction it faces by accreting a particle on its leading edge. The latter process necessarily entails collision with another step (albeit perhaps of zero height). Since opposite facing steps approach each other systematically whereas like facing steps have only a diffusive relative motion, we will focus on collisions of the former pairs. We introduce the notation $P_R(\delta h)$, $P_L(\delta h)$ for the statistical distributions of left and right facing steps (which should in fact be equal). Then we have the equations of motion:

$$\frac{\partial}{\partial n} P_R(x) = \frac{\partial}{\partial x} P_R - 2P_R(x) \sum_y P_L(y) + 2 \sum_y P_R(x+y)P_L(y) \ , \tag{8}$$

and symmetrically for $P_L(x)$. The terms are respectively: increase of δh, loss by collision, and gain by collision with a larger step. For a steady state with $P_R=P_L=P/2$ we then have in a continuum approximation:

$$-2\frac{\partial}{\partial x} \ P(x) \ -2P(x) \ +2 \int_0^\infty dy \ P(x+y)P(y) = 0 \ , \tag{9}$$

which has an exponential solution $P(x)=ke^{-kx}$ with $k=1/2$. The value of k compares fortuitously well with the observed limiting slope of $k_{obs}\sim 0.39$ for the NN model. Thus we have good evidence, both from simulation and analysis, that small steps dominate the fluctuations in height. This fact leads us to suspect that the single-step model should be essentially identical to true ballistic aggregation.

Simulations of the Single-step Model in 2d

A similar set of simulations (with $\ell \leq 512$ and $h_{max} = 5000$) to those discussed above has been carried out for the single step model in 2d. From 15 simulations the result $\nu = 0.332 \pm 0.003$ was obtained for the active zone of deposits grown to heights in the range $0.01h_{max} \leq \bar{h} \leq 0.1h_{max}$. For deposits with heights in the range $0.1h_{max} \leq \bar{h} \leq h_{max}$ the result was $\nu = 0.330 \pm 0.012$. These results suggest that the asymptotic value for the exponent may be exactly 1/3 for this model. We have also determined the dependence of the width of the active zone, ξ, on the strip width ℓ for strips of width 64-1024 lattice units. The effective value for the exponent α is 0.5000 ± 0.0015.

These results strongly suggest that the limiting value for the exponent α may be exactly 1/2 for this model. We conclude that $\nu \simeq 1/3$, $\alpha \simeq 1/2$ and $\gamma \simeq 3/2$ for the single-step model.

Three Dimensional Ballistic Deposition

Simulations of ballistic deposition were carried out on a 3d cubic lattice with periodic boundary conditions in the lateral directions. The surfaces on which the deposition processes was simulated were of size $\ell \times \ell$ lattice units with $8 \leq \ell \leq 512$. The dependence of the width of the active zone, ξ, on ℓ suggests that the asymptotic value for the exponent α (Eq. (2)) may be close to 1/3.

We have also measured the dependence of the width of the active zone (ξ) on the mean height of the active zone using simulations carried out on a 1024 x 1024 surface. The results obtained from this simulation suggest that the exponent ν may have a value close to 1/4. However, in this case, it is difficult to approach the limit $\bar{h} \ll \ell$ and still be at heights which are sufficiently large to see the correct asymptotic behavior (i.e., we need to satisfy $\bar{h} \gg 1$ and $\bar{h} \ll \ell$ simultaneously).

The Three-Dimensional Single-Step Model

A three-dimensional version of the single-step model described above was also investigated. In this model (illustrated in Fig. 4) we start with a checkerboard of raised and lowered sites (Fig. 4a). Positions on the $\ell x \ell$ surface are selected randomly and the height at these positions is raised by two lattice units if all four nearest neighbors have a height which is greater than that of the selected site.

These simulations were carried out on $\ell x \ell x h$ simple cubic lattices with ℓ=8, 16, 32,64, 128, 256 and 512. For the case ℓ=512 lattice units, a total of $2x10^9$ pairs of sites were added ($h \simeq 1.6x10^4 \simeq 4\ell^{4/3}$). For smaller values of ℓ the limit $h \gg \ell^\gamma$ was approached more closely. These simulations gave results which were very similar to those obtained from the ordinary 3d ballistic deposition model and a value of 0.363 ± 0.005 was obtained for the exponent α.

Simulations were also carried out using 1024 x 1024 surfaces to investigate the dependence of the surface height variance ξ on the mean surface height $\bar{h}$. Again, the results obtained were very similar to those for the 3d NN model and an exponent ν with a value of about 1/4 was obtained assuming that the dependence of ξ on $\bar{h}$ is described by equation 2. For mean heights in the range $5 < \bar{h} < 50$ lattice units ν has an effective value of 0.221 ± 0.003 and for $50 < \bar{h} < 500$ lattice units ν= 0.2303 ± 0.0006.

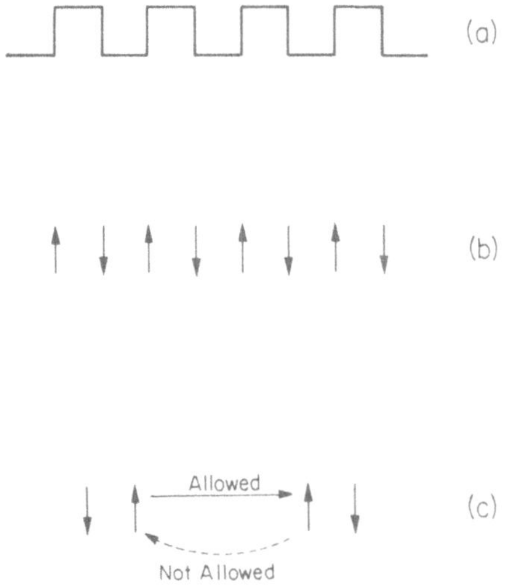

Fig. 3. Spin Model in 2d

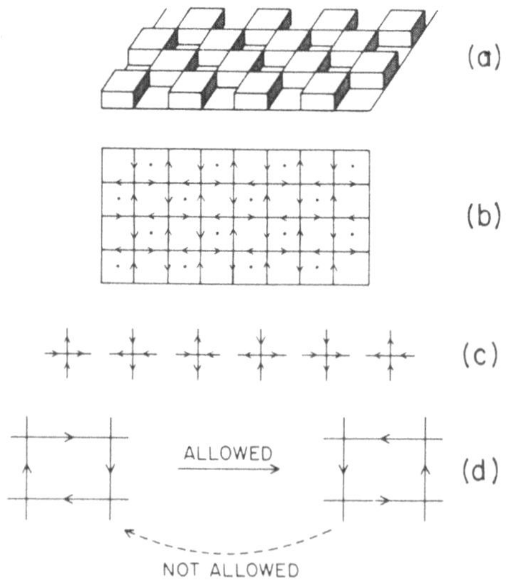

Fig. 4. Spin Model in 3d

Analytic Results for Single-step Models

Since the single-step models both in 2d and 3d have similar scaling behavior to full ballistic aggregation, we are motivated to devise analytical treatments for these simpler processes. In the process we will produce an almost trivial proof that $\alpha=1/2$ in 2d in agreement with simulations and Ref. 12. We will show in the next section that there is a relationship between α and γ, so that we can give a full treatment in this case.

The single-step model in 2d can be mapped onto a spin model with an up-spin representing a step up in the interface and down-spin a step down, see Fig. 3. Growth at a site is represented by a biased nearest neighbor spin exchange with an up-spin moving only to the left and a down-spin to the right. The opposite sense of spin exchange would correspond to a lowering of height which we do not allow. The interface dynamics in the moving frame $\overline{h}=0$ is now a spin exchange problem. The height at site m is given by:

$$h_m = \sum_{i=1}^{m} \sigma_i + h_1 \tag{10}$$

where $\sigma_i = \pm 1$. Also $\sum\sigma_\uparrow = \sum\sigma_\downarrow$ follows from periodic boundary conditions. Now suppose that after many spin exchanges the system approaches the equilibrium of the spin model. In this "long-time" limit we would have:

$$\overline{\sigma_i \sigma_j} = \delta_{ij} \quad . \tag{11}$$

Then we have from Eq. (10):

$$\overline{(h_m - h_n)^2} = |m-n| \quad . \tag{12}$$

This is the definition of a self-affine fractal[5,13,14] with a fractal dimension of 3/2. Also, it follows at once that:

$$\xi^2 = \overline{(h_m - \overline{h})^2} \sim \ell \quad , \tag{13}$$

so that $\alpha = 1/2$.

Now, we can show that the spin system maintains equilibrium, where every configuration such that the magnetization is zero is equally likely from the following observations: a) every configuration can be formed by one deposition in one way for each maximum of its interface. b) every configuration can form, by one deposition, one new configuration for each minimum of its interface. c) the number of minima and maxima are always equal. It follows that if every allowed configuration is equally likely, then every configuration is as likely to be formed as lost, and equilibrium preserves itself.

In 3d the single-step interface can be mapped onto an ice-rule 6-vertex model; see Fig. 4. The initial configuration (Fig. 4a) is mapped onto the vertex configuration Fig. 4b. Looking down on a vertex and sweeping close around it anticlockwise, a step-up is denoted by outwards arrow and a step-down by an inwards one. The allowed vertices are those with equal numbers of incoming and outgoing arrows, as shown in Fig. 4c. Thus the arrows consistute a conserved current, and it is easy to see that the difference in height between two sites is the net transverse current flowing across a line between them. The allowed growth step is to reverse all the arrows of a local cycle (plaquette) from clockwise to anticlockwise (but not vice versa), as shown in Fig. 4d, and may be viewed as a simultaneous horizontal and vertical biased spin exchange each as in the 2d case. The height at site n,m is

$$h(n,m) = \sum_{i=1}^{n} \sigma^{v}_{i,1} + \sum_{j=1}^{m} \sigma^{h}_{n,j} + h(1,1) . \tag{14}$$

where σ^v and σ^h are the vertical and horizontal arrows with values ±1. Once more, if we know the equilibrium arrow correlation function, we can calculate α.

Finding the equilibrium of this vertex model is not as easy as the 2d case. We should note that the ways of forming a configuration are in correspondence with the maxima of its interface (anticlockwise plaquettes) and the ways it can grow with the minima (clockwise plaquettes) as before. Thus if the ensemble of all possible configurations (equally weighted) has predominantly nearly equal number of maxima and minima, then at long times our system probably reaches a good approximation to the equilibrium of our 3d model. Since maxima and minima are features of the local geometry, this assumption is likely to be justified in the limit of a large system.

If we do assume that all configurations of the vertex model are equally likely, then using the result of Sutherland[16] for the arrow correlation in the 6 vertex model, we obtain the result of Beijeren[17] for the body-centered-solid-on-solid model in the limit of infinite temperature:

$$\overline{(h_a - h_b)^2} \sim \ln(R_{ab}) , \tag{15}$$

where R_{ab} is the distance between sites a and b. The result is also obtained for the discrete Gaussian model[18] which we believe to be in the same universality class. We can understand this equation by noting that the long wavevector components of a conserved "current" density are correlated. In fact, if all allowed current configurations are equally likely the correlation function in real space is:

$$\overline{j(o)j(r)} \sim 1/r^2 \tag{16}$$

which is also the arrow correlation result of Sutherland.[16] This logarithmic law corresponds conceptually to α=o and is rather difficult to test; we have seen that our data does not support it.

Kardar et al.[12] obtain α=1/2 for this case as in 2d. Their resul is different from ours (α=o) and from the simulation data ($\alpha \approx 1/3$). We have no explanation of this multiple discrepency.

Continuum Approximation and Scaling

With both simulation and theory suggesting universality for the asymptotic scaling of the interface thickness (at least in d=2) it is natural to look for simplified equations of its motion on the large scale. The observed exponents ν and α are less than unity. Thus ξ/ℓ, $\xi/\overline{h} \to 0$ as $\xi \to \infty$ so that coarse scale derivatives dh/dx exist, and the proof that the step heights are bounded shows that dh/dx still exists even locally. Thus we are lead to consider differential equations. For the NN model we have a growth rate, averaged over temporal noise, given by

$$\langle \frac{\partial h}{\partial n} \rangle = \langle \delta h \rangle_{\text{local spatial average}} \sim \langle \delta h \rangle_o (1 + \varepsilon |\nabla h|^2 \ldots) \tag{17}$$

where $\varepsilon|\nabla h|^2$ is the first term in a gradient expansion. Subtracting the advance of a flat interface, this equation can be rescaled in the weak gradient limit to give the form considered by Kardar et al.,[12]

$$\frac{\partial \tilde{h}}{\partial t} \sim (\nabla \tilde{h})^2 + \text{noise}. \tag{18}$$

Here we observe that if for large scale solutions we can neglect the noise, then there only exist non-trivial solutions to the above with the scaling form

$$\tilde{h}(\vec{x},t) = t^{\nu} f(\vec{x}t^{-\nu/\alpha}) \tag{19}$$

if:

$$\frac{2}{\alpha} = 1 + \frac{1}{\nu} \quad , \tag{20}$$

whereupon $f(\vec{y})$ obeys

$$\nu f(\vec{y}) - \frac{\nu}{\alpha}\vec{y} \cdot \frac{\partial}{\partial y} f(\vec{y}) = [\frac{\partial}{\partial y} f(\vec{y})]^2 \quad . \tag{21}$$

This scaling law is consistent in two dimensions with the results of Kardar et al.[12] and our simulation data. For the single-step model in d=2 where we have α=1/2 explicitly, it determines ν= 1/3 .

For the single-step model in d=3, where we have presented an argument that $\alpha \to 0$, it follows that $\nu \to 0$ also. However $\gamma = 2$. This may ultimately be the clearest prediction to test in this case.

DISCUSSION

In our analysis of the lattice models for ballistic deposition we have assumed that the scaling form found by Family and Vicsek is correct and have attempted to measure the exponents α, ν and γ by approaching as close as possible to the asymptotic limits available to us. In the case of the single-step model the corrections to the simple scaling picture are very much smaller and we believe that our results for this model are quite accurate. In any event, our results from all of the 2d models are consistent with the theoretical work we have presented and that of Kardar et al.[12] (α=1/2, ν=1/3, γ=1.5).

In the case of the three dimensional NN model and the single-step model our results suggest that $\alpha \simeq 1/3$ and $\nu \simeq 1/4$. However, in three dimensions it is even more difficult to approach the asymptotic limits.

Our observation that the distribution of step heights, $N(\delta h,\bar{h})$, in the ballistic deposition model is an exponentially decaying funciton of δh indicates that the surface of 2d ballistic deposits is not a self similar fractal. The fact that large steps are exponentially improbable for the square lattice ballistic deposition model strongly suggest that this model and the single step model should belong to the same universality class. If this is the case we expect to find α=1/2 and ν=1/3 for 2d ballistic deposition. For 3d the situation is much less clear. Neither our analysis nor that of Ref. 12 agrees with our direct simulations. Further work on this case (which is, after all, the one of most direct physical interest) is needed.

ACKNOWLEDGEMENTS

Supported by DoE Grant No. DE-FG02-85ER45189. We would like to thank M. Kardar for a useful discussion. Much of this work was done in collaboration with R.C. Ball and Paul Meakin.

References

1. M. Eden Proc. 4th Berkeley Symp. Math. Stat. Probab. 4, 223 (1961).
2. T.A. Witten and L.M. Sander, Phys. Rev. Lett. 47, 1400 (1982); Phys. Rev. B 27, 1119 (1983).
3. P. Meakin, Phys. Rev. Lett. 51, 1119 (1983).
4. M. Kolb, R. Botet and R. Jullien, Phys. Rev. Lett. 51, 1123 (1983).
5. B.B. Mandelbrot, "The Fractal Geometry of Nature" (Freeman, San Francisco 1982).
6. P. Meakin, J. Colloid and Interface Sci. 105, 240 (1985).
7. D. Bensimon, B. Shraiman and S. Liang, Phys. Lett. 102A, 238 (1984).
8. R.C. Ball and T.A. Witten, Phys. Rev. A 29, 2966 (1984).
9. M. Plischke and Z. Racz, Phys. Rev. Lett. 53, 415 (1984).
10. F. Family and T. Vicsek, J. Phys. A 18, L75 (1985).
11. R. Jullien and R. Botet, Phys. Rev. Lett. 54, 2055 (1985); J. Phys. A 18, 2279 (1985).
12. M. Kardar, G. Parisi and Y.-C. Zhang, Phys. Rev. Lett. 56, 889 (1986).
13. B. Mandelbrot in "Fractals in Physics", L. Pietronero and E. Tossati, eds. (Elsevier, 1986).
14. S. Alexander, Proceedings of Gaithersburg Conference on Transport and Relaxation Processes in Random Materials. M. Schesinger and Y. Klafter, Editors, October 1985.
15. P. Meakin, P. Ramanlal, L.M. Sander and R.C. Ball, Phys. Rev. A., in press.
16. B. Sutherland, Phys. Lett. 26A, 532 (1968).
17. H.V. Beijeren, Phys. Rev. Lett. 38, 993 (1977).
18. S.T. Chui and J.D. Weeks, Phys. Rev. B 14, 4978 (1976).

MONTE CARLO SIMULATION OF LARGE EDEN CLUSTERS ON A CRAY-2

D. Stauffer and John G. Zabolitzky

Supercomputer Institute and School of Physics and Astronomy
University of Minnesota
Minneapolis, MN 55415

INTRODUCTION

One of the simplest mathematical models of growth is the Eden model[1]. Consider an infinite square lattice in two dimensions with all lattice sites empty, except one single occupied site initially. The growth process then may be completely defined by the single sentence, to be applied recursively: Occupy at random one of the empty sites neighbouring an occupied site. In spite of this remarkably simple definition the objects generated by this procedure exhibit astonishingly complex properties.

The Eden clusters are compact objects, mass $\propto$ $(\text{radius})^d$ in d dimensions. The quantity of interest is the thickness W of the surface defined as the rms fluctuation of the distance r of the perimeter sites from the origin of cluster growth, where a perimeter site is any unoccupied site neighbouring an occupied site.

SPHERICAL CLUSTERS IN TWO DIMENSIONS

Let us consider the model exactly as defined above. The surface width then is a function of the cluster mass S, i.e. the number of occupied sites. For large clusters a power law behaviour is expected,

$$W \propto S^x$$

Varying values for the exponent x may be found in the literature[2,3]. Our results[4] are shown in fig. 1. For clusters in the region $S=10^6$ to 10^7 the exponent is about $x=1/4$, i.e. $W \propto \sqrt{R}$, for larger clusters $x=1/2$, i.e. $W \propto R$, as is shown clearly in the inset. These simulations support earlier findings[3] that the anisotropy of the square lattice leads to anisotropic clusters, i.e. diamond shapes. The surface width measured therefore is the intrinsic surface width for smaller clusters, where the anisotropy is not yet felt, whereas for larger clusters the anisotropy contribution overwhelms the instrinsic contribution to the width. Rather large clusters are required since the anisotropy is small.

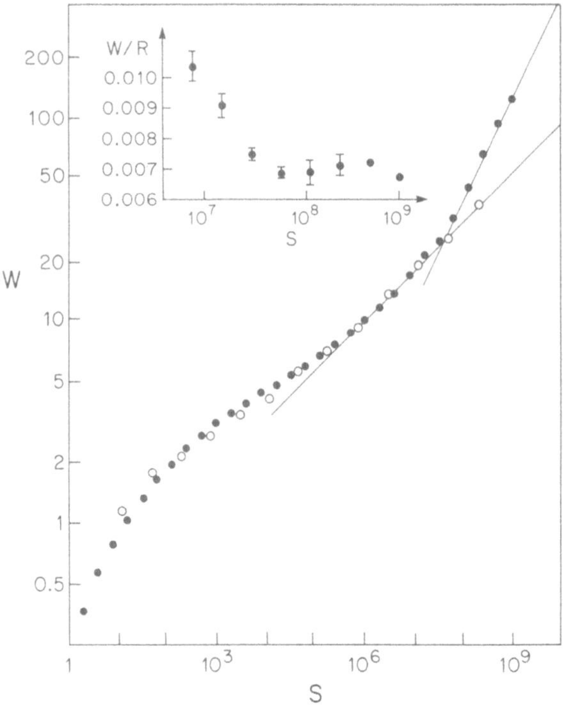

Fig. 1. Simulation results for surface width W as a function of cluster mass S, in two dimensions. Solid dots, sperical clusters. Open circles, rectangular clusters.

FLAT CLUSTERS IN TWO DIMENSIONS

Because of this complication we choose to modify the geometry slightly for further studies. In two dimensions we still have a square lattice, but initially we begin with a line of occupied sites of length L. This line is then allowed to grow in thickness in one direction. Periodic boundary conditions are applied perpendicular to the growth direction. This geometry may be taken to represent a section of surface from an infinite, round cluster - but without the anisotropy problem, since growth occurs only in one direction parallel to a lattice axis. Results from simulations in this geometry are given in fig. 1 as open circles. It is seen clearly that the anisotropy effect has vanished.

A more detailed analysis of these results yields

$$W \propto L^{0.511\pm0.025}$$

which is compatible with the exponent ½. In order to find this behaviour of width W upon line length L, one has to carry the simulation into the "equilibrium" regime, where as a function of simulation time or growth height h the surface width W remains constant.

One may consider also the non-equilibrium regime for fixed line length L, where $1 \ll W(h) \ll W(\infty)$. $W(\infty)$ is the quantity discussed above as a function of L, the equilibrium width for fixed L. Obviously, for given L, there is only a finite amount of simulation time where above inequality is valid. If one is interested in the non-equilibrium behaviour for large h, i.e. large simulation times, one must do these simulation on the large L systems, where the non-equilibrium regime is larger. Again one expects a power-law behaviour[4,5]

$$W \propto h^{\beta}$$

with β=1/3. Results from our simulations are shown in fig. 2. Here we give the apparent exponent β at given height or simulation time. It is seen that enormously large systems are required in order to reach the asymptotic regime. The calculations for fig. 2 used line length L = 1 048 576. Our findings are consistent with the expectations.

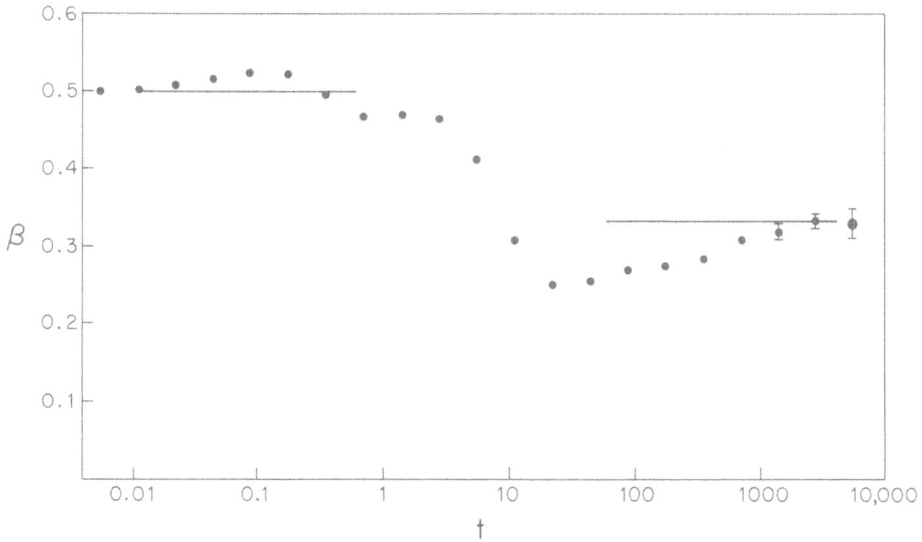

Fig. 2. Dynamic exponent β=d(ln W)/d(ln h) vs. simulation time t$\propto$h, in two dimensions for the square lattice. Strip length L = 1 048 576.

RECTANGULAR CLUSTERS IN HIGHER DIMENSION

Results from simulations in strip geometry for three and four dimensions are given in fig. 3. In three dimensions, one could hope for W $\propto$ $\sqrt{}$ log L. If one looked only at small systems, one might be led to believe that this behaviour were consistent with the data. Considering the last few data points this is not the case. No definite inference can be carried from these data - in spite of these being about the largest Eden simulations possible on a Cray-2.

In four dimensions the surface may be supposed to become smooth, W $\propto$ const. for large L. Again, considering only part of our results (L $\leq$ 32 for example), one may be tempted to see no discrepancy with this theoretical assumption. The larger systems show, however, that the surface width continues to increase with system size L. No definite law has been extracted from this data. It is to be doubted if the asymptotic regime has yet been reached, i.e. if one would not require still much larger systems to obtain asymptotic information. Unfortunately, it does not seem possible to perform such simulations on a Cray-2.

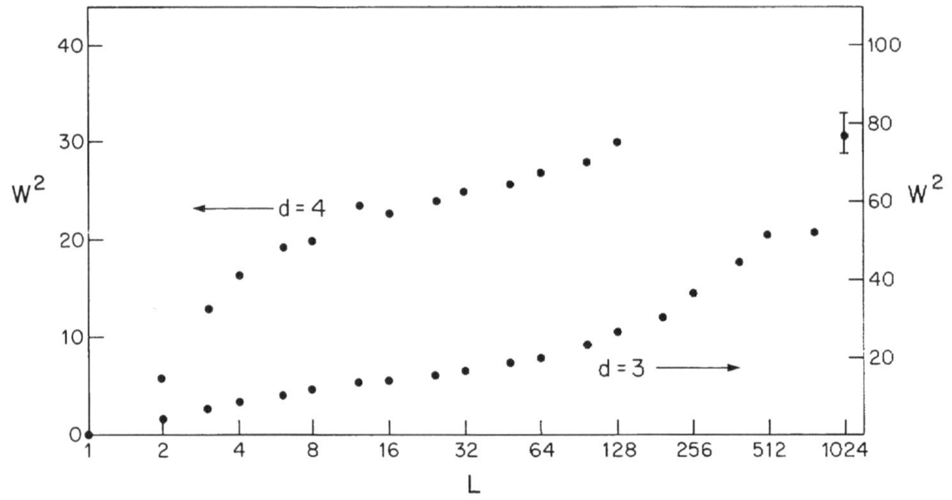

Fig. 3. Equilibrium surface width as function of linear base dimension L, in three and four dimensions.

REFERENCES

1. M. Eden, in "Proceedings of the Fourth Berkeley Symposium on Mathematical Statistics and Probability", F. Neyman, ed., University of California, Berkeley, 1961
2. F. Leyvraz, J. Phys. A18, L941 (1985); D. Stauffer, Phys. Rev. Lett. 41, 1333 (1978); H. P. Peters, D. Stauffer, H. P. Holters, K. Loewenich, Z. Phys. B34, 399 (1979); M. Plischke, Z. Racz, Phys. Rev. Lett. 53, 415 (1984); 54, 2056 (1985); Z. Racz, M. Plischke, Phys. Rev. A31, 985 (1985); R. Jullien, R. Botet, Phys. Rev. Lett. 54, 2055 (1985); J. Phys. A18, 2279 (1985)
3. P. Freche, D. Stauffer, H. E. Stanley, J.Phys. A18, L1163 (1985); R. Hirsch, D. E. Wolf, J. Phys. A19, L251 (1986); D. E. Wolf, unpublished; P. Meakin, R. Botet, R. Jullien, Europhys. Lett., in print
4. J. G. Zabolitzky, D. Stauffer, Phys. Rev. A34, 1523 (1986); D. Stauffer, J. G. Zabolitzky, Phys. Rev. Lett., in print
5. F. Family, T. Vicsek, J. Phys. A18, L75 (1985); M. Kardar, G. Parisi, Y. C. Zhang, Phys. Rev. Lett. 56, 889 (1986)

RENORMALIZATION GROUP METHODS FOR PHASE SEPARATION PROBLEMS

Zhaowei Lai and Gene F. Mazenko

The James Franck Institute and Department of Physics, The University of Chicago
Chicago, Illinois 60637

Oriol T. Valls

School of Physics and Astronomy
University of Minnesota
Minneapolis, Minnesota 55455

The kinetic Ising model with both a non-conserved and a conserved order parameter has been extensively studied as a simple representation of the growth kinetics of binary alloys.[1] When the order parameter is not conserved (NCOP), as in an order-disorder transition in a binary alloy, it is well known that the growth kinetics are governed by the curvature-driven dynamics developed by Lifshitz,[2] Cahn, and Allen[3] (LCA), which lead to a growth law for the typical domain size $L(t) \sim t^{1/2}$, as a function of time, for dimensionality greater than one. For a conserved order parameter, (COP) (phase separation or spinodal decomposition in a binary alloy) it has been commonly asserted that the long-time growth kinetics are controlled by the evaporation-condensation mechanism of Lifshitz and Slyozov[4] which gives $L(t) \sim t^n$ with exponent $n=1/3$ at least when the concentration of one of the two species is small. However, the available results from experiment, theory, or simulation on the growth law for spinodal decomposition in binary alloys give widely differing exponents[5] (in the range 0.07 to 0.25) when they are fitted to a power law. There is no hard evidence for any single power law. This points out to the possibility of having a more complicated time dependence in this problem, which appears to be a power law only when a relatively restricted time interval is considered.

---

[1]For recent reviews of the general problem see: J.D. Gunton, M. San Miguel, and P.S. Sahni, in "Phase Transitions and Critical Phenomena" ed. by C. Domb and J.L. Lebowitz (Academic, London 1983) vol. 8; J.D. Gunton, J. State. Phys. 34, 1019 (1984); K. Binder in "Systems far from equilibrium", vol. 132 of "Lecture Notes in Physics" ed. by L. Garrido (Springer, Heidelberg, 1980).

[2]I.M. Lifshitz, Zh. Eksp. Teor. Fiz. 42, 1354 (1962)[Sov.Phys. JETP 15, 939 (1962)]

[3]S.M. Allen, and J.W. Cahn, Acta. Metall. 27, 1085 (1979); J. Phys.(Paris) Colloq. C7, 54 (1977).

[4]E.M. Lifshitz and V.V. Slyozov, J. Phys. Chem. Solids, 19, 35 (1961).

[5]See Sect. I of Ref. (6) for a review of the literature on these results.

In order to clarify this issue, a renormalization group (RG) method was developed in Ref. (6). This method was implemented by means of Monte Carlo (MC) simulations and applied to the study of the growth kinetics for critical quenches of two-dimensional Ising models with either conserved or non-conserved order parameters. In Ref. (6), the time rescaling parameter, Δ, which relates spatially rescaled correlation functions at different times, was determined by a numerical strategy. Scaling laws were then obtained analytically from the existence and functional form of Δ. In the NCOP case the procedure led to the LCA law, $L(t) \sim t^{1/2}$. For a COP, the temperature dependence of Δ turned out to be of crucial importance. An analysis of this dependence was carried out[6] for the one value (b=2) of the spatial rescaling parameter b only. The analysis of Ref. (6) led to the conclusion that for quenches to low temperatures the long-time growth law was logarithmic for a COP on a lattice. This result is very different from that which is found for the NCOP case and indicates that the associated RG structure is quite different. In the NCOP case there is a single zero-temperature fixed point which controls the long time growth kinetics. In the COP case there are two distinct fixed points. There is a freezing fixed point associated with the behavior for T=0 where the system does not equilibrate and there is another fixed point which is accessed for temperatures between the critical temperature and zero temperature. This means that the limits of long time and low temperature do not commute in this case. Thus the freezing of a system at T=0 is an indication that the temperature is not an irrelevant variable, but it does not tell one about the nature of the growth kinetics at nonzero temperatures. One must do the calculations for T >0 to determine the appropriate RG structure. The freezing behavior was verified in Ref. (7) for a wide variety of concentrations.

The ideas of Ref. (6) were followed up in Ref. (8) and put on a sounder footing, both from a theoretical RG point of view and regarding their practical implementation. Our main result in Ref. (8) is the derivation of a new differential RG equation for the time rescaling parameter Δ. By analytically solving this equation, which is of a standard "textbook" form, one can evaluate Δ from MC data in a way which is both practical and elegant. We illustrated this procedure by presenting extensive results for Δ as a function of b and the quenching temperature for a kinetic Ising model with a COP. For quenches to zero temperature we found that the system freezes. For quenches to nonzero but low temperatures, the characteristic domain size increases with time according to a logarithmic law. For quenches to temperatures near but below the critical temperature it appears that there will be a regime with a power-law growth with an exponent which may be close to that given by the Lifshitz-Slyozov theory (1/3), followed by an eventual long-time crossover to the final logarithmic behavior at long times.

These conclusions seem to be compatible with recent[9] direct Monte Carlo simulations.

Acknowledgements

This work was supported by NSF Grant No. DMR 84-12901 and the Computer Central Facility of the Materials Research Laboratory at the University of Chicago.

[6]G.F. Mazenko, O.T. Valls, and F.C. Zhang, Phys. Rev. B31, 4454 (1985).
[7]G.F. Mazenko and O.T. Valls, Phys. Rev. B33, 1823 (1986).
[8]Z. Lai, G.F. Mazenko, and O.T. Valls, preprint.
[9]D. Huse, preprint and J. Tobochnik, preprint.

MEASURING FRACTALS: COMPARISON OF THEORY AND EXPERIMENT ON THE GLOBAL PROPERTIES OF A STRANGE ATTRACTOR

Leo P. Kadanoff

The Research Institutes
The University of Chicago
Chicago, IL

This talk will be about how to characterize and measure the properties of fractal sets. It will center about an experiment by Libchaber, Heslot, Stavans, and Glazier which sees the onset of chaos via period doubling and via a quasi-periodic behavior in a simple hydrodynamic system. The results of this experiment will be compared with the theory of Feigenbaum, for period doubling, and that of Shenker, Feigenbaum and Kadanoff for the quasi-periodic case. The actual comparison is performed using a novel way of looking at the global properties of the attractor which was first introduced by Parisi and Frisch. The basic device is to study the distribution of densities of points on the attractor. This, in turn, is estimated, following Procaccia, by measuring the time it will take for a recurrence to within a specified distance. This same analysis is applied to the experimental data and also to the theoretical models, giving descriptions of the topological properties of both. The descriptions are compared and the theory and experiment are seen to agree within the errors. The net result is the best quantative evidence to date of the universality of the Feigenbaum mechanism.

References

General

"The Fractal Geometry of Nature," Beniot B. Mandelbrot, W. H. Freeman and Company.

"Fractals, Where's the Physics," Leo P. Kadanoff, Physics Today, February (1986).

"Snatching Chaos from Order," Leo P. Kadanoff, Physics Today, December (1984).

f-α

Thomas C. Halsey, Mogens H. Jensen, Leo P. Kadanoff, Itamar Procaccia, and Boris Shraiman, Phys. Rev. A 33, 1141 (1986).

U. Frisch and G. Parisi, "Turbulence and Predictability of Geophysical Flows and Climate Dynamics, Varenna Summer School LXXXVIII.

R. Benzi, G. Paladin, G. Parisi, and A. Vulpani, J. Phys. A 17, 3521 (1984).

T. C. Halsey, P. Meakin, and I. Procaccia, Phys. Rev. Lett. 56, 854 (1986).

Experiment

M. Jensen, L. Kadanoff, A. Libchaber, I. Procaccia, and J. Stavans, Phys. Rev. Lett. 55, 2798 (1985)......quasi-periodic.
J. Glazier, M. Jensen, A. Libchaber, J. Stavans....period doubling.... in press.

Renormalization Group

D. Bensimon, M. Jensen, L. Kadanoff, Phys. Rev. A 33, 3622 (1986).
L. Kadanoff, J. Stat. Phys., May (1986).

D_q

P. Billingsley, "Ergodic Theory and Information," Wiley, New York, (1965).
A. Renyi, "Probability Theory," North Holland, Amsterdam (1970) pp. 586-595.
P. Grassberger, Phys. Lett. 97A, 227 (1983)....gives application to physical problems.
H.G.E. Hentschel and I. Procaccia, Physica (Amsterdam) 8D, 435 (1983)....describes methods of application to physical problems.

FLUCTUATION AND EXCHANGE IN THE FRACTIONAL QUANTIZED HALL EFFECT

S. T. Chui

Bartol Research Foundation of the Franklin Institute at
University of Delaware, Newark, Delaware 19716

Abstract

It is generally agreed that there is large positional fluctuation in the ground state in the fractional quantized Hall effect. This is illustrated by a series of correlated Gaussian trial wave functions, one of which possesses an energy comparable to the Laughlin wavefunction as well as a gap in the excitation spectrum. The multiparticle exchange integral of this wave function is found to increase as the number of particles exchanged is increased. The shear modulus for this wave function as well as that of the Laughlin type wave function is calculated and is found to be finite. The effect of exchange is illustrated by considering the fractional quantized Hall effect in narrow channels. It is found that for narrow channels a gap exist in the excitation spectrum for even denominator filling factors. As the channel width is increased, this gap goes to zero and then becomes finite again.

Introduction

The fractional quantized Hall effect (FQHE) is an interesting subject in that it illustrates new physical phenomena that has not been encountered previously. Here we shall discuss (1) The question of positional fluctuation, the shear modulus and Goldstone's theorem. (2) The effect of multiparticle exchange.

I Positional fluctuation

Wigner first pointed out that electrons would crystallize into a solid at low densities, contrary to the behaviour of most other materials.[1] This is due to the predominance of the Coulomb potential energy over kinetic energy when the average distance between electrons is large. In the presence of an extremely strong magnetic field, all the electrons would be coerced into having the same kinetic energy of zero-point motion, and only the Coulomb energy would remain effective in governing their distribution in space. A strong magnetic field could, therefore, be expected to facilitate the formation of a crystal from an electron fluid. However, the electrons move around in circles and the x and the y motion are no longer independent. This ultimately lead to a large fluctuation in the position of the electrons. Here we discuss two ways to understand this.

(A) Correlated gaussian trial wavefunction

We have recently studied[2] the effect of exchange and correlations by exploring a series of possible solid wavefunctions. Solids and charge density waves (CDW) were amongst the first ideas examined as a plausible explanation of the FQHE, but were found wanting in the absence of a sizable commensuration energy, at rational filling factors, in the Hartree-Fock approximation,[3] and their ground-state energy estimates were soon superseded by the Laughlin fluid wavefunction[4]. The next improvement is to introduce correlations. In the harmonic approximation, this yields the magnetophonon solid wavefunction [5]. One can improve upon this by going to self-consistent magnetophonons, but a glaring defect remains in that exchange effects have not yet been included. If we try to increase the correlation between the motion of the electrons in some direction, they would tend to correlate less in the other direction. For example, we have looked at wavefunctions $|\psi\rangle$ of the form $|\psi\rangle = \exp(\sum_q V_q \delta z_q \delta z_{-q})|\psi HF\rangle$. Here $z=x+iy$, δz_q is the Fourier component of the displacements from lattice positions with wave vactor q; and $|\psi HF\rangle$ is the Hartree-Fock ground state constructed from Landau orbitals

$g_{\vec{R}}(\vec{r})$ located at the sites $\vec{R}$ of a triangular lattice[6]. We have assumed that the electrons lie in the lowest Landau level, so that functions of z only, and not of its complex conjugate, can appear in the first factor. $\delta z_q \delta z_{-q} = |x_q|^2 - |y_q|^2$+imag.terms. $|x_q|^2$ and $|y_q|^2$ occurs with opposite signs! Hence no optimum correlations can be introduced in both directions simultaneously. However, exchange effects come in to counteract this and keep the electrons apart from each other. In other words, if we are going to increase correlation of electron motion in one direction, the electrons would then come closer together on the average due to decreased correlation in the other direction, and would be able to lower their energy from exchange. The direct Coulomb energy would suffer only a relatively small increase, since the correlations were first optimized with respect to this aspect. This effect, we think, is the reason why fluctuations are so large here.

We found a trial quasi-solid wavefunction which has a slightly lower energy (-0.412) than that of Laughlin's (-0.410). The probability density of this wavefunction looks like the partition function of a 2-D solid at a finite temperature. It only possesses "algebraic" long range translational order, and therefore, is not subject to the dictates of Goldstone's theorem. In fact, the phonons of this wavefunction exhibit a gap of the same magnitude (0.076) as that of the Laughlin type wavefunction (0.106).[7] Our wavefunction provides an interesting example of evading Goldstone's theorem while retaining a solid-like behavior which is a peculiar feature in 2-D. It brings back all the interesting questions of long-range order vs. rigidity and melting in that case. Our wavefunction is stabilized with respect to the magneto-phonon wave function by benefiting from a much larger exchange while retaining a comparable direct energy. This effect is purely of quantum mechanical origin and, to the best of our knowledge, has not been observed in other quantum solids such as the rare-gas crystals. Presumably the difference lies in the absence of a hard core in the present case. We next discuss the more formal aspects of Goldstone's theorem.

(B) Broken symmetry

A puzzle for the fractional quantized Hall effect concerns the nature of the borken symmetry and its relationship with the excitation spectrum. We recently proposed that this broken symmetry is connected with the centers of gyration of the electrons. In the presence of a magnetic field, the electrons move around in circles, it is more appropriate to discuss the long range order of their centers of gyration (C_x, C_y) rather than that of their positions. The possible expectation value of each center ranges from positive to minus infinity. A broken symmetry will be indicated if some kind of fluctuation about the mean of the center of gyration is finite. For all variational wavefunctions that have been studied, the diagonal fluctuation $<(C_{ix}-<C_{ix}>)(C_{ix}-<C_{ix}>)>$ is infinite; the off-diagonal fluctuation $<(C_{ix}-<C_{ix}>)(C_{iy}-<C_{iy}>)>$, which is intimately related to the pase lag of the x and the y motion, remains finite for all particle i. We call this a partially broken symmetry.

The total center of gyration commutes with the Hamiltonian; hence it is a symmetry of the system. $C_{x(y)}$ is the generator for moving the center of gyration in the y(x) direction. Starting from these and the assumption that $C_{x(y)}$ 0> is a well defined function satisfying the boundary conditions, a Goldstone's theorem that implies other degenerate ground states can be proved. Normally the elementary excitation is related to the broken symmetry. The elementary excitation of mementum q discussed by Laughlin[3], by Girvin et al[3] are intimately related. Both of them can be obtained by moving the center of gyration of an electron by a finite amount perpendicular and proportional to q. The situation is similar for the "Phonon" excitation in the solid wavefunction studied by Chui et al. A gap occurs in the excitation spectrum because of the large diagonal fluctuation.

We first review the quantum mechanics of a single charged particle moving in a plane under an external magnetic field first discussed by Kubo[8] and recently rediscovered by different authors[8]. We have cast the results in a gauge covariant form and picked units such that the effective mass, Planck's constant, the charge, and the cyclotron frequency are all unity. We shall also adopt convention that x+iy is written as z, and p, p_x+ip_y, etc.

The Hamiltonian governing this simple system is $H_0=(v+v_+1)/2$ where v=p-a is the velocity. The general solution to the equation of motion is z=C+Rexp(-it),v=-iRexp(-it) where R is the radius of gyration, and C is the center of gyration in complex notation. Eliminating R and t from the solution, we find C=z-iv which is then regarded as the operator definition for C. It is straightforward to show that C commutes with H_0. $[C_x,C_y]=-i$ and hence, $C_{x(y)}$ is the generator for moving the center of gyration in the y(x) direction. The operator $T(s)=\exp[-i\vec{s}\times\vec{C}]$, is a translation operator associated with a displacement $\vec{s}$.

When restricted to the lowest Landau level, the C and C^+ operators are therefore, none other than multiplying by z, and differentiating the polynomial part by z respectively. The operators z and $2\partial_z$ have been discussed by Laughlin and later formalized by Girvin and Jach to be a pair of adjoint operators in Bargmann space[9].

When the interaction between the particles are introduced, the system is still translationally invariant, and the total center of gyration $C=\sum_j C_j$ still commutes with the total Hamiltonian. Moreover, in this system, since $[C_x,C_y]=-i$, a non-zero complex number, it is impossible to have a ground state that has translational symmetry in both x and y directions!

(C) Goldstone's theorem

Two main ideas of broken symmetry are: (a) it is possible to have a ground state that does not possess some of the continuous symmetries that the governing Hamiltonian exhibits, in which case, the ground state would be degenerate; and (b) as a consequence of the broken symmetry we may have low lying excitations, foreclosing the existence of a gap. Following the notation of Itzykson and Zuber[10], consider the commutator of a symmetry generator G and a summetry breaking variable A, viz. $\delta A(t)=\langle 0|[G(t),A]|0\rangle$. Using the Lehmann-Symanizk representation of δA, one obtains

$$\delta A(t)= \sum_n \langle 0|G|n\rangle\langle n|A|0\rangle \exp(-iE_{n0}t)-\langle 0|A|n\rangle\langle n|G|0\rangle \exp(iE_{n0}t). \qquad (1)$$

If the symmetry is broken $\delta A(t)\neq 0$ and $\langle 0|G|n\rangle\langle |A|0\rangle\neq 0$. On the other hand, since G is a symmetry of the Hamiltonian $d\delta A(t)/dt=0$ and

$$0=-i \sum_n [\langle 0|G|n\rangle\langle |A|0\rangle \exp(-iE_{n0}t)+\langle 0|A|n\rangle\langle n|G|0\rangle \exp(iE_{n0}t)]E_{n0} \qquad (2)$$

Hence there exist a state $|n\rangle$ such that $\langle 0|G|n\rangle\langle n|A|0\rangle\neq 0$ for which $E_{n0}=0$. For example, in the case of broken translational symmetry, G is the total momentum and A is the displacement.[11]

For the present case, we identify G as $\sum_j C_{x,j}(t)$ and A as $C_{y,1}$. Just as p_x generates a translation of x, here $\sum_j C_{x,j}(t)$ generates a translation of $C_{y,1}$. If $\langle (C_{ix}-\langle C_{ix}\rangle)(C_{iy}-\langle C_{iy}\rangle)\rangle = X_{ixy}$ is finite and $C_{x(y)}|0\rangle$ is a well defined function satisfying the boundary conditions, then these arguments can be carried over unchanged if we identify G as $\sum_j C_{x,j}(t)$ and A as $C_{y,1}$. Just as p_x generates a translation of x, here $\sum_j C_{x,j}(t)$ generates a translation of $C_{y,1}$. Thus there exist a state $|n\rangle$

degenerate with the ground state and for which $<0|\sum_j C_{x,j}|n><n|C_{y,1}|0>\neq 0$. Let us note, that, for this argument to go through, it is essential to have $<(C_{ix}-<C_{ix}>)(C_{iy}-<C_{iy}>)>=X_{ixy}$ finite.

The above result is consistent with the apparent degeneracy of the ground state emphasized by Thouless and coworkers[12] and seen in finite cluster calculations[13] in the Landau gauge[14]. Haldane[15] have recently pointed out that this degeneracy is due to the center of mass motion. The same state of affair happens in situations with no external magnetic field, in which case it is also possible to separate out center-of-mass motion from the relative motion of the particles. However, the conslusion of Goldstone's theorem and the form of the excitation spectrum is unaffected by this possibility. Both the separability of the center-of-mass motion from the relative motion, and the conclusion of Goldstone's theorem are consequences of the translational invariance of the Hamiltonian.

II The Shear Modulus

If there is large fluctuation in the position of the electrons and no long range positional order, is the system then a fluid[2,16]? As we know from the study of melting in two dimensions, it is possible for a system to possess a finite shear modulus and yet no long range positional order. For the trial wavefunctions studied by Laughlin[4] and by Chui, Ma and Hakim[2], this turn out to be the case[17]. The physics behind the finite shear modulus seems not to be related to dislocations or grain boundaries, however.

Laughlin proposed that the electrons form a "fluid" with ground state wavefunction ψ_0 the probability density of which is proportional to the partition function of a one component plasma at a sufficiently high temperature that this plasma is in a fluid state. It is then assumed that the quantum electron system behaves like this one component plasma and the shear modulus is zero. Whereas in classical physics the shear modulus is basically determined by the partition function, in quantum physics it is determined not just by the ground state wavefunction. The shear modulus is determined by a response function and depends on the properties of the excited states as well. Using the excited states wavefunction proposed by Girvin MacDonald and Platzman[7] we calculated the shear modulus μ from the zero wave-vector limit of the autocorrelation function $G_{tt}(\vec{q},w=0)$ of the shear strain operator of wave-vector $\vec{q}$, $\sigma_q=\sum_j x_j \exp(iqy_j)/\sqrt{N}$, where we have picked the y axis to lie along $\vec{q}$. If a shear stress S of wave-vector $\vec{q}$ is applied, the perturbation Hamiltonian δH is proportional to $Sq\sigma_q/\rho$

where ρ is the density. The resulting strain is proportional to $q<\sigma_q>$ and the shear modulus, defined as the ratio of applied stress to resulting strain is given by $\mu=\lim_{q\to 0}\rho/q^2 G_{tt}(q,w=0)$, where the correlation function G is given by

$$G_{tt}=\sum_i <0\,|\sigma_{-q}\,|qi><qi\,|\sigma_q\,0>/E_{qi} \qquad (3)$$

E_{qi} is the excitation energy and i denotes the branch of the excitation. We found that μ is finite ($0.0056\ e^2/l^3$)! In a sense this is not totally unexpected. When a shear is applied to the electrons, say in the y direction, they move in the x direction as well because of the magnetic field. Hence a compression is automatically developed and the response is different from what one would expect of shears in a zero magnetic field situation.

Chui, Ma and Hakim[2] have recently studied a class of correlated "solid" wavefunctions that possess an energy comparable to that of Laughlin's. Not too unexpectedly, the shear modulus in this case is also finite ($0.006\ e^2/l^3$). In some systems, an excited state labelled by, say the momentum index q, goes continuously to the ground state as q approaches zero; as a consequence of this continuity the properties of the excited states are reflected in the ground state. Under those circumstances it may be possible to decide on the shear modulus just by looking at the ground state. This is not the case here, however. As a consequence of the large fluctuations of the positions of the electrons, the excited states labelled by q does not go continuously to the ground state as q approaches zero[18]. For ordinary solids, phonon excitation energy is linearly proportional to q with the coefficient a function of the shear modulus. There are other wavefunctions whose excitation energies do not depend on q linearly and yet exhibit a finite shear modulus. For example, the magnetophonons[5], obtained from solving the Hamiltonian that one gets by expanding H about a set of lattice points and treating e^2/r_{ij} in the harmonic approximation, exhibits and excitation spectrum proportional to $q^{3/2}$ and a finite shear modulus. The present calculation provides yet another example for which this is violated.

III Exchange

One of the key questions behind the FQHE concerns the apparent stability of the wave function at odd-denominator filling factors. There is a feeling at present that many-particle exchanges may be the cause of this commensuration condition[19]. Common features can be seen in the Laughlin[4] trial wavefunction, in CDW wavefunctions as is discussed by Tosatti and Parrinello[20] and more recently by Chui Ma and Hakim[2], in the path-integral formulation by Kivelson, Kallen, Avoras and Schrieffer[21] as well as in the eigenstates of small clusters of particles[19]. For exchanges involving particles in a loop on a lattice area $A\alpha_0^2$ (α_0 is the lattice constant, at one-third filled, α_0=4.665) invloving $N_{\|}$ particles, the exchange integral has a phase factor equal to $A\alpha_0^2/2$. $A\alpha_0^2/2$ is mod $N_{\|}\pi/\nu$ where $1/\nu$ is the filling factor. Hence when ν is odd, these exchanges enhance the total energy. We have examined this multiparticle exchange and found[2,22] that the commensuration condition holds only for large loops. For a class of correlated Gaussian wave functions studied by us, the evaluation of this exchange can be cast into languages involving the evaluation of grain boundary energies in two dimensions. Exchanges involving a large number of particles are significant in that, for a loop that consists of two long parallel boundaries of area A involving $N_{\|}$ particles, the "energy" cost in creating the loop is negative. More precisely, the overlap integral is given by

$$<\psi|\psi P>=\exp[N_{\|}c-A^2\alpha_0^2/(2iA-4.8N_{\|})]\simeq\exp(N_{\|}c+1A\alpha_0^2/2+1.2\alpha_0^2N_{\|})$$

where we have expanded the exponent in powers of $N_{\|}/A$. c is a non-negative constant. For a collection of such boundaries this formula is still valid except that the "entropy" factor $N_{\|}c$ needs to be modified since the boundaries now exclude each other. The factor of $iA\alpha_0^2/2$ in the exponent provides for the commensuration condition. The sign of the term multiplying $N_{\|}$ is positive. Hence if $N_{\|}/A$ is small, the overlap integral actually increases as $N_{\|}$ is increased. Of course, eventually, the expansion of $N_{\|}/A$ ceases to be valid and the overlap integral starts decreasing. Hence there is a repulsion between the boundaries of a range proportional to the size of the boundary.

The above explanations made specific assumption about the long range order of the ground state, which may not be crucial to the argument. It is thus important to isolate the essential features that are necessary for the effect. An explanation in the Landau gauge is recently suggested by us[19]. This effect can be best illustrated by calculating the eigenstates for narrow channels where, for even-denominator filling factors, the gap oscillates as a function of the channel width[23].

We find that the Hamiltonian usually consists of a difference of a direct and an exchange contribution. Because of the exchange there is an effective attractive interaction between the electrons with a range of the order of the magnetic length, the electrons thus form clusters of a size of the order of megnetic length. For a narrow channel, a one cluster configuration is stable. As L_x is increased. there will be a competition between the 1 and 2 cluster configurations and eventually the ground state becomes the 2 cluster configuration. For example, for Ns=16, the ground state at small L_x corresponds to the single cluster configuration [5,6,7,8,9,10,11,12] with J=68. For L_x=10.25, the ground state corresponds to the two cluster configuration [3,4,5,9,10,11,12,13] with J=67. The energy difference of these two states as a function of L_x is equal to 0.418, 0.238, -0.027, -0.16, -0.16 for L_x=4.76,6.73,8.23,9.51,10.25 respectively.

For a general filling factor the configurations with different number of clusters possess different total y momenta J and hence are not coupled to each other. This is characteristic of the even-denominator situation. As the transverse dimension is increased, the fluctuation in the number of clusters eventually destroy the gap. There are situations such that configurations with different number of clusters occur with the same J and hence are coupled to each other. A gap is produced due to the hybridization of these local minima. This is characteristic of the bulk limit odd-denominator situation.

This difference between the "odd" and "even" denominators can best be illustrated by finite cluster calculations performed under periodic boundary conditions[25]. For a 12 site situation and L_y= 4.3,7.1,7.7,8.7, the gap at half-filled oscillates and is equal to 0.31,0.,0.05, 0.07 respectively whereas at one-third-filled, the gap is equal to 0.15,0.075,0.08,0.08 respectively and is never close to zero. From calculations using the computer renormalization group technique with systematic truncation that we have described previously[16] we get gaps equal to 0.11 and 0 for L_y=8.7,7.1 for 24 site clusters at half-filled, consistent with the 12 site results. For L_y=8.7 and Ns=12, the static structure factor S(q,0) for the ground state possesses a peak of magnitude 1.3 at $q=2\pi/6\alpha$ where a is the intersite spacing, indicating the formation of the "cluster" distribution with three contigous sites occupied and then three empty sites next to it. For large q,S(q,0) approaches 1. For other values of J, we do not see any peak in the structure factor. The magnitude of this peak for the 12 site case is similar to that from finite cluster calculations at one-third filled.

References

1) For a recent discussion, see L. Bonsall and A. A. Maradudin, Phys. Rev. B 15 1959 (1977).

2) S. T. Chui, K. B. Ma and T. Hakim, Phys. Rev. B 33, 7110 (1986)

3) K. Maki and X. Zotos, Phys. Rev. B, 28 4349 (1983)

4) R. B. Laughlin, Phys. Rev. Lett 50

5) A. V. Chaplik Soviet Phys. JETP 35 395 (1972)

6) The unit length and energy are the magnetic length and e^2/(magnetic length) in this paper and will not be explicitly specified from now on. The $g_{\vec{R}}(\vec{r})$ on different $\vec{R}$s are not orthogonal to each other. However, if the $\vec{R}$s are far apart, the overlap will be small.

7) There are many estimates. See, for example, A. H. MacDonald, P. M. Platzman Phys. Rev. Lett. 54 581 (1985). They obtain a gap of 0.106. See also, R. Morf, B. I. Halperin Phys. Rev. B33, 2221, (1986).

8) See, for example, R. Kubo, S. J. Miyake, and N. Hashitsume Solid State Physics Vol. 17 ed. by F. Seitz and D. Turnbull, Academic Press (1965). The relevance of this to the present problem seems to have been first pointed out by V. Emery, unpublished. This has also been discussed recently by various authors. See, for example, A. H. MacDonald, Phys. Rev. B30,3550, (1984).

9) S. M. Girvin and T. Jach Phys. Rev. B29 5617 (1984)

10) C. Itzykson and J. B. Zuber "Wuantum Field Theory" page 520. Published by McGraw Hill Co. (1980) See also, D. Forster, "Hydrodynamic Fluctuations, Broken Symmetry, and Correlation Functions" page 166, published by Benjamin Co. (1975)

11) When there is translation symmetry, the wavefunctions $|G\ 0\rangle$ and $A\ |0\rangle$ are not normalizable and the routine proof of Goldstone's theorem breaks down. However, for both the fluid and solid trial wavefunctions $\langle GA\rangle$ and $\langle AG\rangle$ are finite and non-zero; furthermore, the wavefunctions $G|0\rangle$ and $A\ |0\rangle$ are normalizable. We have hence assume that this is the case in general.

12) R. Tao and D. J. Thouless Phys. Rev. B 28, 1142 (1983), Q. Niu, D. J. Thouless, and U. S. Wu Phys. Rev. B 31 3372 (1985), R. Tao and D. Haldane unpublished.

13)W. P. Su Phys. Rev. B30, 1069 (1984).

14) In that gauge, the Landau orbitals are eigenstates of C_x. Suppose $|1\rangle$ and $|2\rangle$ are two of these states with $\sum_i C_{ix}=c_2,c_2$ respectively. Then $\alpha|1\rangle+b|2\rangle$ and $b|1\rangle-\alpha|2\rangle$ are also energy eigenstates and they are coupled by the operators C_{ix}. Hence they provide for an example of our result. For b or a = 0, the wavefunction $\sum_i C_{y,1}|0\rangle$ is

not normalizable and our argument does not apply.

15) D. Haldane, Phys. Rev. Lett, 55 2095 (1985)

16) S. T. Chui, Phys. Rev. B32, 1436 (1985)

17) S. T. Chui, Phys, Rev, B34, 1409 (1986)

18) S. T. Chui and K. B. Ma unpublished

19) S. T. Chui, Phys. Rev. B32, 8438, (1985)

20) E. Tosatti, M. Parrinello, Lett. nuovo Cimento 36, 289 (1983).

21) S. Kivelson, C. Kallen, D. P. Arovas, J. R. Schrieffer, Phys. Rev. Lett. 58, 873 (1986).

22) S. T. Chui, Phys Rev. B34, 1130 (1986)

23) S. T. Chui, Phys. Rev. Lett 56, (1986) and unpublished

24) For hard wall boundary conditions, it is difficult to predict when the Js for the different cluster configurations will be the same. This is why we have picked periodic b.c.

LINEAR RESPONSE AND THE QUANTIZATION OF THE HALL CONDUCTIVITY

Alpo Kallio, Jari Kinaret and Mauri Puoskari

University of Oulu
Department of Theoretical Physics
Linnanmaa, SF-90570 Oulu, Finland

ABSTRACT

It is shown that the specific behaviour of the density-density response function χ_{nn} required for the quantization of the Hall conductance σ_{xy} of two-dimensional electron system in a strong magnetic field proposed by McMullen [5] can be derived from a variational theory of quantum fluids. The behaviour $\chi_{nn} \sim \nu q^2$ for small wave vector q and for the Hall quantum number ν labelling the plateau, is shown to be a direct consequence of the long range behaviour of the Laughlin's variational wave function for the ground state with filling factors $\nu = 1/m$. It is shown that this implies the behaviour $S(q) \sim \frac{1}{2}q^2$ for the liquid structure factor. Since the small q behaviour of $\chi_{nn}(q) \sim \nu q^2$ is direct consequence of the conservation laws in the 2-D geometry, the criterium $S(q) \sim \frac{1}{2}q^2$ for the many-body wave function at any Hall plateau ν should be valid. The consequences of this to magneto-roton excitations and the sum rules is discussed.

For sufficiently low temperatures the Hall conductance of interacting two-dimensional electrons in a strong magnetic field B shows plateaus at accurately quantized values [1,2]:

$$|\sigma_{xy}| = \nu e^2/h \tag{1}$$

while σ_{xx} vanishes [3]. The Hall quantum numbers ν can be identified with the filling factors of Landau levels, $\nu = 2\pi a^2 n$ where $a = \sqrt{\hbar c/eB}$ and $\omega_c = eB/m^*c$ and n is the uniform density of the electron gas confined to a plane. Integral values of ν come from complete filling of Landau levels and fractional values from fractionally filled Landau levels. The ground state for the Landau levels with the filling factor of the $\nu = 1/m$ where m is an odd integer can be described quite well by Laughlin's wave functions [4]

$$\psi_L = \prod_{i<j} (z_i - z_j)^m \exp\,[-\sum_i |z_i|^2/4],\ m = 3,5,\ldots \tag{2}$$

which leads to a density $n = 1/2\pi a^2 m$.

Recently McMullen [5] has shown that on an ideal Hall plateau the quantized values of the Hall conductance can be expressed in terms of the density-density response function for the electron liquid

$$\sigma_{xy} = -\frac{e^2\omega_c}{2} \lim_{q\to 0} \frac{\partial^2\chi_{nn}(q)/\partial q^2}{1-V_0(q)\chi_{nn}(q)} \tag{3}$$

where V_0 is the Coulomb potential. McMullen then conjectures that the necessary and sufficient conditions for the quantization of σ_{xy} and for $\sigma_{xx} = 0$ are [5]

$$\text{(i)} \lim_{q\to 0} \frac{\chi_{nn}(q)}{q} = 0 \tag{4}$$

$$\text{(ii)} \lim_{q\to 0} \frac{\chi_{nn}(q)}{q^2} = \frac{\nu}{2\pi\hbar\omega_c} \tag{5}$$

The first condition (i) is required for $\sigma_{xx} = 0$ whereas (ii) then gives the quantization rule (1) with the help of eq. (3). In this brief report we show that the features (i) and (ii) follow directly from the Laughlin's wave function (2) by using a general variational linear response theory of quantum fluids develooed recently by us [6,7].

We start from a variational many-body wave function for a single impurity with the field $\vec{B}$ corresponding to the midpoint of a plateau

$$\psi = \prod_{i<j}^{N} R(r_{iA})\, \psi_L \tag{6}$$

Here ψ_L is the host many-body wave function (in this case e.g. the Laughlin wave function) and $R(r_{iA})$ is the impurity-host correlation factor. In order to calculate the static density-density response function

$$\chi_{nn}(q) = \delta n(q)/V_{ext}(q) \tag{7}$$

where $V_{ext}(q)$ and $\delta n(q)$ are the Fourier-transform of the external potential and the density fluctuation, respectively, we will apply the hypernetted chain (HNC) formalism for a mixture [6-9]. In this formalism one can create an external field by taking the limit of zero concentration N_2=1 and infinite mass $m_2\to\infty$ for the second component. Now the wave function of the binary mixture system becomes formally equivalent with the trial function (6) while the interaction V_{12} becomes formally the external field V_{ext} and the partial radial distribution function g_{12} gives the density fluctuation created by the external field in the host system

$$n_1 g_{12}(r) = n\, g_{12}(r) = n(r) \equiv n+\delta(r) \tag{8}$$

The total energy for the Hamiltonian $H = H_0 + \sum_{i=1}^{N} V_{ext}(r)$ is then given by [7]

$$E = N_1\, \varepsilon_0(n_1) + \varepsilon_c \tag{9}$$

where ε_0 is the energy per particle of the host system at the density $n_1 = n \equiv 1/2\pi a^2 m$ described by the Hamiltonian

$$H_0 = \frac{1}{2m^*} \sum_{i=1}^{N} [\vec{p}_i - \frac{e}{c}\vec{A}_i]^2 + \sum_{i<j} \frac{e^2}{r_{ij}} \tag{10}$$

The correlation energy between the host system and the impurity is then [7]

$$\varepsilon_c = \frac{h^2}{2\mu}\int [\nabla \sqrt{n(r)}]^2 d^2r + \int n(r)V_{ext}(r)d^2r$$
$$+ \frac{1}{2}\frac{1}{(2\pi)^2 n}\int x(k)\ \delta^2(k)\ d^2k \qquad (11)$$

Here μ is the reduced mass $\mu = m_1m_2/(m_1+m_2)$, $\delta(k)$ is the Fourier-transform of the density fluctuation

$$\delta(k) = \int e^{i\vec{k}\cdot\vec{r}}\delta(r)d^2r = n\int e^{i\vec{k}\cdot\vec{r}}(g_{12}(r)-1)d^2r \qquad (12)$$

and S(k) is the structure factor of the host system

$$S(k) = 1+n\int e^{i\vec{k}\cdot\vec{r}}(g_{11}(r)-1)d^2r \qquad (13)$$

The auxiliary function x (in ref. 7 it is called χ) is given by [7]

$$x(k) = \frac{1}{(2\pi)n}\int e^{i\vec{k}\cdot\vec{r}}\ x(r)\ d^2r$$
$$= \frac{\hbar^2}{4m^*}k^2\left[\frac{S(k)-1}{S^2(k)}\right]\cdot\left[\frac{m^*}{\mu}S(k)+1\right] \qquad (14)$$

Hence we obtain the Euler-Lagrange equation for the density fluctuation directly by varying ε_c with respect to $\sqrt{n}$ [6,7]

$$-\frac{\hbar^2}{2\mu}\nabla^2\sqrt{n} + [V_{ext}(r) + W(r)]\sqrt{n} = 0 \qquad (15)$$

$$W(r) = \frac{1}{(2\pi)^2 n}\int \delta(k)x(k)\ e^{i\vec{k}\cdot\vec{r}}\ d^2k \qquad (16)$$

The linear response function can be solved in a closed form by going to the uniform limit

$$\sqrt{n(r)} = \sqrt{n}\ (1+\frac{1}{2}\frac{\delta(r)}{n} + ..) \qquad (17)$$

and to the massive impurity limit $\mu \approx m^*$. The result is

$$\delta(k) = -\frac{nV_{ext}(k)}{\frac{\hbar^2k^2}{4m^*}\left[1 - \frac{(S^2(n)-1)}{S^2(k)}\right]} \qquad (18)$$

Hence the linear response function is given by [6]

$$\chi_{nn}(k) = -\frac{4m^*}{\hbar^2}\ n\ \frac{S^2(k)}{k^2} \qquad (19)$$

For liquid ^{4}He the formula (19) turns out to be exact in the limit $k \approx 0$.

To use it in our present problem we need the small wave vector behaviour of the structure factor. The radial distribution function for the Laughlin wave function can be obtained via the hypernetted chain (HNC) equation [4,10]

$$g(r) = e^{\ln r^{2m}+N(r)+B(r)} \tag{20}$$

where the nodal function N(r) is given by the Orstein-Zernicke relation

$$N(k) = \frac{1}{(2\pi)^2 n} \int e^{i\vec{k}\cdot\vec{r}}\, N(r)\, d^2r = \frac{[S(k)-1]^2}{S(k)} \tag{21}$$

and B(r) is the bridge function, which takes account the elementary diagrams.

The bridge function B(r) is a short-range function [10] (especially at m = 1 it has a gaussian behaviour). Since the radial distribution function has to remain finite for large r, the singular behaviour of the factor $\ln r^{2m}$ must be cancelled by a corresponding term in N(r). This requires that the term $S^{-1}(k)$ in eq. (21) cancels the Fourier-transform of $\ln r^{2m}$, which is $-4\pi mn/k^2$. Hence the structure factor behaves as

$$S \approx \frac{1}{4\pi mn} k^2 = \frac{1}{2} k^2 \tag{22}$$

This result is obviously exact in the limit when k goes to zero. Once the quadratic behaviour of S eq. (22) is established the two conjectures of McMullen, eqs. (4) and (5) follow then from the linear response formula (19) since for small wave vectors

$$\chi_{nn}(k) \approx \frac{1}{2\pi\hbar\omega_c} \frac{k^2}{m} \tag{23}$$

Hence we have shown that the quantization of the Hall conductance, with filling factors of the form $\nu = 1/m$ at least, follows directly from the Laughlin's ground state wave functions. MacDonald et al. [13] have suggested the pair-correlation function for the filling factor $\nu = 2/5$

$$g_{2/5} \approx \frac{3}{4} g_{1/3} + \frac{1}{4} g_1 \tag{24}$$

In this case the small-k behaviour of S(k) agrees with eq. (23), since $S_1(k) \approx \frac{1}{2}k^2$. Finally since both formulas (3) and (19) are valid also for more general fractional states $\nu = p/m$ we can make prediction for the small q behaviour of their liquid structure factor

$$S(q) \cong \frac{q^2}{4\pi\nu^{-1}\rho_\nu} = \frac{1}{2} q^2 \tag{25}$$

This should also be valid for normal QHE when ν is an integer. Clearly general proof here would require detailed wave function to be known.

One can show that the behaviour (22) for the structure factor remains unchanged also when one tries to improve the Laughlin variational wave function by multiplying it with a real two-body factor

$$\psi = \prod_{i<j} e^{U(rij)} \psi_L \tag{26}$$

When we use the HNC equation to relate the radial distribution function $g_2(r_{12})$ to the wave function (26) and optimise the ground state energy, $\delta E/\delta g(r) = 0$, we obtain an Euler-Lagrange equation for $\sqrt{g}$ [11]

$$\left[-\frac{\hbar^2}{m^*}\nabla^2 + \frac{e^2}{r} + W_0(r) + W_m(r) + W_3(r)\right]\sqrt{g}(r) = 0 \tag{27}$$

where the induced potentials W are

$$W_0(r) = -\frac{\hbar^2}{4m^*}\frac{1}{(2\pi)^2 n}\int e^{i\vec{k}\cdot\vec{r}}\, k^2\,\frac{(S-1)^2}{S^2}(2S+1)\,d^2k \tag{28a}$$

$$W_m(r) = \frac{\hbar^2}{m^*}\left[\frac{m^2}{r^2} + \frac{m}{2r}\frac{dN}{dr}\right] - \frac{\hbar^2}{2m^*}\frac{m}{(2\pi)^2 n}\int e^{i\vec{k}\cdot\vec{r}}\,\xi_k\,\frac{S^2-1}{S^2}\,d^2k \tag{28b}$$

$$W_3(r) = \frac{\hbar^2}{2m^*}\frac{m}{(2\pi)^2 n}\int e^{i\vec{k}\cdot\vec{r}}\left[\frac{S^2-1}{S^2} - \frac{S^2-1}{S}\right]\xi_k\,d^2k + \Delta W_3 \tag{28c}$$

Here

$$\xi_k = n\int e^{i\vec{k}\cdot\vec{r}}\left(\frac{1}{r}\frac{dg}{dr}\right)d^2r$$

and ΔW_3 indicates additional three-body terms which are irrelevant in our discussion here. Now we see that the term $(S^2-1)\xi_k/S^2 \sim k^{-4}$ in W_m is cancelled by the first term in W_3 since $\xi(k=0)=2\pi n$. In order to obtain a self-consistant solution for $\sqrt{g}$ also the term $k^2/S^2 \sim k^{-2}$ in W_0 has to be cancelled. This is done by the second term in W_3 when the small wave vector behaviour of S(k) obeys eq. (22). Hence we can argue that eq. (25) is true quite generally even for the exact wave function, not only for the Laughlin's wave function. The solution of eq. (27) would enable one to calculate the energy shift due to the effect of higher Landau levels not contained in the Laughlin wave function.

Finally the real impurity problem in FQHE can also be treated with the linear response theory using the idea of refs. 6-7. All the derivations of ref. 5 remain unchanged if also the external field due to the impurities is added to the δv_{ext}. The theory breaks down when the impurity concentration exceeds the limit beyond which the linear response theory no longer works.

In the derivation above we did not make any assumption about the excitation spectrum of the system. Using the projection technique onto the lowest Landau states Girvin, MacDonald and Platzman have recently [14] calculated the magneto-roton spectrum for FQHE which reduces to magneto-plasmon spectrum for integer Hall effect $\nu=1$. Their derivation is based on the analogy between QHE and the Feynman phonon-roton spectrum in liquid ^{4}He. For $\nu < 1$ their excited state wave function is

$$\psi_{\vec{k}} = \bar{\rho}_k\,\psi_\nu \tag{29}$$

where $\bar{\rho}_k$ is the projected density operator

$$\bar{\rho}_k = \sum_{j<\nu}\exp\left[-ik\frac{\partial}{\partial z_j}\right]\exp\left[-\frac{ik^*}{2}z_j\right] \tag{30}$$

where $k=k_x+ik_y$ is the wave vector in complex $\vec{k}$-plane and ψ_ν is the ground state wave function made solely of single-particle orbitals belonging to the lowest Landau level. The resulting excitation spectrum is of the form [14]

$$\varepsilon(k) \equiv \Delta(k) = \frac{\bar{f}(k)}{\bar{S}(k)} \tag{31}$$

where $\bar{f} = N^{-1} \langle\psi_\nu|\bar{\rho}_k^\dagger [\bar{H},\bar{\rho}_k]|\psi_\nu\rangle$

is the oscillator strength and $\bar{S}(k) = S_\nu(k) - (1-e^{-\frac{1}{2}k^2}) = S_\nu(k)-S_1(k)$ is the projected liquid structure function, $S_\nu(k)$ being the liquid structure function for ψ_ν. For $\nu=1$ one obtains the Feynman spectrum,

$$\Delta(k) = \frac{\hbar^2}{2m}\frac{k^2}{S_1(k)} = \varepsilon_F(k) \tag{32}$$

Unlike in liquid ^{4}He one obtains here a gap at $k=0$, $\Delta=\hbar\omega$. In this magnetoplasmon mode the wave function (29) is a linear combination of particle-hole excitations and the mode corresponds to the dipole states in nuclei. These high lying states have been seen experimentally in nuclei.

For the fractional states $\nu<1$, the states ψ_ν and $\bar{\rho}_k\psi_\nu$ are degenerate without Coulomb force and disorder. The Coulomb force separates the magneto-roton excitations from the ground state and one obtains a gap at $k=0$ of the order $e^2/\varepsilon a$. The dispersion curve has a "roton"-minimum Δ at finite $k=k_R$. This is interpreted in Ref. 14 to be precursor for an instability towards a Wigner crystalline state where k_R determines the respective Wigner crystal size. For fractions $\nu=1/m$ with increasing m the value of the gap goes down but never collapses to zero without the disorder whose effect is treated by the same authors [15] and independently by Gold [16] with the memory function formalism of Götze and Wölfle [17]. The disorder lowers the roton minimum. Also a percolation model has been proposed [18].

No calculation of the excitation spectrum exists for the hierarchial states based on the Laughlin state $\nu=1/m$. However in the light of discussions above their static structure factor behaves like $S(k) \sim \frac{1}{2}k^2$. Hence for them too $\bar{S}(k) \sim k^4$ and $\bar{f}(k) \sim k^4$. One then obtains quantitatively the same excitation curve for the hierarchial states $\nu=p/q$. Presumably the roton gap for these states is smaller and hence more susceptible for a collapse due to the disorder. As for the sum rules, the single-mode approximation (SMA) of Ref. 14 gives a dynamic structure factor

$$S(k,\omega) = \bar{S}(k)\delta(\omega-\Delta(k)) \tag{33}$$

which does not satisfy the sum rules. Since in the case of FQHE both modes can exist one may take them both infinitely sharp and approximate the energy of the plasmon mode by Feynman spectrum (32) one obtains the following two mode approximation

$$S(\vec{k},\omega) = \bar{S}(k)\ \delta(\omega-\Delta) + S_1(k)\ \delta(\omega-\omega_F). \tag{34}$$

This gives for the sum rules the following

$$S(k) = \int_0^\infty d\omega\ S(k,\omega) = \bar{S}(k) + S_1(k) \tag{35}$$

$$\int_0^\infty \omega\ d\omega\ S(k,\omega) = \frac{k^2}{2m} + \bar{S}(k)\ \Delta(k) \tag{36}$$

$$\chi(k) = -2n \int_0^\infty \frac{d\omega}{\omega} S(k,\omega) = -2n \frac{\bar{S}(k)}{\Delta(k)} - \frac{4mn}{k^2} S_1^2(k)$$

The f - sum rule of the second line is satisfied up to terms of order k^4 and the third line produces a linear response function $\chi(k)$ which for small k is in agreement with our result in eq. (19) due to the second term. At finite k, the response function is determined by the first term which produces the main peak of $\chi(k)$.

As a conclusion one finds that quantization of conductivity at the midpoint of a plateau is determined by the small-k behaviour of the linear response function $\chi(k)$. This in turn requires the existence of the plasmon mode also for the FQHE. We therefore believe that the existence of plateaus is connected with the plasmon mode in both cases of FQHE and QHE. This remains a conjecture until a proper disorder theory is presented which then would set limit to the accuracy of the plateaus for both σ_{xx} σ_{xy} and their temperature dependence. This may or may not be the memory function formalism mentioned above. Finally there seems to be no need for fractionally charged quasiparticles nor do we need any degeneracy for the Laughlin states at $\nu=1/m$, however any candidate for the ground state wave function should produce a liquid structure factor with small-k behaviour $S(k) \sim \frac{1}{2}k^2$. Therefore the meaning of ν in the quantization formula (1) is the filling factor of the lowest Landau state without any mystical fractional charges connected to it.

REFERENCES

1. K. von Klitzing, G. Dorda and M. Pepper, Phys. Rev. Lett. 45, 494 (1980).
2. H.L. Störmer, A. Chang, D.C. Tsui, J.C.M. Hwang, A.C. Gossard and W. Wiegmann, Phys. Rev. Lett. 50, 1953 (1983); D.C. Tsui, H.L. Störmer, A.C. Gossard, Phys. Rev. Lett. 48, 1559 (1982).
3. M.E. Cage, B.F. Field, R.F. Dziuba, S.M. Girvin, A.C. Gossard and D.C. Tsui, Phys. Rev. B30, 2286 (1984).
4. R.B. Laughlin, Phys. Rev. Lett. 50, 1395 (1983).
5. T. McMullen, Phys. Rev. B32, 1415 (1985).
6. A. Kallio, M. Puoskari, L. Lantto, P. Pietiläinen and V. Halonen, Lecture Notes in Physics, vol. 198, p. 210, edited by H. Kümmel and M.L. Ristig (Springer-Verlag, Berlin, 1984).
7. P. Pietiläinen and A. Kallio, Phys. Rev. B27, 224 (1983).
8. J.C. Owen, Phys. Rev. Lett. 47, 5861 (1981).
9. T. Chakraborty, Phys. Rev. B25, 3177 (1982).
10. J.M. Caillol, D. Levesque, J.J. Weis and J.P. Hansen, J. Stat. Phys. 28, 235 (1982).
11. A. Kallio, J. Kinaret, P. Pollari and M. Puoskari, Proceedings of IX International Workshop on Condensed Matter Theories, San Francisco, 1985.
12. Qian Niu, D.J. Thouless and Yong-Ski Wu, Phys. Rev. B31, 3372 (1981).
13. A.H. MacDonald, G.C. Aers and M.W.C. Dharma-wardana, Phys. Rev. B31, 5529 (1985).
14. S.M. Girvin, A.H. MacDonald and P.M. Platzman, Phys Rev. Lett. 54, 581 (1985); Phys. Rev. B33, 2481 (1986).
15. P.M. Platzman, S.M. Girvin and A.H. MacDonald, Phys. Rev. B32, 8458 (1985).
16. A. Gold, Europhys. Lett. 1, 241 (1986).
17. W. Götze and P. Wölfle, Phys. Rev. B6, 1226 (1972).
18. R.F. Kazarinov and Serge Luryi, Phys. Rev. B25, 7626 (1982).

WHAT UNDERLIES THE ANDERSON HAMILTONIAN?

D. D. Koelling*

Materials Science Division
Argonne National Laboratory
9700 South Cass Avenue
Argonne, Illinois 60439

ABSTRACT

The Anderson Hamiltonian characterizes a model in which local orbitals interact with each other through Coulomb interactions and with a bath of itinerant states through hybridization. It has been utilized with considerable success to discuss phenomena ranging from the Kondo problem to the occurrence of two peaks in the photoemission spectra of Ce compounds. Normally, it is studied by examining the content of its solutions. Of comparable significance, however, is the determination of the various parameters entering the model. In trying to find a prescription for the parameters of the model, one simultaneously develops some understanding of the states involved and studies whether the model incorporates enough of the essential physics. One useful tactic is to compare this model with other models and ask how different effects are accommodated. Here, I will compare results from the model with atomic, band, and modified band calculations for f-electron materials. In light of the complexity of the problems posed, it should be appreciated that I will concentrate more on posing questions than on providing answers.

My interest in the actual content of the Anderson Hamiltonian [1] was kindled by its utilization to interpret the photoemission spectra of Ce compounds. Beyond my natural disinclination to accept claims that this (or any) model was the answer, I was quite concerned that it did not appear to incorporate effects that seemed to be significant. Specifically, it is difficult to see the relation of this model to the multiple-screening-mode model [2,3] which is a natural interpretation of our band calculations. Yet the Anderson model can represent the experimental situation reasonably well. One could always casually dismiss its success by saying that it is mere curve fitting, with model parameters that have been assigned phenomenologically [4]. But that would be a bit too cavalier in light of the added fact that f-counts derived from it correlate at least qualitatively with those found in band calculations. Thus arises the question of just what can be

*Work supported by the U. S. Department of Energy, BES-Materials Sciences, under contract #W-31-109-ENG-38.

represented in a model Hamiltonian of the Anderson form. Such an examination needs to be done very carefully if it is to be useful. Consider a different example: that of the spin Hamiltonian. If one has not been properly introduced to it, one would naturally assume that the spin Hamiltonian is nothing more than a phenomenological model. Yet it can be related to utilization of the spin-coordinates to describe what is occurring in the associated underlying spatial wave function. (Normal usage is, of course, as a phenomenological model.) Let us examine the Anderson Hamiltonian in a similar fashion. To do this, we first repesent the computations as a surrogate for the experiment and then consider how this model Hamiltonian might represent the effects observed. A proviso is in order: it is not so much my intent to present solutions as to pose questions.

Selected Density Functional Theory

Density functional theory is based on the observation that the **ground state properties** (and the total energy in particular) are functionals of the density ($E=E[\rho]$). There is, of course, the problem that one does not know these functionals and they must be approximated, but when the intra-atomic repulsion U is truly the dominant factor, this will not be critical. This direct Coulomb interaction (which is explicitly included) will far overshadow any uncertainty in our knowledge of the exchange-correlation functionals. Writing the density as a sum of single particle amplitudes,

$$\rho = \sum \psi_k^* \psi_k$$

the ψ_k must satisfy single-particle-like equations of the form

$$(t + V)\, \psi_k = e_k \psi_k .$$

Here, t is a single-particle kinetic-energy operator and V incorporates the self-consistent Hartree field plus the functional derivative of the exchange-correlation functional (usually evaluated in the local density approximation -- but that is not of concern here). The total energy is not a simple sum of the eigenvalues e_k but must include the effects of Coulomb double counting and the involved functional form of the exchange correlation functional. The eigenvalue e_k is a functional of all the other occupied orbitals and will reflect changes occurring in those orbitals. A particularly useful observation that will be exploited extensively is the relation [5]

$$e_k = dE/dn_k ,$$

which states that the eigenvalue is the local derivative of the energy. (This energy is the analytic continuation of the total energy, which is strictly valid only at integral occupation values rather than the more physical energy of a statistical ensemble. Consequently, questions of discontinuous behaviour do not enter this discussion.) To study the photoemission process, we must break slightly with the formal theory, which limits itself to the ground state or to ensembles of states. Making that leap of faith, we nonetheless will be accounting for the major Coulomb interactions, including those that are responsible both for screening and for the intra-atomic repulsion U. As the occupation number n_k is changed, the eigenvalue e_k will change to reflect the changes occurring in the entire (model) many-body system. In essence, this theory is mapping the energetics of a many-body system onto a set of one-particle eigenvalues. A natural consequence is that as one excites the system, these eigenvalues will **change**! If the excitation is spatially delocalized, the change will be negligible.

When it is not, one can integrate on the occupation number from one down to zero to obtain the removal energy. However, as the eigenvalue is very nearly a linear function of the occupation number, one of several simple things can be done: one can calculate at the half integral value (which is the standard transition state prescription) or, as will be useful here, one can use the average of the eigenvalues obtained with the orbital fully occupied and completely empty.

An Atomic Case Study

To make this a bit more concrete, consider the results of a series calculations on cerium and uranium atoms involving transitions between f- and d-orbitals. The total energy can be written

$$E_T = e_f^o\, n_f + U_f n_f^2/2 + e_d^o\, n_d + U_d n_d^2/2 + X_{df} n_d n_f + E_o(n_d, n_f),$$

where E_o represents the response of the s electrons and the core to changes in the d and f occupation. With the assumption that this dependency can also be adequately expanded to quadratic order, we may redefine the parameters appearing in the remainder of the expression to include these additional terms (relaxation effects) and treat E_{oo} as constant. The eigenvalues are then the derivative of this expression (I -- as well as many others -- have checked that this is really so [6]):

$$e_f = e_f^o + U_f n_f + X_{df}\, n_d$$

$$e_d = e_d^o + U_d n_d + X_{df} n_f$$

For the uranium atom, we can make the approximation $U_d \sim 0$ and consider $(f^3 d^2 s^1) \rightarrow (f^{2.5} d^{2.5} s^1)$ to find $U_f = 3.4$ eV and $X_{df} = 0.7$ eV. This leaves about 1-1.5 eV for solid state screening. With the tripositive ion, one finds $U_f \sim 25.6$ eV and $X_{df} \sim 7.3$ eV more nearly akin to the unscreened result. In the case of cerium, we do not make the $U_d \sim 0$ approximation and we find $U_f \sim 12$ eV, $U_d \sim 6$ eV, and $X_{df} \sim 7$ eV. These parameterizations represent the behaviour of the eigenvalues reasonably well. Note now that if we fix $n_d = 1$ and look at the dependence of e_f on n_f,

$$e_f(1) = e_f^o + U_f + X_{df},$$

$$e_f(0) = e_f^o + X_{df},$$

and the removal energy is

$$\Delta_f = e_f^o + U_f/2 + X_{df}.$$

Because $U_f > 0$, Δ_f will appear at greater binding energy than the eigenvalue $e_f(1)$. Again, such careful consideration is required for local orbitals as a price for mapping many-body behaviour onto a set of single-particle-like quantities. The band supercell calculations [7-9] to be discussed presently indicate that, for the Ce compounds, an f to d transition is more appropriate:

$$\Delta_{fd} = e_f^o - e_d^o \quad U_f/2 - 3U_d/2 + X_{df},$$

which differs from the ground state eigenvalue difference by

$$\delta_{relax.} = U_f/2 - U_d/2 + X_{df}$$

for $(f^1d^1s^2)\rightarrow(f^0d^2s^2)$. This is actually not the best case because this is a very large energy. It is reduced when the atom is compressed into a solid and the s state is converted to a d state. Nonetheless, it does demonstrate the difficulty of reaching the f^0 state -- which is a crucial point in mixed-valence discussions.

Multiple Screening Mode Model in a Solid

To discuss photoemission from the f-states on a Ce atom in a solid, the model we have applied is based on the following assertions:

1. Transitions from the f-orbitals are local excitations -- no matter whether the ground state wave function is or is not local. This requires that we deal with a special excited (i.e., impurity) atom. This is done by forming a supercell consisting of the excited atom surrounding "host" atoms, repeated as a periodic array.

2. The **Total Energy** difference must be determined to find the excitation energy. This is, after all, a many-body problem. Although there are some formal fine points to worry about in what we do, we at least include the major direct Coulomb interactions -- including the U term of the Anderson Hamiltonian! We usually -- but not always -- utilize eigenvalue arguments like the ones used above to get at this total energy difference.

3. The transitions occur into channels which are the different possible screening states -- i.e., excited states of the many-body problem. These must be postulated. Those that we consider are as follows:

a. A fully (f-) screened peak which will occur very near the eigenvalue in the ground state potential because the potential is changed very little changed as a consequence of the similarity between the screening charge and the charge of the removed electron. (We believe the ESCA-type transitions and band f- transitions to be very improbable.) Because the ground state eigenvalues occur at or very near the Fermi energy, these transitions will occur very near but not necessarily exactly there as well. The fact that the screening arises from the remaining f-orbitals can provide an explanation of the observation of the spin-orbit splitting in that peak [10].

b. A poorly screened state which is created by forcing an f-hole and then letting the system relax. The hole is screened by a Ce d-electron and the binding energy appears about 2.5 eV below the fully screened peak.

c. An "unscreened" peak (perhaps seen in CeN) which could appear 5-7.5 eV below the fully screened peak. This energy is obtained in the renormalized atom calculation [11] (as the eigenvalue), partly as a consequence of the use of Hartree-Fock formalism in the renormalized atom calculations and partly as a consequence of the local unhybridized form assumed for the f-orbital.

Several results are obtained from this model. The peak separation between the fully and the partially screened peak agrees very well with those observed for Ce materials; this finding gives some credibility to the results. The poorly screened peak appears to be very well described as a d-screened f-hole. One thus sees approximately an atomic transition, which explains the fact that the peak separation is roughly constant (it does increase slightly as one proceeds down the isoelectronic pnictide series). The f-eigenvalues calculated in the f^0 state are very flat bands, indicating that the hybridization has been "turned off" in that state. The one exception in the series is CeN, where the strong admixture into the nitrogen p-bands indicates that the state is actually unstable. This is experimentally manifested by the absence of a second peak in CeN. Apparently, a similar situation occurs in the uranium materials. One important additional observation needs to be made concerning the ground state alone: the f-density always consists of two spin-orbit split atomic-like levels just above the Fermi energy, with a tail, pulled below that energy by the hybridization, containing roughly one f-electron for the cerium materials and roughly 2.5 electrons for the uranium materials. This is a result of the calculation -- not an input -- and will need to be rationalized with the Anderson model parameter E_f.

Relation to Anderson Model

Now let us examine how these features might appear in the Anderson model. Anderson [1] originally proposed this model Hamiltonian to study local moment formation of transition metal impurities in metals. The essence of the problem, which this rather schematic model is intended to capture twofold is: (1) The almost core-like d- or f-orbitals are strongly correlated (H_{corr}) by their strong intra-atomic Coulomb repulsion -- with the small radial extent greatly enhancing the size of $\langle 1/|r_1 - r_2|\rangle$. (2) Because these orbitals are not for a free atom in a vacuum, they must couple with the surrounding sea of electrons (H_{cf}). The model is better tuned to the Kondo problem than the photoemission problem because the former does not involve a net change in the number of local orbitals. The form of the Hamiltonian is

$$H = H_c + H_f + H_{cf} + H_{corr}.$$

H_c represents an unperturbed conduction electron sea:

$$H_c = \sum e_K \, n_{k\sigma}.$$

Because the focus is on the local orbitals, this sea of conduction electrons is viewed as a rather structureless bath which can be treated as nearly free electrons -- Anderson suggested Orthogonalized Plane Waves (OPWs). H_f represents our "free" local orbitals which, although they were d-orbitals for Anderson, will now be our f-orbitals:

$$H_f = E_f \, (n_+ + n_-).$$

The form for a single orbital with spin degeneracy is used for notational simplicity. The generalization to the degenerate case is normally utilized, but only f^o, f^1, and f^2 are taken as significant in the treatment of Ce materials (and the comparable hole states for Yb). As a consequence, it is reasonable to avoid introducing the additional exchange parameters in H_{corr} below. The presence of spin-orbit coupling, producing an $f_{5/2}$ - $f_{7/2}$ splitting, is actually the more important consideration complicating the selection of lowest-lying levels. It is to be noted here that, from the form of the Hamiltonian, E_f is the change in total energy as the number of f-electrons is changed from zero to one if there is no hybridization. This will be significant in the comparisons to be made below. H_{cf} is the single-particle-like hybridization term:

$$H_{cf} = \sum \, (V_{kf} \, c_{k\sigma}^* \, c_{k\sigma} + \text{Hermitian conj.}).$$

V_{kf} embodies the Coulomb and exchange interactions between the local orbital and the member of the conduction electron sea. The correlation term is

$$H_{corr} = K(n_+ + n_-)(n_+ + n_-)/2 \quad - J(n_+n_+ + n_-n_-)/2 \sim Un_+n_- .$$

It is the size of U that provides the first and most dramatic hint that one needs to consider the precise nature of the underlying structure represented by this model. If one simply uses atomic or ionic wave functions to estimate U, one arrives at values two to ten times larger than those derived from the analysis of experiment.

Clearly, this Hamiltonian is for an impurity problem, which is a reasonable model for the photoemission process involving the f-orbital since the excited atom becomes an impurity in the lattice. Because this is an impurity model, it is reasonable to make a unitary transformation on the conduction sea from Bloch (i.e., k) states to site-centered levels. With this mapping, one focuses more on $\rho(e)$ than on the $e_{k\sigma}$ themselves. This expediency of mapping k into $\{e,n\}$ has been utilized by Gunnarsson and Schoenhamer [4] in their extensive analysis of the model. To actually carry it through in practice is a rather more subtle problem because, as in the problem of Wannier functions, there is the matter of the optimal choice of phases [12].

The form of this model assumes the underlying states to be orthogonal. One might think that in light of the core-like nature of the 4f orbital, this would not be difficult to arrange. However, the conduction electrons do have $\ell=3$ character that penetrates at least as far as the tail region of the 4f-orbital. One can realize this either by observing that a plane wave will have $\ell=3$ character associated with a spherical Bessel function or by considering the Lowdin α-expansion of the orbitals from the neighboring atoms about the Ce site. Is this non-orthogonality not likely to be the source of the concept of two types of f-electrons? Anderson clearly thought so in the case of the d-states [1]. Orthogonality can always be arranged by a suitable transformation of states. It has long been known that orthogonalization weakens the hybridization interaction. So practioners of the combined interpolation schemes [13] perform the orthogonalization step first and only then set up the hybridization term. Within the Anderson model, this property is the basis for Zitkova and Rivier's observation that non-orthogonality effects narrowed the resonance [14]. This transformation to an orthogonal basis set is also the origin of the direct $k \rightarrow k'$ scattering which they observe and is also mentioned by Harrison [15]. The majority of the non-orthogonality effects can be absorbed into a simple redefinition of the parameters. Only this scattering is not so accommodated -- at least in the Bloch representation -- but I suspect that it could be similarly incorporated as merely a renormalized $\rho(e)$ and $V(e)$ in the site-specific representation. One way to explore this idea is to perform the extraction of this information from the non-orthogonal Slater-Koster band parameterization rather than the orthogonal version used for YbP [16].

It is a stringent approximation that the conduction electrons states are taken to be the same independent of the condition of the local orbitals. If the set of states is sufficiently complete, the response can be described by changes in occupation alone. But perhaps a better description is achieved by considering the modifications of the states produced in response to changes in local (i.e., f-) orbital occupation at the site. This can be illustrated by considering Anderson's suggestion that one use OPWs. Because the 4f orbital is roughly the same size as the 4d and 5s,5p orbitals in Ce, these orbitals will be rather strongly interacting. These core orbitals will be changed according to whether the 4f is occupied or not. If these orbitals are changed, then certainly the OPWs will be changed as well. Although this effect has not been incorporated in models of valence electron spectroscopies, it has for the core electron spectroscopies [17-19]. It is precisely the significance of this effect which is at issue in the discussions of the interpretation of L_{III} absorption. And it enters as well in the comparison of the Anderson model with the multiple-screening-mode interpretation for the two-peak structure.

It is the omission of the direct Couloumb interactions involving the conduction electron sea (and the core electrons) that gives the Anderson model its simplicity. But these effects must then be absorbed in the remaining terms. Is it possible? This is the crucial question because this interaction is the driving term for screening and as such is needed to relate the Anderson mode to the multiple-screening-mode model [2-3]. The band structure results [7-9] indicate that the major screening process is generally the population of an additional d- like orbital. Forming a d-screened f-hole excitation may well be the key to transforming away the direct Coulomb interaction, but this would definitely modify all the parameters. To illustrate, returning to the atomic case, one could write n_d as a linear function of n_f (expanding about $n_f=1$ for Ce) and rewrite E_T and Δ_{fd} as a function of n_f alone. To do this for the solid, one must assume that the atom never makes a transition to a different charged state: the fully screened f^1 and the partially (i.e., d-) screened f^o state both have the same charge. But what of the f^2 state? That would appear to be properly described as well, because an anti-screening occurs when an extra f-electron is added [20]. This type of approach would accommodate the fact that in the atom it costs less energy to add an extra f-electron than an extra d-electron. Such an approach would allow one to appropriately incorporate the core relaxations as well. The remaining screening effects on the f-orbitals might well be incorporated by defining the appropriate "medium" of s- and p-electrons. This scheme has the nice feature that it incorporates the relative stability of the peak separation by tying it to an f to d transition which is only slightly modified by the solid state environment.

Now, however, what is the effect on the hybridization term of the Hamiltonian? Note that the redefined f^o state can scatter the conduction electron and will also require additional effort to maintain the orthogonalization. Is it possible to transform these effects such that they appear in the $\rho(e)$? Even if one ends up with modifications of the $V(e)$ and $\rho(e)$, this would not result in dramatic changes because the photoemission does not appear to depend critically on this structure. However, there is the problem that the explanation used for the non-existence of a second peak in CeN is the strong coupling of the f^1 state to the conduction bands when the system is in the f^o state. How could this effect appear in the modified model? The best way to proceed at this point is to attempt to further refine the definitions suggested above, get the problem into a second quantized form rather than the qualitative c-number discussion given, and then attempt a transformation to the form of the Anderson Hamiltonian -- probably with the introduction of a few averages along the way. This is a necessary program to understand the application of an Anderson-like model for the Ce compounds.

It is interesting to note that the situation is somewhat different in the Yb compounds. In the Yb materials, one must transform to a hole language, but one then has the complementary problem, except that now the f-orbitals are much more localized owing to the lanthanide contraction. The consequence is that the hybridization is weaker and the model is more directly applicable. There may be core relaxation effects but not the d-orbital screening. This would be consistent with the idea that the heavy lanthanides are less tolerant of d-states than the lighter lanthanides [21]. The effect is due to the much smaller size of the cores in the heavier lanthanides. This can best be illustrated by some recent computer experiments. We set up the f-orbitals as core states such that we could control the f-count. The band states were forced to be orthogonal to the f-states by only including variational freedom for the 5f states. Calculations were then done for two assumed configurations, f^{13} and f^{14} (each is a sort of supercell consisting entirely of impurity atoms). When the conduction states were examined, very nearly rigid-band behaviour was found, in which the densities of states for the two calculations were almost identical between the two Fermi energies found, and corresponded to the one additional electron. The implication is that the Anderson model is more easily applied to the Yb compounds. That is why we have begun studies on the Yb systems first [16] rather than on the Ce compounds.

Two general observations should be made. (1) It was noted earlier that the principal peaks of the f-character always appear above the Fermi energy in the band calculations for Ce compounds. This is an immediate consequence of applying Fermi statistics. Yet the Anderson model is only interesting if E_f is below the Fermi energy. This apparent discrepancy is resolved if one notes that E_f is the change in total energy occurring when the f-count is decreased or increased by one in the absence of hybridization. Thus E_F is more nearly to Δ_{fd} in the case of Ce and Δ_f in the case of Yb ($f^{13} \rightarrow f^{14}$). These terms include relaxation and screening effects. The band eigenvalue position is merely a local derivative and not a representative of what happens when the occupation changes by one. Thus E_f is not placed by these peaks in the f density of states but is displaced by an amount demonstrated above for the atomic case. (2) In the Anderson model, the hybridization mixes the two spin states [1]. A similar feature appears in the band calculations for the j=5/2 and j=7/2 states. In a strongly hybridized material, the two spin-orbit split peaks in the density of states will occur well above the Fermi energy with long tails extending down into the occupied states. There will be a large component of $f_{7/2}$ character. As the hybridization becomes weaker, the $f_{5/2}$ derived peak will fall to the Fermi energy, which then occurs in the leading edge of that peak, and the $f_{7/2}$ component will not be occupied. Consequently, the fractional $f_{7/2}$ occupation is a good index of the strength of hybridization.

This discussion really ends with questions rather than conclusions. Suggestions have been made for relating the Anderson model to the multiple-screening-mode model. In the process of doing so, the difficulties in defining the parameters of the model to make them amenable to calculation become apparent. Nonetheless, one would like to try to make that connection, since it would more clearly define the mechanisms acting in these "highly correlated" materials. The problem appears more tractable in the case of the Yb materials than the Ce materials. In the Yb materials, one can work with something close to "bare" particles, whereas in the Ce materials one is committed to defining a set of quasi-particles such as a d-screened f hole, for example. If one succeeds in doing so, will one be able to justify using the same entities for the low-energy (i.e., low-temperature) properties? Given that hybridization and the intra-atomic interaction of the local orbitals are the major significant concepts, it is at least as important to explore how one might be able to transform the entire many-body problem to such a simplified form that a more detailed solution could be sought. But the screening of a charge is also an important concept in this problem. Utilizing a multiple-screening-mode model, one can relionalize a very large body of systematics for core excitations [22]. Unfortunately, charge screening is also a difficult problem not well represented by a linear theory. It would be nice if one were able to schematize it in some way such as folding its effects into the parameters of the Anderson model. Further research in this area would be most significant and useful, and might well lead to breakthroughs in the understanding of mixed-valent and heavy Fermion systems.

References

1. P. W. Anderson, Phys. Rev. 124, 41 (1961)
2 J. C. Fuggle, M. Campanga, Z. Zolnierek, P. Lasser, and A. Plateau, Phys. Rev. Lett. 45, 1597 (1980)
3. S. Hufner, Z. Phys. B58, 1 (1984); S. Hufner and P. Steiner, Z. Phys. B46, 37 (1982)
4. O. Gunnarsson, K. Schoenhammer, J. C. Fuggle, F. U. Hillebrecht, J.-M. Estava, R. C. Karnatak, and B. Hillebrand, Phys. Rev. B28, 7330 (1983); O. Gunnarsson and K. Schoenhammer, Phys. Rev. B28, 4315 (1983)
5. J. Janak, Phys. Rev. B18, 7165 (1978)
6. D. D. Koelling, Rep. Prog. Phys. 44, 139 (1981)
7. M. R. Norman, D. D. Koelling, A. J. Freeman, H. J. F. Jansen, B. I. Min, T. Oguchi, and Ling Ye, Phys. Rev. Lett. 53, 1673 (1984)
8. M. R. Norman, D. D. Koelling, and A. J. Freeman, Phys. Rev. B31, 6251 (1985)
9. M. R. Norman, D. D. Koelling, and A. J. Freeman, Phys. Rev. B32, 7748 (1985)
10. F. Patthey, B. Delley, W.-D. Schneider, and Y. Baer, Phys. Rev. Lett. 55, 1518 (1985); the screening interpretation is due to M. R. Norman (unpublished results).
11. J. F. Herbst, D. N. Lowy, and R. E. Watson, Phys. Rev. B6, 1913 (1972); J. F. Herbst, R. E. Watson, and J. F. Wilkins, Phys. Rev. B13, 1439 (1976)
12. U. Fano, Phys. Rev. Lett. 31, 234 (1973); G. Strinati and U. Fano, J. Math Phys. 17, 434 (1976); G. Strinati, Phys. Rev. B18, 4096 and 4104 (1978)
13. L. Hodges, H. Ehrenreich, and N. D. Lang, Phys. Rev. 153, 505 (1966); F. M. Mueller, Phys. Rev. 153, 659 (1967)
14. J. Zitkova and N. Rivier, J. Phys. C 3, L71 (1970)
15. W. A. Harrison, Solid State Theory, Dover, New York (1980), pp. 480-485
16. R. Monnier, L. Degiorgi, and D. D. Koelling, Phys. Rev. Lett. 56, 2744 (1986)
17. B. Delley and H. Beck, J. Phys. C 17, 4971 (1984)
18. E. Wuilloud, B. Delley, W.-D. Schneider, and Y. Baer, Phys. Rev. Lett. 53, 202 (1984)
19. A. Kotani and J. C. Parlebas, J. Physique 46, 77 (1985); T. Jo and A. Kotani, Solid State Commun. 54, 451 (1985)
20. M. R. Norman, Phys. Rev. B31, 6261 (1985)
21. J. C. Duthie and D. G. Pettifor, Phys. Rev. Lett. 38, 564 (1977)
22. B. W. Veal and A. P. Paulikus, Phys. Rev. Lett. 51, 1995 (1983); Phys. Rev. B31, 5399 (1985); B. W. Veal, D. E. Ellis, and D. J. Lam, Phys. Rev. B32, 5391 (1985).

THE ANDERSON LATTICE AND UNIVERSAL PROPERTIES OF HEAVY FERMION SYSTEMS

Assa Auerbach and K. Levin

James Franck Institute, University of Chicago, Chicago, Illinois 60637

ABSTRACT

Using the Kondo Boson - 1/N expansion, we solve for the Fermi liquid properties of the Anderson lattice at low temperatures. The Kondo limit of this model is shown to necessarily induce large mass enhancements $m^*/m \gg 1$, and generate a low lying energy scale $\bar{T}_K \propto (m^*/m)^{-1}$, which dominates the dynamics of this heavy Fermi liquid. In particular, our calculation leads to the following predictions: (1) The specific heat $C_V = \gamma T \propto T/\bar{T}_K$ with corrections $\Delta C_V = (T/\bar{T}_K)^3 \log(T/\bar{T}_K)$ (2) The zero temperature spin susceptibility $\chi \propto 1/\bar{T}_K$, and (3) the resistivity $\rho \propto (T/\bar{T}_K)^2$. We analyze recent pressure dependent C_V, χ and ρ/T^2 measurements on UPt_3 to confirm the scaling of these quantities with a single strongly pressure dependent energy scale. The universality of these relations is supported by evidence of systematic trends throughout the entire class of heavy fermion compounds.

1. INTRODUCTION

The class of heavy electron materials poses a new challenge for condensed matter theorists, where traditional "tools of the trade" seem unable to provide a link between the underlying microscopic physics and the Fermi liquid phenomena seen in experiments[1]. Strong two-body interactions U between valence electrons, are present at the rare-earth sites. When U is large these cannot simply be treated by standard perturbative expansions, and complications reminiscent of those of the Kondo impurity problem arise. In particular, difficulties are encountered in applying e.g. the local density approximation[2], an important and cherished tool of band structure calculations. Crude estimates of the scale of U in uranium and cerium show it to be of order electron-volts, which necessarily implies that the independent valence and conduction electron picture is a doubtful zeroth order approximation for these materials.

It is the purpose of this paper to explain the origin of Fermi liquid properties in the heavy fermion compounds. We use a simple microscopic model, and predict universal features which are common to most of the materials for which large m^*/m values are observed. We derive a consistent Fermi liquid theory for heavy fermions from the Anderson lattice model (AL). Although no unambigous *ab-initio* calculation of the parameters of the AL has yet been provided[2], it is based on an intuitive real-space picture which seems to capture the important underlying physics. This model is the translationally invariant generalization of the Anderson impurity (AI) model which has successfully been used to explain the Kondo effect.

Experimentally, heavy fermions exhibit the following properties[3]:
The high temperature (T) regime, where the susceptibility follows a 1/T Curie-Weiss law, has the signatures of localized and uncompensated valence spins fluctuating at the rare-earth sites, which weakly interact with the conduction electrons. The low temperature phase, on the other hand, exhibits coherent compensation of the spins and becomes a paramagnetic Fermi liquid of

vanishing electrical resistivity. At even lower temperatures many of the systems undergo further transitions into magnetic or superconducting ground states. In recent years, much of the interest in heavy fermions has been fueled by the peculiar nature of superconductivity in e.g. UPt_3, UBe_{13}, and $CeCu_2Si_2$, however it is clear that these phenomena cannot be fully explained without a detailed understanding of the "normal state" Fermi liquid interactions.

The Fermi liquid properties of heavy fermions as previously deduced by several authors[4,5] can be summarized as follows:

1. The low T specific heat C_V is linear in T, and the coefficient γ, measures the quasiparticle density of states at the Fermi level or their "mass" m^*. The zero temperature susceptibility χ measures m^*/m (m is the conduction electron band mass) as well as the Landau parameter A_0^a, which quantifies the exchange interaction. An anomalously large value of m^*/m in heavy fermions is recognized as a signature of a paramagnetic Fermi liquid with an extremely small Fermi energy $\tilde{T}_K$. The Wilson ratio $\chi/\gamma=1-A_0^a$ can be used, (in the absence of large moment renormalization and spin orbit coupling) to directly measure A_0^a. In Fig. 1 the values of γ and χ of several typical heavy electron compounds are plotted. We can deduce from Fig. 1, taken from Ref. 1, that A_0^a in most compounds is not large, and is systematically negative.

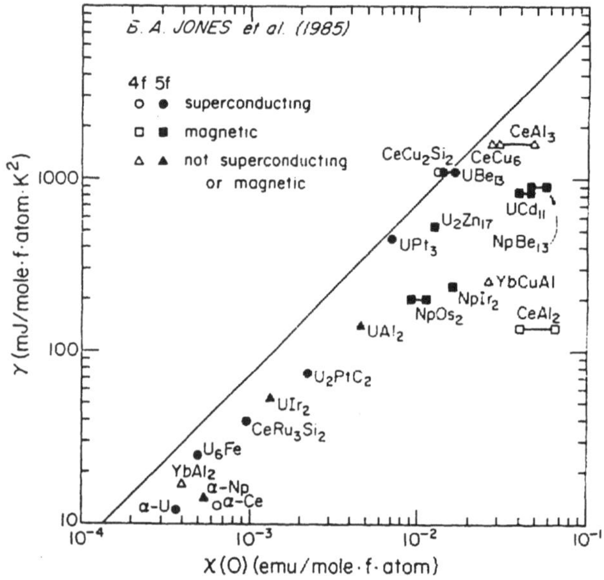

Fig. 1. The specific heat coefficient γ versus the zero temperatue susceptibility $\chi(0)$ for some heavy electron systems, from ref. 1. The solid line represents the free electron gas Wilson ratio $R \equiv \chi/g^2\gamma=1$. The data suggests that $A_0^a \equiv 1-R$ is systematically negative and of order unity.

2. The electronic compressibility is unrenormalized by the large interaction U. This implies that the density response function is suppressed by a large value of F_0^s, so that $1-A_0^s=O(m^*/m)^{-1}$ This behavior derives from the suppression of the valence charge fluctuations and the fact that the chemical potential rides with the filling factor[4]. Also it has been pointed out that in contrast to single component galilean invariant Fermi liquids (such as ^{3}He), here the relation $m^*/m=1+F_1^s/3$ is invalid.

3. The dynamical many-body interactions, (formally represented by a 4-point vertex function Γ), are manifested also in the temperature dependence of the thermodynamic and transport data. The sharp downturn of C_V/T with increasing T, and the T^2 resistivity rise reported in many of the materials, imply that the frequency dependence of Γ is dominated by a low energy scale. In a previous publication[5] we have pointed out that a single energy scale, $\tilde{T}_K$, is inferred from χ, the coefficient of the T^2 term in the resistivity A, and $T^3\log T$ coefficient δ of the specific heat. This is supported by the universal scaling of $\chi \propto \gamma$, $A \propto \gamma^2$ and $\delta \propto \gamma^3$ as seen both in systematic trends throughout the heavy fermion

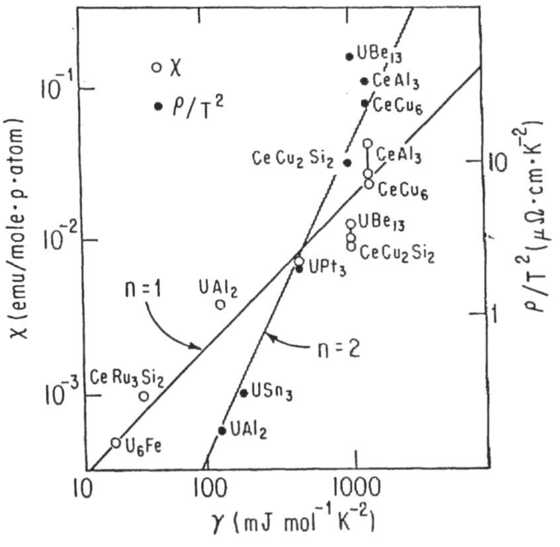

Fig. 2. Universal ratios in the heavy Fermion compounds. χ/γ data are from Ref. 1 and A = ρ/T^2 data are from Ref. 14. The solid lines are theoretical results summarized in Eq. (5.1).

class (Fig.2.), and detailed pressure dependences in UPt_3 (Fig. 4.). By comparison, we show that such scaling does not occur in liquid ^3He, by compiling the pressure dependence of the respective coefficients in Fig. 5.

In section 2 we introduce the large-U AL model and discuss its bare input parameters. The Kondo limit of the AL is defined. Since, unfortunately, no exact solutions are available for the AL problem (in contrast to the AI problem), we resort to the asymptotic approximation of the Kondo-boson (KB) 1/N expansion, where N is the valence degeneracy[6]. The interaction term in the limit of $U \to \infty$ is handled by introducing Colemans' Kondo-Boson (KB) fields[7] (alias "Slave-Bosons"), and a functional integral formalism developed by Read and Newns[8] is applied to enforce the local constraints on the KB and valence occupation. The analogous expansion has been recently studied in the AI problem and compared to exact results[9]. It proved to be successful in calculating the Wilson ratio, as well as continuosly interpolating between the "asymptotically free" local moment phase and the "local Fermi liquid" ground state. Our analysis is separated into two levels: the O(1) - "mean field theory", and the O(1/N) - "fluctuations" respectively.

In section 3 we discuss the mean field theory. Here, the saddle-point variational equations are presented and solved, which results in an effective non-interacting renormalized band structure for the fermion quasiparticles. The KB fields are static (c-numbers) and constant in space. An important consequence of the mean field solution is the emergence of a new energy scale which is exponentially dependent on the parameters of the AL. We define the Kondo limit and show that it necessarily yields a large mass enhancement $m^*/m \gg 1$. The other consequence of this limit is that the average valence occupation becomes very close to an integer, and its fluctuations are greatly suppressed. Most of the so-called many body effects on γ and χ are already included at this level. The mean field solution thus allows us to replace some of the bare AL input parameters by the experimentally observed mass enhancement. The density of states structure at the Fermi level defines the Kondo-lattice temperature $\overline{T}_K$, which for heavy fermions is of order $5-50K$. Since the Debye temperature and other "non universal" features such as the valence charge fluctuation energies and crystal field splittings are typically at much higher energy scales, the Fermi liquid behavior in the absence of incipient instabilities is predicted to be dominated by $\overline{T}_K$.

The mean field theory, however, does not include the quasiparticle interactions. Therefore, to proceed it is necessary to calculate the gaussian fluctuations of the KB fields, that is to say the Kondo-Boson propagator $D(q,\omega)$. In section 4 we perform this calculation in the "radial gauge" of Read and Newns[8]. This allows us to obtain the leading order vertex function and self energy, and the correction to the Wilson ratio. Using the microscopic prescription of Ref. 10 the Landau parameters are determined. Physically, the interactions are mediated by coherent hybridization fluctuations screened by quasiparticle-hole excitations, with characteristic frequency scale of $\bar{T}_K$. The KB thus gives rise to a large $(T/\bar{T}_K)^2$ term in the dc resistivity, and a $(T/\bar{T}_K)^3\log(T/\bar{T}_K)$ contribution to the specific heat. The Fermi liquid properties 1 - 3 are thus derived.

In section 5 we summarize our main results, and show that they agree with recently measured pressure dependence in[11,12] γ, χ, and the resistivity[13] data, and also explain the remarkable universal behavior throughout the heavy fermion class[14]. The data also serves to exclude alternate interpretations such as that of ferromagnetic spin fluctuations.

We conclude with a brief discussion of the role a more realistic band structure plays in producing antiferromagnetic spin fluctuations, and the nature of superconducting instability. Both of them invoke resummation of our O(1/N) theory.

2. THE ANDERSON LATTICE MODEL

The Anderson lattice (AL) hamiltonian in second quantized notation is given by:

$$H^{AL} = \sum_{\mathbf{k},m}\varepsilon_{\mathbf{k}}c^{\dagger}_{\mathbf{k},m}c_{\mathbf{k},m} + \varepsilon_f^0\sum_{\mathbf{k},m}f^{\dagger}_{\mathbf{k},m}f_{\mathbf{k},m} + \frac{V}{\sqrt{N}}\sum_{i,m}(f^{\dagger}_{im}c_{im}+c^{\dagger}_{im}f_{im}) \tag{2.1}$$

$$+\,U\sum_{imm'}f^{\dagger}_{im}f_{im}f^{\dagger}_{im'}f_{im'}\ ,$$

where $\varepsilon_{\mathbf{k}}$ and ε_f^0 are the conduction and dispersionless valence band energies respectively. The label i denotes the Wannier state at site $\mathbf{r}_i$. The N-fold degeneracy of both bands is labelled by m, $|m|\le(N-1)/2$. V is the local hybridization matrix element which in this simplified version is taken to be independent of $\mathbf{k},m$. The large local Coulomb repulsion is parametrized by U. The band structure $\varepsilon_{\mathbf{k}}$ defines the bare density of states ρ_c, and the fermi surface at $\varepsilon_{\mathbf{k}}=\mu_0$. The bare chemical potential μ_0 is determined by the total (valence plus conduction) electron density

$$N_e=\int_{-\infty}^{\mu_0} d\varepsilon\rho_c(\varepsilon)\ .$$

In the case of large U, i.e $U\gg\varepsilon_f^0,\mu_0$, we can proceed by introducing the Kondo-Boson (KB) fields of Coleman at each lattice site[7], and replacing the 4-fermion term by a constraint on the total f-electron and KB occupation. This results in the following path integral representation of the AL partition function:($\hbar$=1,β=1/T)

$$Z_{AL} = \int D\,\lambda b^*bc^*cf^*f\exp\left[-\int_0^\beta d\tau\,(L_{AL}(\tau)+i\sum_{im}\lambda_i(f^*_{im}f_{im}+\frac{1}{N}b^*_ib_i-Q_0))\right]\ , \tag{2.2a}$$

where,

$$L_{AL} = \sum_{\mathbf{k}m}\{\,c^*_{\mathbf{k}m}(\partial_\tau+\varepsilon_{\mathbf{k}})c_{\mathbf{k}m}+f^*_{\mathbf{k}m}(\partial_\tau+\varepsilon_f^0)f_{\mathbf{k}m}\,\} + \sum_i b^*_i\partial_\tau b_i$$

$$+\,\frac{V}{\sqrt{N}}\sum_{im}\{c^*_{im}f_{im}b^*_i + b_if^*_{im}c_{im}\}\ , \tag{2.2b}$$

Here, c_{im} and f_{im} are grassman variables, and b_i are the KB complex fields. The integrations over the Lagrange multiplier fields $\lambda_i(\tau)$ impose the local constraints of $n_f+n_b=Q_0$ at all times and sites, where n_α denotes the number operator of particle α. Q_0 is kept as a fixed parameter (instead of $Q_0=1/N$) in order to define a true N-independent mean field theory.

At this stage we find it useful to apply the Read and Newns[8] (RN) time dependent local gauge transformation which acts simultaneously on the Bose field $b_i\equiv r_i\exp(i\phi_i)$ and the fields f_i:

$$f_i\rightarrow f_i'=e^{-i\phi_i}f_i\ \ ;\ \ \lambda_i\rightarrow\lambda_i'+\dot{\phi}_i \tag{2.3a}$$

Let us rescale the radial coordinate by

$$r_i \to r_i' = \frac{V}{\sqrt{N}} r_i . \tag{2.3b}$$

Rewriting the lagrangian in terms of the primed coordinates and using the Bose periodicity condition of $\phi(\beta)=\phi(0)$ we are left with two real fields r' and λ', and the zero mode ϕ is made redundant as it contributes only to an overall constant of Z. The RN transformation thus eliminates infrared divergencies which are known to plague perturbation theory and 1/N expansions. Price is paid, however, since the KB in the radial gauge looses its physical meaning as a particle operator and becomes a "noise field" in the functional integral formalism. Henceforth we shall drop the primes on the f', λ', and r' fields.

Since the lagrangian is bilinear in the fermion fields, it is possible to integrate them out exactly and arrive at a Bose path integral with an effective action S:

$$Z_{AL} = \exp[-\beta F] = \int D\, r^2 \lambda \ \exp\left[-N \ \beta S([r],[\lambda])\right] \tag{2.4a}$$

$$S = -\frac{1}{2\beta N}\mathrm{Tr}_k \log \det \begin{bmatrix} (ik_0-\varepsilon_k)\delta_{kk'} & r_q\,\delta_{k'k+q} \\ r_q\,\delta_{k'k+q} & (ik_0-\varepsilon_f^0)\delta_{kk'}+i\lambda_q\delta_{k',k+q} \end{bmatrix}$$

$$+\, i\sum_{kq}\frac{\lambda_q}{V^2} r_k r_{-k-q} - Q_0\lambda_q\delta_{q0}, \tag{2.4b}$$

Here, k and q denote the usual Fermi and Bose four-vectors (implicitly including the label m for the fermions). The zeroth components k_0 , q_0, are the Matsubara frequencies $(2n+1)\pi T$, and $2n\pi T$ respectively.

Eq. (2.4a) is used to generate an asymptotic expansion for the free energy F, using 1/N as the small parameter (analogous to $\hbar$ of the semiclassical approximation). By adding source terms to the lagrangian in the form $j_{kk'}\alpha_k^*\alpha_{k'}$ for $\alpha_k=(c_k f_k)$, the electronic correlation functions are determined as functional derivatives of $F[j]$. At low temperatures Eq. (2.4a) is dominated by a non trivial saddle point $\bar{r},\bar{\lambda}\neq 0$, which constitutes the parameters of the mean field theory discussed in the next section. By expanding Eq. (2.4b) around the saddle point and performing the gaussian integration, we generate the higher order terms in 1/N.

3. MEAN FIELD THEORY

The mean field theory $N=\infty$ of the AL has already been amply discussed in the literature[15-18]. It bears close resemblence to the Hartree approximation in other many-body problems such as the Coulomb gas and the Hubbard model. As a variational estimate of the ground state, the same theory has been also derived using other approaches such as a generalized Gutzwiller approximation of Rice and Ueda[17,4], for which the relation to the KB theory has been recently explored[19]. In essence, the Bose fields are replaced by their expectation values and an effective single particle band theory is obtained. Here we shall define the mean field parameters and band structure in terms of the saddle point KB fields given by the variational equations:

$$\frac{\delta S}{\delta \mathbf{r}}\Big|_{\mathbf{r}=\bar{r}} = 0 \tag{3.1}$$

where $\mathbf{r} \equiv (r,\lambda)$ is a vector notation for the two real KB fields. By expanding the logarithm in Eq. (4), it is easy to verify that (in the absence of source currents) $\bar{r} \propto \delta_{q,0}$, i.e. the saddle point fields are constants in space-time. The mean field parameters $r_0=r$ and $\varepsilon_f=\varepsilon_f^0+i\bar{\lambda}$, represent the effective c-f hybridization and renormalized f-level respectively. They determine the renormalized bands $E^{\pm}$, which are separated by a gap at ε_f.

$$E_k^{\pm} = \frac{\varepsilon_k+\varepsilon_f}{2} \pm \sqrt{\left(\frac{\varepsilon_k-\varepsilon_f}{2}\right)^2+r_0^2} \equiv \varepsilon_f \pm r_0 \mathrm{ctg}\theta(\varepsilon_k)^{\pm 1} , \tag{3.2}$$

which defines the function $\theta(\varepsilon) = \theta_{E^-}$. The quasiparticles $\alpha^{\pm}$ of these bands are given by the θ dependent coherence factors:

$$\begin{aligned} \alpha_{k,m}^{+} &= \cos\theta f_{k,m} + \sin\theta c_{k,m} \\ \alpha_{k,m}^{-} &= -\sin\theta f_{k,m} + \cos\theta c_{k,m} \end{aligned} \tag{3.3}$$

The Fermi level at $T=0$ lies on the bottom band and equals $\mu=E^{-}(\mu_0)$. The mass enhancement at the Fermi level is $m^*/m = \sin^{-2}\theta_\mu$. The mean field parameters are determined by the implicit equations

$$\frac{dS}{dr_0} = 2r_0\{ \int_{\tan\theta_0}^{\tan\theta_\mu} d\tan\theta \, \rho_c / \tan\theta + \frac{\varepsilon_f - \varepsilon_f^0}{V^2} \} = 0 \ , \tag{3.4a}$$

$$\frac{dS}{d\varepsilon_f} = -r_0 \int_{\tan\theta_0}^{\tan\theta_\mu} d\tan\theta \, \rho_c/(\tan\theta)^2 - Q_0 \langle n_f \rangle = 0 \ , \tag{3.4b}$$

where,

$$\langle n_f \rangle = 1 - r_0^2/(Q_0 V^2) \ . \tag{3.4c}$$

In heavy fermion systems we are interested in a specific limit of the AL model, the Kondo limit, where $J \equiv \rho_c V^2/(\varepsilon_f - \varepsilon_f^0) \ll 1$. For small Q_0 the slow variations of $\rho_c(\varepsilon)$ away from the Fermi surface at $\varepsilon=\mu_0$ can be ignored. Eqs.(3.4) to leading order in $(m^*/m)^{-1}$ and Q_0 yield:

$$m^*/m \approx \frac{r_0^2}{(\mu_0 - \varepsilon_f)^2} = \frac{Q_0 \langle n_f \rangle}{\rho_c \mu_0} \exp[1/J] \gg 1, \tag{3.5a}$$

and

$$\overline{T}_K \equiv \varepsilon_f - \mu \approx \frac{Q_0 \langle n_f \rangle}{\rho_c} (m^*/m)^{-1} \tag{3.5b}$$

$\overline{T}_K$ is the Kondo Lattice energy scale which measures the width of the renormalized density of states structure and characterizes the heavy Fermi liquid. Since it is exponentially dependent on $(-1/J)$, $\overline{T}_K$ is expected to be much smaller than any bare energy scale (e.g. μ_0, ρ_c^{-1}, V), and thus for small J:

$$1 - \langle n_f \rangle \approx \overline{T}_K/(\rho_c V^2) \ll 1 \ . \tag{3.6}$$

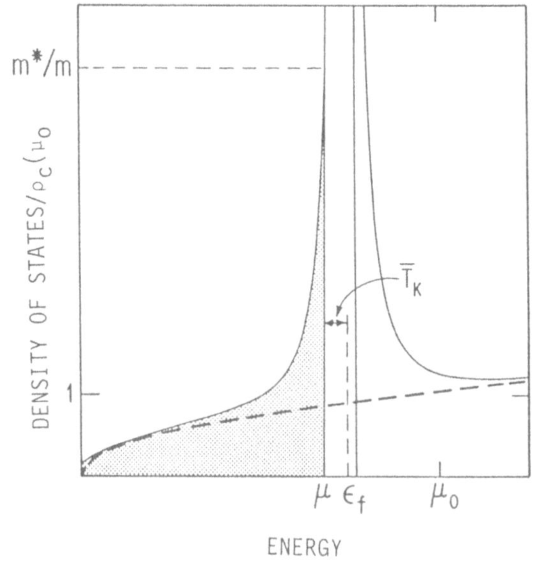

Fig. 3. The spherical band AL model, mean field $N=\infty$ level. The dashed line in the bare conduction band density of states. The solid lines is the renormalized mean field band structure. Shaded area is the occupied Fermi sea at zero temperature.

Eqs. (3.5) and (3.6) prove that the mean field large mass enhancement is necessarily correlated with the near integer valence occupation, and, as will later be shown, suppression of f-charge fluctuations. In Fig. 3 we plot the density of states for a heavy fermion system with a spherical conduction band. The large peak of the renormalized density of states at the Fermi level is a direct consequence of the local f-charge constraint, and it confirms the idea (supported by the "dense Kondo system" approaches[1]) that that the individual Kondo resonances overlap and form a narrow band of mostly f-character.

The mean field theory of the AL replaces two bare parameters ε_f^0 and V, by the renormalized ones r_0 and ε_f. These, in turn, can be replaced by the more physical parameters m^*/m and $<n_f>$. In the Kondo limit, one parameter, namely $<n_f>$, drops out.

In the presence of a magnetic field h, the mean field free energy is given by

$$F^0 = \frac{1}{(2\beta)} \mathrm{Tr} \log G^0(h) \; , \tag{3.7a}$$

where the z-component of the spin is coupled with a magnetic moment g to the static magnetic field. The mean field Greens function is given by:

$$G^0(h) = \left[ik_0 - E_k^{\alpha} - h g m\right]^{-1} \delta_{km\alpha, k'm'\alpha'} \; . \tag{3.7b}$$

The linear temperature coefficient of the specific heat $\gamma^0 = -d^2F^0/dT^2|_{T=0}$ and the zero temperature susceptibitity $\chi^0 = -d^2F^0/dh^2|_{h=0}$ are readily determined:

$$\gamma^0 = \frac{\pi^2 N}{3}(m^*/m)\rho_c \;\; ; \;\; \chi^0 = g^2 N \frac{(N^2-1)}{12}(m^*/m)\rho_c \; . \tag{3.8}$$

It can be shown that field and temperature variations of the saddle point parameters r_0, ε_f, and μ only contribute to higher F^0 derivatives of even order. The Wilson ratio R at the mean field level is thus:

$$R \equiv (2\pi/g)^2/(N^2-1)\, \chi^0/\gamma^0 = 1. \tag{3.9}$$

which is expected of course, since no interactions between quasiparticles have yet been included.

4. FLUCTUATIONS AND INTERACTIONS

We expand the effective action S (Eq. 2.4b) of the partition function in the presence of source currents, to second order in the Bose fields, and obtain (up to overall constants[8]):

$$Z[j] = e^{(-\beta F^0[j])} \int D\,\delta r \exp\left[-\frac{N\beta}{2}\sum_{qq'} \frac{\delta^2 S}{\delta r_q \delta r'_{q'}}\Big|_{r=\bar{r}_j} \delta r_q \delta r_{q'} + O(\delta r^3)\right] . \tag{4.1}$$

$\mathbf{r}_j$ satisfies Eq. (3.1) in the presence of an arbitrary source current j. Performing the gaussian integration in (4.1) yields

$$F^{(2)}[j] = -\frac{1}{\beta}\log Z^{(2)}$$

$$= \frac{1}{(2\beta)}\mathrm{Tr}_k \log\; G(j, \mathbf{r}_j) - \frac{1}{(2\beta)}\mathrm{Tr}_q \log\det D(j, \mathbf{r}_j) + O(1/N) \tag{4.2}$$

where G^0 is the mean field Greens function (with the source term as the self energy) and $D_{r,r'}$ is the KB propagator $<\delta r_r, \delta r_{r'}>$. $D_{r,r'}$ is given by the RPA sum of bubble diagrams, such that

$$D(q) = \frac{-1}{N}(\Pi(q) + \Pi^0)^{-1} \;\; ; \;\; \Pi^0 = 2\begin{bmatrix} 1/J & \sqrt{\hat{\chi}_c} \\ \sqrt{\hat{\chi}_c} & 0 \end{bmatrix} , \tag{4.3a}$$

where Π^0 is the "unscreened vertex" in which $\hat{\chi}_c \equiv [(1-<n_f>)\,Q_0/r_0]^2$. $\hat{\chi}_c$ can be shown to represent the f-charge fluctuations at the gaussian level, and following the discussion of the previous section, it is seen to vanish in the Kondo limit of large m^*/m. The bubble diagram Π is given by:

$$\Pi_{r,r'} = -\beta^{-1} \sum_{k,\alpha\alpha'=\pm} C_r^{\alpha\alpha'}(\theta_k, \theta_{k+q}) C_{r'}^{\alpha'\alpha}(\theta_{k+q}, \theta_k) G_{k,\alpha}^0 G_{k+q,\alpha}^0 \; . \tag{4.3b}$$

The C's are the quasiparticle-boson vertices arising from the orthogonal transformation from the $(c\ f)$ basis to the (+,–) bands:

$$C_1^{--}=\sin(\theta_k+\theta_{k+q})\quad C_1^{-+}=\cos(\theta_k+\theta_{k+q})\quad C_1^{++}=C_1^{--} \tag{4.3c}$$

$$C_2^{--}=i\cos\theta_k\cos\theta_{k+q}\quad C_2^{-+}=-i\cos\theta_k\sin\theta_{k+q}\quad C_2^{++}=i\sin\theta_k\sin\theta_{k+q}\ .$$

For j=0,the functions in Π contain interband and intraband terms, the latter being very similar in their low (q,ω) behavior to the familiar Lindhard functions or polarization insertions in the electron gas. It can also be verified that $\lim_{q,\omega\to 0}\det(-D)$ and $\lim_{q,\omega\to 0}\mathrm{Tr}(-D)$ are positive, which ensures the stability of the mean field solution[7]. The explicit factor of 1/N in Eq. (4.3a) provides us with the small parameter of the RPA or gaussian approximation, since all the corrections either involve higher powers of D, or do not contain the maximal number of internal bubbles which reduces their contribution by factors of 1/N.

We can now derive the electronic response and correlation function, to leading order in 1/N, by functionally differentiating $F^{(2)}[j]$ in Eq. (4.2). In particular it is straightforward to read off the quasiparticle self energy Σ and irreducible vertex function Γ from the first and second derivatives respectively[20]. The quasi particles interact via the exchange of a Kondo Boson propagator. This propagator can be physically interpreted as an effective hybridization fluctuation which is strongly screened by the quasiparticle density response Π. The effective mass correction is given by $d\Sigma/dE_k^-$ at T=0:

$$\frac{\delta m^*}{m}=-\frac{d}{dE_k^-}\sum_{q,rr',\alpha}\int_0^\infty\frac{d\omega}{\pi}D_{rr'}^{im}(q,\omega)C_r^{-\alpha}(\theta_k,\theta_{k+q})C_{r'}^{\alpha-}(\theta_{k+q},\theta_k)$$

$$\times\left[\frac{1+\Theta(\mu-E_{k+q}^{\alpha})}{E_k^--E_{k+q}^{\alpha}-\omega}-\frac{\Theta(\mu-E_{k+q}^{\alpha})}{E_k^--E_{k+q}^{\alpha}+\omega}\right]_{E_k^-=\mu}, \tag{4.4}$$

where $D^{im}=\lim_{\eta\to 0}[D(\omega+i\eta)-D(\omega-i\eta)]$. Γ contains two contributions to the leading order in 1/N:

$$\Gamma^{\alpha\beta;\alpha'\beta'}(k\ m,k+q\ m;k'\ m',k'-q\ m')=\sum_{r,r'}C_r^{\alpha\beta}(\theta_k,\theta_{k+q})C_{r'}^{\alpha'\beta'}(\theta_{k'},\theta_{k'-q})\,D_{r,r'}(q)\quad\left[\equiv\Gamma^{dir}\right]$$

$$-\delta_{m,m'}\sum_{r,r'}C_r^{\alpha\alpha'}(\theta_k,\theta_{k'-q})C_{r'}^{\beta\beta'}(\theta_{k+q},\theta_{k'})\,D_{r,r'}(k'-k-q)\quad\left[\equiv-\Gamma^{exch}\right]+O(1/N^2) \tag{4.5}$$

We can obtain the Landau scattering amplitudes $\{A_l^{s,a}\}$ following the microscopic prescription of Ref.[10]. Here, "s" and "a" denote the generalized symmetric and antisymmetric channels respectively. One considers the $\omega\to 0$ limit of Γ evaluated on the spherical Fermi surface i.e ($E_k=E_{k'}=\mu$), and projects it onto Legendre polynomials, P_l, such that:

$$A_l^s=\rho_c(m^*/m)\,N\,\delta_{l,0}\lim_{|q|,\frac{\omega}{|q|}\to 0}\Gamma^{dir}+A_l^a \tag{4.6}$$

$$A_l^a=-\,\rho_c(m^*/m)\,(2l+1)/2k_f^2\int_0^{2k_f}d\kappa\,\Gamma^{exch}(\kappa,0)\,P_l(1-\kappa^2/2k_f^2)\,\kappa$$

where k_f is the Fermi wave vector ($\varepsilon_{k_f}=\mu_0$). Eq. (4.5) guarantees that the forward scattering sum rule is automatically satisfied. To leading order in 1/N there are no renormalization factors or other corrections to Eqs. (4.4) and (4.5). It should be stressed that in this Fermi liquid theory the "bare" particles are the heavy mean field quasiparticles, and thus $\delta m^*/m$ in Eq. (5) is *not* large. Using Eqs. (4.5) and (4.6) we computed the $\{A_l^{s,a}\}$ numerically. We found that D(q,0) is slowly varying, and higher moments decrease rapidly with l. Considerable simplification arises when a parabolic band structure is used for $\varepsilon(k)\ \alpha\ |k|^2$ since angular integrations may then be done analytically, leaving us with just a 1-dimensional numerical integration. Here we neglect Umklapp processes and set $\hat{\chi}_c$=0. We find that the l=0,1 parameters are (up to relative corrections of $O(1/N,\ m^*/m^{-1})$:

$$A_0^a=\frac{-1.000}{N}+\frac{0.08}{N}(Q_0/\mu_0)\ ;\ A_0^s=1.000\ ;\ A_1^s=A_1^a=\frac{-.12}{N}(Q_0/\mu_0)\ . \tag{4.7}$$

A direct differentiation of the free energy in Eq. (4.2) with respect to temperature and magnetic field yields χ and γ. The mean field contributions of the first term were calculated in Eqs. (3.8) and (3.9). The 1/N corrections arise from the term Tr ln det$D$. After some algebra it follows that the susceptibility and specific heat are renormalized such that

$$\chi = \chi^0(1+\delta m^*/m - A_0^a + O(1/N^2)) , \qquad (4.8)$$

and similarly:

$$\gamma = \gamma^0(1+\delta m^*/m + O(1/N^2)) . \qquad (4.9)$$

These are known Fermi liquid identities related to spin and charge conservation. Their direct verification lends further support to our assignment of Landau parameters in Eq. (4.6). The result for A_0^s is intimately related to the constraint imposed on the f-charge fluctuations by setting $\tilde{\chi}_c=0$. It is important to note that this theory is not galillean invariant (because of hybridization between bands of largely different curvature or "masses"). Therefore the relation $1+\delta m^*/m \doteq (1-A_1^s/3)^{-1}$ is incorrect as could be checked against Eq. (4.4). These observations are consistent with Fermi liquid properties listed under point 2 in the introduction.

In addition to the correction to γ, there exists a specific heat correction ΔC_V analogous to the paramagnon $T^3 \log T$ contribution in liquid ^{3}He. Our analysis follows Ref. 21, where $D_{rr'}$ replaces the RPA susceptibility that mediates the spin fluctuations. We find:

$$\Delta C_V = \delta\, T^3 \log\left[\frac{T}{\tilde{T}_K}\right] + O((T/\tilde{T}_K)^3) \ , \ ; \delta = a \left[\frac{T}{\tilde{T}_K}\right]^3 , \qquad (4.10)$$

where a is a positive number close to unity. The contributions to C_v from higher powers of temperature are dominated by the variation of the mean field parameters r_0, ε_f, and μ, with characteristic energy scale $\tilde{T}_K$.

Using this approach we are also able to estimate the T^2 coefficient of the low temperature resistivity $\rho = AT^2$. We follow the analogous paramagnon calculation[14], and determine A by evaluating $\partial D^{im}/\partial\omega$, at $\omega=0$.The result is:

$$\rho = A\ T^2 + O(T^3)\ ;\ A = \rho_{max}\,(1/\lambda\tilde{T}_K)^2 , \qquad (4.11)$$

where $\rho_{max} = \hbar/(e^2 k_f N^2) = 100\text{–}300\mu\Omega\text{cm}$ and where λ is a Fermi surface geometric factor of order unity.

5. SUMMARY AND DISCUSSION

In heavy fermion materials where crystal field splittings are larger than $\tilde{T}_K \approx 5 - 50\,K$, the valence degeneracy is not large ($N=2$ and $Q_0=1/2$). The present 1/N expansion, while it may not be quantitatively accurate in detail, is nevertheless a viable systematic description of the Fermi liquid properties which has many satisfying features including:

1) The local f-charge constraint is imposed at each order, and Fermi liquid identities and sum rules are satisfied.
2) The model we have used is generic and depends on a minimal set of microscopic parameters. These are the bare band structure, and the Kondo lattice temperature $\tilde{T}_K$.
3) Translational invariance is ab-initio built into the theory. This is in contrast to "interacting impurities" approaches where more sophisticated resummation schemes are needed to recover coherence effects.

We find that the Fermi liquid parameters in Eq. (4.7) for a spherical band structure are given by $A_0^a \approx 1/N$ and $A_0^s \approx 1$, and the higher moments are smaller than $1/N$. Also a $\delta\, T^3 \log T$ contribution to the specific heat and a AT^2 resistivity term are calculated. These last two effects are similarly obtained in paramagnon theories. For this reason paramagnon models have been extensively used to explain both the normal state and the superconductivity in UPt_3 in analogy to ^{3}He. However, the degree of independent evidence for an incipient ferromagnetic instability in UPt_3 remains controversial[5]. The origin of the $T^3 \log$ T behavior in this KB theory is the non analytic low $(|q|,\omega)$ behavior of Π via the ratio $\omega/|q|$. The presence of such a term is not surprising, since it is a general property of theories with an RPA-like boson mediating the interactions. On the other hand, it should be stressed that unlike the paramagnon mechanism, this behavior does not derive solely from the spin fluctuation channel, as seen by the relative magnitudes of A_0^a and A_0^s.

Our results in Eqs. (4.8), (4.9), (4.10) and (4.11), can best be summarized by the simple proportionality relations which are obtained between γ, χ, A and δ.

$$\chi \propto \gamma\ ;\ A \propto \gamma^2\ ;\ \delta \propto \gamma^3 . \qquad (5.1)$$

In Eq.(3.5) we observe that m^*/m strongly depends on the value of the Kondo coupling J, which is expected to increase with pressure. This implies that $\overline{T}_K$ should dramatically increase with applied pressure, and thus explains why the Gruneisen parameter $\Gamma_\gamma = d\log\gamma/d\log V$ is observed to be anomalously large in comparison to typical non heavy metals ($\Gamma_\gamma \approx 57$ in[11] UPt_3). Thus, the pressure dependence is a useful probe to the relations (5.1). In UPt_3, γ can vary under pressure by 40%. As shown in Fig. 4, our predictions appear to be well confirmed by experiments. Also universal relations between γ, χ and A for many different heavy fermions seem to correlate remarkably well with Eq. (5.1). This analysis of the data raises doubts about the validity of paramagnon models, for which relations (5.1) are not expected to be valid as happens in liquid ^3He. In fact, large deviations from these relations are found for the analogous experimental measurements in ^3He. These are plotted in Fig. 5.

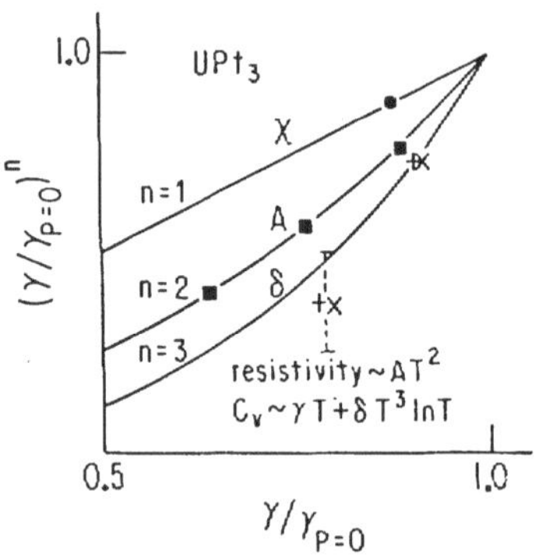

Fig. 4. Scaling of thermodynamic and transport coefficients with pressure dependent γ. (For the latter see Ref. 11). χ are from Ref. 12, resistivity from Ref. 13. The symbols + and x correspond respectively to the coeffient δ of $T^3\ln T$ term in C_V and the coefficient ε of T^{-3} term in C_v (from Ref. 11). The solid lines are theoretical results summarized in Eq. (5.1).

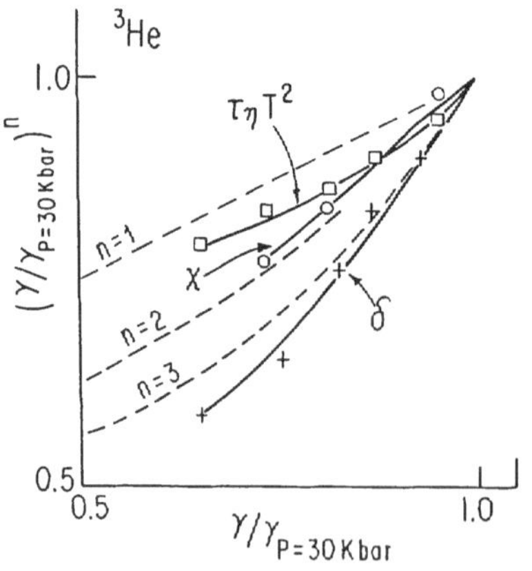

Fig. 5. Scaling of χ, δ, and $A = \tau_\eta T^2$ with pressure dependent γ in liquid ^3He. Here, τ_η is the quasiparticle lifetime as measured by the viscosity. In contrast to Fig. 2., it is evident that the relations (5.1) are not applicable to this Fermi liquid, in which ferromagnetic spin fluctuations are assumed to be important.

It is worth mentioning at this point that the mechanism leading to superconductivity at e.g. $T_c \approx .5K$ in UPt_3 is still unresolved. The participation of the heavy quasiparticles in the formation of cooper pairs is inferred from the large specific heat jump at T_c. The possibility that Kondo-Boson mediated pairing drives the superconductivity in some of the heavy fermion compounds may be the most interesting extension of this theory. The l=2 Landau parameters in Eq. (4.6) were found to be attractive. A simplistic deduction[22,23] of the transition temperature and order parameter symmetry from the Landau (ω=0) limit of the vertex function would yield d-wave pairing. However, it might not be applicable as in ^3He, since it uses a spherical Fermi surface and it assumes that the frequency cut-off scale in Γ is much smaller than the characteristic variations in the electronic energies. In the KB theory both relevant frequency scales appear to be of order T_K.

There is widespread evidence for antiferromagnetic correlations[24], in the heavy fermion materials. Because these fluctuations sometimes lead to spin density wave instabilities, this raises questions about the relation between the two instabilities and their implications on the symmetry of the order parameter. The present O(1/N) level of course is insufficient to provide answers related to multiple scattering of quasiparticles which thus requires an infinite resummation scheme. We are currently investigating such schemes in relation to both channels of instability. The analogy to theories of itinerant antiferromagnetism suggests that the KB interaction, with sufficient Fermi surface nesting, is sufficient to produce such an instability. To provide a detailed understanding of different materials it is necessary of course to generalize the spherical band AL model and include a realistic band structure with the correct crystal symmetries.

We thank P. Coleman, G. Crabtree, R. Dunlap, G.S. Grest, G. Mazenko, P. Schlottmann and S. Shenker for useful discussions. This work was supported by NSF DMR-84-20187 and MRL grant NSF-DMR-82-6892. K.L. also thanks Argonne National Laboratory for their support and hospitality.

References

1. For a theoretical review see P.A. Lee, T.M. Rice, J.W. Serene, L.J. Sham and J.W. Wilkins, Comm. in Cond. Matt. Phys. **12**, 98 (1986) and references therein. Fig. 1 displays the γ *vs.* χ data.
2. See e.g. M.R. Norman, and also D.D. Koelling, these preceedings.
3. For an experimental review see G.R. Stewart, Rev. Mod. Phys. **56**, 755 (1984).
4. T.M. Rice and K. Ueda, Phys. Rev. Lett. **55**, 995 (1985); C. M. Varma, Phys. Rev. Lett. **55**, 2723 (1985).
5. A. Auerbach and K. Levin, Phys. Rev. **B34**, 3524 (1986).
6. A. Auerbach and K. Levin, Phys. Rev. Lett. **57**, 877 (1986); The O(1/N) calculation has been independently carried out in the cartesian KB coordinates and the same results obtained by A. Millis and P.A. Lee, Phys. Rev. B, to be published.
7. P. Coleman, Phys. Rev. B **29**, 3035 (1985); Proc. 8 Tanaguchi Symp. on Mixed Valence, S. St. Phys. **62**, Springer-Verlag (1985).
8. N. Read and D.M.Newns, J. Phys. **C16**, 3273 (1983); N. Read, J. Phys. **C18**, 2651 (1985).
9. P. Coleman and N. Andrei, U.C.S.B. preprint # ITP-85-108.
10. A.A. Abrikosov, L.P. Gorkov and I.E. Dzyaloshinski, "Methods of Quantum Field Theory in Statistical Physics", Dover N.Y. (1975).
11. G.E. Brodale, R.A. Fisher, Norman E. Phillips, G.R. Stewart and A.L. Giorgi (preprint).
12. J.J. M. Franse, P.H. Frings, A. Menovsky and A. de Visser, Physica **130B**, 180 (1985).
13. J.O. Willis, J.D. Thompson, Z. Fisk, A. deVisser, J.J.M. Franse and A. Menovsky, Phys. Rev. B **31**, 1654 (1985).
14. K. Kadowaki and S.B Woods, Sol. St. Comm. **58**, 507 (1986).
15. P.F. de-Chatel, Sol. St. Comm. **41**, 853 (1982).
16. N. Read and D.M. Newns, Sol. St. Comm. **52**, 993 (1984).
17. T.M. Rice and K. Ueda, Phys. Rev. B (in press); D. Baeriswyl, C. Gros and T.M. Rice, ETH-Hongerberg preprint.
18. B.H Brandow, Phys. Rev. B **33**, 215 (1986).

19. G. Kotliar and A.E. Ruckenstein, Phys. Rev. Lett. **57**, 1362 (1986).

20. D. Amit, Field Theory, Renormalization Group And Critical Phenomena, McGraw-Hill, (1978)

21. S. Doniach and S. Engelsberg, Phys. Rev. Lett. **17**, 750 (1966); W.F. Brinkman and S. Engelsberg Phys. Rev. **169**, (1968); E. Riedel, Zeit. Phys. **210**, 403 (1968).

22. R. Jullien, M.T. Beal-Monod and B. Coqblin, Phys. Rev. B **9**, 1441 (1974).

23. B.R. Patton and A. Zaringhalam, Phys. Lett. **55A**, 95 (1975); K. Levin and O.T. Valls, Phys. Rev. B **20**, 105 (1979).

24. G. Aeppli, A. Goldman, G. Shirane, E. Bucher, M.-Ch. Lux-Steiner, preprint.

25. J. C. Wheatley, Rev. Mod. Phys. **47**, 415 (1975).

EXTENDED COUPLED CLUSTER METHOD: QUANTUM MANY-BODY THEORY MADE CLASSICAL

J. Arponen*, R.F. Bishop† and E. Pajanne‡

*Department of Theoretical Physics, University of Helsinki
Siltavuorenpenger 20C, 00170 Helsinki, Finland

†Department of Mathematics
University of Manchester Institute of Science and Technology
P.O. Box 88, Manchester M60 1QD, England

‡Research Institute for Theoretical Physics
Siltavuorenpenger 20C, 00170 Helsinki, Finland

1. INTRODUCTION

We focus attention in this paper on how the general quantum many-body problem can be cast in the form of a variational principle for a specified action functional. After some preliminary discussion in Section 2 concerning the algebra of the many-body operators and the development of a convenient shorthand notation to describe it, we show in Section 3 how each of the configuration-interaction (CI)[1] method, the normal coupled cluster method (CCM),[2-6] and an extended version of the CCM,[7,8] can be derived by specific parametrisations of the ground-state bra and ket wavefunctions in the action functional. In each case we make contact and comparison with time-independent perturbation theory, and we discuss the various "tree-diagram" structures that emerge in each case.

As is widely appreciated by now, the CI method contains unlinked diagrams for the energy, no generalised time-ordering (g.t.o.) properties, and suffers accordingly from the size-extensivity problem.[6] By contrast, the normal CCM takes account of the linked-cluster theorem,[9,10] has a well-known g.t.o. structure ("backwards in time") which leads to connected diagrams for the energy of the normal tree structure (in which each link is a group of particle or particle-hole lines parametrised by some configuration-space index ξ), and does not suffer from the size-extensivity problem. Although evaluation of the energy does not require the bra ground state, it is needed for the expectation values of other operators, and in the normal CCM the operators parametrising the bra ground state are not linked. Practical problems can thereby still arise as for the CI method. Finally, it is shown how maximal use is made of the linked-cluster theorem in the extended CCM, and how each of the amplitudes $\sigma(\xi)$ and $\tilde{\sigma}(\xi)$ which now completely specify the bra and ket ground states, is fully linked and is hence quasi-local in the sense of obeying the usual cluster property. We point out how the extended CCM (ECCM) can also be cast in the form of generalised tree diagram structures that now have a g.t.o. property both forwards as well as backwards in time.

In view of the above properties, the remainder of this paper is concerned with the ECCM. Just as the normal CCM is intimately connected with a similarity transformation,[2] as is by now well-known, so the ECCM contains a double similarity transformation. In Section 4 we show how various matrix elements involving the double similarity transform of an arbitrary operator can be related to the functional derivatives with respect to the basic amplitudes $\sigma,\tilde{\sigma}$ of the average-value functional for the original operator. We hence show how the average values of arbitrary operator products can be evaluated. These results are applied in Section 5 to the very important case of operator commutators, and in so doing we show very explicitly how a set of generalised Poisson brackets naturally arises, and how thereby the general many-body ground-state problem can be formally mapped exactly onto the classical hamiltonian mechanics for the many-body, classical (c-number), quasi-local, configuration-space fields $\sigma(\xi)$ and $\tilde{\sigma}(\xi)$. This discussion is extended in Section 6 to show how a generalised mean-field theory can be exactly associated with the original many-body quantum theory, in terms of a set of generalised coherent states defined in some (fictitious) boson Hilbert space $\mathcal{H}^B$, which can itself be associated with the original Hilbert space $\mathcal{H}$. In this way we develop an exact correspondence between the generalised coherent states in $\mathcal{H}^B$ and the states representable in $\mathcal{H}$ via the ECCM as previously described. The quantum many-body problem in $\mathcal{H}$ is thus exactly "bosonised" into a (semiclassical) effective boson theory in $\mathcal{H}^B$ which is to be treated at the (generalised) mean-field level only. Finally, we compare this new "bosonisation" procedure with previous such methods in Section 7, where further extensions and applications of the ECCM are also briefly discussed.

2. OPERATOR ALGEBRA IN THE MANY-BODY HILBERT SPACE

It is typical of many quantum-mechanical calculations that the construction of states belonging to the full Hilbert space $\mathcal{H}$, is based on some initial or model state $|\Phi\rangle$. This is often, but not necessarily chosen to be some suitable state that the system would otherwise be in when (some part of) the interactions are turned off. We start here with such a state $|\Phi\rangle$, and assume furthermore that the algebra of all operators in $\mathcal{H}$ is spanned by the two subalgebras of creation and destruction operators defined with respect to the given model state $|\Phi\rangle$. We assume further that these two subalgebras and the state $|\Phi\rangle$ are *cyclic* in the sense that all of the ket states in $\mathcal{H}$ can be constructed from linear combinations of the states reached by operating on $|\Phi\rangle$ with the elements of the creation operator subalgebra; and similarly for the bra states with respect to the state $\langle\Phi|$.

We introduce a very convenient short-hand notation for the general creation and annihilation operators, $C^{\dagger}(\xi)$ and $C(\xi)$ respectively, where the label ξ is intended to represent a subset of any suitable complete set of general configuration-space indices. As an example, if $|\Phi\rangle$ is chosen to be the vacuum of a many-boson system, so that $a_i|\Phi\rangle = 0$ for all single-boson destruction operators a_i, then we may choose

$$C^{\dagger}(\xi) \rightarrow \prod_{i=1}^{m} (n_i!)^{-\frac{1}{2}} (a_i^{\dagger})^{n_i} \tag{1a}$$

and the configuration index $\xi \rightarrow \{n_i\}$ is a shorthand for any such set of integers $(n_1, n_2, \cdots n_m)$. Alternatively, in real space we may write

$$C^{\dagger}(\xi) \rightarrow (m!)^{-\frac{1}{2}} \prod_{i=1}^{m} a^{\dagger}(\vec{x}_i) \tag{1b}$$

and the configuration index $\xi \rightarrow (\vec{x}_1,\vec{x}_2, \cdots \vec{x}_m)$. In any case we assume that with any such appropriate configuration space, the identity operator I in $\mathcal{H}$ can be resolved as,

$$
\begin{aligned}
I &= |\Phi\rangle\langle\Phi| + \int' d\xi\ C^{\dagger}(\xi)|\Phi\rangle\langle\Phi|C(\xi) \\
&= \int d\xi\ C^{\dagger}(\xi)|\Phi\rangle\langle\Phi|C(\xi) \quad ,
\end{aligned}
\tag{2}
$$

in terms of a normalised set of creation and destruction operators,

$$
\langle\Phi|C(\xi)C^{\dagger}(\xi')|\Phi\rangle = \delta(\xi - \xi') \ , \tag{3}
$$

and where the integral over ξ in Eq.(2) must as usual be interpreted as a sum for discrete configuration-space labels. We also point out our convention in Eq.(2) that a prime on the integration symbol indicates the restriction that we include only those creation operators $C^{\dagger}(\xi)$ which create at least one particle. Conversely, an integration with no prime also includes the model state itself, that is $C^{\dagger}(0) \equiv I$ is included.

The reader should take careful note of our shorthand notation now, since it is quite vital for later formal developments. Without this compact notation, our later formulae would become very complicated. Worse still, they would be system-dependent. We note that our formalism is very general, and the extension to systems of fermions or spin-algebraic systems etc. is, at least in principle, straightforward. *However, for the purposes of the remainder of this paper, we shall henceforth restrict ourselves to bosonic systems.*

As indicated previously, we now assume that an arbitrary ket state $|K\rangle$ in $\mathcal{H}$ can be written as

$$
|K\rangle = \int d\xi\ k(\xi)C^{\dagger}(\xi)|\Phi\rangle \quad ; \quad k(\xi) = \langle\Phi|C(\xi)|K\rangle \tag{4}
$$

and similarly for an arbitrary bra state.

For later use we will also find it very useful to develop a notation for compounding configuration indices. We shall denote these by the ordinary symbols of addition and subtraction, e.g. $(\xi + \xi')$ and $(\xi - \xi')$, but it is clear from our prior discussion that these cannot possibly be interpreted in the usual arithmetical sense. Instead, we give the very specific definitions,

$$
\begin{aligned}
C^{\dagger}(\xi+\xi') &\equiv C^{\dagger}(\xi)C^{\dagger}(\xi') \quad ; \quad C(\xi+\xi') = C(\xi')C(\xi) \\
C^{\dagger}(\xi-\xi')|\Phi\rangle &\equiv C(\xi')C^{\dagger}(\xi)|\Phi\rangle; \quad \langle\Phi|C(\xi-\xi') \equiv \langle\Phi|C(\xi)C^{\dagger}(\xi') \quad .
\end{aligned}
\tag{5}
$$

Since the operator $C^{\dagger}(\xi-\xi')$ is thus, more specifically, defined to be the creation part (with respect to $|\Phi\rangle$) of the full contraction of the product $C(\xi')C^{\dagger}(\xi)$, it is clear that it is non-zero only when the index set ξ' is a proper subset of the index set ξ. We may similarly then define compounded expansion coefficients,

$$
\begin{aligned}
k(\xi+\xi') &\equiv \langle\Phi|C(\xi)C(\xi')\int d\eta\ k(\eta)C^{\dagger}(\eta)|\Phi\rangle = \langle\Phi|C(\xi+\xi')|K\rangle \quad , \\
k(\xi-\xi') &\equiv \langle\Phi|C(\xi)C^{\dagger}(\xi')\int d\eta\ k(\eta)C^{\dagger}(\eta)|\Phi\rangle = \langle\Phi|C(\xi-\xi')|K\rangle \quad .
\end{aligned}
\tag{6}
$$

The reader should be warned, in case of later temptation, that these operations of addition or subtraction on the configuration indices are generally neither associative nor commutative, e.g. $k(\xi+(\xi'-\xi'')) \neq k((\xi+\xi')-\xi'')$;

although $k(\xi-(\xi'+\xi'')) = k((\xi-\xi')-\xi'')$. Finally, we note that a very useful form of Wick's theorem (for bosons) can be proved which, in our notation, reads as

$$C(\xi')C^{\dagger}(\xi) = \int d\eta C^{\dagger}(\xi-\eta)C(\xi'-\eta) \quad . \tag{7}$$

3. DYNAMIC VARIATIONAL PRINCIPLE AND GENERALISED TREE DIAGRAMS

For later purposes we shall not wish to restrict ourselves to hamiltonians, or their relevant subsequent transforms, that are necessarily hermitian. Hence, with complete generality we denote the ground-state bra and ket states of the exact many-body hamiltonian H as $\langle\Psi'|$ and $|\Psi\rangle$, where

$$H|\Psi\rangle = E|\Psi\rangle \quad ; \quad \langle\Psi'|H = E\langle\Psi'| \quad , \tag{8}$$

corresponding to ground-state energy E. If either the hamiltonian depends on time or if the system is not in equilibrium, we must use the time-dependent Schrödinger equations instead of Eq.(8). In turn, these can be formulated in terms of a *dynamic variational principle* based on an action-like (henceforth referred to as the action) functional

$$\mathcal{A} = \mathcal{A}[\Psi,\Psi'] \equiv \int dt \langle\Psi'(t)|(i\partial/\partial t - H)|\Psi(t)\rangle \quad . \tag{9}$$

Stationarity of $\mathcal{A}$ with respect to all variations in the independent states $|\Psi\rangle$ and $\langle\Psi'|$ (subject only to the vanishing of $|\delta\Psi\rangle$ and $\langle\delta\Psi'|$ at the implied end-points of the time-integration in Eq.(9) -- usually $t \to \pm\infty$), then gives the correct Schrödinger equations of motion,

$$\frac{\delta\mathcal{A}}{\delta\Psi'} = 0 \Rightarrow i\frac{\partial}{\partial t}|\Psi(t)\rangle = H|\Psi(t)\rangle ; \frac{\delta\mathcal{A}}{\delta\Psi} = 0 \Rightarrow -i\frac{\partial}{\partial t}\langle\Psi'(t)| = \langle\Psi'(t)|H \quad . \tag{10}$$

We now show how various *parametrisations* of $\langle\Psi'|$ and $|\Psi\rangle$ lead to purely algebraic ways to generate many-body structures (diagrams) that can very usefully be cast in the form of *generalised tree diagrams*. In particular, we compare and contrast three different (potentially exact) many-body formalisms: (i) the configuration-interaction (CI) method[1], (ii) the normal coupled cluster method (CCM),[2-6] and (iii) the extended coupled cluster method (ECCM).[7,8] We shall show how the ECCM has many formal advantages over the normal CCM, just as the normal CCM has many merits over the CI method, as is by now very well known.[4,6] In the later Sections we then discuss the ECCM and some of its applications in more detail.

3·1. Configuration-Interaction Representation

In the CI method, $|\Psi\rangle$ and $\langle\Psi'|$ are parametrised as,

$$|\Psi\rangle = F|\Phi\rangle = \int d\xi\sigma_1(\xi)C^{\dagger}(\xi)|\Phi\rangle \; ; \; \langle\Psi'| = \langle\Phi|\tilde{F} = \int d\xi\tilde{\sigma}_1(\xi)\langle\Phi|C(\xi) \quad . \tag{11}$$

It is then easy to show that the action functional of Eq.(9) becomes,

$$\begin{aligned}\mathcal{A}[\sigma_1,\tilde{\sigma}_1] \equiv \mathcal{A}_1 &= \int dt \left\{ i\int d\xi\tilde{\sigma}_1(\xi)\dot{\sigma}_1(\xi) - \bar{H}[\sigma_1,\tilde{\sigma}_1] \right\} \\ &= \int dt \left\{ -i\int d\xi\dot{\tilde{\sigma}}_1(\xi)\sigma_1(\xi) - \bar{H}[\sigma_1,\tilde{\sigma}_1] \right\}\end{aligned} \tag{12}$$

where $\bar{H} \equiv \langle\Psi'|H|\Psi\rangle$ is specified as

$$\bar{H}_1 \equiv \bar{H}[\sigma_1,\tilde{\sigma}_1] \equiv \langle\Psi'|H|\Psi\rangle = \int d\xi \int d\xi' \langle\xi|H|\xi'\rangle \tilde{\sigma}_1(\xi)\sigma_1(\xi') ,$$
$$\langle\xi|H|\xi'\rangle = \langle\Phi|C(\xi)HC^\dagger(\xi')|\Phi\rangle \quad . \tag{13}$$

Requiring $\mathcal{A}_1$ to be stationary with respect to arbitrary small changes in $\tilde{\sigma}_1(\xi)$ and $\sigma_1(\xi')$, then yields respectively the dynamical equations,

$$i\dot{\sigma}_1(\xi) = \frac{\delta\bar{H}_1}{\delta\tilde{\sigma}_1(\xi)} \quad ; \quad i\dot{\tilde{\sigma}}_1(\xi) = -\frac{\delta\bar{H}_1}{\delta\sigma_1(\xi)} \quad . \tag{14}$$

The ground state is clearly given by the (stationary in time) equilibrium point,

$$\frac{\delta\bar{H}_1}{\delta\sigma_1(\xi)} = 0 = \frac{\delta\bar{H}_1}{\delta\tilde{\sigma}_1(\xi)} \quad . \tag{15}$$

If the hamiltonian can be split into a sum of a one-body operator (kinetic energy) part T and an interaction part V, $H = T+V$, and furthermore if the configuration space has been chosen so that the states $C^\dagger(\xi)|\Phi\rangle$ are eigenstates of T, then it is easily shown that

$$\langle\Psi'|T|\Psi\rangle \equiv \bar{T}[\sigma_1,\tilde{\sigma}_1] \equiv T_1 = \int d\xi\, E(\xi)\tilde{\sigma}_1(\xi)\sigma_1(\xi) + \text{const.}, \tag{16}$$

where $E(\xi)$ is the extra kinetic energy of the configuration $C^\dagger(\xi)|\Phi\rangle$ with respect to the model state $|\Phi\rangle$. Equation (15) may then be cast in the form,

$$\sigma_1(\xi) = -\frac{1}{E(\xi)}\frac{\delta\bar{V}_1}{\delta\tilde{\sigma}_1(\xi)} \quad ; \quad \tilde{\sigma}_1(\xi) = -\frac{1}{E(\xi)}\frac{\delta\bar{V}_1}{\delta\sigma_1(\xi)} \quad , \tag{17}$$

where the factors $E(\xi)$ appear in the usual guise of "energy denominators". Equations (17) may be regarded as the Dyson equations for the functions $\sigma_1(\xi)$ and $\tilde{\sigma}_1(\xi)$. Their solution by iteration leads to a set of terms which can be recognised as Goldstone diagrams.

The resulting CI equations, while simple, are known to have a very serious drawback for applications to many-body systems. This hinges on the fact that the diagrams for both $\sigma_1(\xi)$ and $\tilde{\sigma}_1(\xi)$ contain *disconnected* pieces. A corollary is that both sets of amplitudes are *strongly non-local* in the sense that they do not possess the cluster property, e.g. $\sigma_1(\vec{x}_1,\vec{x}_2,\vec{x}_3) \not\rightarrow 0$ as $\vec{x}_3$ is removed very far from both $\vec{x}_1$ and $\vec{x}_2$. Although the CI method is in principle exact, in practice it needs to be truncated, and the disconnected (unlinked) nature of the amplitudes then leads to the well-known "size-consistency" or "size-extensivity" problem.[6] A further, more minor, drawback to the method is that there is no manifest normalisation built into it, viz. $\langle\Psi'|\Psi\rangle \neq 1$ necessarily (although Eq.(10) does guarantee that $(d/dt)\langle\Psi'|\Psi\rangle = 0$). Nevertheless, the CI method has often been used in few-body applications, especially to problems in the realm of quantum chemistry.

3·2. Normal Coupled Cluster Representation

The roots of the coupled cluster method (CCM) date back to Hubbard[9] who realised that the problems associated with the disconnected nature of the operator F in Eq.(11) could be rectified by writing F in the form $F = \exp(S)$, where S contains only linked terms; and whence follows the linked cluster theorem of Goldstone[10] for the ground-state energy. So long as we are only interested in the ground state energy E, the static Schrödinger equation (8) may be written in the form,[2]

$$e^{-S}He^{S}|\Phi\rangle = E|\Phi\rangle \quad . \tag{18}$$

The overlap of Eq.(18) with $\langle\Phi|$ gives E itself, while the overlaps with the states $\langle\Phi|C(\xi)$ give the normal ground-state CCM hierarchy of equations that fully determine the operator S. When we need expectation values of operators other than H, it becomes necessary also to involve the state $\langle\Psi'|$. In the normal CCM, the parametrisation of the bra and ket ground states is given as,

$$\begin{aligned} |\Psi\rangle &= e^{S}|\Phi\rangle \;; \quad S = \int' d\xi\sigma_2(\xi)C^{\dagger}(\xi) \\ \langle\Psi'| &= \langle\Phi|\Omega e^{-S}; \quad \Omega = 1 + \int' d\xi\tilde{\sigma}_2(\xi)C(\xi) \quad , \end{aligned} \tag{19}$$

which may be compared with the equivalent CI expressions (11). It is immediately clear from Eq.(19) that our ground states are manifestly normalised, $\langle\Psi'|\Psi\rangle = 1$. Inserting from Eq.(19) into the action functional of Eq.(9) leads to the expression,

$$\mathcal{A}[\sigma_2,\tilde{\sigma}_2] \equiv \mathcal{A}_2 = \int dt \left\{ i\int' d\xi\tilde{\sigma}_2(\xi)\dot{\sigma}_2(\xi) - \bar{H}[\sigma_2,\tilde{\sigma}_2] \right\} \;; \tag{20}$$

$$\bar{H}[\sigma_2,\tilde{\sigma}_2] \equiv \bar{H}_2 \equiv \langle\Psi'|H|\Psi\rangle = \langle\Phi|\Omega e^{-S}He^{S}|\Phi\rangle \quad , \tag{21}$$

and where we note that the fact that the creation and annihilation subalgebras are Abelian makes the taking of the relevant derivatives above very easy.

We may now define an *average-value functional* for an arbitrary operator Θ as,

$$\begin{aligned} \langle\Psi'|\Theta|\Psi\rangle &\equiv \bar{\Theta}[\sigma_2,\tilde{\sigma}_2] \equiv \bar{\Theta}_2 = \langle\Phi|\Omega e^{-S}\Theta e^{S}|\Phi\rangle \\ &= \sum_{n=o}^{N} \frac{1}{n!}\langle\Phi|\Omega[\;\cdots\;[[\Theta,S],S],\;\cdots\;S]|\Phi\rangle \end{aligned} \tag{22}$$

where in the well-known nested commutator expansion employed in Eq.(22), the operator S appears n times. If Θ is a symmetrised sum of j-body operators, then the upper limit on the sum becomes $N = 2j$. Formally we may write,

$$\begin{aligned} \bar{\Theta}[\sigma_2,\tilde{\sigma}_2] = \sum_{m=o}^{1}\sum_{n=o}^{N}\frac{1}{m!n!}\int d\xi_1\int' d\xi_1'\cdots\int' d\xi_n'\langle\xi_m|\Theta|\xi_1'\;\cdots\;\xi_n'\rangle \\ \times\,\tilde{\sigma}_2(\xi_1)\sigma_2(\xi_1')\cdots\sigma_2(\xi_n'), \end{aligned} \tag{23}$$

where, by comparison with Eq.(22), the matrix elements may be written schematically as

$$\langle\xi_1|\Theta|\xi_1'\;\cdots\;\xi_n'\rangle = \langle\Phi|C(\xi_1)\Theta C^{\dagger}(\xi_1')\;\cdots\;C^{\dagger}(\xi_n')|\Phi\rangle_{\mathcal{L}} \quad , \tag{24}$$

and where the suffix $\mathcal{L}$ on these matrix elements indicates the very definite linked structure implied by Eq.(22).

With this notation, the stationarity of $\mathcal{A}_2$ with respect to small variations in $\sigma_2(\xi)$ and $\tilde{\sigma}_2(\xi)$ then yields dynamical equations of precisely the same form as in the CI equations (14), but with all (CI) indices 1 replaced by (normal CCM) indices 2. At the equilibrium point, the ground state is again similarly given as in Eq.(15), which equations now reduce to the form,

$$\langle\Phi|C(\xi)e^{-S}He^{S}|\Phi\rangle = 0 \quad ; \quad \xi \neq 0$$
$$\langle\Phi|\Omega[e^{-S}He^{S}, C^{\dagger}(\xi)]|\Phi\rangle = 0 \quad , \tag{25}$$

which determine the static amplitudes $\sigma_2(\xi)$ and $\tilde{\sigma}_2(\xi)$. We find that the Dyson equations for $\sigma_2, \tilde{\sigma}_2$ can again be put in similar form to the CI equations (17), but a big difference now arises from the replacement $\overline{V}_1 \to \overline{V}_2$. Thus in the present case the resulting diagrams for the energy are fully connected, as indeed are the diagrams for $\sigma_2(\xi)$ as expected from our earlier remarks on the linked-cluster theorem. However, $\tilde{\sigma}_2(\xi)$ still contains disconnected terms, and remains problematic. Of course if we restrict ourselves to the energy, then the amplitudes $\tilde{\sigma}_2(\xi)$ are not needed, as already mentioned. However, problems still remain for the expectation values of arbitrary operators.

Very related to the above discussion is the concept of *"generalised time ordering"* (g.t.o.), which is a useful tool for classifying and combining classes of Goldstone diagrams.[11,7] This technique is based on the factorisation property of disjoint sets of legs of a Goldstone diagram, and leads to the factorisation of corresponding energy denominators across such legs when all permitted time orderings are included. The normal CCM generates diagrams for the energy with a g.t.o. in the "downward" direction (i.e. backwards in time) only. The diagrams for the expectation value of the hamiltonian can thereby be represented by what we now call *"normal g.t.o. trees"* or *"normal CCM trees"*. These are diagrams which "branch out" in the downward direction only (-- thus resembling the root system of a real tree rather than the visible tree structure!). As explained elsewhere,[7] each link or branch (i.e., root!) in such a diagram corresponds to a definite set of particle/hole lines associated with the configuration ξ of the corresponding state $C^{\dagger}(\xi)|\Phi\rangle$. If such a (downward) tree diagram for the energy (the totality of which give the normal CCM expression), is divided into two by cutting one link, the lower part (which constitutes a diagram associated with some amplitude $\sigma_2(\xi)$) will always be linked, whereas the upper part (which constitutes a diagram for the corresponding amplitude $\tilde{\sigma}_2(\xi)$) may be unlinked. For further details we refer the interested reader to the by now quite extensive literature on the normal CCM and its applications.[4-6,12-15]

3·3. Extended Coupled Cluster Representation

For reasons already alluded to above, the normal CCM has been almost wholly concerned with energy calculations. There are virtually no calculations involving average-value properties of other operators. In such cases, the unlinked nature of the operator Ω (or equivalently its amplitudes $\tilde{\sigma}_2(\xi)$) may well lead to computational difficulties in the case of practical (i.e., truncated) calculations. Just as in passing from the CI method to the normal CCM we cured the $\sigma_1(\xi)$ of their disconnectedness by their replacement with $\sigma_2(\xi)$, so the extended CCM (ECCM) aims to cure the remaining disconnectedness in $\tilde{\sigma}_2(\xi)$. This is achieved by the following ECCM parametrisation,

$$|\Psi\rangle = e^{S}|\Phi\rangle \quad ; \quad S = \int' d\xi\, s(\xi)C^{\dagger}(\xi)$$
$$\langle\Psi'| = \langle\Phi|e^{S''}e^{-S} \quad ; \quad S'' = \int' d\xi\, \tilde{\sigma}_3(\xi)C(\xi) \quad , \tag{26}$$

which may be compared with its normal CCM counterpart in Eq.(19). Just as in the normal CCM, the coefficients $\tilde{\sigma}_2(\xi)$ are the (unlinked) average values of the creation operators, $\tilde{\sigma}_2(\xi) = \langle\Psi'|C^{\dagger}(\xi)|\Psi\rangle$, so now it can be shown that the linked parts of these averages are precisely the new amplitudes $\tilde{\sigma}_3(\xi)$. Again, the ground states are trivially observed to be manifestly normalised, $\langle\Psi'|\Psi\rangle = 1$. An average-value functional of an arbitrary operator Θ now takes the form,

$$\overline{\Theta} \equiv \langle\Psi'|\Theta|\Psi\rangle = \langle\Phi|\hat{\Theta}|\Phi\rangle \quad , \tag{27}$$

in terms of the (doubly) similarity-transformed operator $\hat{\Theta}$,

$$\hat{\Theta} \equiv e^{S''}e^{-S}\Theta e^{S}e^{-S''} \quad . \tag{28}$$

Finally in the ECCM, it turns out to be very convenient to define a new operator Σ,

$$\Sigma \equiv \int' d\xi\, \sigma_3(\xi)C^\dagger(\xi) \quad ; \quad \sigma_3(\xi) \equiv \langle\Phi|C(\xi)e^{S''}S|\Phi\rangle \quad , \tag{29}$$

which has an inverse transformation,

$$S|\Phi\rangle = e^{-S''}\Sigma|\Phi\rangle \quad ; \quad s(\xi) = \langle\Phi|C(\xi)e^{-S''}\Sigma|\Phi\rangle \quad . \tag{30}$$

Insertion of the above parametrisation into the action functional of Eq.(9) gives again a form analogous to that in Eqs.(12) and (20),

$$\mathcal{A}[\sigma_3,\tilde{\sigma}_3] \equiv \mathcal{A}_3 = \int dt\, \{i\int' d\xi \tilde{\sigma}_3(\xi)\dot{\sigma}_3(\xi) - \overline{H}[\sigma_3,\tilde{\sigma}_3]\} \quad . \tag{31}$$

The variational principle again leads to the dynamic equations,

$$i\dot{\sigma}_3(\xi) = \frac{\delta\overline{H}_3}{\delta\tilde{\sigma}_3(\xi)} \quad ; \quad i\dot{\tilde{\sigma}}_3(\xi) = -\frac{\delta\overline{H}_3}{\delta\sigma_3(\xi)} \quad . \tag{32}$$

The expression for $\overline{H}_3 \equiv \overline{H}[\sigma_3,\tilde{\sigma}_3]$ is now more complicated than before. The general form of an arbitrary average-value functional $\overline{\Theta}$ can now be shown to be given by,

$$\overline{\Theta}[\sigma_3,\tilde{\sigma}_3] \equiv \overline{\Theta}_3 = \sum_{m,n} \frac{1}{m!n!} \int' d\xi_1 \cdots \int' d\xi_m \int' d\xi'_1 \cdots \int' d\xi'_n \langle\xi_1\cdots\xi_m|\Theta|\xi'_1\cdots\xi'_n\rangle$$
$$\times\, \tilde{\sigma}(\xi_1)\cdots\tilde{\sigma}(\xi_m)\sigma(\xi'_1)\cdots\sigma(\xi'_n) \quad , \tag{33}$$

in which the matrix elements, expressed schematically by the expression,

$$\langle\xi_1\cdots\xi_m|\Theta|\xi'_1\cdots\xi'_n\rangle = \langle\Phi|C(\xi_1)\cdots C(\xi_m)\Theta C^\dagger(\xi'_1)\cdots C^\dagger(\xi'_n)|\Phi\rangle_{\mathcal{DL}} \tag{34}$$

have a very definite linked ($\mathcal{DL}$) structure, as can be found by explicit construction, and as explained in detail elsewhere.[7] We note here only that the matrix elements of Eq.(34) are not all independent, in the sense that one may find definite recursion relations between them. Particularly useful in this respect is the identity,

$$\frac{\delta\overline{\Theta}_3}{\delta\tilde{\sigma}_3(\xi+\xi')} = \frac{\delta^2\overline{\Theta}_3}{\delta\tilde{\sigma}_3(\xi)\delta\tilde{\sigma}_3(\xi')} + \int' d\eta \left[\sigma_3(\xi+\eta)\frac{\delta^2\overline{\Theta}_3}{\delta\sigma_3(\eta)\delta\tilde{\sigma}_3(\xi')} + \frac{\delta^2\overline{\Theta}_3}{\delta\tilde{\sigma}_3(\xi)\delta\sigma_3(\eta)}\sigma_3(\eta+\xi')\right] + \int' d\eta \int' d\eta' \sigma_3(\xi+\eta)\frac{\delta^2\overline{\Theta}_3}{\delta\sigma_3(\eta)\delta\sigma_3(\eta')}\sigma_3(\eta'+\xi') \quad . \tag{35}$$

Such relations as Eq.(35) in the ECCM are most easily proven by starting with the average value functional $\overline{\Theta}$ written as a functional of the operators S'' and S; and then making the change of 'variables' to the operators S'' and Σ in the usual partial differential sense, as discussed further in Section 4.

As usual, the energy functional $\bar{H} \equiv \langle\Psi'|H|\Psi\rangle$ now plays the particularly important role, through Eq.(32), of determining the dynamics of the amplitudes $\sigma_3(\xi)$ and $\tilde{\sigma}_3(\xi')$, and their equilibrium values. The kinetic energy operator T has an average value functional which, after some algebra, can be expressed in terms of these amplitudes as,

$$\bar{T}[\sigma_3,\tilde{\sigma}_3] = \langle\Phi|S''T\Sigma|\Phi\rangle = \text{const.} + \int' d\xi E(\xi)\tilde{\sigma}_3(\xi)\sigma_3(\xi) \quad . \tag{36}$$

Comparison with Eq.(16), and the discussion following it, again shows that at equilibrium the values of $\sigma_3,\tilde{\sigma}_3$ represent the contributions of definite classes of Goldstone diagrams. These may again be conveniently classified in terms of *extended g.t.o. trees.*[7] By comparison with the normal CCM, these extended CCM tree diagrams now branch upwards as well as downwards at any vertex. The average value functional for an arbitrary operator may then uniquely be expressed in terms of such extended g.t.o. tree diagrams. A consequence is that if any such extended tree diagram for the energy is divided into two by cutting any single link, both the top part (which is a diagram for $\tilde{\sigma}_3$) and the bottom part (which is a diagram for σ_3) now separately remain connected diagrams. In this way, now *all* of our basic amplitudes $\sigma_3(\xi)$ and $\tilde{\sigma}_3(\xi)$ in the ECCM are quasi-local in the sense that they obey the *cluster property*, namely that if any subset of the particles incorporated in the configuration-space labelling ξ is removed infinitely far from the remainder, the amplitude goes to zero.

In the remainder of this paper we now deal only with the extended CCM and investigate some of its consequences and uses. We therefore now drop the suffices 3, and henceforth rename $\tilde{\sigma}_3(\xi) \rightarrow \tilde{\sigma}(\xi)$; $\sigma_3(\xi) \rightarrow \sigma(\xi)$.

4. FUNCTIONAL DERIVATIVES AND MATRIX ELEMENTS

Neither the normal nor extended version of the CCM is manifestly hermitian. In the ECCM with which we are now concerned, this ultimately derives from the fact that the (double) similarity transformations which generate the ket and bra ground states are not unitary. Thus the formalism is actually a *biorthogonal* formulation of the many-body problem, and the functional $\bar{\Theta}$ of Eq.(27), for an arbitrary operator Θ, represents the real expectation value functional, in view of the definite normalisation $\langle\Psi'|\Psi\rangle = 1$.

For further development of the ECCM formalism, we will have need for various matrix elements involving the doubly similarity-transformed operator $\hat{\Theta}$ of Eq.(28), and we now point out and exploit the intimate connection between such matrix elements and the functional derivatives of the average-value functional $\bar{\Theta}$. It is clear from our earlier discussion that it is easiest to compute $\bar{\Theta}$ in the first place as a functional of the operators S and S'',

$$\bar{\Theta} = \langle\Phi|e^{S''}e^{-S}\Theta e^{S}|\Phi\rangle \quad , \tag{37}$$

where we have used Eqs.(27) and (28), and the fact that S'' is built only from annihilation operators, as in Eq.(26). From Eq.(37) we may thus express the average-value functional in the form $\bar{\Theta} = \bar{\bar{\Theta}}[s,\tilde{\sigma}]$, using Eqs.(26), where the double-bar notation simply reminds us of the functional arguments. It is then trivial to see directly from Eq.(37) that the first-order functional derivatives in this representation are given by,

$$\frac{\delta\bar{\bar{\Theta}}}{\delta\tilde{\sigma}(\xi)} = \langle\Phi|C(\xi)\hat{\Theta}|\Phi\rangle \quad ; \quad \frac{\delta\bar{\bar{\Theta}}}{\delta s(\xi)} = \langle\Phi|e^{S''}e^{-S}[\Theta,C^{\dagger}(\xi)]e^{S}|\Phi\rangle \quad . \tag{38}$$

Finally, we may "change variables" from the operators S,S'' to Σ,S'' using

Eq.(29) to give equivalently $\bar{\Theta} \rightarrow \bar{\Theta}[\sigma,\tilde{\sigma}] \equiv \bar{\bar{\Theta}}[s,\tilde{\sigma}]$. Making use of the usual chain rule of partial differentiation, we readily find,

$$\frac{\delta\bar{\bar{\Theta}}}{\delta\tilde{\sigma}(\xi)} = \frac{\delta\bar{\Theta}}{\delta\tilde{\sigma}(\xi)} + \int' d\xi'\sigma(\xi+\xi')\frac{\delta\bar{\Theta}}{\delta\sigma(\xi')} \quad ; \quad \frac{\delta\bar{\bar{\Theta}}}{\delta s(\xi)} = \int' d\xi'\omega(\xi-\xi')\frac{\delta\bar{\Theta}}{\delta\sigma(\xi')} \quad , \tag{39}$$

where we have used the notation of Eqs.(5) and (6) to compound the configuration space indices; and where the amplitude $\omega(\xi)$ and a comparable amplitude $\bar{\omega}(\xi)$ needed later, are defined as,

$$\omega(\xi) \equiv \langle\Phi|e^{S''}C^{\dagger}(\xi)|\Phi\rangle \quad ; \quad \bar{\omega}(\xi) \equiv \langle\Phi|e^{-S''}C^{\dagger}(\xi)|\Phi\rangle \quad . \tag{40}$$

These latter amplitudes are easily seen to obey the orthogonality relations,

$$\int d\eta\bar{\omega}(\xi-\eta)\omega(\eta-\xi') = \delta(\xi-\xi') = \int d\eta\omega(\xi-\eta)\bar{\omega}(\eta-\xi') \quad . \tag{41}$$

We also note in passing that our previous Eq.(35) is proven by making repeated use of Eq.(39) together with the trivial relation $\delta^2\bar{\Theta}/\delta\tilde{\sigma}(\xi)\delta\tilde{\sigma}(\xi') = \delta\bar{\Theta}/\delta\tilde{\sigma}(\xi+\xi')$, which follows immediately from Eq.(38).

The combination of Eqs.(38) and (39) immediately gives that for an arbitrary operator A,

$$\langle\Phi|C(\xi)\hat{A}|\Phi\rangle = \frac{\delta\bar{A}}{\delta\tilde{\sigma}(\xi)} + \int' d\xi'\sigma(\xi+\xi')\frac{\delta\bar{A}}{\delta\sigma(\xi')} \quad ; \quad \xi \neq 0 \quad . \tag{42}$$

Relatively straightforward algebra using the above relations and judicious insertions of the identity operator from Eq.(2), also leads to other comparable expressions involving matrix elements of the operator $\hat{A}$ being able to be written in terms of functional derivatives of the average-value functional $\bar{A} = \bar{A}[\sigma,\tilde{\sigma}]$. As a further example, we simply quote the result (valid for $\xi \neq 0$),

$$\langle\Phi|\hat{A}C^{\dagger}(\xi)|\Phi\rangle = \frac{\delta\bar{A}}{\delta\sigma(\xi)} + \int' d\eta\frac{\delta\bar{A}}{\delta\tilde{\sigma}(\eta)}L(\eta\xi) + \int' d\eta\int' d\eta'\frac{\delta\bar{A}}{\delta\sigma(\eta)}\sigma(\eta+\eta')L(\eta'\xi) \quad , \tag{43}$$

where the function $L(\xi\xi') = L(\xi'\xi)$ is defined as,

$$L(\xi\xi') = \int d\eta\int d\eta'\omega(\eta+\eta')\bar{\omega}(\xi-\eta)\bar{\omega}(\xi'-\eta') = \langle\Phi|e^{S''}C^{\dagger}(\xi)C^{\dagger}(\xi')|\Phi\rangle \quad . \tag{44}$$

In a similar fashion one may evaluate higher matrix elements in terms of higher-order functional derivatives, e.g. the matrix element $\langle\Phi|C(\xi)\hat{A}C^{\dagger}(\xi')|\Phi\rangle$ will involve second-order derivatives. Evaluation of these higher elements is often facilitated by a judicious use of Wick's theorem in the form of Eq.(7).

4·1. Average Values of Operator Products

As a nice example of the above techniques we now turn to the very important topic of the representation in the ECCM of operator products. It is clear that products of operators transform under the double similarity transformation of Eq.(28) into the corresponding product of transformed operators, and therefore that their average values can be represented in the ECCM as,

$$\langle\Psi'|\Theta_1\Theta_2 \cdots \Theta_n|\Psi\rangle = \langle\Phi|\hat{\Theta}_1\hat{\Theta}_2 \cdots \hat{\Theta}_n|\Phi\rangle \quad . \tag{45}$$

If, on the right-hand side of Eq.(45), one inserts the identity operator from Eq.(2) between each adjacent pair of operators, we may thereby express such average values in terms of first or second functional derivatives of the average values of the individual operators. For present purposes we restrict ourselves to what is in any case the most important example of the product of two operators. In this case we need only first-order derivatives, and Eqs. (42) and (43) suffice. In this way we may easily evaluate the average value $\overline{AB} \equiv \langle\Psi'|AB|\Psi\rangle \equiv \langle AB\rangle$ as,

$$\overline{AB} = \overline{A}\cdot\overline{B} + \int' d\xi \int' d\xi' \left[\frac{\delta\overline{A}}{\delta\sigma(\xi)} \frac{\delta\overline{B}}{\delta\sigma(\xi')} X_{11}(\xi\xi') + \frac{\delta\overline{A}}{\delta\sigma(\xi)} \frac{\delta\overline{B}}{\delta\tilde{\sigma}(\xi')} X_{12}(\xi\xi') + \frac{\delta\overline{A}}{\delta\tilde{\sigma}(\xi)} \frac{\delta\overline{B}}{\delta\sigma(\xi')} X_{21}(\xi\xi') + \frac{\delta\overline{A}}{\delta\tilde{\sigma}(\xi)} \frac{\delta\overline{B}}{\delta\tilde{\sigma}(\xi')} X_{22}(\xi\xi') \right] \tag{46}$$

where the X-coefficients are given by the expressions

$$X_{11}(\xi\xi') = X_{11}(\xi'\xi) = \sigma(\xi+\xi') + \int' d\eta \int' d\eta' \sigma(\xi+\eta) L(\eta\eta') \sigma(\eta'+\xi'),$$
$$X_{22}(\xi\xi') = X_{22}(\xi'\xi) = L(\xi\xi'), \tag{47}$$
$$X_{12}(\xi\xi') = \delta(\xi-\xi') + \int' d\eta \sigma(\xi+\eta) L(\eta\xi') \; ; \; X_{21}(\xi\xi') = \int' d\eta L(\xi\eta) \sigma(\eta+\xi') .$$

5. COMMUTATORS AND GENERALISED POISSON BRACKETS

As a particularly important application of the results of Section 4·1, we now consider the expectation value of the *commutator* of two operators. In view of the high degree of symmetry exhibited in Eq.(47) by the X-coefficients, there is a considerable consequent simplification in this regard. We find,

$$\langle\Psi'|AB - BA|\Psi\rangle \equiv \langle[A,B]\rangle = i\{\overline{A},\overline{B}\} \quad , \tag{48}$$

where the *generalised Poisson bracket* $\{\overline{A},\overline{B}\}$ is defined as,

$$i\{\overline{A},\overline{B}\} \equiv \int' d\xi \left[\frac{\delta\overline{A}}{\delta\sigma(\xi)} \frac{\delta\overline{B}}{\delta\tilde{\sigma}(\xi)} - \frac{\delta\overline{B}}{\delta\sigma(\xi)} \frac{\delta\overline{A}}{\delta\tilde{\sigma}(\xi)} \right] \quad . \tag{49}$$

We can make these results even more suggestive by choosing as new basic (quasi-local) field variables, the *generalised fields* $\phi(\xi)$ and their canonically conjugate *generalised momentum densities* $\pi(\xi)$, defined as,

$$\phi(\xi) \equiv 2^{-\frac{1}{2}}[\sigma(\xi) + \tilde{\sigma}(\xi)] \quad ; \quad \pi(\xi) \equiv 2^{-\frac{1}{2}} i[\tilde{\sigma}(\xi) - \sigma(\xi)] \quad . \tag{50}$$

In this way we can re-define our average values $\langle A\rangle \equiv \overline{A} \rightarrow \overline{A}[\phi,\pi]$, in terms of which Eq.(49) becomes

$$\{\overline{A},\overline{B}\} \equiv \int' d\xi \left[\frac{\delta\overline{A}}{\delta\phi(\xi)} \frac{\delta\overline{B}}{\delta\pi(\xi)} - \frac{\delta\overline{A}}{\delta\pi(\xi)} \frac{\delta\overline{B}}{\delta\phi(\xi)} \right] \tag{51}$$

The equations of motion (32) for the amplitudes $\sigma,\tilde{\sigma}$ are thereby re-cast into the form,

$$\dot{\phi}(\xi) = \frac{\delta\overline{H}}{\delta\pi(\xi)} = \{\phi(\xi),\overline{H}\} \quad ; \quad \dot{\pi}(\xi) = -\frac{\delta\overline{H}}{\delta\phi(\xi)} = \{\pi(\xi),\overline{H}\} \quad . \tag{52}$$

Finally, for an arbitrary, intrinsically time-dependent operator $A(t)$, it is easy to show using Eqs.(52) that the equation of motion for its average-value functional $\langle A\rangle = \overline{A}[\phi,\pi;t]$ is,

$$\frac{d\overline{A}}{dt} = \left\langle \frac{\partial A}{\partial t} \right\rangle + \{\overline{A},\overline{H}\} = \left\langle \frac{\partial A}{\partial t} \right\rangle + \frac{1}{i} \langle [A,H] \rangle \quad . \tag{53}$$

Equation (53) shows both that the equation of motion is indeed the proper quantum-mechanical one, and that the connection to classical physics arises through a very well-defined, suitably generalised version of the *correspondence principle*.

We are thus led to the very important result that the whole of our quantum many-body problem has formally been exactly mapped onto the classical Hamiltonian mechanics for the (c-number) quasi-local fields $\phi(\xi)$ and $\pi(\xi)$ which are themselves functions in the many-body configuration space labelled by the indices ξ. In this way we can take over (or suitably extend) the whole of the classical formalism to describe (*exactly*, in principle, if no truncations are made) the quantum many-body system. In particular we can make easy contact with such things as *conservation laws* and the associated *sum rules*, through the corresponding Noether currents.

6. EXACT BOSONISATION AND GENERALISED COHERENT STATES

It comes as no surprise that the ECCM when truncated at its lowest obvious level of approximation (-- namely, the so-called SUB1 approximation wherein in the expansions (26) and (29) for S'' and Σ, the 'sums over configurations' are restricted to one-body (i.e. one particle or one particle-hole) configurations ξ), is precisely equal to the *mean field theory* or semiclassical approximation.[7] For the bosonic systems which have mainly been emphasised here, this is precisely the ordinary *coherent-state approximation* (while for fermionic systems it is just the Hartree-Fock approximation). We briefly remind the reader how the mean field approximation for bosons can be expressed in terms of the (classical) atomic or ordinary coherent states of Glauber.[16,17] In terms of a complete set of our original single-boson creation operators in $\mathcal{H}$, say $a^{\dagger}(\vec{x})$, (and see the discussion surrounding Eqs. (1a,b)), the Glauber coherent states are defined as

$$|\gamma\rangle \;=\; e^{\Gamma}|\Phi\rangle \quad ; \quad \Gamma \;=\; \int d\vec{x}[\phi(\vec{x})a^{\dagger}(\vec{x}) - \phi^{*}(\vec{x})a(\vec{x})] = -\Gamma^{\dagger} \quad , \tag{54}$$

where $\phi(\vec{x})$ and its complex conjugate $\phi^{*}(\vec{x})$ are scalar (i.e. c-number) fields, and with $|\Phi\rangle$ the vacuum, $a(\vec{x})|\Phi\rangle = 0$, as before. It is straightforward to show that these states are eigenstates of the destruction operator $a(\vec{x})$,

$$a(\vec{x})e^{\Gamma}|\Phi\rangle \;=\; \phi(\vec{x})e^{\Gamma}|\Phi\rangle \quad ; \quad \langle\Phi|e^{-\Gamma}a^{\dagger}(\vec{x}) \;=\; \phi^{*}(\vec{x})\langle\Phi|e^{-\Gamma} \quad . \tag{55}$$

We can then also use Eq.(55) to show rather easily that the expectation value of a suitably normal-ordered arbitrary operator $\Theta = :\Theta[a,a^{\dagger}]:$ in these coherent states is just given by replacing the field operators $a(\vec{x}), a^{\dagger}(\vec{x})$ in the functional by their c-number coherent-state expectation values $\phi(\vec{x})$, $\phi^{*}(\vec{x})$ respectively:

$$\langle\Theta\rangle_{\Gamma} \;\equiv\; \langle\Phi|e^{-\Gamma} : \Theta[a,a^{\dagger}] : e^{\Gamma}|\Phi\rangle \;=\; \Theta[\phi,\phi^{*}] \quad . \tag{56}$$

In particular, the static mean field approximation is obtained by minimising $\langle H\rangle_{\Gamma}$ with respect to the one-body fields $\phi(\vec{x}), \phi^{*}(\vec{x})$.

In the light of the discussion in the previous Section, it is now natural to enquire whether our entire formalism can also be *exactly* recast as a *generalised mean field theory* in terms of a well-defined set of (fictitious) *ideal coherent bosons*, whose appropriate expectation values (defined suitably in their own Hilbert space) equal the (many-body) "classical" fields $\sigma, \tilde{\sigma}$. The basic configuration space operators $C(\xi)$, $C^{\dagger}(\xi)$ are definitely not candidates for the annihilation and creation operators associated with these ideal bosons, both because their expectation values are generally much more complicated functionals of $\sigma(\xi)$ and $\tilde{\sigma}(\xi)$, and because their commutation relations are also generally not those of ideal bosons. Instead, we simply attempt to map the original many-body Hilbert space $\mathcal{H}$ onto some fictitious boson Hilbert space $\mathcal{H}^{B}$. In view of the biorthogonal nature of our previous formulation, we now postulate the existence in $\mathcal{H}^{B}$ of vacuum states $|\Phi_{B}\rangle$ and $\langle\Phi'_{B}|$, and ideal boson operators $A(\xi)$, $\tilde{A}(\xi)$ associated with each configuration ξ, such that

$$[A(\xi),A(\xi')] = 0 = [\tilde{A}(\xi),\tilde{A}(\xi')] \quad ; \quad [A(\xi),\tilde{A}(\xi')] = \delta(\xi-\xi') ,$$
$$A(\xi)|\Phi_B\rangle = 0 = \langle\Phi'_B|\tilde{A}(\xi) \quad . \tag{57}$$

Furthermore, for every operator Θ in $\mathcal{H}$ we now associate its *boson image* Θ^B in $\mathcal{H}^B$, defined as:

$$\Theta^B \equiv \sum_{m,n} \frac{1}{m!n!} \int' d\xi_1 \cdots \int' d\xi_m \int' d\xi'_1 \cdots \int' d\xi'_n \langle\xi_1 \cdots \xi_m|\Theta|\xi'_1 \cdots \xi'_n\rangle$$
$$\times \tilde{A}(\xi_1) \cdots \tilde{A}(\xi_m) A(\xi'_1) \cdots A(\xi'_n) , \tag{58}$$

by analogy to Eq.(33), and with matrix elements exactly as specified in Eq. (34). By analogy with Eq.(54) it seems reasonable to consider a set of *generalised coherent states* (or, more properly, bi-coherent states) in $\mathcal{H}^B$, defined as

$$|\Psi_B\rangle \equiv e^G|\Phi_B\rangle \quad ; \quad \langle\Psi'_B| \equiv \langle\Phi'_B|e^{-G}$$
$$G \equiv \int' d\xi [\sigma(\xi)\tilde{A}(\xi) - \tilde{\sigma}(\xi)A(\xi)] \quad , \tag{59}$$

which are to be interpreted as the images in $\mathcal{H}^B$ of the corresponding states $|\Psi\rangle$ and $\langle\Psi'|$ in $\mathcal{H}$. The generalised coherent-state expectation value of Θ^B can be shown, by precise analogy with our previous atomic coherent state result of Eq.(56), to be identically equal to the earlier expression in Eq. (33),

$$\langle\Phi'_B|e^{-G}\Theta^B e^G|\Phi_B\rangle = \bar{\Theta}[\sigma,\tilde{\sigma}] \quad . \tag{60}$$

Finally, we are led to consider the generalised coherent-state action functional $\mathcal{A}^B$, defined in $\mathcal{H}^B$ to be the image of the action $\mathcal{A}$ of Eq.(9) in $\mathcal{H}$,

$$\mathcal{A}^B = \int dt \, \langle\Phi'_B|e^{-G(t)}(i\partial/\partial t - H^B)e^{G(t)}|\Phi_B\rangle \quad . \tag{61}$$

By making use of the result,

$$e^{-G}\frac{\partial}{\partial t}e^G = \int' d\xi\{\dot{\sigma}(\xi)\tilde{A}(\xi) - \dot{\tilde{\sigma}}(\xi)A(\xi) + \tfrac{1}{2}[\dot{\sigma}(\xi)\tilde{\sigma}(\xi) - \sigma(\xi)\dot{\tilde{\sigma}}(\xi)]\} \quad , \tag{62}$$

which follows from Eqs.(59) and (57), we finally find the result that $\mathcal{A}^B$ has an identical form to our earlier results in Eqs.(12) and (20), and most particularly in Eq.(31);

$$\mathcal{A}^B = \int dt \, \{i\int' d\xi \tilde{\sigma}(\xi)\dot{\sigma}(\xi) - \bar{H}[\sigma,\tilde{\sigma}]\} \quad . \tag{63}$$

Hence, a variational principle in $\mathcal{H}^B$ applied to the action $\mathcal{A}^B$, *exactly* reproduces our previous *exact* equations of motion (32) in $\mathcal{H}$ for the ECCM. These results are discussed further in Section 7.

7. SUMMARY AND DISCUSSION

We have shown how a variational principle for our action functional of Eq.(9) enables quantum many-body theory to be written in the form of classical hamiltonian mechanics for the many-body (c-number) configuration-space amplitudes $\sigma,\tilde{\sigma}$ for each of the CI method, the normal CCM and the extended CCM. Only in the latter case are all of these fields quasi-local in the sense of obeying the cluster property. In any realistic calculation, each of these methods must be truncated, e.g. in the so-called SUBn approximation by restricting the configuration indices ξ to at most n particles (or particle-hole pairs). In the CI method, the limbs of the associated CI trees would need to be very thick (large n), unless the interaction is so weak that low-order perturbation theory suffices. By contrast, the normal and extended CCM always perform such various infinite-order summations of Goldstone diagrams, even for low n, that the CCM trees need not be so thick for good energy results.

However, the results are still in principle quite dependent on the choice of model state $|\Phi\rangle$. The normal (ground-state) CCM may miss altogether a phase of the system with some broken symmetry not built into $|\Phi\rangle$. One of us has shown[18] how this may be overcome within the normal CCM by constructing new model states $|\Phi'\rangle$, using a combination of the excited-state CCM formalism of Emrich[15] to search for *"de-excited states"* of lower energy than in the ground-state formalism discussed here, and the *maximum-overlap stability criterion* of Kümmel.[19] In the extended CCM, the hope is that the limbs of the generalised trees can again be relatively slim for good approximations, but by contrast the hope has been expressed[7] that the model state $|\Phi\rangle$ can remain the naive (symmetry-conserving) vacuum state even in the broken-symmetry phase. This is certainly true at the fermion SUB1 level, for example, where the normal SUB1 approximation does not produce the correct deformed Hartree-Fock state, whereas the extended CCM at SUB1 level does. The clear hope is that since the ECCM easily permits symmetry-breaking to be incorporated by introducing suitable symmetry-breaking amplitudes $\sigma, \tilde{\sigma}$ with a *given* $|\Phi\rangle$, it will find future applications to such *topological excitations* as vortices in liquid helium. Our formalism also permits applications to *non-equilibrium phenomena*, and to nonlinear behaviour far from equilibrium.

We turn now to the implications of our new bosonisation procedure discussed in Section 6. We note that there is a very long tradition in the bosonisation of spin-algebraic or fermionic systems. As examples of the genre we mention the methods of Holstein and Primakoff,[20] Dyson,[21] Schwinger,[22] and others;[23,24] and refer the reader to the review by Garbaczewski.[25] In most previous procedures the philosophy has been to establish an exact equivalence between the Lie algebra of the spin operators or suitably chosen pairs of fermion operators, and the Lie algebra of canonical boson operators in an ideal boson space. It usually transpires that the boson Hilbert space is too large, in the sense that physically realisable states in the original Hilbert space map only onto a *subspace* of the boson image space. The aim is that with a judicious choice of mapping, the low-lying collective excitations of the original system might be treated (semi-) classically or nearly so in the mapped space, in the sense that as a zeroth approximation the ideal bosons can be treated as non-interacting. Further, one hopes that to go beyond zeroth order, the residual interactions (in the usual quantum-mechanical sense) are weak enough to be easily treated by conventional means in the boson space. Generally, however, the boson interactions are far from trivial, and the formalism is still a relatively complicated theory of interacting bosons in a projected subspace of an ideal boson space.

By contrast, our own bosonisation procedure differs in at least three ways. In the first place, no effort is made to preserve the Lie algebra. Secondly, whereas earlier bosonisation schemes have been applied only to fermion or spin systems, ours may equally be applied to boson systems. Our bosonisation procedure transforms the original boson (or other) theory to an effective boson theory which is to be used at mean-field level (with only classical interactions between the bosons). Clearly, therefore, it cannot be bosonised any further! Thirdly, whereas in the earlier bosonisation schemes, the physical states map onto a subspace of the image space $\mathcal{H}^B$, here the generalised coherent states form only a *subset* of the states in $\mathcal{H}^B$. In this way we lose *the superposition principle* in the mapped physical space. The ultimate reason for this is the strongly coherent form adopted for our ground-state trial wavefunctions in Eqs.(26), which means that only the ground state and those adiabatically excited states which are strongly collective, are easily so describable.

With regard to this latter point, it is quite possible to develop further the ground-state ECCM to include excited states in very much the same spirit as Emrich[15] has developed the normal CCM. The upshot is that the *excited states* and their *excitation energies* can be found by solving a linear eigenvalue problem to diagonalise an operator built from second-order functional derivatives of $\bar{H}$, evaluated at the equilibrium point. Likewise, our time-dependent formalism enables us to treat the same problem via the dynamics of small oscillations around the stationary equilibrium. This dynamics is again governed by an effective hamiltonian obtained by linearising the equations of motion (32) around the stationary point. In this way we can build up a set of *generalised random phase approximation* (RPA) equations in configuration space, with two main differences from the ordinary RPA.[26] Firstly, the configuration indices ξ are arbitrary, and not restricted to just one particle (or particle-hole pair) as in ordinary RPA; and secondly, when no such truncations are made the resulting generalised RPA equations are *exact*.

Further details and applications of both the present work and the extensions mentioned above, will be published elsewhere.[27,28]

REFERENCES

1. R. K. Nesbet, Phys.Rev. 109:1632 (1958).
2. F. Coester, Nucl.Phys. 7:421 (1958).
3. F. Coester and H. Kümmel, Nucl.Phys. 17:477 (1960).
4. H. Kümmel, K. H. Lührmann and J. G. Zabolitzky, Phys.Reports 36C:1 (1978).
5. V. Kvasnička, V. Laurinc and S. Biskupič, Phys.Reports 90C:160 (1982).
6. H. Kümmel, in: "Nucleon-Nucleon Interaction and Nuclear Many-Body Problems", S. S. Wu and T. T. S. Kuo (eds.), World Scientific, Singapore (1984), p.46.
7. J. Arponen, Ann.Phys. (N.Y.) 151:311 (1983).
8. J. Arponen and E. Pajanne, in: "Recent Progress in Many-Body Theories", H. Kümmel and M. L. Ristig (eds.), Lecture Notes in Physics Vol 198, Springer-Verlag, Berlin (1984), p.319.
9. J. Hubbard, Proc.Roy.Soc. London A240:539 (1957).
10. J. Goldstone, Proc.Roy.Soc. London A239:267 (1957).
11. B. H. Brandow, Phys.Rev. 152:863 (1966); Rev.Mod.Phys. 39:771 (1967); Ann.Phys. (N.Y.) 57:214 (1970).
12. R. F. Bishop and K. H. Lührmann, Phys.Rev. B17:3757 (1978); ibid. 26:5523 (1982).
13. R. F. Bishop, in: "Nucleon-Nucleon Interaction and Nuclear Many-Body Problems", S. S. Wu and T. T. S. Kuo (eds.), World Scientific, Singapore (1984), p.604.
14. K. Szalewicz, J. G. Zabolitzky, B. Jeziorski and H. J. Monkhorst, J.Chem.Phys. 81:2723 (1984).
15. K. Emrich, Nucl.Phys. A351:379, 397 (1981).
16. R. J. Glauber, Phys.Rev.Lett. 10:84 (1963); Phys.Rev. 130:2529 (1963); Phys.Rev. 131:2766 (1963).
17. J. R. Klauder and B.-S. Skagerstam, "Coherent States - Applications in Physics and Mathematical Physics", World Scientific, Singapore (1985).
18. R. F. Bishop, in: "Recent Progress in Many-Body Theories", P. J. Siemens and R. A. Smith (eds.), Lecture Notes in Physics, Springer-Verlag Berlin (1986).
19. H. Kümmel, Nucl.Phys. A317:199 (1979).
20. T. Holstein and H. Primakoff, Phys.Rev. 58:1098 (1940).
21. F. J. Dyson, Phys.Rev. 102:1217, 1230 (1956).
22. J. Schwinger, in: "Quantum Theory of Angular Momentum", L. C. Biedenharn and H. van Dam (eds.), Academic Press, New York (1956) p.229.
23. T. Marumori, M. Yamamura and A. Tokunaga, Prog.Theor.Phys. 31:1009 (1964).
24. D. Janssen, F. Dönau, S. Frauendorf and R. V. Jolos, Nucl.Phys. A172:145 (1971).
25. P. Garbaczewski, Phys.Reports 36C:65 (1978).

26. D. Bohm and D. Pines, Phys.Rev. 82:625 (1951); ibid.92:609 (1953); D. Pines, Phys.Rev. 92:626 (1953).
27. J. Arponen, R. F. Bishop and E. Pajanne, the succeeding article in this volume.
28. J. Arponen, R. F. Bishop, E. Pajanne and N. I. Robinson, to be published.

ON AN EFFECTIVE GAUGE FIELD DESCRIPTION OF A POSITRON IMPURITY IN POLARIZABLE MEDIA

J. Arponen*, R.F. Bishop**, and E. Pajanne***

*Department of Theoretical Physics, University of Helsinki
**Department of Mathematics, UMIST, Manchester
***Research Institute of Theoretical Physics
University of Helsinki

Abstract

The local polarization around a positron impurity is described by a unitary operator, which defines a dynamical gauge field in interaction with the particle. We study the relation of this gauge field to the elementary collective excitations of the medium. We make contact with the generalized coherent bosonization scheme recently introduced in the extended coupled cluster theory, which suggests a definite parametrization of the polarization unitary operator in terms of a double similarity transformation. We derive the exact equations for the wavefunction and for the CCM amplitudes and show that they satisfy the conservation laws.

Introduction

The technique of positron annihilation at low energies has in the last 3 decades become much used in studying the properties of condensed materials, e.g. metals and their alloys[1]. In order to obtain a reliable interpretation for the experimental findings it is of great interest to find a first-principles theoretical description for the system composed of a positron, or generally a charged impurity, embedded in an electron medium. Although there exist rather advanced theoretical techniques e.g. for the case of a homogeneous electron gas, the extension of such formalisms to the general inhomogeneous case presents considerable difficulties. The two-component density-functional theory[2] (DFT) can be used to formulate the concept of the positron wavefunction[3-4], but the theory still has problems of two kinds: 1) There is no systematic way to go beyond the local density approximation (LDA); 2) Since the DFT wavefunction (i.e. the Slater determinant of the DFT orbitals) is not the true wavefunction, the amount of information, which can be extracted, is rather limited.

The problem considered in the present article has a very general nature. It is a particular and relatively simple example of the field-theoretical one-body problem, and the present treatment is accordingly strongly field-theoretical in spirit. It was shown elsewhere[5] and will be demonstrated more thoroughly in the present article that the theory of an impurity particle in a polarizable medium can be formulated as a special kind of gauge field theory, where one expresses the polarization of the electron medium in terms of an internal gauge field A_μ. Our approach will be applicable not only to homogeneous electron systems, but also to the treatment of such inhomogeneous problems as e.g. surfaces and the localization problem, as well as to dynamical and transient problems.

The basic problem then consists of choosing a convenient parametrization describing the abstract internal gauge field operator and thus the displacement or polarization of the medium. In this article we shall show that a very convenient framework for the present problem is afforded by the well known coupled cluster method (CCM), which has been developed by Coester, Kümmel and coworkers in a series of fundamental papers[6-8]. We shall use the extended variational version of the CCM (ECCM)[9-10], which allows e.g. a consistent treatment of the dynamics and of the average values of physical observables in terms of the linked-cluster quasi-local subsystem correlation amplitudes.

General definition of the impurity wave function

We start with a general formulation of the impurity wavefunction. For the sake of simplicity we assume the interaction between the electron system and the positron to be spin-independent and thus omit the spin of the positron. As in our previous article (Arponen and Pajanne[5]), we define the state of the whole system, consisting of a positron embedded in the interacting electron medium, to be

$$|\Psi_1\rangle = \int d^3r \, b_r^\dagger \, \chi(rt)U(rt)|\Psi_0\rangle \, , \qquad (1)$$

where $b_r{}^\dagger$ is the positron creation field operator at the (three-dimensional) position vector r, $|\Psi_0\rangle$ is the true ground state of the electron system without the positron, U(rt) is a unitary operator acting in the electron Hilbert space and $\chi(rt)$ can be interpreted to be the positron wave function, because $\langle\Psi_1|b_r^\dagger \, b_r|\Psi_1\rangle = |\chi(rt)|^2$.

In the coordinate-space representation the many-body wavefunction corresponding to the state (1) is of the form $\Psi(r;x_1x_2...x_N) = \chi(r)\psi_r(x_1x_2...x_N)$, where r is the positron coordinate and the $\{x_i\}_{i=1,..,N}$ are the electron coordinates, and the electron wavefunction $\psi_r(x_1x_2...x_N)$ is required to be normalized for each r. Due to the strong screening correlations and electron density enhancement around the positron the electron wavefunction $\psi_r(x_1x_2...x_N)$ depends on the coordinate r in a crucial way, and therefore χ is to be understood as a quasi-wavefunction. In the independent particle model (IPM) the r-dependence of ψ_r is ignored and the electron wavefunction further represented as a Slater determinant, for which reason e.g. the positron annihilation rate in metals becomes drastically underestimated.

Next we define the internal gauge field $A_\mu(rt) = (A_0(rt), -\vec{A}(rt))$ through

$$A_\mu(rt) = iU^\dagger(rt)(\partial_\mu U(rt)) . \tag{2}$$

The covariant derivative $D_\mu = (D_0, -\vec{D})$ will then read

$$D_\mu = \partial_\mu - iA_\mu = \partial_\mu + U^\dagger(\partial_\mu U). \tag{3}$$

The time development of the state $|\Psi_1\rangle$ is given by the usual Schrödinger equation, which after premultiplication with $b_r U^\dagger(rt)$ can now be written as

$$i\partial_t \chi(rt)\ |\Psi_0\rangle = \left[-\frac{1}{2M}\vec{D}(r)^2 - A_0(rt) - ZV^c(r) + U^\dagger(rt)H^e U(rt)\right] \chi(rt)|\Psi_0\rangle \tag{4}$$

Here M and Z are the mass and the charge number of the impurity , V^c is the Coulomb potential operator (including the induced part) at the impurity position, and H^e is the Hamiltonian of the interacting pure electron system. Taking the scalar product of (4) with $\langle\Psi_0|$, we get the Schrödinger equation for $\chi(rt)$

$$i\partial_t \chi(rt) = \frac{1}{2M}\left[-\nabla_r^2 - 2i\vec{\alpha}(rt)\cdot\nabla_r\right] \chi(rt) + V^{tot}(rt)\ \chi(rt). \tag{5}$$

The total potential felt by the positron, V^{tot}, is

$$\begin{aligned} V^{tot}(rt) &= \frac{1}{2M}\left\{\langle\Psi_0|\vec{A}(rt)\cdot\vec{A}(rt)|\Psi_0\rangle - i\nabla_r\cdot\vec{\alpha}(rt)\right\} - \alpha_0(rt) \\ &\quad - Z\int d^3x\ v(r-x)\langle\Psi_0|U^\dagger(rt)\rho^e(x)U(rt)|\Psi_0\rangle + Ze\phi_{ext}(rt) \\ &\quad + \langle\Psi_0|U^\dagger(rt)\ H^e\ U(rt)|\Psi_0\rangle . \end{aligned} \tag{6}$$

In this equation $v(r-x) = e^2/4\pi\varepsilon_0|r-x|$ is the Coulomb potential, $\rho^e(r)$ is the total electron number density operator, and $\phi_{ext}(rt)$ is the external electrostatic potential. The third term in (6) is the interaction potential ZV^c of the impurity with the electron system, and finally, $\alpha_\mu(rt) = (\alpha_0(rt), -\vec{\alpha}(rt))$ is the average of the gauge field A_μ

$$\alpha_\mu(rt) = \langle\Psi_0|A_\mu(rt)|\Psi_0\rangle = i\langle\Psi_0|U^\dagger(rt)\partial_\mu U(rt)|\Psi_0\rangle. \tag{7}$$

The total potential in (6) looks formally simple. However, in reality it is very complicated; besides containing terms that arise from interaction with the unperturbed electron system, it also implicitly depends on the wavefunction χ itself, because the self-consistent polarization operator $U(rt)$ is coupled to χ by the requirement that the projections of equation (4) to other states than $|\Psi_0\rangle$ must also be satisfied. The equations (4)-(5) thus form a coupled set of equations in which χ depends on U and U depends on χ in an intricate fashion.

Internal gauge field operator and path-ordered phase factor

From the definition in equation (2) we easily derive for the internal gauge field operator A_μ the following expression

$$\partial_\mu A_\nu - \partial_\nu A_\mu - i\,[A_\mu, A_\nu] = 0\ , \tag{8}$$

or in component form

$$\partial_t \vec{A} + \nabla A_0 - i\,[A_0, \vec{A}] = 0 \quad \text{and} \tag{9}$$

$$\nabla \times \vec{A} + i\vec{A} \times \vec{A} = 0\ . \tag{10}$$

As a side remark we should mention that in the relativistic non-Abelian gauge field theories the gauge fields A_μ^a typically contain extra indices (a) related to the group structure of an internal symmetry group, and the commutator in equation (8) also involves the group commutator.

Furthermore, one can rewrite (2) to give a differential equation for U(rt) :

$$\nabla U(rt) = iU(rt)\vec{A}(rt)\ . \tag{11}$$

This can be integrated along some path in coordinate space to give U(rt) as a path-ordered phase factor

$$U(rt) = P\,\exp\Big\{-i\int_r^\infty d\vec{s}\cdot\vec{A}(st)\Big\}\ , \tag{12}$$

Here we have assumed that the impurity is localized in some finite region of real space for which reason the operator U(rt) at any given time t can be assumed to satisfy the boundary condition $U(rt) \to I$ as $r \to \infty$.

Since the operator field U(rt) is unique by definition, it follows that the integral in (12) cannot depend on the chosen particular path; therefore e.g. integration around any closed path gives the identity operator I. Nevertheless, path ordering in expression (12) is vital, because we cannot assume that the gauge field operators $\vec{A}(rt)$ commute with each other at different points.

On the basis of equation (12), the unitary operator U(rt), as well as the hermitean conjugate

$$U^\dagger(rt) = P\,\exp\Big\{-i\int_\infty^r d\vec{s}\cdot\vec{A}(st)\Big\}$$

can be considered as functionals of the abstract operator-valued gauge field A_μ. The functional dependence is local in time, but not in space. Thus all the constituents of our theory, for example the potential $V^{tot}(rt)$

in the Schrödinger equation (5), are functionals of the gauge field A_μ.

If the system experiences a nonzero classical external electromagnetic field $A_\mu(rt)^{em}$, we have to modify the above notation and write for the total gauge field A^{tot},

$$A_\mu \to A_\mu^{tot} = A_\mu + eA_\mu^{em}, \tag{13}$$

where +e is the positron charge (Z=1). Because the external field is now just a c-number field we can transform equations (9)-(10) with the familiar definitions for the magnetic induction $\vec{B}^{em}$ and electric field $\vec{E}^{em}$ to read as

$$\nabla \times \vec{A}^{tot} + i\vec{A}^{tot} \times \vec{A}^{tot} = e\vec{B}^{em} \quad \text{and} \tag{14}$$

$$\partial_t \vec{A}^{tot} + \nabla A_0^{tot} - i[A_0^{tot}, \vec{A}^{tot}] = - e\vec{E}^{em} \quad . \tag{15}$$

If we consider external gauge transformations, which leave the external fields $\vec{E}$ and $\vec{B}$ invariant, we may assume the unitary operators U(rt) and thus the internal gauge fields A_μ invariant. Then these gauge transformations will be associated with the phases of the wavefunction χ in standard fashion.

Parametrization by expS theory using effective-action formalism

As one notices from equation (1) the definition of the state $|\Psi_1\rangle$ contains the ground state of the interacting electron system $|\Psi_0\rangle$ modified by the local deformation U caused by the impurity. The question now arises as to what would be the optimal way to parametrize the internal gauge field A_μ that describes the polarization of the electron medium. One way to do this systematically is to apply the ideas of the extended coupled cluster method (EECM) of references[9-11] to this system. We start by defining a ket-state $|\Psi_1\rangle$ and a bra-state $\langle\Psi_1'|$ for the system,

$$|\Psi_1\rangle = \int d^3r \, b_r^\dagger \, \chi(rt) \, e^{S(rt)} \, e^{-S''(rt)} |\Phi\rangle \tag{16}$$

$$\langle\Psi_1'| = \int d^3r \, \langle\Phi| e^{S''(rt)} \, e^{-S(rt)} \, \tilde{\chi}(rt) b_r . \tag{17}$$

Here $|\Phi\rangle$ is an independent particle model (IPM) state. The amplitudes S and S" depend explicitly on the point r, where the impurity is located. This differs now from the previous definitions in equation (1) in the sense that the amplitudes S and S" are now assumed to give both the correlation between the electrons as well as the local enhancement of charge around the positron. This means that in the limit $r \to \infty$, i.e. far out of the positron, both S and S" (or more precisely, the components S_i and S''_i, see below) should approach their values for the unperturbed electron system. Furthermore, we have now, because the ECCM approach makes use of a double similarity transform exp(S)exp(-S"), a gauge field A_μ that is defined as

$$A_\mu = ie^{S''}e^{-S}\,\partial_\mu(e^{S}e^{-S''})$$

$$= ie^{S''}(\partial_\mu S)e^{-S''} - i(\partial_\mu S'') \qquad , \qquad (18)$$

which is generally no longer hermitean.

We point out that the particular choice in equations (16)-(17) corresponds to definite gauge-fixing conditions for the internal gauge fields. This comes from the fact that the operators S and S" are genuine creation and destruction operators and do not contain any c-number terms. Therefore we have no freedom to impose further restrictions on the internal gauge fields such as e.g. the Coulomb-gauge condition $\nabla\cdot\langle\vec{A}\rangle = 0$. One consequence of the present gauge-fixing parametrization is that the internal gauge field necessarily appears e.g. in the definition of the positron current (see next section) even if no external electromagnetic fields are present. Because the similarity transformations are not unitary, the wavefunctions χ and $\tilde{\chi}$ in this fixed internal gauge are not each other's complex conjugates.

The equations of motion for the positron wave function and for the amplitudes S and S" can be obtained in a concise[9-11] manner by applying a variational principle to the action functional

$$\mathcal{A} = \int dt \int d^3r \int d^3r' \langle\Phi| e^{S''(rt)} e^{-S(rt)} \tilde{\chi}(rt) b_r \left[i\partial_t - H\right] b^\dagger_{r'} \chi(r't) e^{S(r't)} e^{-S''(r't)} |\Phi\rangle$$

$$\mathcal{A} = \mathcal{A}_0 - \int dt\, \bar{H} \quad ; \quad \bar{H} = \langle\Psi_1'|H|\Psi_1\rangle \qquad (19)$$

The most convenient parametrization from the point of view of diagram expansions involves a change from the original partial amplitudes $\{S_i, S''_i\}$ to new amplitudes $\{\sigma_i, \tilde{\sigma}_i\}$, and we write [9-11]

$$S(rt) = \sum' S_i(rt) C_i^\dagger \;\rightarrow\; \Sigma = \sum' \sigma_i(rt) C_i^\dagger \qquad (20)$$

$$S''(rt) = \sum' S''_i(rt) C_i \;\rightarrow\; \tilde{\Sigma} = \sum' \tilde{\sigma}_i(rt) C_i \;. \qquad (21)$$

Here C_i and $C_i^\dagger$ are normalized even configuration operators, and the primed sum means that at least one particle-hole pair is annihilated or created ($i \neq 0$). These new linked-cluster amplitudes σ_i and $\tilde{\sigma}_i$ are now (quasi)local fields in the configuration space and they are given in terms of the amplitudes S and S" (Arponen et al.[10-11])

$$\sigma_i(rt) = \langle\Phi|C_i e^{S''} S|\Phi\rangle \;, \qquad (22)$$

$$\tilde{\sigma}_i(rt) = S''_i \;, \qquad (23)$$

$$S_i = \sum' \sigma_j\, \bar{\omega}_{j-i} \;, \qquad (24)$$

where $\bar{\omega}_{j-i}$ is the functional

$$\bar{\omega}_{j-i} = \langle\Phi|C_i e^{-S''} C_j^\dagger|\Phi\rangle \ . \tag{25}$$

We postpone the precise description of the index set $\{i\}$ and the summation rules and, for the moment, keep the formalism completely general and thus applicable to an arbitary medium. For the general calculation rules we must refer to the original sources [10-11]. Using the above definitions one can now easily calculate the functional $\mathcal{A}_0$,

$$\mathcal{A}_0 = \iint dt d^3r \ \{i\tilde{\chi}\dot{\chi} + i\tilde{\chi}\chi\dot{\sigma}_0 - i\tilde{\chi}\chi \textstyle\sum' \dot{\tilde{\sigma}}_j \sigma_j\} \ , \text{ where} \tag{26}$$

$$\sigma_0(rt) = \langle\Phi|e^{S''} S|\Phi\rangle = - \textstyle\sum' \bar{\omega}_i(rt)\sigma_i(rt) \ . \tag{27}$$

The functional derivatives of $\mathcal{A}_0$ with respect to the parameters are

$$\frac{\delta\mathcal{A}_0}{\delta\tilde{\chi}(rt)} = i\dot{\chi}(rt) + [i\dot{\sigma}_0 - i\textstyle\sum' \dot{\tilde{\sigma}}_j\sigma_j] \ \chi \ , \tag{28a}$$

$$\frac{\delta\mathcal{A}_0}{\delta\chi(rt)} = -i\dot{\tilde{\chi}} + [i\dot{\sigma}_0 - i\textstyle\sum' \dot{\tilde{\sigma}}_j\sigma_j] \ \tilde{\chi} \ , \tag{28b}$$

$$\frac{\delta\mathcal{A}_0}{\delta\tilde{\sigma}_i(rt)} = i\tilde{\chi}\chi\dot{\sigma}_i + i(\dot{\tilde{\chi}}\chi + \tilde{\chi}\dot{\chi})(\sigma_i - S_i) \ , \tag{28c}$$

$$\frac{\delta\mathcal{A}_0}{\delta\sigma_i(rt)} = -i\tilde{\chi}\chi\dot{\tilde{\sigma}}_i + i(\dot{\tilde{\chi}}\chi + \tilde{\chi}\dot{\chi})\bar{\omega}_i \ . \tag{28d}$$

In the above equations and often later we have omitted the argument r, which is common to all the amplitudes σ_i, $\tilde{\sigma}_i$ and $\bar{\omega}_i$. The expressions above are formally local with respect to the space point r, which appears just as an external common parameter.

Average values of operators

Let us now present the average values of some operators of the theory that will be needed later. We focus first on the positron density matrix, for which we may write

$$\begin{aligned} \langle b_r^\dagger b_{r'}\rangle &= \langle\Psi_1'|b_r^\dagger b_{r'}|\Psi_1\rangle \\ &= \tilde{\chi}(rt)\chi(r't) \ K(r,r') \quad , \text{ where} \end{aligned} \tag{29}$$

$$K(r,r') = \langle\Phi|\hat{K}(r,r')|\Phi\rangle \quad , \text{ and} \tag{30}$$

$$\begin{aligned} \hat{K}(r,r') &= e^{S''(rt)} e^{-S(rt)} e^{S(r't)} e^{S''(r't)} \\ &= P \exp[-i\textstyle\int_{r'}^{r} d\vec{s}\cdot\vec{A}(s)] \end{aligned}$$

From the fact that $K(r,r) = 1$ it follows at once that the impurity density is given by $\langle b_r^\dagger b_r\rangle = \tilde{\chi}(rt)\chi(rt)$. The true positron current density is now obtained from the above equations

$$\vec{J}^p(r) = \frac{i}{2M}\,(\nabla_r - \nabla_{r'})\,\langle b_r^\dagger\, b_{r'}\rangle\Big|_{r=r'}$$

$$= \frac{i}{2M}\left[(\nabla\tilde{\chi})\chi - \tilde{\chi}(\nabla\chi)\right] + \frac{1}{M}\,\tilde{\chi}\chi\,\vec{\alpha}\,, \qquad (31)$$

where $\vec{\alpha} = i\nabla_r\, K(r,r')|_{r=r'} = \langle\Phi|\vec{A}(rt)|\Phi\rangle$. The last term $\tilde{\chi}\chi\vec{\alpha}/M$ arises from the polarization of the electron medium and from our specific gauge-fixing conditions.

The positron kinetic energy $\langle T^p\rangle$, which is an interesting quantity in positron physics, is found to be

$$\langle T^p\rangle = \frac{1}{2M}\int d^3r\; \nabla_r\cdot\nabla_{r'}\,\langle b_r^\dagger\, b_{r'}\rangle\Big|_{r=r'}$$

$$= \frac{1}{2M}\int d^3r\,\left\{ \nabla\tilde{\chi}\cdot\nabla\chi + i\vec{\alpha}\cdot\left[(\nabla\tilde{\chi})\chi - \tilde{\chi}(\nabla\chi)\right] + \vec{\alpha}^2\,\tilde{\chi}\chi + \Delta\,\tilde{\chi}\chi \right\} \qquad (32)$$

Here the average of the gauge field, $\vec{\alpha}(rt)$, and the fluctuation of the gauge field around its average, Δ, are given as :

$$\vec{\alpha}(rt) = \langle\Phi|\vec{A}(rt)|\Phi\rangle$$

$$= i\,\sum{}'\left\{ \bar{\omega}_j\nabla\sigma_j + (\sigma_j - S_j)\nabla\tilde{\sigma}_j \right\} \quad \text{and} \qquad (33)$$

$$\Delta(rt) = \langle\Phi|\vec{A}(rt)\cdot\vec{A}(rt)|\Phi\rangle - \langle\Phi|\vec{A}(rt)|\Phi\rangle^2$$

$$= \nabla_r\cdot\nabla_{r'}\,K(r,r')\Big|_{r=r'} - \vec{\alpha}(rt)^2$$

$$= -\sum{}'\sum{}' L_{ij}\nabla\sigma_i\cdot\nabla\sigma_j + \sum{}'\sum{}'\left[\delta_{ij} + 2\sum{}' L_{ik}\sigma_{k+j}\right]\nabla\sigma_i\cdot\nabla\tilde{\sigma}_j$$

$$- \sum{}'\sum{}'\left[\sigma_{i+j} + \sum{}'\sum{}'\sigma_{i+k}\,L_{kl}\,\sigma_{l+j}\right]\nabla\tilde{\sigma}_i\cdot\nabla\tilde{\sigma}_j\,. \qquad (34)$$

The quantity S_i is defined in equation (24) and L_{ij} is

$$L_{ij} = L_{ji} = \sum{}'\;\omega_{k+l}\,\bar{\omega}_{i-k}\,\bar{\omega}_{j-l}\,. \qquad (35)$$

As pointed out by Arponen et al.[10] these coefficients ($\tilde{L}_{ij}$) have a well-understood diagrammatic interpretation in terms of the $\tilde{\sigma}$'s.

The additional expectation values that one needs to calculate the total potential of the positron V^{tot} in (40a) are

$$\langle V^{ep}\rangle = -Z\iint d^3x\,d^3y\;\tilde{\chi}(x)\chi(x)v(x-y)\langle\rho^e(y)_x\rangle, \qquad (36)$$

$$\langle H^e\rangle = \int d^3r\;\tilde{\chi}(r)\chi(r)\langle H_r^e\rangle\,, \qquad (37)$$

$$\alpha_0(rt) = \langle\Phi|A_0(rt)|\Phi\rangle = -i\sum{}'[\dot{\sigma}_i\tilde{\omega}_i + \dot{\tilde{\sigma}}_i(\sigma_i - S_i)] \quad , \tag{38}$$

where we use the notation

$$\langle H^e_r\rangle = \langle\Phi|e^{S''(r)}e^{-S(r)}\, H^e\, e^{S(r)}e^{-S''(r)}|\Phi\rangle \tag{39}$$

$$\langle\rho^e(y)_x\rangle = \langle\Phi|e^{S''(x)}e^{-S(x)}\rho^e(y)e^{S(x)}e^{-S''(x)}|\Phi\rangle$$

Equations of motion for the wave functions

The equations of motion for the wave functions $\tilde{\chi}$ and χ are straightforwardly obtained by writing down the stationary conditions for the action functional in equation (19) with respect to the variations $\delta\tilde{\chi}$ and $\delta\chi$. Without going into details we only give here the resulting Schrödinger equations, which are

$$i\dot{\chi}(rt) = -\frac{1}{2M}\nabla^2\chi(rt) - \frac{i}{M}\vec{\alpha}(rt)\cdot\nabla\chi(rt) + V^{tot}(r)\,\chi(rt), \tag{40a}$$

$$-i\dot{\tilde{\chi}}(rt) = -\frac{1}{2M}\nabla^2\tilde{\chi}(rt) + \frac{i}{M}\vec{\alpha}(rt)\cdot\nabla\tilde{\chi}(rt) + \tilde{V}^{tot}(r)\,\tilde{\chi}(rt). \tag{40b}$$

The total potential V^{tot} is

$$V^{tot}(r) = -\frac{i}{2M}\nabla\cdot\vec{\alpha}(rt) + \frac{(\vec{\alpha}^2 + \Delta)}{2M} - \alpha_0(rt) - Z\int d^3y\; v(r-y)\langle\rho^e(y)_r\rangle + \langle H^e_r\rangle\;, \tag{41}$$

and $\tilde{V}^{tot}$ is identical with V^{tot}, except for the change in sign of the first term on the right-hand side of equation (41).

The total potential in (41) seems to contain the time-derivatives $\dot{\sigma}$ and $\dot{\tilde{\sigma}}$ through $\alpha_0(rt)$ (cf. equation (38)), but these can be eliminated with the use of the respective equations of motion for σ and $\tilde{\sigma}$ presented in the following section (cf. equation (53)) , and thus the Hamiltonians appearing in the Schrödinger equations (40a,b) for χ and $\tilde{\chi}$ are truly independent of time derivatives. The problem at hand thus reverts to finding a reasonable approximation for the sets of parameters $\{\sigma_i,\tilde{\sigma}_i\}$.

If we now combine the two equations (40) in an obvious way, and note the definition in equation (31), we obtain the familiar continuity equation for the positron density $\rho^p = \tilde{\chi}\chi$,

$$\frac{\partial\rho^p(rt)}{\partial t} + \nabla\cdot\vec{J}^p(rt) = 0\;. \tag{42}$$

Equations of motion for the CCM amplitudes σ and $\tilde{\sigma}$

The calculation of the equations of motion for the CCM amplitudes proceeds analogously to the ideas presented in the previous section by finding the stationary conditions for the action functional in (19) with respect to variations $\delta\sigma_i$ and $\delta\tilde{\sigma}_j$. In this case, however, the needed functional derivatives of H with respect to σ_i and $\tilde{\sigma}_j$ are rather cumbersome, although in principle straightforward to calculate. The variation δH can be written in the following form

$$\delta H = \int d^3r\, \vec{J}^p(rt)\cdot\delta\vec{\alpha}(rt) + \frac{1}{2M}\int d^3r\, \rho^p(rt)\delta\Delta(rt)$$

$$- Z\int\int d^3r d^3x\, \rho^p(r)v(r-x)\langle\delta\rho^e(x)\rangle_r + \int d^3r\, \rho^p(r)\langle\delta H^e_r\rangle \qquad (43)$$

Now we only have to express the variations $\delta\vec{\alpha}$, $\delta\Delta$, $\langle\delta\rho^e\rangle$ and $\langle\delta H^e\rangle$ in terms of the variations $\delta\sigma_i$ and $\delta\tilde{\sigma}_i$ of the basic amplitudes and to use (43) in conjunction with the results in equations (28c) and (28d). Observing the continuity equation (42), these two equations (28c,d) are written in the form

$$\frac{\delta\mathcal{A}_0}{\delta\tilde{\sigma}_i} = i\rho^p\,\dot{\sigma}_i + i[s_i - \sigma_i]\,\nabla\cdot\vec{J}^p \qquad (44)$$

$$\frac{\delta\mathcal{A}_0}{\delta\sigma_i} = -i\rho^p\,\dot{\tilde{\sigma}}_i - i\bar{\omega}_i\,\nabla\cdot\vec{J}^p \qquad (45)$$

In the actual calculations there will occur partial cancellations between $\delta\mathcal{A}_0$ and the first term on the right-hand side of equation (43), because we find for the functional derivatives of this term :

$$\frac{\delta}{\delta\tilde{\sigma}_i}\{\int d^3r\,\vec{J}^p\cdot\delta\vec{\alpha}\} = i(s_i - \sigma_i)\nabla\cdot\vec{J}^p - i\vec{J}^p\cdot\nabla\sigma_i \qquad (46)$$

$$\frac{\delta}{\delta\sigma_i}\{\int d^3r\,\vec{J}^p\cdot\delta\vec{\alpha}\} = -i\bar{\omega}_i\nabla\cdot\vec{J}^p + i\vec{J}^p\cdot\nabla\tilde{\sigma}_i \qquad (47)$$

The equation of motion e.g. for the amplitude σ_i is now obtained by taking the functional derivative of $\mathcal{A}$ with respect to $\tilde{\sigma}_i$. Taking notice of (43), (44) and (46) we finally have :

$$i\rho^p\frac{d}{dt}\sigma_i = \frac{\delta}{\delta\tilde{\sigma}_i}\bar{K} \quad , \text{ where} \qquad (48)$$

$$\bar{K} = \langle V^{ep}\rangle + \langle H^e\rangle + \frac{1}{2M}\int d^3r\,\rho^p\Delta \quad , \qquad (49)$$

where the various terms are given in (34), (36) and (37). What appears in (48) is the convective time derivative

$$\frac{d}{dt} = \frac{\partial}{\partial t} + \frac{\vec{J}^p}{\rho^p}\cdot\nabla \quad , \qquad (50)$$

which describes the time rate of change for an observer moving with the positron. The positron velocity field is

$$\vec{v}(r) = \frac{\vec{J}^p(r)}{\rho^p(r)} = \frac{i}{2M} \nabla \log(\frac{\tilde{\chi}}{\chi}) + \frac{1}{M} \vec{\alpha} \tag{51}$$

Analogous to equation (48) we have for $\tilde{\sigma}_i$:

$$-i\rho^p \frac{d}{dt} \tilde{\sigma}_i = \frac{\delta}{\delta\sigma_i} \overline{K} \tag{52}$$

We can use equations (48), (52) and (33) to eliminate the apparent time-dependence of $\alpha_0(rt)$ in Eq. (38) and obtain

$$\alpha_0(rt) = + \frac{1}{\rho^p} \vec{J}^p \cdot \vec{\alpha} - \frac{1}{\rho^p} \sum{}' \{ \bar{\omega}_i \frac{\delta\overline{K}}{\delta\tilde{\sigma}_i} + (S_i - \sigma_i) \frac{\delta\overline{K}}{\delta\sigma_i} \} \quad , \tag{53}$$

which now guarantees that the Hamiltonians in equations (40a-b) are truly independent of time derivatives. We stress that all the equations obtained above are exact. To be able to calculate the functional derivatives of $\overline{K}$, we must further specify the Hamiltonian of the medium, as will be discussed later on.

Connection to the Lee-Low-Pines transformation

In addition to the present impurity problem there exist many others, like e.g. the polaron problem and the meson-nucleus system, which all have typical common features. Therefore a method introduced for one problem often can be applied also in the other cases. A particularly convenient trick of transforming to the impurity rest coordinate system was introduced for the polaron problem by Lee et al. (LLP)[12]. Two of the present authors used this transformation earlier for the problem of a positron in an homogeneous electron gas[13]. The LLP transformation is performed by the unitary operator

$$U_{LLP} = \exp[i(\vec{P} - \vec{P}^e)\cdot\vec{r}^p] \, , \tag{54}$$

where $\vec{P}^e$ is the total-momentum operator of the medium (electron gas), $\vec{r}^p$ the position operator of the impurity, and $\vec{P}$ is a constant vector, which represents the total momentum of the translationally invariant eigenstate of the total system. The main effects of this transformation are: 1) The impurity coordinate $\vec{r}^p$ becomes a "dead" variable and the impurity can be regarded to be fixed at e.g. $\vec{r}^p = 0$, 2) The Hamiltonian is reduced to describe only the medium in the presence of the fixed impurity center, but it obtains new recoil terms describing the impurity-induced interactions between the degrees of freedom of the medium.

The Lee-Low-Pines transformation to the impurity-centric description is a global one, and assumes global translational symmetry. Looking carefully at the equations of motion of the present article, it can be seen that the representations (1) or (16)-(17) for the total wavefunction actually perform a local (or differential) LLP transformation in which no global symmetry needs to be assumed. The transformed Hamiltonian is essentially the K of equation (49), and the term containing the fluctuation Δ corresponds to the recoil terms. If the impurity density $\rho^p(r) = \tilde{\chi}(r)\chi(r)$ is constant, it factorizes away from the equations of motion (48), (52) for σ and $\tilde{\sigma}$, and the impurity position parameter r remains a constant external parameter. If, however, the density ρ^p is not constant, like in the cases of a real lattice, nonstationary state, or spontaneous trapping, new terms containing $\nabla\log(\rho^p)$ arise in the equations for σ ,$\tilde{\sigma}$ from the variations of the recoil term. These new terms cannot be obtained - at least easily - by the global LLP transformation.

It is interesting to note that from the point of view of the medium the positron enters only through the hydrodynamic variables ρ^p and $\vec{J}^p$ describing classical average flow. The same is true in the opposite direction; the quantities affecting the impurity wavefunctions χ, $\tilde{\chi}$ can be interpreted hydrodynamically to be due to the physical fields and fluctuations describing the medium. The relative phases of the basic amplitudes $\chi,\tilde{\chi}$ or $\sigma,\tilde{\sigma}$ are combined within each subsystem (i.e. the impurity vs. the medium) into such physical quantities, which have classical meaning.

As indicated above, the differential LLP transformation is adequate to describe spontaneous breaking of translational symmetry like in the case of spontaneous trapping. There have been suggestions that positron trapping might occur in low-density electron gas, but to our knowledge the present expectation is that it does not happen in the metallic density regime.

<u>Generalized coherent bosonization and collective eigenmodes of the medium</u>

Several authors have used the CCM to study the homogeneous electron gas[14-19]. The extended version of the theory (ECCM), however, has not yet been applied to this problem, and therefore we shall remain rather sketchy about the general formalism, and devote more attention to a few limiting cases.

As is pointed out in references 10-11, the exponential similarity transformations of the ECCM theory can be regarded as a definite bosonization scheme, where the guidelines are sought from a careful analysis of the structure of perturbation diagrams. In contrast to the conventional bosonization schemes (like e.g. the Sawada bosons [13,17]), which are based on isomorphisms of the Lie algebras, the ECCM bosonization is rather aimed at entirely eliminating the quantal interactions of the bosons to the ultimate extent that the bosons become classical in a definite sense[10-11].

The collective eigenmodes are found by expanding the average value of the Hamiltonian in powers of $\delta\sigma_i$, $\delta\tilde{\sigma}_i$ around the stable ground state, and by diagonalizing the leading (second-order) terms of the expansion. The result is formally a "Ginzburg-Landau" Hamiltonian

$$H[\sigma_i^{\circ} + \delta\sigma_i, \tilde{\sigma}_i^{\circ} + \delta\tilde{\sigma}_i] = H[\sigma_i^{\circ},\tilde{\sigma}_i^{\circ}] + \delta H$$

$$= E^{\circ} + \sum E_\ell \tilde{C}_\ell C_\ell + (\text{ terms with } CCC,\ CCCC,\dots) . \qquad (55)$$

The deviations $\delta\sigma_i$, $\tilde{\delta\sigma}_i$ are linear functions of the normal-mode amplitudes C_ℓ, $\tilde{C}_\ell$ with coefficients that are given by the eigenvectors of the dynamical matrix[10]. The dynamics is still given by the equations

$$i\dot{C}_\ell = \delta H/\delta\tilde{C}_\ell \quad , \; i\dot{\tilde{C}}_\ell = -\delta H/\delta C_\ell \qquad , \qquad (56)$$

and the amplitudes C, $\tilde{C}$ can be considered to be the average values of ideal boson operators in the (bi-)coherent states of an ideal boson Hilbert space.

The average values of all other operators can as well be expanded in powers of C and $\tilde{C}$. It is obvious that the leading (quadratic) terms of H determine the linear response of the system to arbitrary small perturbations. The higher-order terms (CCC, CCCC, ..) are related to large perturbations and non-linear phenomena. By the aid of the Poisson-bracket relations of Ref. 10 it is easy to show that the above formalism exactly satisfies all conservation laws not only in the linear but also in the non-linear regime.

Coming back to the positron impurity problem, we find that the deviations C and $\tilde{C}$ become functions of the positron position r. For instance, the electronic energy becomes

$$H^e \rightarrow H^e_r = E^{\circ} + \sum E_\ell \tilde{C}_\ell(r) C_\ell(r) + (\text{ higher terms }). \qquad (57)$$

Assuming that the medium would be in its ground state in the absence of the impurity, the amplitudes then behave, for a fixed index ℓ, as $C_\ell(r),\tilde{C}_\ell(r) \rightarrow 0$ in the physical sense as $r \rightarrow \infty$. For an accurate description of the electron density enhancement the higher-order terms of the Hamiltonian, however, turn out to be important in the metallic density regime and at lower densities[1,5,13].

The role of the interaction (V^{ep}) is vital, because it is the driving term of the polarization of the medium; without it there would be no r-dependence in the amplitudes $\sigma_i(r)$ and $\tilde{\sigma}_i(r)$, and the recoil Δ would be simply zero, and $\langle H^e\rangle = E^{\circ}$. The full expression

$$V^{ep} = -Z\int\int d^3r d^3x\ \tilde{\chi}(r)\chi(r)v(r-x)\langle\Phi| e^{S''(r)} e^{-S(r)}\ a^\dagger_x a_x\ e^{S(r)}|\Phi\rangle,$$

where $a_x^\dagger$, a_x are the electron field operators, gives rise to the diagrams of Fig.1. Analogous diagram representations can be drawn for $\langle H^e\rangle$ and the recoil term. In constructing such expressions it will be useful to express the average value of a product of operators in terms of low-order functional derivatives of the average values of the factors[10].

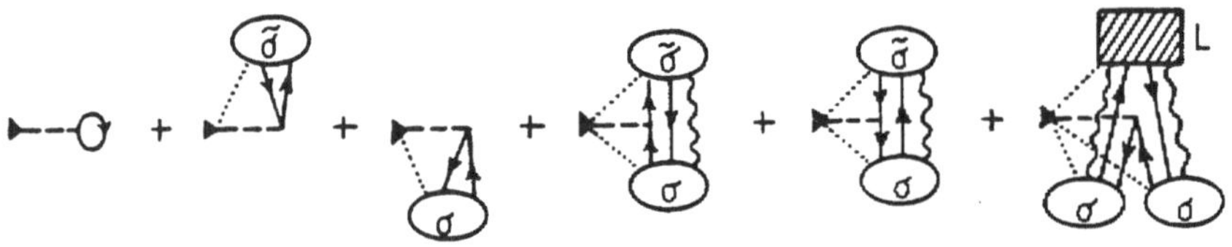

Fig.1 The ECCM diagrams for the electron-positron interaction. The small black triangle corresponds to the positron density factor $\tilde{\chi}(r)\chi(r)$; the dashed line is the Coulomb interaction, and the dotted lines are a mnemonic showing that the amplitudes σ, $\tilde{\sigma}$ and L depend on r. The wavy lines denote other possible electron-hole pairs.

Summary and discussion

The general formalism presented above is an exact description for the positron motion in the polarizable medium. It satisfies the conservation laws, like e.g. the continuity equation, and the total energy conservation. For example, the slowing down of a fast positron in this picture takes place through emission of "classical" dissipative wave motion, or elastic waves, which disappear to infinity in a large system.

In the present article we have not considered specific approximations to the exact equations. The simplest truncation, SUB1[9], would correspond to the time-dependent Hartree-Fock approximation. This approximation becomes exact in the uniform limit, which is the limit of high electron density. The proper eigenexcitations are then the Sawada bosons, which in the low-momentum regime can be treated classically and thus identified as the generalized coherent bosons. The other interesting limit is that of low electron density. The extreme case is the two-body problem where the ground state is the bound positronium (Ps) atom. Also this case is accurately treated by the SUB1 truncation. Now the non-linearity of the equations must be fully considered in contrast to the uniform limit where the linear-response treatment is essentially sufficient.

As was emphasized earlier, all the observable properties of the system are in the ECCM method expressible through the quasi-local, linked-cluster subsystem correlation amplitudes $\{\sigma,\tilde{\sigma}\}$, which amplitudes may be interpreted as the average values of generalized coherent boson operators acting in a fictious ideal boson Hilbert space. The formalism preserves all the microscopic information in contrast to e.g. the density functional theory, where the obtained DFT orbital functions do not have close bearing to the true correlated many-body wavefunction.

The present representation of the many-body wavefunction was shown above to correspond to a local or differential form of the well-known Lee-Low-Pines transformation into the impurity-centric coordinate system.

As a consequence of this, the equations for the medium contain recoil energy terms, which correspond to impurity-induced interactions between the degrees of freedom of the medium. The advantage of a local description is that it can be used in a conceptually simple way for translationally non-invariant systems like e.g. real metals and localization problems.

The formalism leads to separate, but coupled equations of motion for the impurity and for the medium. From the point of view of each of the subsystems, i.e. the impurity and the medium, the other subsystem enters only through classical or hydrodynamical variables. For the impurity system the theory provides a hydrodynamically complete and consistent description, in which all the conservation laws are fully satisfied.

References

1. For review, see "Positron Annihilation", ed. by P.C.Jain, R.M.Singru & K.P.Gopinathan, 1985, World Scientific Publ. Co., Singapore.
2. See e.g. "Theory of the Inhomogeneous Electron Gas", ed. by S.Lundqvist and N.H.March, 1983, Plenum Press, New York.
3. B.Chakraborty, Phys.Rev. B24: 7423 (1981).
4. B.Chakraborty and R.W.Siegel, Phys.Rev. B27: 4535 (1983).
5. J.Arponen and E.Pajanne, 1985, in Ref.1
6. F.Coester, Nucl. Phys. 7: 421 (1958).
7. F.Coester and H.Kümmel, Nucl. Phys. 17: 477 (1960).
8. For further references, see e.g. H.Kümmel, K.H.Lührmann and J.G.Zabolitzky, Phys. Rep. 36C: 1 (1978).
9. J.Arponen, Ann.Phys. 151: 311 (1983).
10. J.Arponen, R.F.Bishop, E.Pajanne and N.Robinson, to be published.
11. J.Arponen, R.F.Bishop and E.Pajanne, this volume.
12. T.D.Lee, F.E.Low and D.Pines, Phys. Rev. 90: 297 (1953).
13. J.Arponen and E.Pajanne, Ann. Phys. 121: 343 (1979).
14. D.L.Freeman, Phys. Rev. B15:5512 (1977).
15. R.F.Bishop and K.H.Lührmann, Phys. Rev. B17: 3757 (1978).
16. R.F.Bishop and K.H.Lührmann, Phys. Rev. B26: 5523 (1982).
17. J.Arponen and E.Pajanne, J.Phys. C15: 2665 (1982).
18. E.Pajanne and J.Arponen, J.Phys. C15: 2683 (1982).
19. K.Emrich and J.G.Zabolitzky, Phys. Rev. B30: 2049 (1984).

A TEMPERATURE DEPENDENT COUPLED CLUSTER METHOD

M. Altenbokum, K. Emrich, and H. Kümmel
Institut für Theoretische Physik II
Ruhr-Universität Bochum

J.G. Zabolitzky
Supercomputer Institute
University of Minnesota

The Bloch equation for the density operator is written down in terms of coupled cluster amplitudes. The partition function for the Lipkin-Meshkov-Glick model is computed approximately and compared with exact results.

I. INTRODUCTION

At the last many body conference the protagonists of the coupled cluster method (CCM) [1] have been asked to invent a temperature dependent version. At that time we had such a theory, but due to some mistake the results on the model by Lipkin et. al (LMG model) did not look encouraging. Meanwhile we have corrected this and therefore we now present the theory together with some first results.

The major problem encountered in approaching thermodynamics with CCM is the fact that the standard version of CCM is treating physics asymmetrically. That is to say, writing the wave function as $\psi^{g} = \exp \underline{S}^{g} \cdot \phi$ with ϕ as Slater determinant, we project only on the l.h.s. with a complete orthonormal set of particle-hole states $\phi_{p\ell}$ obtaining

$$\langle \phi_{p\ell} | e^{-\underline{S}^{g}} \underline{H} e^{\underline{S}^{g}} \phi \rangle = 0. \tag{1.1}$$

There is no projection on Φ_{ph} states from the right. Therefore there is no simple relation to expectation values or variational methods. Only more recently an "extended CCM" has been introduced by Arponen et al. which is trying to remedy this situation [2]. So far no large scale tests of this method exist, however.

Statistical mechanics is based on expectation values. This implies the use of matrix elements of all kinds - not just those with excited states on the left and ground states on the right as in (1.1). This naturally leads to an Ansatz with exponentials on both sides of the density operator as described below.

II. CCM FORM OF BLOCH EQUATION

Let

$$\underline{\varrho} = \frac{1}{Z} \sum_n |n\rangle e^{-E_n \beta} \langle n| \tag{2.1}$$

be the exact density matrix with

$$Z = \sum_n e^{-E_n \beta} \tag{2.2}$$

as partition function. Here $|n\rangle / E_n$ are the exact eigenfunctions / energies of the Hamiltonian $\underline{H}$.
We are making the Ansatz

$$\underline{\varrho} = e^{\tilde{S}} \underline{\tilde{\varrho}} e^{\tilde{S}^+} \tag{2.3}$$

Here

$$\underline{\tilde{S}}(\beta) = S_0(\beta) + \underline{S}(\beta) \ , \ \underline{S}(\beta) = \sum_n \underline{S}_n(\beta), \tag{2.4}$$

$$\underline{S}_n(\beta) = \frac{1}{n!} \sum_{\substack{\nu_1 \dots \nu_n \\ \varrho_1 \dots \varrho_n}} \underline{a}^+_{\varrho_1} \dots \underline{a}^+_{\varrho_n} \underline{a}_{\nu_n} \dots \underline{a}_{\nu_1} \langle \varrho_1 \dots \varrho_n | S_n(\beta) | \nu_1 \dots \nu_n \rangle \tag{2.5}$$

$S_0(\beta)$ is a number needed to normalize

$$Tr(\underline{\varrho}) = 1 . \tag{2.6}$$

$\varrho_i, \tau_i / \nu_i, \mu_i$ label non-occupied/occupied states with respect to a reference state

$$\phi = \prod_{i}^{N} \underline{a}^{+}_{\nu_i} \, \phi_{vac} \, . \tag{2.7}$$

$\underline{a}^{+}_{\alpha} / \underline{a}_{\beta}$ are the usual creation/annihilation operators. $\underline{\tilde{\varrho}}$ is not a "free" or independent particle (IP) density matrix. It is easily seen that the $\langle \varrho_1 .. | S_1 | \nu_1 .. \rangle$ do not suffice to describe an arbitrary density matrix. Therefore we make the Ansatz

$$\underline{\tilde{\varrho}} = \sum_{n,m} \sum_{\substack{\varrho_1 .. \varrho_n \\ \nu_1 .. \nu_n}} \sum_{\substack{\tau_1 .. \tau_m \\ \mu_1 .. \mu_m}} |\phi^{\varrho_1 .. \varrho_n}_{\nu_1 .. \nu_n}\rangle \, e^{-\frac{1}{2}\left[\sum^{n}(\varepsilon_{\varrho_i} - \varepsilon_{\nu_i}) + \sum^{m}(\varepsilon_{\tau_i} - \varepsilon_{\mu_i})\right]}$$

$$\times \frac{1}{n!\,m!} \langle \varrho_1 .. \varrho_n \, \nu_1 ... \nu_n | e^{C_{mn}} | \tau_1 .. \tau_m \mu_1 .. \mu_m \rangle \langle \phi^{\tau_1 .. \tau_m}_{\mu_1 .. \mu_m} | \, , \tag{2.8}$$

with

$$\langle \varrho_1 .. \varrho_n | C_{n0} | \nu_1 .. \nu_n \rangle = \langle \mu_1 .. \mu_m | C_{0m} | \tau_1 .. \tau_m \rangle = 0 \, . \tag{2.9}$$

Here the

$$|\phi^{\varrho_1 .. \varrho_n}_{\nu_1 ... \nu_n}\rangle = \underline{a}^{+}_{\varrho_1} .. \underline{a}^{+}_{\varrho_n} \underline{a}_{\nu_n} .. \underline{a}_{\nu_1} \, \phi \tag{2.10}$$

are n particle-hole states. Clearly, without C_{nm} $\underline{\tilde{\varrho}}$ would be the independent particle density matrix $\underline{\varrho}_{IP}$. On the other hand, just these operators C_{nm} introduce the elements missing in (2.3) with $\underline{\tilde{\varrho}} \rightarrow \underline{\varrho}_{IP}$. Now in addition to the expS-type excitations on both (right and left) sides we allow for more general ones.

We recall that the ground state is written [1] as

$$\psi^{g} = e^{\underline{S}^{g}} \phi \, . \tag{2.11}$$

Excited states are of the form [3]

$$\psi^{e}_{m} = \underline{S}^{e}_{m} \, e^{\underline{S}^{g}} \phi \, , \tag{2.12}$$

with

$$\underline{S}^{e}_{m} = \sum_{n \geq m} \underline{S}^{e}_{m,n} \, . \tag{2.13}$$

Each state ψ^e_m is labelled by a certain linear combination of m particle-hole states $\phi^{\rho_1..\rho_m}_{\nu_1..\nu_m}$. $\underline{S}^e_{m,n}$ creates n particle-hole pairs like the $\underline{S}_n$ of (2.5). Thus already in the $\beta \to \infty$ (T=0) case there occur amplitudes $\exp\underline{S} \cdot \phi^{\rho_1..\rho_n}_{\nu_1..\nu_n}$. In other words, ϕ and all $\phi^{\rho_1..\rho_n}_{\nu_1..\nu_n}$ must occur on both (left and right) sides of $\underline{\rho}$ (not only those with n = m (as in $\underline{\rho}_{IP}$). Introducing exp C_{nm} is just a convenient parametrization. (2.9) avoids overcounting since contributions with a ϕ on one side are already taken care of by the $\underline{S}_n$. After these remarks it is not surprising that later the $\underline{S}_n$ will turn out to be related to $\underline{S}^g_n$, whereas the C_{nm} are related to the $\underline{S}^e_m$. There are some more motivations for using the Ansatz (2.3) with (2.8): it will turn out that in this way the equations for $\underline{S}_n$ not only are very similar to the ground state equations. They also are decoupled from the equations involving the C_{nm}. Furthermore it is hoped that the C_{nm} amplitudes are small such that one can do rather primitive approximations for them. We have found some numerical evidence for this being true. No convincing general arguments for their smallness seem to exist, however.

The Bloch equation in the form

$$\left(\underline{H} + \frac{d}{d\beta}\right) e^{\underline{\tilde{S}}} \underline{\rho}\, e^{\underline{\tilde{S}}^+} = 0 \tag{2.14}$$

can now be made explicit in the standard way.

i) By projecting on both sides with ϕ we obtain

$$\frac{dS_0}{d\beta} = -\frac{1}{2}\langle \phi | \underline{H}\, e^{\underline{S}} \phi \rangle . \tag{2.15}$$

ii) By projecting from the left with $\phi^{\rho_1..\rho_n}_{\nu_1..\nu_n}$ and from the right with ϕ we obtain

$$\frac{d}{d\beta}\langle \rho_1..\rho_n | S_n | \nu_1..\nu_n \rangle = -\langle \phi^{\rho_1..\rho_n}_{\nu_1..\nu_n} | e^{-\underline{S}} \underline{H}\, e^{\underline{S}} \phi \rangle , \tag{2.16}$$

iii) By projecting with particle-hole states on both sides we obtain

$$\langle \phi^{\rho_1..\rho_n}_{\nu_1..\nu_n} | e^{-\underline{S}} \left(\underline{H} + \frac{d}{d\beta}\right) e^{\underline{\tilde{S}}} \underline{\tilde{\rho}}\, e^{\underline{\tilde{S}}^+} \phi^{\tau_1 ... \tau_m}_{\mu_1..\mu_m} \rangle = 0 , \tag{2.17}$$

or

$$\sum_{n=1}^{m+2} \langle \phi^{\varrho_1\ldots\varrho_m}_{\nu_1\ldots\nu_m} | e^{-\underline{S}} [\underline{H}, \underline{Q}^e_n] e^{\underline{S}} \phi \rangle \tag{2.18}$$

$$= - \langle \phi^{\varrho_1\ldots\varrho_m}_{\nu_1\ldots\nu_m} | \frac{d\underline{Q}^e_m}{d\beta} \phi \rangle .$$

Here we have introduced (m > 0)

$$\underline{Q}^e_m = \tilde{\underline{\varrho}}\, e^{\underline{S}^+} e^{\underline{S}^g} \underline{S}^e_m , \tag{2.19}$$

replacing the amplitudes C_{nm}. The reason for this replacement is rather evident from the low temperature limit: in this case equations (2.16) with $d\langle|S_n|\rangle/d\beta = 0$ become the well known standard CCM equations for the ground state. If we replace $\underline{Q}^e_n$ by $\underline{S}^e_n$ (2.18) with $d\underline{Q}^e_m/d\beta = 0$ turn out to be the equations for excited states [3]. Note also

$$\underline{\varrho} \xrightarrow[\beta\to\infty]{} |\psi^g\rangle\langle\psi^g| = \frac{e^{\underline{S}^g} |\phi\rangle\langle\phi| e^{\underline{S}^{g+}}}{\langle\psi^g|\psi^g\rangle} , \tag{2.20}$$

i.e. in the zero temperature limit $\underline{\varrho}$ is just the ground state projection operator. Thus necessarily $\underline{S}_n(\beta) \to \underline{S}^g_n$, $\underline{Q}^e_m \to \underline{S}^e_m$ and - if we put the ground state energy equal to zero - $dS_0/d\beta \to 0$. Thus it is no surprise that the boundary conditions going with the differential equations (2.15), (2.16), and (2.18) can be put in the form

$$\frac{d\underline{S}_n}{d\beta} \xrightarrow[\beta\to\infty]{} 0 , \quad \underline{Q}^e_m \xrightarrow[\beta\to\infty]{} \underline{S}^e_m . \tag{2.21}$$

Thus we have obtained a rigorous set (2.15), (2.16), and (2.18) of differential equations with initial conditions (2.21) determining $\underline{\varrho}$.

III. PARTITION FUNCTION OF THE LMG MODEL

The model by Lipkin et al. [4] is a two level model with N distinguishable particles. Let the upper (ϱ) and lower (ν) levels have the distance 1 such that $\varepsilon_\varrho - \varepsilon_\nu \equiv \varepsilon = 1$. The Hamiltonian is

$$\underline{H} = \frac{1}{2} \sum_i^N \left(\underline{a}^+_{\varrho_i} \underline{a}_{\varrho_i} - \underline{a}^+_{\nu_i} \underline{a}_{\nu_i} \right) + \frac{K}{2} \left[\left(\sum^N \underline{a}^+_{\varrho_i} \underline{a}_{\nu_i} \right)^2 + \left(\sum^N \underline{a}^+_{\nu_i} \underline{a}_{\varrho_i} \right)^2 \right]. \tag{3.1}$$

It's spectrum can be obtained analytically. Therefore the partition function can be obtained numerically [5]. We have solved the set of equations including only S_1, S_2, Q_1^e. Fortunately, in this model these quantities are just numbers. Thus the equations look rather simple:

$$\frac{dS_0}{d\beta} = -\frac{1}{2}\left[N(N-1)KS_2 - \frac{N}{2}\right] . \qquad (3.2)$$

$$S_1 + S_1 S_2 K(N^2-4N+3) - S_1^3 K(N-1) = -\frac{dS_1}{d\beta} . \qquad (3.3)$$

$$K(N^2-7N+9)S_2^2 + 2S_2 + (6N-12)^2 K S_1 S_2 + K S_1^4 + K = \frac{dS_2}{d\beta} . \qquad (3.4)$$

$$\{1 - K(N-1)[3S_1^2 + (N-3)S_2]\} Q_1^e(\beta) = -\frac{d}{d\beta} Q_1^e(\beta) . \qquad (3.5)$$

The solution of the equations (3.2) for S_0 and (3.5) for Q_1^e is trivial once S_1 and S_2 have been determined from (3.3) and

Table 1. $-\ln(\ln Z)/\beta$ for N = 2o, KN = 0.1

β	exact values	CCM (S_2, Q_1 approx.)
0.001	$-2.6285\ 10^3$	$-2.6284\ 10^3$
0.01	$-2.6220\ 10^2$	$-2.6219\ 10^2$
0.1	$-2.5563\ 10^1$	$-2.5562\ 10^1$
0.5	-4.4985	-4.4992
1.0	-1.8348	-1.8353
5.0	$4,0176\ 10^{-1}$	-3.9968
10.0	$7.0066\ 10^{-1}$	-6.9939
50.0	$9.4030\ 10^{-1}$	-9.3889
100.0	$9.6822\ 10^{-1}$	-9.6751

(3.4). Note that there is substantial freedom in the choice of the reference state and the single particle energies going with it. Indeed initially we had problems here since we used naively ε from the zero order Hamiltonian. This did lead to unsatisfactory results. Thus we finally decided to do an improved calculation including S_2^g and S_4^g (only even S_n^g occur) and S_1^e, S_3^e and S_5^e (only odd S_n^e occur) to obtain a better ε. We arrived at the solutions $S_1(\beta) = 0$ and

$$S_2 = K \frac{e^{-2\sqrt{\Delta}\beta} - 1}{1+\sqrt{\Delta} - (1-\sqrt{\Delta})e^{-2\sqrt{\Delta}\beta}}, \tag{3.6}$$

with

$$\Delta = 1 - K\gamma, \quad \gamma = K(N^2 - 7N + 9). \tag{3.7}$$

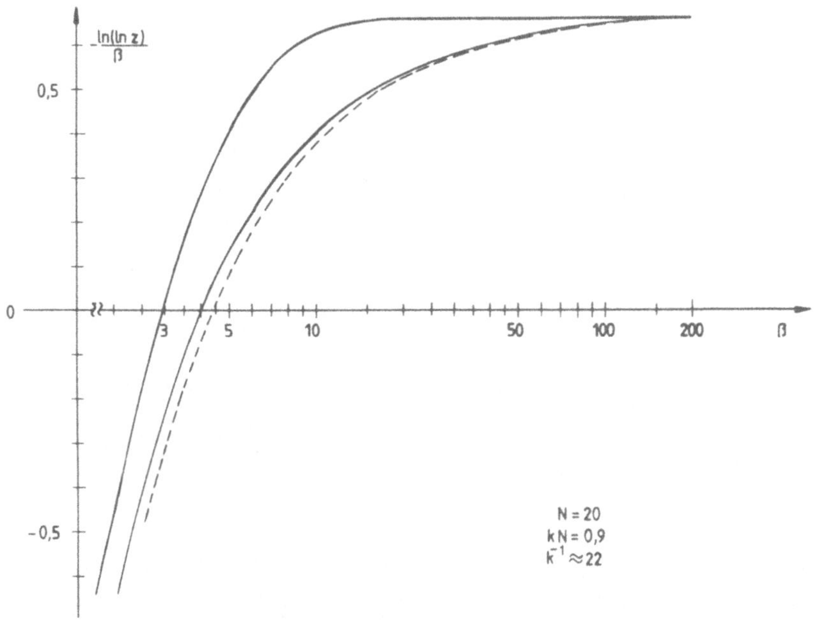

Fig. 1: Results for the strong coupling limit

The computation of the partition function then is a lengthy but straight forward affair. We represent the results for two extreme cases, i.e. for weak coupling (table) and strong coupling (figure). It is seen that in the first case the agreement with the exact partition function is very good, whereas in the latter one only the high and low temperature limits come out well. We remark in passing that the influence of the C_{11} term indeed is small (Putting $C_{11} = 0$ means $\langle g | Q_1^e | v \rangle = e^{-\varepsilon\beta}$).

IV. SUMMARY

The extension of the CCM to finite temperature once again is in line with the experience made before with CCM: the principles are simple and straightforward and therefore suggest systematical truncation schemes. In practice the equations turn out to be rather complicated. This is why we did only a very low order calculation within the LMG model. This could be done fairly easily. From the results so obtained we guess that a higher order approximation would allow us to go both higher with the coupling strength as well as away from the high or low temperature limit. Whether or not this method will be practical for realistic systems remains to be seen.

REFERENCES

1. H. Kümmel, K.H. Lührmann, J.G. Zabolitzky, Phys.Rep.36C, 1(1978)
2. J. Arponen, E. Pajanne, International Conference on Recent Progress in Many Body Theories, San Francisco, 1985, to published.
3. K. Emrich, Nucl.Phys.A351, 379 and 397 (1981)
4. H. Lipkin, N.Meshkow, A.J. Glick, Nucl.Phys.A622,188(1965)
5. J.G. Zabolitzky, unpublished

CBF DESCRIPTION OF LIGHT NUCLEI

M.C. Boscá and R. Guardiola

Departamento de Física Nuclear. Univ. de Granada

18071 Granada (Spain)

INTRODUCTION

The Correlated Basis Function theory (CBF) was invented many years ago by E. Feenberg (1) with the objective of studying the ground state and excitations of strongly interacting quantum systems. The underlying idea of the theory is the use of a set of correlated states

$$|\Psi_m> = F\,|\Phi_m> \qquad (1)$$

which constitute the basis where the hamiltonian will be diagonalized. In eq. (1) the functions Φ_m are a complete set of model states, e.g., Slater determinants of plane waves in the case of extended systems or shell model states in the case of nuclei. On the other hand, F is a correlation operator which main role is to account properly for the strongly repulsive character of the interaction at short distances.

The eigenstates of the system are obtained by solving the generalized eigenvalue problem

$$< \Psi_m \mid H \mid \Psi_n > = E < \Psi_m \mid \Psi_n > \qquad (2)$$

Note that the correlated states are not orthonormal because of the presence of the correlation factor.

This generalized eigenvalue problem is solved in the following two ways:

i) By means of a (non-orthogonal) perturbation expansion, in the case of extended systems, or
ii) By direct diagonalization of eq. (2) when the basis is discrete, like in the case of nuclei.

The correlation factor is usually chosen of the Jastrow form, i.e., involving only two-particle correlations. It can be of the state independent type

$$F = \prod_{i<j} f(r_{ij}) \qquad (3)$$

or of the state dependent type (i.e., spin/isospin dependent)

$$F = S\{\prod_{i<j} f(r_{ij}, \sigma_i, \sigma_j, \tau_i, \tau_j)\} \qquad (4)$$

where S is a symmetrizer operator so as to ensure the antisymmetry of the correlated state, or finally of the independent pair form

$$F = \prod_{i<j} f(r_{ij}) \prod'_{i<j} \{1 + h(r_{ij},\sigma_i,\sigma_j,\tau_i,\tau_j)\} \qquad (5)$$

where the prime on the second product means that when expanding this product only terms which commute have to be maintained. In this manner the complex problem of antisymmetrizing is avoided.

The two-particle correlation $f(r_{ij})$ has an structure directly related to the two-body interaction. The actual form of its radial dependence is determined variationally, both from a parametrized form or from Euler equations corresponding to a second order expansion in the medium.

STUDY OF NUCLEI

This work deals with p-shell nuclei, from ^4He to ^{16}O. For this study we have considered the interactions known as Reid V6 and Reid V8 (2), which have the general structure

$$V = V_C(r) + V_S(r)(\sigma_1\sigma_2) + V_I(r)(\tau_1\tau_2) + V_{SI}(r)(\sigma_1\sigma_2)(\tau_1\tau_2)$$
$$+ V_T(r)S_{12} + V_{TI}(r)(\tau_1\tau_2)S_{12} + V_{LS}(r)(\ell_{12}(\sigma_1+\sigma_2)) + \qquad (6)$$
$$+ V_{LSI}(r)(\tau_1\tau_2)(\ell_{12}(\sigma_1+\sigma_2))$$

Equation (6) correspond to the V8 form, and when the spin-orbit terms are eliminated then we have the V6 form. The value of the radial functions is related to the Reid soft core interaction components. In both cases we will consider a V6 form for the two-particle correlation.

In order to determine the effect of the operatorial terms, both in the potential and in the correlation, in nuclei, one can follow two approaches. The first method consists in using in a first step a state independent correlation, and subsequently letting the configuration mixing mechanism to account for the operatorial dependence. The main motivation of this approach is the fact that the repulsive part of the various radial components of the potential is esentially independent of the spin/isospin channel. However, this method requires the diagonalization in a very large shell model basis, just because of the very important contribution of the tensor part of the interaction.

A way of dealing with this large basis problem was devised by Clark and Krotscheck (3) with the name of CBF-RPA approximation. This method follows three steps:

i) Use the Fermi Hypernetted Chain (FHNC) theory in nuclear matter at a density appropriate to the nucleus, to determine an effective interaction. This step is carried out by using state independent correlations.

ii) Renormalize this effective interaction so as to match the CBF perturbative corrections in nuclear matter, and

iii) Use this renormalized interaction as an input to a conventional RPA or TD algorithm.

This method has been applied to the determination of the odd-parity levels of ^{16}O (4), as well as the gamma transition rates.

The second method to account for operatorial effects consists in using from the very beginning state-dependent correlations, and appropriate cluster expansions valid for finite systems (5). This method involves the calculation of matrix elements of the type

$$h^{(n)}_{pq} = \langle \Phi_p \mid F^{(n)} H^{(n)} F^{(n)} \mid \Phi_q \rangle$$
$$n^{(n)}_{pq} = \langle \Phi_p \mid F^{(n)} \quad F^{(n)} \mid \Phi_q \rangle \tag{7}$$

where $F^{(n)}$ and $H^{(n)}$ represent, respectively, the correlation and the hamiltonian of the n-particle subsystem. This procedure requires high precision calculations to properly account for the (approximate) cancelation of diagrams which will be unlinked or irreducible in the limit of infinite nuclear systems. From the matrix elements given in eq. (7) the A-body matrix elements are obtained by means of a cluster expansion (5), the ones of Van Kampen multiplicative type being the more appropriate.

Convergence properties of these expansions are not well known. Empirically we have shown that the main contribution comes from the first and second terms of the expansion, whereas the fourth term almost cancels the contribution of the third term (6). This convergence is controlled by the smallness parameter

$$K_S = \langle \Phi \mid f^2 - 1 \mid \Phi \rangle \tag{8}$$

which is related to the wound volume of Brueckner theory. To speed up the convergence of the expansions it is convenient tu put a constraint in the correlation so that K_S is null or is a very small quantity (7), and consider only up to second order in the expansion. This is known as LOCV, lowest order constrained variation, and is the method we use in this work. Our constraint was K_S=0.

We should finally mention that variational Montecarlo method has been used to compute the diagonal matrix elements in medium light nuclei, in particular in ^{16}O, using state dependent correlations of the independent pair form (8). This method is esentially a generalization of the procedure used in very light systems (9).

SPECTROSCOPY OF P-SHELL NUCLEI

We have determined the ground state and excited levels of p-shell nuclei, from ^{4}He to ^{16}O, using the LOCV method with V6 and V8 interactions.

At second order the relevant matrix elements are given by the following expressions

$$N_{pp} = 1$$

$$H_{pp} = A \langle \Phi_p \mid T_1 \mid \Phi_p \rangle + \tfrac{1}{2} A (A-1) \langle \Phi_p \mid \mathcal{V}_{12} \mid \Phi_p \rangle$$

$$N_{pq} = \frac{1}{2} A (A-1) < \Phi_p \mid f^2_{12} \mid \Phi_q > \tag{9}$$

$$H_{pq} = \frac{1}{2} A (A-1) < \Phi_p \mid \mathcal{V}_{12} \mid \Phi_q > + \frac{1}{2} (H_{pp}+H_{qq}) N_{pq}$$

where $\mathcal{V}$ is a tamed potential given by (10)

$$\mathcal{V}_{12} = \frac{1}{2} \left[f_{12}, \left[T_1+T_2 , f_{12}\right]\right]+f_{12} V_{12} f_{12} \tag{10}$$

In eqs. (9) the correlated states have been normalized. Note the particular form of the non-diagonal matrix elements, which depend also on the diagonal matrix elements and on the overlap term. This dependence is important to have the correct cluster behaviour.

Explicit evaluation of eq. (10) gives rise to a long formula involving all potential components, all correlation componentes as well as their gradients and the nuclear density, both direct and exchange parts. The final expression has again a V8 structure.

In closed shell nuclei (He and O) tensor and spin-orbit components of the effective interaction do not contribute to the energy expectation value. Note, however, that the central part and the spin/isospin components of the effective interaction do depend on all operatorial components of both, potential and correlation. In these cases there results a reasonably simple formula for the diagonal matrix elements. This idea has been extended to all other nuclei, so that the calculations have been carried out along the following steps:

i) Obtain an average value for the diagonal matrix elements by scaling appropriately the terms linear and quadratic in the number of particles of the p-shell. These expectation matrix elements then depend only on the direct and exchange parts of the nuclear density.

ii) Derive and solve Euler-Lagrange equations for the two-body correlations by imposing a healing condition at a healing distance d, i.e., the correlation goes to 1 at d with zero derivative (2). This last statement has to be interpreted in operatorial sense. At this point the calculation depends only on two parameters, the harmonic oscillator parameter which determines the single particle wave functions and the value of the healing distance d. However, because of the sequential condition constraint, $K_S=0$, both variables are related and there remains only a free parameter, which is variationally fixed.

After completion of this step we have both the correlation functions and a first estimate for the ground state energy.

iii) Correct the previous averaged results by allowing the particles to distribute freely in the p 1/2 and p 3/2 shells. This is exactly a configuration interaction calculation within the p-shell.

iv) Determine the excited levels by allowing for 1p-1h or 2p-2h excitations. In this form there result levels with different or with the same parity than the ground state, respectively. We have limited this configuration mixing so that only particles from the p-shell are promoted to the next s-d shell. Even with that limitation in some of the middle of the shell nuclei the dimensionality of the basis turned out to be too large for our computing facilities.

Some of the obtained results are shown in the figures which follow. In Figure 1 we shown the calculation of the ground state energies for the

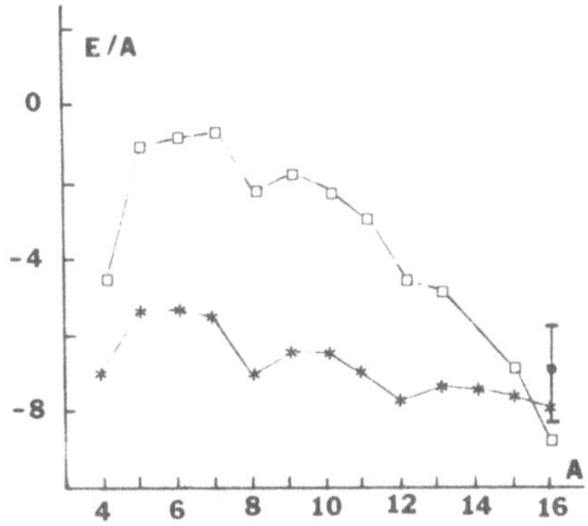

<u>Figure 1.</u> Energy per nucleon as a function of the mass number for the V6 interaction. The squares are the results of our work at step iii). The stars are experimental values and the dot with error bars corresponds to the variational Montecarlo calculation of Ref. (8).

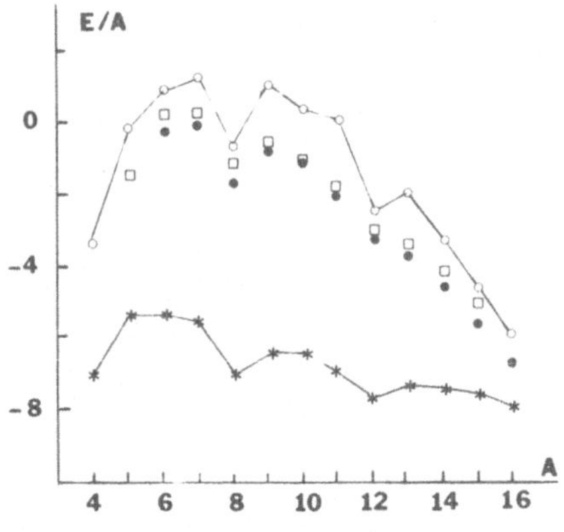

<u>Figure 2.</u> Same as Fig. 1 but for V8 interaction. Open circles correspond to average calculation. Squares include p-shell configuration mixing and black dots correspond to 2p-2h calculation. The stars are the experimental results. Energies are given in MeV.

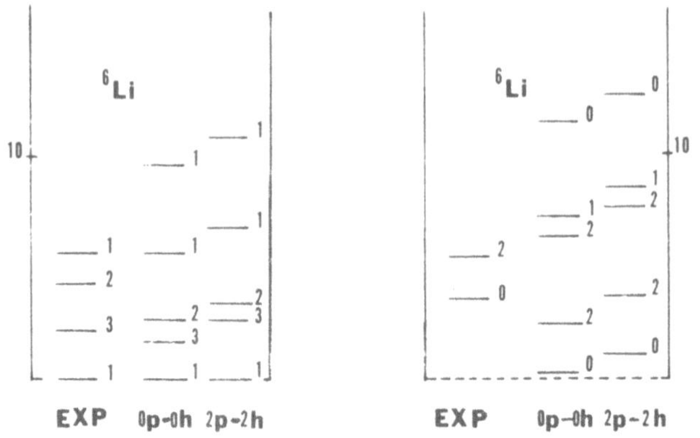

Figure 3. Low lying even parity levels in MeV of ^{6}Li for the V8 interaction. The left part corresponds to isospin 0, and the right part to isospin 1. The harmonic oscillator parameter used is $\hbar\omega$=15 MeV and corresponds to the variational minimum for the g.s.

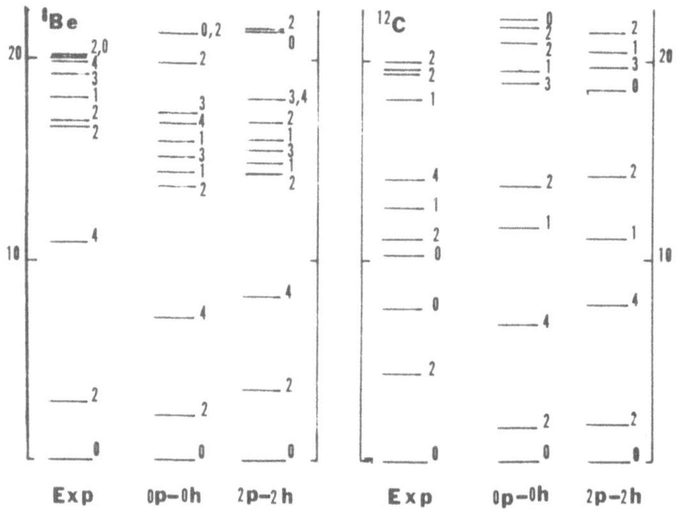

Figure 4. Low lying, even parity isospin 0 levels in MeV corresponding to V8 interaction. Left part is for Berillium ($\hbar\omega$ =20 MeV) and right part is for Carbon ($\hbar\omega$=20 MeV)

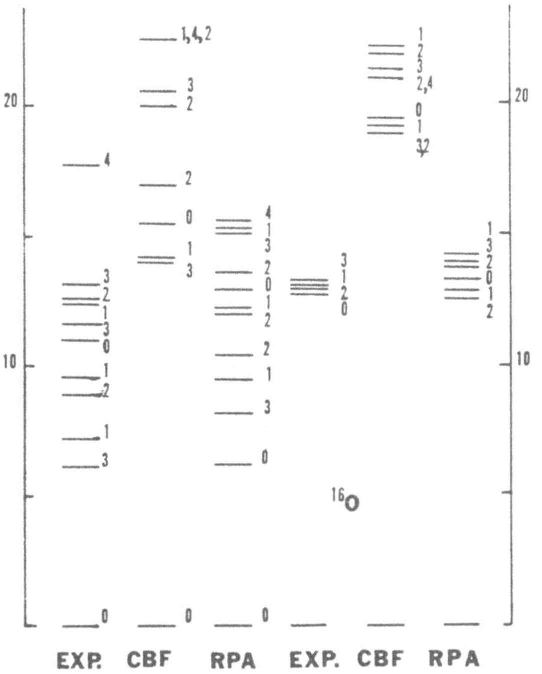

Figure 5. Odd parity levels of ^{16}O with isospin 0 (left) and isospin 1 (right). EXP refers to the experimental levels, CBF ($\hbar\omega$=15 MeV) to our calculation and RPA ($\hbar\omega$=12.2 MeV) to the CBF-RPA calculation of Ref. (4). Computed with potential V8.

V6 interaction in the step iii) described above, i.e. with an average correlation but allowing for a configuration interaction within the p-shell. The main interest of this figure is the comparison with the Montecarlo variational calculation(8) in oxigen, being in reasonable agreement with our result. We should notice however that the two-body correlations used in these two calculations are different, but the effect of this difference should not be very large.

In Figure 2 we show a panorama of the various steps followed in the calculation of ground state energies, as well as the comparison with the experimental results. The figure shows the importance of both the appropriate coupling within the p-shell (0p-0h calculation) and the 2p-2h excitations which give a contribution close to 1 MeV per nucleon. It should be mentioned that the 2p-2h corrections for the middle-of-the-shell nuclei do not include all excitations because of the large dimensionality of the space. With regard to this figure we observe a qualitative agreement with the experimental values, showing all the structure of the shell, but our calculations are clearly very far from the experimental results. The lack of agreement should be atributed to the poorness of the interaction.

Finally, the rest of the figures correspond to selected levels of various nuclei of the shell. Again many of the qualitative features of the spectra are reflected in our results, and the role of 2p-2h excitations is noticeable.

Figure 5 corresponds to the odd-parity levels of ^{16}O, and our results (CBF) are compared with the experiment and with the CBF-RPA calculation of Ref. (4). In this case we have not used the harmonic oscillator parameter corresponding to the variational minimum of the ground state, because it turned out to be very large (25 MeV) and consequently the single-particle energies and consequently the 1p-1h excitations where pushed up too much. Our results differ very much from the CBF-RPA calculations, the main reason being that we only included 1p-1h excitations. The next basis states of the shell model would correspond to 3p-3h states in s-d shell, and with analogy with the calculation of 2p-2h corrections, one should expect quite large effects associated to 3p-3h states in these odd-parity levels.

ACKNOWLEDGEMENTS

This work has been supported by the CAICYT (Spain). The authors acknowledge fruitful suggestions from R.F. Bishop and J.W. Clark, as well as the collaboration of A. Polls in the earlier stages of this work. One of us (RG) is grateful to the Joint U.S.-Spain Commitee in Science and Technology for financial support.

REFERENCES

1. E. Feenberg, Theory of Quantum Fluids, Academic Press (New York 1969).
2. V.R. Pandharipande and R.B. Wiringa, Rev. Mod. Phys. 51 (1979) 821.
3. E. Krotscheck and J.W. Clark, Nucl. Phys. A333 (1980) 77.
4. J.W. Clark, E. Krotscheck and B. Schwesinger, "Nucleon matter particle-hole force and a correlated RPA theory of ^{16}O," Preprint Washington Un. 1985.
5. J.W. Clark, Springer Lecture Notes in PHysics 138 (1981) 184, R. Guardiola, Nucl. PHys. A384 (1982) 143.
6. E. Buendia and R. Guardiola, in preparation.
7. V.R. Pandharipande, Nucl. Phys. A234 (1974) 237; J.M. Irvine, G.S. Mani, V.F.E. Pucknful, M. Vallieres, and F. Yacici, Ann. Phys. 102 (1976) 129; R.F. Bishop, C. Howes, J.M. Irvine, and M. Modarres, J. Phkys. G4 (1985) 2105.
8. J. carlson, and M.H. Kalos, Phys. Rev. C32 L1985) 2105.
9. J. Lomnitz-Adler, V.R. Pandharipande, and R.A. Smith, Nucl. Phys. A361 (1981) 399; J. Carlson, R.B. Wiringa, and V.R. Pandharipande, Nucl. Phys. A371 (1981) 301.
10. M.L. ristig, W.J. Ter-Louw, and J.W. Clark, Phys. Rev. C3 (1971), 1504; C5 (1972) 695.

INDEX